Introduction to Business Statistics

Special Seventh Edition

Ronald M. Weiers

CENGAGE Learning®

Australia • Brazil • Japan • Korea • Mexico • Singapore • Spain • United Kingdom • United States

CENGAGE Learning

Introduction to Business Statistics: Special Seventh Edition

Executive Editors:
Maureen Staudt
Michael Stranz

Senior Project Development Manager:
Linda deStefano

Marketing Specialist:
Courtney Sheldon

Senior Production/Manufacturing Manager:
Donna M. Brown

Production Editorial Manager:
Kim Fry

Sr. Rights Acquisition Account Manager:
Todd Osborne

Printed in the United States of America

Introduction to Business Statistics, 7th Edition
Ronald M. Weiers

ISBN-13: 978-1-285-10486-7

ISBN-10: 1-285-10486-2

Cengage Learning
5191 Natorp Boulevard
Mason, Ohio 45040
USA

Cengage Learning is a leading provider of customized learning solutions with office locations around the globe, including Singapore, the United Kingdom, Australia, Mexico, Brazil, and Japan. Locate your local office at:
international.cengage.com/region.

Cengage Learning products are represented in Canada by Nelson Education, Ltd.
For your lifelong learning solutions, visit **www.cengage.com/custom.**
Visit our corporate website at **www.cengage.com.**

Table of Contents

CHAPTER 1

A Preview of Business Statistics

STATISTICS CAN ENTERTAIN, ENLIGHTEN, ALARM

Today's statistics applications range from the inane to the highly germane. Sometimes statistics provides nothing more than entertainment—e.g., a study found that 31% of U.S. adults have re-gifted a present for the holidays.[1] Regarding an actual entertainer, other studies found that the public's "favorable" rating for actor Tom Cruise had dropped from 58% to 35% between 2005 and 2006.[2]

On the other hand, statistical descriptors can be highly relevant to such important matters as corporate ethics and employee privacy. For example, 5% of workers say they use the Internet too much at work, and that decreases their productivity.[3] In the governmental area, U.S. census data can mean millions of dollars to big cities. According to the Los Angeles city council, that city will have lost over $180 million in federal aid because the 2000 census had allegedly missed 76,800 residents, most of whom were urban, minority, and poor.[4]

At a deadly extreme, statistics can also describe the growing toll on persons living near or downwind of Chernobyl, site of the world's worst nuclear accident. Just 10 years following this 1986 disaster, cancer rates in the fallout zone had already nearly doubled, and researchers are now concerned about the possibility of even higher rates with the greater passage of time.[5] In general, statistics can be useful in examining any geographic "cluster" of disease incidence, helping us to decide whether the higher incidence could be due simply to chance variation, or whether some environmental agent or pollutant may have played a role.

[1]Source: Michelle Healy and Veronica Salazar, "Thanks... Again," *USA Today*, December 23, 2008, p. 1D.
[2]Source: Susan Wloszczyna, "In Public's Eyes, Tom's Less of a Top Gun," *USA Today*, May 10, 2006, p. 1D.
[3]Source: Jae Yang and Marcy Mullins, "Internet Usage's Impact on Productivity," *USA Today*, March 21, 2006, p. 1B.
[4]Source: cd13.com. Letter from Los Angeles City Council to U.S. House of Representatives, April 11, 2006.
[5]Allison M. Heinrichs, "Study to Examine Breast Cancer in Europeans," *Pittsburgh Tribune-Review*, April 23, 2006; and world-nuclear.org/info/chernobyl, April 2009.

Anticipating coming attractions

1.1 INTRODUCTION

Timely Topic, Tattered Image

At this point in your college career, toxic dumping, armed robbery, fortune telling, and professional wrestling may all have more positive images than **business statistics.** If so, this isn't unusual, since many students approach the subject believing that it will be either difficult or irrelevant. In a study of 105 beginning students' attitudes toward statistics, 56% either strongly or moderately agreed with the statement, "I am afraid of statistics."[6] (Sorry to have tricked you like that, but you've just been introduced to a statistic, one that you'll undoubtedly agree is neither difficult nor irrelevant.)

Having recognized such possibly negative first impressions, let's go on to discuss statistics in a more positive light. First, regarding ease of learning, the only thing this book assumes is that you have a basic knowledge of algebra. Anything else you need will be introduced and explained as we go along. Next, in terms of relevance, consider the unfortunates of Figure 1.1 and how just the slight change of a single statistic might have considerably influenced each individual's fortune.

What Is Business Statistics?

Briefly defined, business statistics can be described as *the collection, summarization, analysis, and reporting of numerical findings relevant to a business decision or situation.* Naturally, given the great diversity of business itself, it's not surprising that statistics can be applied to many kinds of business settings. We will be examining a wide spectrum of such applications and settings. Regardless of your eventual career destination, whether it be accounting or marketing, finance or politics, information science or human resource management, you'll find the statistical techniques explained here are supported by examples and problems relevant to your own field.

For the Consumer as Well as the Practitioner

As a businessperson, you may find yourself involved with statistics in at least one of the following ways: (1) as a practitioner collecting, analyzing, and presenting

FIGURE 1.1

Some have the notion that statistics can be irrelevant. As the plight of these individuals suggests, nothing could be further from the truth.

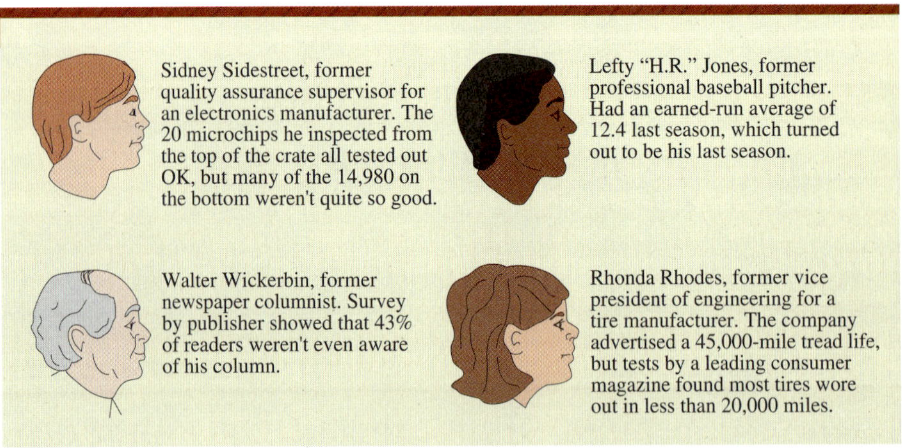

[6]Source: Eleanor W. Jordan and Donna F. Stroup, "The Image of Statistics," *Collegiate News and Views,* Spring 1984, p. 11.

findings based on statistical data or (2) as a consumer of statistical claims and findings offered by others, some of whom may be either incompetent or unethical.

As you might expect, the primary orientation of this text will be toward the "how-to," or practitioner, dimension of business statistics. After finishing this book, you should be both proficient and conversant in most of the popular techniques used in statistical data collection, analysis, and reporting. As a secondary goal, this book will help you protect yourself and your company as a statistical consumer. In particular, it's important that you be able to deal with individuals who arrive at your office bearing statistical advice. Chances are, they'll be one of the following:

1. **Dr. Goodstat** The good doctor has painstakingly employed the correct methodology for the situation and has objectively analyzed and reported on the information he's collected. Trust him, he's OK.
2. **Stanley Stumbler** Stanley means well, but doesn't fully understand what he's doing. He may have innocently employed an improper methodology and arrived at conclusions that are incorrect. In accepting his findings, you may join Stanley in flying blind.
3. **Dr. Unethicus** This character knows what he's doing, but uses his knowledge to sell you findings that he knows aren't true. In short, he places his own selfish interests ahead of both scientific objectivity and your informational needs. He varies his modus operandi and is sometimes difficult to catch. One result is inevitable: When you accept his findings, he wins and you lose.

STATISTICS: YESTERDAY AND TODAY (1.2)

Yesterday

Although statistical data have been collected for thousands of years, very early efforts typically involved simply counting people or possessions to facilitate taxation. This record-keeping and enumeration function remained dominant well into the 20th century, as this 1925 observation on the role of statistics in the commercial and political world of that time indicates:

It is coming to be the rule to use statistics and to think statistically. The larger business units not only have their own statistical departments in which they collect and interpret facts about their own affairs, but they themselves are consumers of statistics collected by others. The trade press and government documents are largely statistical in character, and this is necessarily so, since only by the use of statistics can the affairs of business and of state be intelligently conducted.

Business needs a record of its past history with respect to sales, costs, sources of materials, market facilities, etc. Its condition, thus reflected, is used to measure progress, financial standing, and economic growth. A record of business changes—of its rise and decline and of the sequence of forces influencing it—is necessary for estimating future developments.[7]

Note the brief reference to "estimating future developments" in the preceding quotation. In 1925, this observation was especially pertinent because a transition was in process. Statistics was being transformed from a relatively passive record keeper

[7]Source: Horace Secrist, *An Introduction to Statistical Methods,* rev. ed. New York: Macmillan Company, 1925, p. 1.

and descriptor to an increasingly active and useful business tool, which would influence decisions and enable inferences to be drawn from sample information.

Today

Today, statistics and its applications are an integral part of our lives. In such diverse settings as politics, medicine, education, business, and the legal arena, human activities are both measured and guided by statistics.

Our behavior in the marketplace generates sales statistics that, in turn, help companies make decisions on products to be retained, dropped, or modified. Likewise, auto insurance firms collect data on age, vehicle type, and accidents, and these statistics guide the companies toward charging extremely high premiums for teenagers who own or drive high-powered cars like the Chevrolet Corvette. In turn, the higher premiums influence human behavior by making it more difficult for teens to own or drive such cars. The following are additional examples where statistics are either guiding or measuring human activities.

- Well beyond simply counting how many people live in the United States, the U.S. Census Bureau uses sampling to collect extensive information on income, housing, transportation, occupation, and other characteristics of the populace. The Bureau used to do this by means of a "long form" sent to 1 in 6 Americans every 10 years. Today, the same questions are asked in a 67-question monthly survey that is received by a total of about 3 million households each year. The resulting data are more recent and more useful than the decennial sampling formerly employed, and the data have a vital effect on billions of dollars in business decisions and federal funding.[8]
- According to the International Dairy Foods Association, ice cream and related frozen desserts are consumed by more than 90% of the households in the United States. The most popular flavor is vanilla, which accounts for 30% of sales. Chocolate is a distant second, at 10% of sales.[9]
- On average, U.S. stores lose $35 million each day to shoplifters. The problem becomes even worse when the national economy is weak, and more than 10 million people have been detained for shoplifting in the past five years.[10]

Throughout this text, we will be examining the multifaceted role of statistics as a descriptor of information, a tool for analysis, a means of reaching conclusions, and an aid to decision making. In the next section, after introducing the concept of descriptive versus inferential statistics, we'll present further examples of the relevance of statistics in today's world.

EXERCISES

1.1 What was the primary use of statistics in ancient times?

1.2 In what ways can business statistics be useful in today's business environment?

[8]Source: Haya El Nasser, "Rolling Survey for 2010 Census Keeps Data Up to Date," *USA Today*, January 17, 2005, p. 4A.
[9]Source: idfa.org, June 19, 2009.
[10]Source: shopliftingprevention.org, June 19, 2009.

DESCRIPTIVE VERSUS INFERENTIAL STATISTICS (1.3)

As we have seen, statistics can refer to a set of individual numbers or numerical facts, or to general or specific statistical techniques. A further breakdown of the subject is possible, depending on whether the emphasis is on (1) simply describing the characteristics of a set of data or (2) proceeding from data characteristics to making generalizations, estimates, forecasts, or other judgments based on the data. The former is referred to as **descriptive statistics,** while the latter is called **inferential statistics.** As you might expect, both approaches are vital in today's business world.

Descriptive Statistics

In descriptive statistics, we simply summarize and describe the data we've collected. For example, upon looking around your class, you may find that 35% of your fellow students are wearing Casio watches. If so, the figure "35%" is a descriptive statistic. You are not attempting to suggest that 35% of all college students in the United States, or even at your school, wear Casio watches. You're merely describing the data that you've recorded. In the year 1900, the U.S. Postal Service operated 76,688 post offices, compared to just 27,276 in 2007.[11] In 2008, the 1.12 billion common shares of McDonald's Corporation each received a $1.63 dividend on net income of $3.76 per common share.[12] Table 1.1 (page 6) provides additional examples of descriptive statistics. Chapters 2 and 3 will present a number of popular visual and statistical approaches to expressing the data we or others have collected. For now, however, just remember that descriptive statistics are used only to summarize or describe.

Inferential Statistics

In inferential statistics, sometimes referred to as *inductive* statistics, we go beyond mere description of the data and arrive at *inferences* regarding the phenomenon or phenomena for which sample data were obtained. For example, based partially on an examination of the viewing behavior of several thousand television households, the ABC television network may decide to cancel a prime-time television program. In so doing, the network is assuming that millions of other viewers across the nation are also watching competing programs.

Political pollsters are among the heavy users of inferential statistics, typically questioning between 1000 and 2000 voters in an effort to predict the voting behavior of millions of citizens on election day. If you've followed recent presidential elections, you may have noticed that, although they contact only a relatively small number of voters, the pollsters are quite often "on the money" in predicting both the winners and their margins of victory. This accuracy, and the fact that it's not simply luck, is one of the things that make inferential statistics a fascinating and useful topic. (For more examples of the relevance and variety of inferential statistics, refer to Table 1.1.) As you might expect, much of this text will be devoted to the concept and methods of inferential statistics.

[11]Source: Bureau of the Census, U.S. Department of Commerce, *Statistical Abstract of the United States 2009*, p. 690.
[12]Source: McDonald's Corporation, Inc., *2008 Annual Report.*

TABLE 1.1

Some examples of descriptive and inferential statistics.

Descriptive Statistics

- According to the Bureau of the Census, there are 2.2 million U.S. households with a single father and one or more children younger than 18. [p. 1A]
- There have been 82 confirmed or suspected suicides among active-duty service personnel this year, compared to 51 for the same period in 2008. [p. 1A]
- The number of mutual funds peaked at 8305 in 2001, but the combination of bear markets and mergers and acquisitions has driven the number of funds down to 8011. [p. 1B]
- Since March 4, 2009, there have been 190,000 mortgage modifications through President Obama's relief plan; 396,724 homes in payment default; and 607,974 homes in either foreclosure or auction proceedings. [p. 1A]

Inferential Statistics

- In observing a sample of nurses and other healthcare workers who were likely infected with the swine flu, researchers found that only half routinely wore gloves when dealing with patients. [p. 3A]
- In a Zagat survey of diners, Outback Steakhouse had the top-rated steaks in the full-service restaurant category. [p. 7A]
- Survey results revealed that 26% of thirsty golfers order a sports drink when they finish their round and head for the clubhouse. [p. 1C]
- In a survey of U.S. motorists, 33% said their favorite American roadside store was South of the Border, in South Carolina. [p. 1D]

Source: *USA Today*, June 19, 2009. The page references are shown in brackets.

Key Terms for Inferential Statistics

In surveying the political choices of a small number of eligible voters, political pollsters are using a **sample** of voters selected from the **population** of all eligible voters. Based on the results observed in the sample, the researchers then proceed to make inferences on the political choices likely to exist in this larger population of eligible voters. A sample result (e.g., 46% of the sample favor Charles Grady for president) is referred to as a **sample statistic** and is used in an attempt to estimate the corresponding **population parameter** (e.g., the actual, but unknown, national percentage of voters who favor Mr. Grady). These and other important terms from inferential statistics may be defined as follows:

- **Population** Sometimes referred to as the *universe*, this is the entire set of people or objects of interest. It could be all adult citizens in the United States, all commercial pilots employed by domestic airlines, or every roller bearing ever produced by the Timken Company.

A population may refer to things as well as people. Before beginning a study, it is important to clearly define the population involved. For example, in a given study, a retailer may decide to define "customer" as all those who enter her store between 9 A.M. and 5 P.M. next Wednesday.

- **Sample** This is a smaller number (a *subset*) of the people or objects that exist within the larger population. The retailer in the preceding definition may decide to select her sample by choosing every 10th person entering the store between 9 A.M. and 5 P.M. next Wednesday.

A sample is said to be *representative* if its members tend to have the same characteristics (e.g., voting preference, shopping behavior, age, income, educational level) as the population from which they were selected. For example, if 45% of the population consists of female shoppers, we would like our sample to also include 45% females. When a sample is so large as to include all members of the population, it is referred to as a complete *census*.

- **Statistic** This is a measured characteristic of the sample. For example, our retailer may find that 73% of the sample members rate the store as having higher-quality merchandise than the competitor across the street. The sample statistic can be a measure of *typicalness* or central tendency, such as the mean, median, mode, or proportion, or it may be a measure of *spread* or dispersion, such as the range and standard deviation:

The *sample mean* is the arithmetic average of the data. This is the sum of the data divided by the number of values. For example, the mean of $4, $3, and $8 can be calculated as ($4 + $3 + $8)/3, or $5.

The *sample median* is the midpoint of the data. The median of $4, $3, and $8 would be $4, since it has just as many values above it as below it.

The *sample mode* is the value that is most frequently observed. If the data consist of the numbers 12, 15, 10, 15, 18, and 21, the mode would be 15 because it occurs more often than any other value.

The *sample proportion* is simply a percentage expressed as a decimal fraction. For example, if 75.2% is converted into a proportion, it becomes 0.752.

The *sample range* is the difference between the highest and lowest values. For example, the range for $4, $3, and $8 is ($8 − $3), or $5.

The *sample standard deviation,* another measure of dispersion, is obtained by applying a standard formula to the sample values. The formula for the standard deviation is covered in Chapter 3, as are more detailed definitions and examples of the other measures of central tendency and dispersion.

- **Parameter** This is a numerical characteristic of the population. If we were to take a complete census of the population, the parameter could actually be measured. As discussed earlier, however, this is grossly impractical for most business research. The purpose of the sample statistic is to estimate the value of the corresponding population parameter (e.g., the sample mean is used to estimate the population mean). Typical parameters include the population mean, median, proportion, and standard deviation. As with sample statistics, these will be discussed in Chapter 3.

  For our retailer, the actual percentage of the population who rate her store's merchandise as being of higher quality is unknown. (This unknown quantity is the parameter in this case.) However, she may use the sample statistic (73%) as an estimate of what this percentage would have been had she taken the time, expense, and inconvenience to conduct a census of all customers on the day of the study.

EXERCISES

1.3 What is the difference between descriptive statistics and inferential statistics? Which branch is involved when a state senator surveys some of her constituents in order to obtain guidance on how she should vote on a piece of legislation?

1.4 In 2008, Piedmont Natural Gas Corporation sold 50.4 million cubic feet of gas to residential customers, an increase of 3.7% over the previous year. Does this information represent descriptive statistics or inferential statistics? Why? Source: Piedmont Natural Gas Corporation, *Annual Report 2008*, p. 16.

1.5 An article in a fitness magazine described a study that compared the cardiovascular responses of 20 adult subjects for exercises on a treadmill, on a minitrampoline, and jogging in place on a carpeted surface. Researchers found average heart rates were significantly less on the minitrampoline than for the treadmill and stationary jogging. Does this information represent descriptive statistics or inferential statistics? Why?

(1.4) TYPES OF VARIABLES AND SCALES OF MEASUREMENT

Qualitative Variables

Some of the variables associated with people or objects are **qualitative** in nature, indicating that the person or object belongs in a category. For example: (1) you are either male or female; (2) you have either consumed Dad's Root Beer within the past week or you have not; (3) your next television set will be either color or black and white; and (4) your hair is likely to be brown, black, red, blonde, or gray. While some qualitative variables have only two categories, others may have three or more. Qualitative variables, also referred to as *attributes,* typically involve counting how many people or objects fall into each category.

In expressing results involving qualitative variables, we describe the percentage or the number of persons or objects falling into each of the possible categories. For example, we may find that 35% of grade-school children interviewed recognize a photograph of Ronald McDonald, while 65% do not. Likewise, some of the children may have eaten a Big Mac hamburger at one time or another, while others have not.

Quantitative Variables

Quantitative variables enable us to determine *how much* of something is possessed, not just whether it is possessed. There are two types of quantitative variables: discrete and continuous.

Discrete quantitative variables can take on only certain values along an interval, with the possible values having gaps between them. Examples of discrete quantitative variables would be the number of employees on the payroll of a manufacturing firm, the number of patrons attending a theatrical performance, or the number of defectives in a production sample. Discrete variables in business statistics usually consist of observations that we can count and often have integer values. Fractional values are also possible, however. For example, in observing the number of gallons of milk that shoppers buy during a trip to a U.S. supermarket, the possible values will be 0.25, 0.50, 0.75, 1.00, 1.25, 1.50, and so on. This is because milk is typically sold in 1-quart containers as well as gallons. A shopper will not be able to purchase a container of milk labeled "0.835 gallons." The distinguishing feature of discrete variables is that gaps exist between the possible values.

Continuous quantitative variables can take on a value at any point along an interval. For example, the volume of liquid in a water tower could be any quantity between zero and its capacity when full. At a given moment, there might be 325,125 gallons, 325,125.41 gallons, or even 325,125.413927 gallons, depending on the accuracy with which the volume can be measured. The possible values that could be taken on would have no gaps between them. Other examples of continuous quantitative variables are the weight of a coal truck, the Dow Jones Industrial Average, the driving distance from your school to your home town, and the temperature outside as you're reading this book. The exact values each of these variables could take on would have no gaps between them.

Scales of Measurement

Assigning a numerical value to a variable is a process called *measurement.* For example, we might look at the thermometer and observe a reading of 72.5 degrees Fahrenheit or examine a box of lightbulbs and find that 3 are broken. The numbers 72.5 and 3 would constitute measurements. When a variable is measured, the result will be in one of the four levels, or *scales,* of measurement—nominal, ordinal, interval, or ratio—summarized in Figure 1.2. The scale to which the measurements belong will be important in determining appropriate methods for data description and analysis.

The Nominal Scale

The **nominal scale** uses numbers only for the purpose of identifying membership in a group or category. Computer statistical analysis is greatly facilitated by the use of numbers instead of names. For example, Entergy Corporation lists four types of domestic electric customers.[13] In its computer records, the company might use "1" to identify residential customers, "2" for commercial customers, "3" for industrial customers, and "4" for government customers. Aside from identification, these numbers have no arithmetic meaning.

The Ordinal Scale

In the **ordinal scale,** numbers represent "greater than" or "less than" measurements, such as preferences or rankings. For example, consider the following

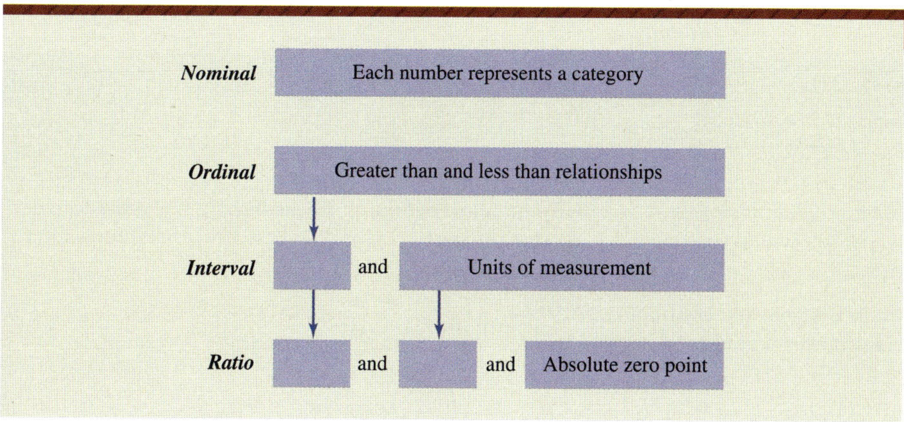

FIGURE 1.2

The methods through which statistical data can be analyzed depend on the scale of measurement of the data. Each of the four scales has its own characteristics.

[13]Source: Entergy Corporation, *2008 Annual Report.*

Women's Tennis Association singles rankings for female tennis players:[14]

1. Dinara Safina
2. Serena Williams
3. Venus Williams
4. Elena Dementieva

In the ordinal scale, numbers are viewed in terms of rank (i.e., greater than, less than), but do not represent distances between objects. For example, we cannot say that the distance between Dinara Safina and Serena Williams is the same as the distance between Serena Williams and Venus Williams. This is because the ordinal scale has no unit of measurement.

The Interval Scale

The **interval scale** not only includes "greater than" and "less than" relationships, but also has a unit of measurement that permits us to describe *how much more or less* one object possesses than another. The Fahrenheit temperature scale represents an interval scale of measurement. We not only know that 90 degrees Fahrenheit is hotter than 70 degrees, and that 70 degrees is hotter than 60 degrees, but can also state that the distance between 90 and 70 is twice the distance between 70 and 60. This is because degree markings serve as the unit of measurement.

In an interval scale, the unit of measurement is arbitrary, and there is no absolute zero level where *none* of a given characteristic is present. Thus, multiples of measured values are not meaningful—e.g., 2 degrees Fahrenheit is not twice as warm as 1 degree. On questionnaire items like the following, business research practitioners typically treat the data as interval scale since the same physical and numerical distances exist between alternatives:

	[ ]	[ ]	[ ]	[ ]	[ ]
Kmart prices are	1	2	3	4	5
	low				high

The Ratio Scale

The **ratio scale** is similar to the interval scale, but has an absolute zero and multiples are meaningful. Election votes, natural gas consumption, return on investment, the speed of a production line, and FedEx Corporation's average daily delivery of 6,900,000 packages during 2008[15] are all examples of the ratio scale of measurement.

EXERCISES

1.6 What is the difference between a qualitative variable and a quantitative variable? When would each be appropriate?

1.7 What is the difference between discrete and continuous variables? Under what circumstances would each be applicable?

1.8 The Acme School of Locksmithing has been accredited for the past 15 years. Discuss how this information might be interpreted as a

a. qualitative variable.
b. quantitative variable.

[14]Source: ESPN.com, June 19, 2009.
[15]Source: FedEx Corporation, *2008 Annual Report,* p. 26.

1.9 Jeff Bowlen, a labor relations expert, has collected information on strikes in various industries.

a. Jeff says, "Industry A has been harder hit by strikes than Industry B." In what scale of measurement is this information? Why?
b. Industry C has lost 10.8 days per worker, while Industry D has lost 14.5 days per worker. In what scale of measurement is this information? Why?

1.10 The Snowbird Ski Lodge attracts skiers from several New England states. For each of the following scales of measurement, provide one example of information that might be relevant to the lodge's business.

a. Nominal b. Ordinal
c. Interval d. Ratio

STATISTICS IN BUSINESS DECISIONS (1.5)

One aspect of business in which statistics plays an especially vital role is decision making. Every year, U.S. businesses risk billions of dollars in important decisions involving plant expansions, new product development, personnel selection, quality assurance, production techniques, supplier choices, and many others. These decisions almost always involve an element of uncertainty. Competitors, government, technology, and the social and economic environment, along with sometimes capricious consumers and voters, constitute largely uncontrollable factors that can sometimes foil the best-laid plans.

Prior to making decisions, companies often collect information through a series of steps called the *research process*. The steps include: (1) defining the problem in specific terms that can be answered by research, (2) deciding on the type of data required, (3) determining through what means the data will be obtained, (4) planning for the collection of data and, if necessary, selection of a sample, (5) collecting and analyzing the data, (6) drawing conclusions and reporting the findings, and (7) following through with decisions that take the findings into consideration. Business and survey research, discussed more fully in Chapter 4, provides both descriptive and inferential statistics that can improve business decisions in many kinds of situations.

EXERCISES

1.11 Restaurants sometimes provide "customer reaction" cards so that customers can evaluate their dining experience at the establishment. What kinds of decisions might be made on the basis of this information?

1.12 What kinds of statistical data might a burglar alarm company employ in trying to convince urban homeowners to purchase its product?

BUSINESS STATISTICS: TOOLS VERSUS TRICKS (1.6)

The techniques of business statistics are valuable tools for the enhancement of business operations and success. Appropriately, the major emphasis of this text will be to acquaint you with these techniques and to develop your proficiency in using them and interpreting their results.

On the other hand, as suggested earlier, these same techniques can be abused for personal or corporate gain. Improperly used, statistics can become an effective weapon with which to persuade or manipulate others into beliefs or behaviors that we'd like them to adopt. Note too that, even when they are not intentionally misused, the results of statistical research and analyses can depend a lot on when and how they were conducted, as Statistics in Action 1.1 shows.

Unlike many other pursuits, such as defusing torpedoes, climbing mountains, or wrestling alligators, improper actions in business statistics can sometimes work in your favor. (As embezzlers know, this can also be true in accounting.) Naturally, we don't expect that you'll use your knowledge of statistics to manipulate unknowing customers and colleagues, but you should be aware of how *others* may be using statistics in an attempt to manipulate *you.* Remember that one of the key goals of this text is to make you an informed consumer of statistical information generated by others. In general, when you are presented with statistical data or conclusions that have been generated by others, you should ask yourself this key question: *Who carried out this study and analyzed the data, and what benefits do they stand to gain from the conclusions reached?*

EXERCISES

1.13 The text claims that a company or organization might actually benefit when one of its employees uses statistics incorrectly. How can this be?

1.14 The headline of an article in your daily newspaper begins "Research Study Reveals. . . ." As a statistics student who wishes to avoid accepting biased results, what single question should be foremost in your mind as you begin reading the article?

1.7 SUMMARY

Business statistics can be defined as the collection, summarization, analysis, and reporting of numerical findings relevant to a business decision or situation. As businesspersons and citizens, we are involved with statistics either as practitioners or as consumers of statistical claims and findings offered by others. Very early statistical efforts primarily involved counting people or possessions for taxation purposes. More recently, statistical methods have been applied in all facets of business as a tool for analysis and reporting, for reaching conclusions based on observed data, and as an aid to decision making.

Statistics can be divided into two branches: descriptive and inferential. Descriptive statistics focuses on summarizing and describing data that have been collected. Inferential statistics goes beyond mere description and, based on sample data, seeks to reach conclusions or make predictions regarding the population from which the sample was drawn. The population is the entire set of all people or objects of interest, with the sample being a subset of this group. A sample is said to be representative if its members tend to have the same characteristics as the larger population. A census involves measuring all people or objects in the population.

The sample statistic is a characteristic of the sample that is measured; it is often a mean, median, mode, proportion, or a measure of variability such as the range or standard deviation. The population parameter is the population characteristic that the sample statistic attempts to estimate.

Variables can be either qualitative or quantitative. Qualitative variables indicate whether a person or object possesses a given attribute, while quantitative

STATISTICS IN ACTION 1.1

High Stakes on the Interstate: Cell Phones and Accidents

Do cell phones contribute to auto accidents? The National Safety Council says they do, and has called for a nationwide ban on cell phone use while driving. A number of research studies support this view. In one preliminary study, the researchers randomly selected 100 New York motorists who had been in an accident and 100 who had not. Those who had been in an accident were 30% more likely to have a cell phone. In another study, published in *The New England Journal of Medicine,* researchers found that cell phone use while driving quadrupled the chance of having an accident, a risk increase comparable to driving with one's blood alcohol level at the legal limit.

The Cellular Telecommunications Industry Association has a natural stake in this issue. There are more than 270 million cell phone subscribers, tens of thousands are signing up daily, and a high percentage of subscribers use their phones while driving. The association would have a vested interest in dismissing accident studies such as the ones above as limited, flawed, and having research shortcomings.

One thing is certain: More research is on the way. It will be performed by objective researchers as well as by individuals with a vested interest in the results. Future studies, their sponsors, and the interpretation of their results will play an important role in the safety of our highways and the economic vitality of our cellular phone industry.

Sources: "Survey: Car Phone Users Run Higher Risk of Crashes," *Indiana Gazette,* March 19, 1996, p. 10; "Ban Car Phones?", *USA Today,* April 27, 2000, p. 16A; "Get Off the Cell Phone," *Pittsburgh Tribune-Review,* January 29, 2000, p. A6; and "Safety Council Urges Ban on Cell Phone Use While Driving," cnn.com, June 19, 2009.

variables express how much of an attribute is possessed. Discrete quantitative variables can take on only certain values along an interval, with the possible values having gaps between them, while continuous quantitative variables can take on a value at any point along an interval.

When a variable is measured, a numerical value is assigned to it, and the result will be in one of four levels, or scales, of measurement—nominal, ordinal, interval, or ratio. The scale to which the measurements belong will be important in determining appropriate methods for data description and analysis.

By helping to reduce the uncertainty posed by largely uncontrollable factors, such as competitors, government, technology, the social and economic environment, and often unpredictable consumers and voters, statistics plays a vital role in business decision making. Although statistics is a valuable tool in business, its techniques can be abused or misused for personal or corporate gain. This makes it especially important for businesspersons to be informed consumers of statistical claims and findings.

CHAPTER EXERCISES

1.15 A research firm observes that men are twice as likely as women to watch the Super Bowl on television. Does this information represent descriptive statistics or inferential statistics? Why?

1.16 For each of the following, indicate whether the appropriate variable would be qualitative or quantitative. If you identify the variable as quantitative, indicate whether it would be discrete or continuous.

a. Whether you own a Panasonic television set
b. Your status as either a full-time or a part-time student
c. The number of people who attended your school's graduation last year
d. The price of your most recent haircut
e. Sam's travel time from his dorm to the student union
f. The number of students on campus who belong to a social fraternity or sorority

1.17 What kinds of statistical data play a role in an auto insurance firm's decision on the annual premium you'll pay for your policy?

1.18 For each of the following, indicate the scale of measurement that best describes the information.

a. In 2008, Dell Corporation had approximately 78,000 employees. Source: Fortune, May 4, 2009, p. F-48.

b. *USA Today* reports that the previous day's highest temperature in the United States was 105 degrees in Death Valley, California. Source: *USA Today*, June 19, 2009, p. 12A.
c. An individual respondent answers "yes" when asked if TV contributes to violence in the United States.
d. In a comparison test of family sedans, a magazine rates the Toyota Camry higher than the VW Passat.

1.19 Most undergraduate business students will not go on to become actual practitioners of statistical research and analysis. Considering this fact, why should such individuals bother to become familiar with business statistics?

1.20 Bill scored 1200 on the Scholastic Aptitude Test and entered college as a physics major. As a freshman, he changed to business because he thought it was more interesting. Because he made the dean's list last semester, his parents gave him $30 to buy a new Casio calculator. For this situation, identify at least one piece of information in the

a. nominal scale of measurement.
b. ordinal scale of measurement.
c. interval scale of measurement.
d. ratio scale of measurement.

1.21 Roger Amster teaches an English course in which 40 students are enrolled. After yesterday's class, Roger questioned the 5 students who always sit in the back of the classroom. Three of the 5 said "yes" when asked if they would like *A Tale of Two Cities* as the next class reading assignment.

a. Identify the population and the sample in this situation.
b. Is this likely to be a representative sample? If not, why not?

1.22 In studying the performance of the company's stock investments over the past year, the research manager of a mutual fund company finds that only 43% of the stocks returned more than the rate that had been expected at the beginning of the year.

a. Could this information be viewed as representing the nominal scale of measurement? If so, explain your reasoning. If not, why not?
b. Could this information be viewed as representing the ratio scale of measurement? If so, explain your reasoning. If not, why not?

CHAPTER 2

Visual Description of Data

"USA SNAPSHOTS" SET THE STANDARD

When it comes to creative visual displays to summarize data, hardly anything on the planet comes close to *USA Today* and its "USA Snapshots" that appear in the lower-left portion of the front page of each of the four sections of the newspaper. Whether it's "A look at statistics that shape the nation" (section A), "your finances" (section B), "the sports world" (section C), or "our lives" (section D), the visual is apt to be both informative and entertaining.

For example, when the imaginative folks who create "USA Snapshots" get their hands on some numbers, we can expect that practically any related object that happens to be round may end up becoming a pie chart, or that any relevant entity that's rectangular may find itself relegated to duty as a bar chart. An example of this creativity can be seen later in the chapter, in Figure 2.3.

If you're one of the many millions who read *USA Today,* chances are you'll notice a lot of other lively, resourceful approaches to the visual description of information. Complementing their extensive daily fare of news, editorials, and many other items that we all expect a good daily newspaper to present, *USA Today* and the "USA Snapshot" editors set the standard when it comes to reminding us that statistics can be as interesting as they are relevant.

Visualizing the data

Elamul Hoque/Rod Steele/Getty Images

LEARNING OBJECTIVES

After reading this chapter, you should be able to:

- Construct a frequency distribution and a histogram.
- Construct relative and cumulative frequency distributions.
- Construct a stem-and-leaf diagram to represent data.
- Visually represent data by using graphs and charts.
- Construct a dotplot and a scatter diagram.
- Construct contingency tables.

(2.1) INTRODUCTION

When data have been collected, they are of little use until they have been organized and represented in a form that helps us understand the information contained. In this chapter, we'll discuss how raw data are converted to frequency distributions and visual displays that provide us with a "big picture" of the information collected.

By so organizing the data, we can better identify trends, patterns, and other characteristics that would not be apparent during a simple shuffle through a pile of questionnaires or other data collection forms. Such summarization also helps us compare data that have been collected at different points in time, by different researchers, or from different sources. It can be very difficult to reach conclusions unless we simplify the mass of numbers contained in the original data.

As we discussed in Chapter 1, variables are either quantitative or qualitative. In turn, the appropriate methods for representing the data will depend on whether the variable is quantitative or qualitative. The frequency distribution, histogram, stem-and-leaf display, dotplot, and scatter diagram techniques of this chapter are applicable to quantitative data, while the contingency table is used primarily for counts involving qualitative data.

(2.2) THE FREQUENCY DISTRIBUTION AND THE HISTOGRAM

Raw data have not been manipulated or treated in any way beyond their original collection. As such, they will not be arranged or organized in any meaningful manner. When the data are quantitative, two of the ways we can address this problem are the frequency distribution and the histogram. The **frequency distribution** is a table that divides the data values into classes and shows the number of observed values that fall into each class. By converting data to a frequency distribution, we gain a perspective that helps us see the forest instead of the individual trees. A more visual representation, the **histogram** describes a frequency distribution by using a series of adjacent rectangles, each of which has a length that is proportional to the frequency of the observations within the range of values it represents. In either case, we have summarized the raw data in a condensed form that can be readily understood and easily interpreted.

The Frequency Distribution

We'll discuss the frequency distribution in the context of a research study that involves both safety and fuel-efficiency implications. Data are the speeds (miles per hour) of 105 vehicles observed along a section of highway where both accidents and fuel-inefficient speeds have been a problem.

EXAMPLE

Raw Data and Frequency Distribution

Part A of Table 2.1 lists the raw data consisting of measured speeds (mph) of 105 vehicles along a section of highway. There was a wide variety of speeds, and these data values are contained in data file **CX02SPEE**. If we want to learn more from this information by visually summarizing it, one of the ways is to construct a frequency distribution like the one shown in part B of the table.

EXAMPLEEXAMP

TABLE 2.1

Raw data and frequency distribution for observed speeds of 105 vehicles.

A. Raw Data

54.2	58.7	71.2	69.7	51.7	75.6	53.7	57.0	82.5	76.8
62.1	67.4	64.6	70.5	48.4	69.0	65.2	64.1	65.8	56.9
82.0	62.0	54.3	73.2	74.5	73.6	76.2	55.1	66.9	62.4
70.7	68.3	62.8	83.5	54.8	68.3	69.4	59.2	60.9	60.4
60.2	75.4	55.4	56.3	77.1	61.2	50.8	67.1	70.6	60.7
56.0	80.7	80.1	56.2	70.1	63.7	70.9	54.6	61.6	63.2
72.2	84.0	76.8	61.6	61.1	80.8	58.3	52.8	74.3	71.4
63.2	57.8	61.9	75.8	80.8	57.3	62.6	75.1	67.6	78.0
52.2	57.6	61.1	66.9	88.5	66.7	61.2	73.9	79.1	70.2
65.2	61.3	69.5	72.8	57.5	71.4	64.7	78.4	67.6	76.3
78.6	66.8	71.1	58.9	61.1					

B. Frequency Distribution (Number of Motorists in Each Category)

Speed (mph)	Number of Motorists
45–under 50	1
50–under 55	9
55–under 60	15
60–under 65	24
65–under 70	17
70–under 75	17
75–under 80	13
80–under 85	8
85–under 90	1

LEEXAMPLEEXAMPLEEXAMPLEEXAMPLE

Key Terms

In generating the frequency distribution in part B of Table 2.1, several judgmental decisions were involved, but there is no single "correct" frequency distribution for a given set of data. There are a number of guidelines for constructing a frequency distribution. Before discussing these rules of thumb and their application, we'll first define a few key terms upon which they rely:

Class Each category of the frequency distribution.

Frequency The number of data values falling within each class.

Class limits The boundaries for each class. These determine which data values are assigned to that class.

Class interval The width of each class. This is the difference between the lower limit of the class and the lower limit of the next higher class. When a frequency distribution is to have equally wide classes, the approximate width of each class is

$$\text{Approximate class width} = \frac{\text{Largest value in raw data} - \text{Smallest value in raw data}}{\text{Number of classes desired}}$$

Class mark The midpoint of each class. This is midway between the upper and lower class limits.

Guidelines for the Frequency Distribution

In constructing a frequency distribution for a given set of data, the following guidelines should be observed:

1. The set of classes must be **mutually exclusive** (i.e., a given data value can fall into only one class). There should be no overlap between classes, and limits such as the following would be inappropriate:

   Not allowed, since a value of 60 *could fit into either class:* 55–60, 60–65

   Not allowed, since there's an overlap between the classes: 50–under 55, 53–under 58

2. The set of classes must be **exhaustive** (i.e., include all possible data values). No data values should fall outside the range covered by the frequency distribution.
3. If possible, the classes should have equal widths. Unequal class widths make it difficult to interpret both frequency distributions and their graphical presentations.
4. Selecting the number of classes to use is a subjective process. If we have too few classes, important characteristics of the data may be buried within the small number of categories. If there are too many classes, many categories will contain either zero or a small number of values. In general, about 5 to 15 classes will be suitable.
5. Whenever possible, class widths should be round numbers (e.g., 5, 10, 25, 50, 100). For the highway speed data, selecting a width of 2.3 mph for each class would enhance neither the visual attractiveness nor the information value of the frequency distribution.
6. If possible, avoid using **open-end classes**. These are classes with either no lower limit or no upper limit—e.g., 85 mph or more. Such classes may not always be avoidable, however, since some data may include just a few values that are either very high or very low compared to the others.

The frequency distribution in part B of Table 2.1 was the result of applying the preceding guidelines. Illustrating the key terms introduced earlier, we will refer to the "50–under 55" class of the distribution:

- **Class limits** 50–under 55. All values are at least 50, but less than 55.
- **Frequency** 9. The number of motorists with a speed in this category.
- **Class interval** 5. The difference between the lower class limit and that of the next higher class, or 55 minus 50.
- **Class mark** 52.5. The midpoint of the interval; this can be calculated as the lower limit plus half the width of the interval, or 50 + (0.5)(5.0) = 52.5.

Relative and Cumulative Frequency Distributions

Relative Frequency Distribution. Another useful approach to data expression is the **relative frequency distribution,** which describes the proportion or percentage of data values that fall within each category. The relative frequency distribution for the speed data is shown in Table 2.2; for example, of the 105 motorists, 15 of them (14.3%) were in the 55–under 60 class.

Relative frequencies can be useful in comparing two groups of unequal size, since the actual frequencies would tend to be greater for each class within the larger group than for a class in the smaller one. For example, if a frequency distribution of incomes for 100 physicians is compared with a frequency distribution for 500 business executives, more executives than physicians would be likely to fall into a given class. Relative frequency distributions would convert the groups to the same *size:* 100 percentage points each. Relative frequencies will play an important role in our discussion of probabilities in Chapter 5.

Cumulative Frequency Distribution. Another approach to the frequency distribution is to list the number of observations that are within or below each of the classes. This is known as a **cumulative frequency distribution.** When cumulative frequencies are divided by the total number of observations, the result is a *cumulative relative frequency distribution.* The "Cumulative Relative Frequency (%)" column in Table 2.2 shows the cumulative relative frequencies for the speed data in Table 2.1. Examining this column, we can readily see that 62.85% of the motorists had a speed less than 70 mph.

Cumulative percentages can also operate in the other direction (i.e., "greater than or within"). Based on Table 2.2, we can determine that 90.48% of the 105 motorists had a speed of at least 55 mph.

TABLE 2.2

Frequencies, relative frequencies, cumulative frequencies, and cumulative relative frequencies for the speed data of Table 2.1.

Speed (mph)	Number of Motorists	Relative Frequency (%)	Cumulative Frequency	Cumulative Relative Frequency (%)
45–under 50	1	0.95	1	0.95
50–under 55	9	8.57	10	9.52
55–under 60	15	14.29	25	23.81
60–under 65	24	22.86	49	46.66
65–under 70	17	16.19	66	62.85
70–under 75	17	16.19	83	79.05
75–under 80	13	12.38	96	91.43
80–under 85	8	7.62	104	99.05
85–under 90	1	0.95	105	100.00

The Histogram

The **histogram** describes a frequency distribution by using a series of adjacent rectangles, each of which has a length proportional to either the frequency or the relative frequency of the class it represents. The histogram in part (a) of Figure 2.1 is based on the speed-measurement data summarized in Table 2.1. The lower class limits (e.g., 45 mph, 50 mph, 55 mph, and so on) have been used in constructing the horizontal axis of the histogram.

The tallest rectangle in part (a) of Figure 2.1 is associated with the 60–under 65 class of Table 2.1, identifying this as the class having the greatest number of observations. The relative heights of the rectangles visually demonstrate how the frequencies tend to drop off as we proceed from the 60–under 65 class to the 65–under 70 class and higher.

The Frequency Polygon

Closely related to the histogram, the **frequency polygon** consists of line segments connecting the points formed by the intersections of the class marks with the class frequencies. Relative frequencies or percentages may also be used in constructing the figure. Empty classes are included at each end so the curve will intersect the horizontal axis. For the speed-measurement data in Table 2.1, these are the 40–under 45 and 90–under 95 classes. (*Note:* Had this been a distribution for which the first nonempty class was "0 but under 5," the empty class at the left would have been "−5 but under 0.") The frequency polygon for the speed-measurement data is shown in part (b) of Figure 2.1.

FIGURE 2.1

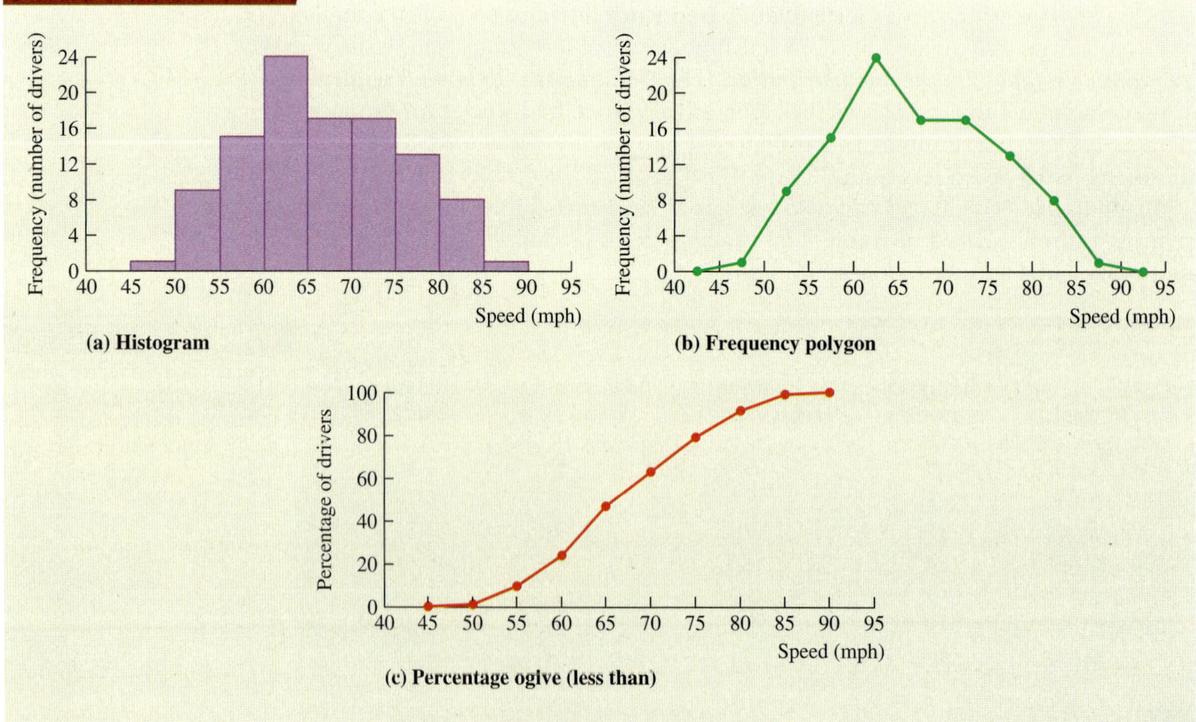

Histogram, frequency polygon, and percentage ogive for the speed-measurement data summarized in Table 2.1

Compared to the histogram, the frequency polygon is more realistic in that the number of observations increases or decreases more gradually across the various classes. The two endpoints make the diagram more complete by allowing the frequencies to taper off to zero at both ends.

Related to the frequency polygon is the **ogive,** a graphical display providing cumulative values for frequencies, relative frequencies, or percentages. These values can be either "greater than" or "less than." The ogive diagram in part (c) of Figure 2.1 shows the percentage of observations that are less than the upper limit of each class.

We can use the computer to generate a histogram as well as the underlying frequency distribution on which the histogram is based. Computer Solutions 2.1 describes the procedures and shows the results when we apply Excel and Minitab in constructing a histogram for the speed data in Table 2.1.

COMPUTER 2.1 SOLUTIONS

The Histogram

EXCEL

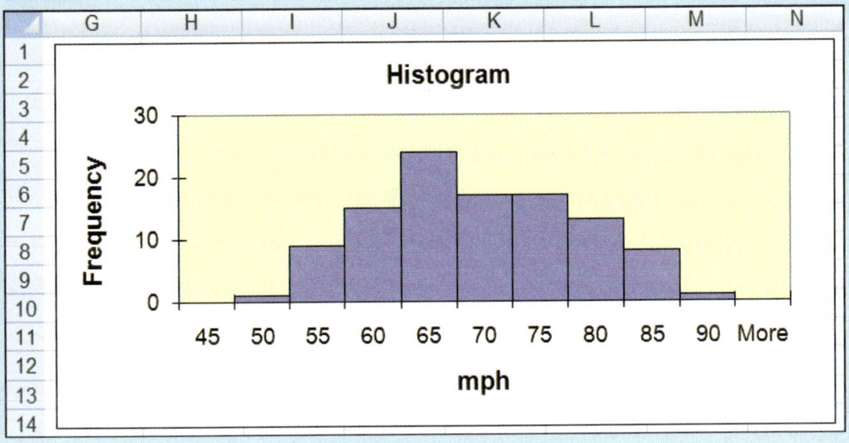

1. Open the Excel data file **CX02SPEE**. The name of the variable (Speed) is in cell A1, and the 105 speeds are in the cells immediately below.
2. Type **Bin** into cell C1. Enter the bin cutoffs (45 to 90, in multiples of 5) into C2:C11. (Alternatively, you can skip this step if you want Excel to generate its default frequency distribution.)
3. From the **Data** ribbon, click **Data Analysis** in the rightmost menu section. Within **Analysis Tools,** click **Histogram.** Click **OK.**
4. Enter the data range **(A1:A106)** into the **Input Range** box. If you entered the bin cutoffs as described in step 2, enter the bin range **(C1:C11)** into the **Bin Range** box. Click to place a check mark in the **Labels** box. (This is because each variable has its name in the first cell of its block.) Select **Output Range** and enter where the output is to begin—this will be cell **E1.**
5. Click to place a check mark into the **Chart Output** box. Click **OK.**
6. Within the chart, click on the word **Bin.** Click again, and type in **mph.** Right-click on any one of the bars in the chart. In the **Format Data Series** menu, click on the **Gap Width** slider and move it leftward to the **No Gap** position. Click **Close.** You can further improve the appearance by clicking on the chart and changing fonts, item locations, such as the key in the lower right, or the background color of the display. In the printout shown here, we have also enlarged the display and moved it slightly to the left.

(continued)

MINITAB

1. Open the Minitab data file **CX02SPEE**. The name of the variable (Speed) is in column C1, and the 105 speeds are in the cells immediately below.

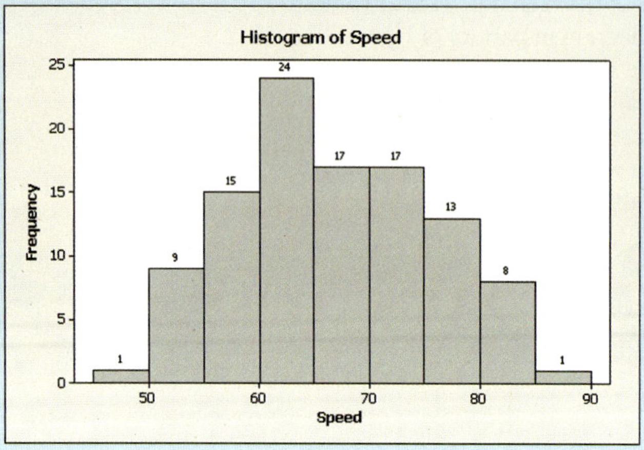

2. Click **Graph.** Click **Histogram.** Select **Simple** and click **OK.** Within the **Histogram** menu, indicate the column to be graphed by entering **C1** into the **Graph variables** box.
3. Click the **Labels** button, then click the **Data Labels** tab. Select **Use y-value labels.** (This will show the counts at the top of the bars.) Click **OK.** Click **OK.**
4. On the graph that appears, double-click on any one of the numbers on the horizontal axis. Click the **Binning** tab. In the **Interval Type** submenu, select **Cutpoint.** In the **Interval Definition** submenu, select **Midpoint/Cutpoint positions** and enter **45:90/5** into the box. (This provides intervals from 45 to 90, with the width of each interval being 5.) Click **OK.**

NOTE

Excel and Minitab differ slightly in how they describe the classes in a frequency distribution. If you use the defaults in these programs, the frequency distributions may differ slightly whenever a data point happens to have exactly the same value as one of the upper limits because

1. Excel includes the *upper* limit but not the *lower* limit; the Excel bin value of "60" in Computer Solutions 2.1 represents values that are "more than 55, but not more than 60." A particular speed (x) will be in Excel's interval if $55 < x \leq 60$.
2. Minitab includes the *lower* limit, but not the *upper* limit, so a category of 55–60 is "at least 55, but less than 60." Thus, a particular speed (x) will be in Minitab's 55–60 interval if $55 \leq x < 60$.

This was not a problem with the speed data and Computer Solutions 2.1. However, if it had been, and we wanted the Excel frequency distribution to be the same as Minitab's, we could have simply used Excel bin values that were very slightly below those of Minitab's upper limits—for example, a bin value of 44.99 instead of 45.00, 49.99 instead of 50, 54.99 instead of 55, and so on.

EXERCISES

2.1 What is a frequency distribution? What benefits does it offer in the summarization and reporting of data values?

2.2 Generally, how do we go about deciding how many classes to use in a frequency distribution?

2.3 The National Safety Council reports the following age breakdown for licensed drivers in the United States. Source: Bureau of the Census, *Statistical Abstract of the United States 2009,* p. 682.

Age (Years)	Licensed Drivers (Millions)
under 20	10.72
20–under 25	16.84
25–under 35	40.78
35–under 45	45.59
45–under 55	36.70
55–under 65	23.32
65 or over	28.76

Identify the following for the 35–under 45 class: (a) frequency, (b) upper and lower limits, (c) width, and (d) midpoint.

2.4 Using the frequency distribution in Exercise 2.3, identify the following for the 25–under 35 class: (a) frequency, (b) upper and lower limits, (c) width, and (d) midpoint.

2.5 The National Center for Health Statistics reports the following age breakdown of deaths in the United States during 2006. Source: *The New York Times Almanac 2009,* p. 392.

Age (Years)	Number of Deaths (Thousands)
under 15	39.80
15–under 25	34.23
25–under 35	41.93
35–under 45	84.79
45–under 55	183.53
55–under 65	275.30
65–under 75	398.36
75 or over	1389.83

Identify the following for the 45–under 55 class: (a) frequency, (b) upper and lower limits, (c) width, and (d) midpoint.

2.6 Using the frequency distribution in Exercise 2.5, identify the following for the 15–under 25 class: (a) frequency, (b) upper and lower limits, (c) width, and (d) midpoint.

2.7 What is meant by the statement that the set of classes in a frequency distribution must be mutually exclusive and exhaustive?

2.8 For commercial banks in each state, the U.S. Federal Deposit Insurance Corporation has listed their total deposits (billions of dollars) as follows. Source: Bureau of the Census, *Statistical Abstract of the United States 2009,* p. 722.

	Deposits		Deposits		Deposits
AL	74.3	LA	73.0	OH	209.1
AK	6.9	ME	20.0	OK	57.4
AZ	80.0	MD	95.0	OR	47.9
AR	47.4	MA	180.8	PA	259.4
CA	751.0	MI	154.7	RI	25.5
CO	81.3	MN	106.2	SC	64.1
CT	81.0	MS	41.6	SD	63.3
DE	160.2	MO	102.6	TN	107.2
FL	373.9	MT	14.6	TX	450.0
GA	177.9	NE	36.0	UT	181.7
HI	26.8	NV	188.1	VT	9.9
ID	17.7	NH	21.5	VA	182.2
IL	338.9	NJ	222.5	WA	105.7
IN	88.6	NM	21.9	WV	25.9
IA	56.6	NY	722.8	WI	109.7
KS	54.0	NC	206.3	WY	10.6
KY	63.2	ND	13.9		

Construct a frequency distribution and a histogram for these data.

2.9 The accompanying data describe the hourly wage rates (dollars per hour) for 30 employees of an electronics firm:

22.66	24.39	17.31	21.02	21.61	20.97	18.58	16.61
19.74	21.57	20.56	22.16	20.16	18.97	22.64	19.62
22.05	22.03	17.09	24.60	23.82	17.80	16.28	19.34
22.22	19.49	22.27	18.20	19.29	20.43		

Construct a frequency distribution and a histogram for these data.

2.10 The following performance scores have been recorded for 25 job applicants who have taken a preemployment aptitude test administered by the company to which they applied:

66.6	75.4	66.7	59.2	78.5	80.8	79.9	87.0	94.1
70.2	92.8	86.9	92.8	66.8	65.3	100.8	76.2	87.8
71.0	92.9	97.3	82.5	78.5	72.0	76.2		

Construct a frequency distribution and a histogram for these data.

2.11 During his career in the NHL, hockey great Wayne Gretzky had the following season-total goals for each of his 20 seasons. Source: *The World Almanac and Book of Facts 2000,* p. 954.

51	55	92	71	87	73	52	62	40	54
40	41	31	16	38	11	23	25	23	9

Construct a frequency distribution and a histogram for these data.

2.12 According to the U.S. Department of Agriculture, the distribution of U.S. farms according to value of annual sales is as follows. Source: Bureau of the Census, *Statistical Abstract of the United States 2006,* p. 546.

Annual Sales	Farms (Thousands)
under \$10,000	1262
\$10,000–under \$25,000	256
\$25,000–under \$50,000	158
\$50,000–under \$100,000	140
\$100,000–under \$250,000	159
\$250,000–under \$500,000	82
\$500,000–under \$1,000,000	42
\$1,000,000 or more	29

Convert this information to a
a. Relative frequency distribution.
b. Cumulative frequency distribution showing "less than or within" frequencies.

2.13 Convert the distribution in Exercise 2.3 to a
a. Relative frequency distribution.
b. Cumulative frequency distribution showing "less than or within" frequencies.

2.14 Convert the distribution in Exercise 2.8 to a
a. Relative frequency distribution.
b. Cumulative relative frequency distribution showing "greater than or within" relative frequencies.

2.15 Using the frequency distribution obtained in Exercise 2.8, convert the information to a "less-than" ogive.

2.16 For the frequency distribution constructed in Exercise 2.11, convert the information to a "less-than" ogive.

(DATA SET) *Note:* Exercises 2.17–2.19 require a computer and statistical software.

2.17 The current values of the stock portfolios for 80 clients of an investment counselor are as listed in data file **XR02017**. Use your computer statistical software to generate a frequency distribution and histogram describing this information. Do any of the portfolio values seem to be especially large or small compared to the others?

2.18 One of the ways to evaluate the ease of navigating a corporate website is to measure how long it takes to reach the link where you can view or download the company's most recent annual report. In randomly selecting and visiting 80 corporate websites, an investor needed an average of 39.9 seconds to reach the point where the company's annual report could be viewed or downloaded. The 80 times are in data file **XR02018**. Generate a frequency distribution and histogram describing this information. Comment on whether any of the corporate sites seemed to require especially long or short times for the annual report to be found.

2.19 A company executive has read with interest the finding that the average U.S. office worker receives about 60 e-mails per day. Assume that an executive, wishing to replicate this study within her own corporation, directs information technology personnel to find out the number of e-mails each of a sample of 100 office workers received yesterday, with the results as provided in the data file **XR02019**. Generate a frequency distribution and histogram describing this information and comment on the extent to which some workers appeared to be receiving an especially high or low number of e-mails. Source: emailstatcenter.com, June 20, 2009.

2.3 THE STEM-AND-LEAF DISPLAY AND THE DOTPLOT

In this section, we examine two additional methods of showing how the data values are distributed: the stem-and-leaf display and the dotplot. Each of the techniques can be carried out either manually or with the computer and statistical software.

The Stem-and-Leaf Display

The **stem-and-leaf display,** a variant of the frequency distribution, uses a subset of the original digits as class descriptors. The technique is best explained through a few examples. The raw data are the numbers of Congressional bills vetoed during the administrations of seven U.S. presidents, from Johnson to Clinton.[1]

	Johnson	Nixon	Ford	Carter	Reagan	Bush	Clinton
Vetoes	30	43	66	31	78	44	38

In *stem-and-leaf* terms, we could describe these data as follows:

Stem (10's Digit)	Leaf (1's Digit)
3|018	(represents 30, 31, and 38)
4|34	(represents 43 and 44)
5|	(no data values in the 50s)
6|6	(represents 66)
7|8	(represents 78)

The figure to the left of the divider (|) is the **stem,** and the digits to the right are referred to as **leaves.** By using the digits in the data values, we have identified five different categories (30s, 40s, 50s, 60s, and 70s) and can see that there are three data values in the 30s, two in the 40s, one in the 60s, and one in the 70s.

Like the frequency distribution, the stem-and-leaf display allows us to quickly see how the data are arranged. For example, none of these presidents vetoed more than 44 bills, with the exceptions of Presidents Ford and Reagan, who vetoed 66 and 78, respectively. Compared with the frequency distribution, the stem-and-leaf display provides more detail, since it can describe the individual data values as well as show how many are in each group, or stem. Computer Solutions 2.2 shows the procedure and results when Excel and Minitab are used to generate a stem-and-leaf display based on the speed data of Table 2.1.

When the stem-and-leaf display is computer generated, the result may vary slightly from the previous example of the presidential vetoes. For example, Minitab may do the following:

1. Break each stem into two or more lines, each of which will have leaves in a given range. In Computer Solutions 2.2, speeds in the 50s are broken into two lines. The first line includes those with a leaf that is 0 to 4 (i.e., speeds from 50.x to 54.x), the second covers those for which the leaf is from 5 to 9 (i.e., speeds from 55.x to 59.x).
2. Include stem-and-leaf figures for outliers. An *outlier* is a data value very distant from most of the others. When outliers are present, they can be shown separately in a "HI" or a "LO" portion to avoid stretching the display and ending up with a large number of stems that have no leaves.

The stem-and-leaf display shows just two figures for each data value. For example, the data value 1475 could be represented as

1|4 (1000's digit and 100's digit)

or

14|7 (100's digit and 10's digit)

or

147|5 (10's digit and 1's digit)

In a given stem-and-leaf display, only one of these alternatives could be used because (1) a given data value can be represented only once in a display, and

[1]Source: Bureau of the Census, *Statistical Abstract of the United States 2009,* p. 247.

COMPUTER 2.2 SOLUTIONS

The Stem-and-Leaf Display

EXCEL

	A	B	C	D	E	F
1	**Stem & Leaf Display**					
2						
3	**Stems**	**Leaves**				
4	4	->8				
5	5	->012234444556666777778889				
6	6	->00001111111112222233344455566667777889999				
7	7	->000000111122333445555666678889				
8	8	->000022348				

1. Open the Excel data file **CX02SPEE**. Click on **A1** and drag to **A106** to select the label and data values in cells **A1:A106.** From the **Add-Ins** ribbon, click **Data Analysis Plus** in the leftmost menu section.
2. In the **Data Analysis Plus** menu, click **Stem and Leaf Display.** Click **OK.** Ensure that the **Input Range** box contains the range you specified in step 1. In the **Increment** menu box, select **10** (because the 10's digit is the stem for the display). Click **Labels.** Click **OK.**

MINITAB

```
Stem-and-Leaf Display: Speed
Stem-and-leaf of Speed  N  = 105
Leaf Unit = 1.0

  1    4  8
 10    5  012234444
 25    5  556666777778889
 49    6  000011111111122222333444
(17)   6  55566667777889999
 39    7  00000011112233344
 22    7  5555666678889
  9    8  00002234
  1    8  8
```

1. Open the Minitab data file **CX02SPEE**. The variable (Speed) is in column C1.
2. Click **Graph.** Click **Stem-And-Leaf.** Enter **C1** into the **Variables** box. Click **OK.**

(2) all of the values must be expressed in terms of the same stem digit and the same leaf digit. If we were to deviate from either of these rules, the resulting display would be meaningless.

When data values are in decimal form, such as 3.4, 2.5, and 4.1, the stem-and-leaf method can still be used. In expressing these numbers, the stem would be the 1's digit, and each leaf would correspond to the first number to the right of the decimal point. For example, the number 3.4 would be converted to a stem of 3 and a leaf of 4. The location of the decimal point would have to be considered during interpretation of the display.

COMPUTER 2.3 SOLUTIONS

The Dotplot

EXCEL

Excel does not currently provide the dotplot display.

MINITAB

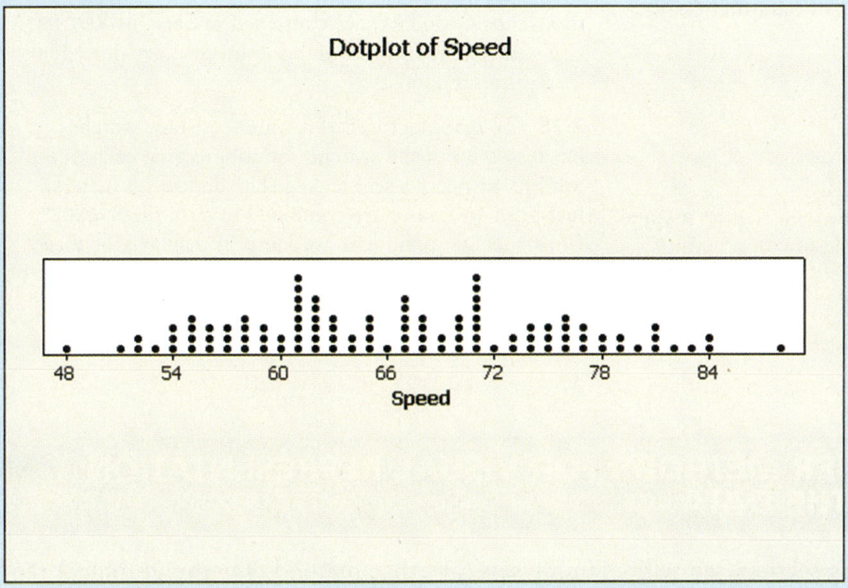

1. Open the Minitab data file **CX02SPEE**. The variable (Speed) is in column C1.
2. Click **Graph.** Click **Dotplot.** Select **One Y Simple.** Click **OK.** Enter **C1** into the **Variables** box. Click **OK.**

The Dotplot

The **dotplot** displays each data value as a dot and allows us to readily see the shape of the distribution as well as the high and low values. Computer Solutions 2.3 shows the procedure and the result when we apply Minitab's dotplot feature to the speed data shown previously in Table 2.1.

EXERCISES

2.20 Construct a stem-and-leaf display for the following data:

15	64	15	34	75	24
81	67	19	25	48	57
69	62	41	46	35	27
72	64	48	51	77	71
21	20	26	42	83	38

2.21 Construct a stem-and-leaf display for the following data:

11	22	28	32	41	46	52	56	59
13	24	28	33	41	46	52	57	60
17	24	29	33	42	46	53	57	61
19	24	30	37	43	48	53	57	61
19	26	30	39	43	49	54	59	63

2.22 In the following stem-and-leaf display for a set of two-digit integers, the stem is the 10's digit, and each leaf is the 1's digit. What is the original set of data?

2|002278
3|011359
4|1344
5|47

2.23 In the following stem-and-leaf display for a set of three-digit integers, the stem is the 100's digit, and each leaf is the 10's digit. Is it possible to determine the exact values in the original data from this display? If so, list the data values. If not, provide a set of data that could have led to this display.

6|133579
7|00257
8|66899
9|0003568

(DATA SET) *Note:* Exercises 2.24–2.28 require a computer and statistical software capable of constructing stem-and-leaf and dotplot displays.

2.24 Construct stem-and-leaf and dotplot displays based on the investment portfolio values in the data file (**XR02017**) for Exercise 2.17.

2.25 Construct stem-and-leaf and dotplot displays based on the access times in the data file (**XR02018**) for Exercise 2.18.

2.26 Construct stem-and-leaf and dotplot displays based on the e-mail receipts in the data file (**XR02019**) for Exercise 2.19.

2.27 City tax officials have just finished appraising 60 homes near the downtown area. The appraisals (in thousands of dollars) are contained in data file **XR02027**. Construct stem-and-leaf and dotplot displays describing the appraisal data.

2.28 The manufacturer of a water-resistant watch has tested 80 of the watches by submerging each one until its protective seal leaked. The depths (in meters) just prior to failure are contained in data file **XR02028**. Construct stem-and-leaf and dotplot displays for these data.

(2.4) OTHER METHODS FOR VISUAL REPRESENTATION OF THE DATA

In this section, we will examine several other methods for the graphical representation of data, then discuss some of the ways in which graphs and charts can be used (by either the unwary or the unscrupulous) to deceive the reader or viewer. We will also provide several Computer Solutions to guide you in using Excel and Minitab to generate some of the more common graphical presentations. These are just some of the more popular approaches. There are many other possibilities.

The Bar Chart

Like the histogram, the **bar chart** represents frequencies according to the relative lengths of a set of rectangles, but it differs in two respects from the histogram: (1) the histogram is used in representing quantitative data, while the bar chart represents qualitative data; and (2) adjacent rectangles in the histogram share a common side, while those in the bar chart have a gap between them. Computer Solutions 2.4 shows the Excel and Minitab procedures for generating a bar chart describing the number of U.S. nonrecreational airplane pilot certificates in each of four categories.[2] The relevant data file is **CX02AIR**.

[2]Source: General Aviation Manufacturers Association, *General Aviation Statistical Databook 2008 Edition*, p. 38.

The Line Graph

The **line graph** is capable of simultaneously showing values of two quantitative variables (y, or vertical axis, and x, or horizontal axis); it consists of linear segments connecting points observed or measured for each variable. When x represents time, the result is a time series view of the y variable. Even more information can be presented if two or more y variables are graphed together. Computer Solutions 2.5 shows the Excel and Minitab procedures for generating a line chart showing annual net income for McDonald's Corporation from 2003 through 2008.[3] The pertinent data file is CX02MCD.

COMPUTER 2.4 SOLUTIONS

The Bar Chart

EXCEL

	A	B	C	D
1	Nonrecreational Airplane Pilot Certificates			
2				
3	Certificate Type		Number (thousands)	
4	Student		80.989	
5	Private		222.596	
6	Commercial		124.746	
7	Transport		146.838	
8				
9				
10				
11				
12				
13				

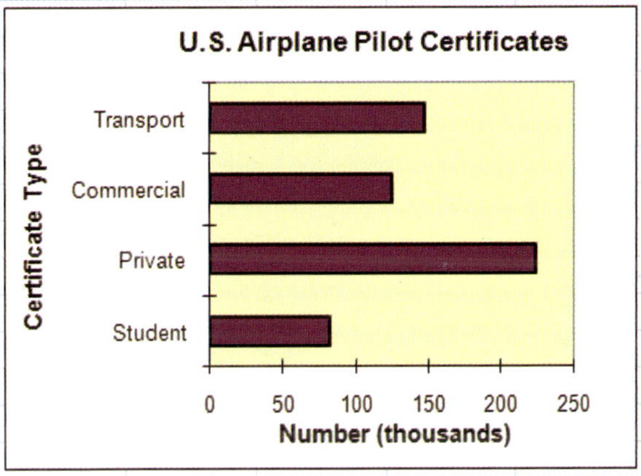

1. Open the Excel data file CX02AIR. The labels, **Certificate Type** and **Number (thousands),** have already been entered into **A3** and **C3,** respectively. The name for the first category has already been entered into **A4,** and its corresponding numerical value (thousands of pilots) has been entered into **C4.** We have continued downward, entering category names into column A and corresponding numerical values into column C, until all category names and their values have been entered into columns A and C.
2. Click on **A4** and drag to **C7** to select cells **A4:C7.**
3. From the **Insert** ribbon, click **Bar** from the **Charts** submenu. From the next menu, click the first choice **(Clustered Bar)** from the **2-D Bar** section.
4. To further improve the appearance, click on the chart and drag the borders to expand it vertically and horizontally. If you don't want the little "Legend" box that originally appears, just right-click within it, then click **Delete.** Additional editing can include insertion of the desired labels: Double-click within the chart area. From the **Layout** ribbon, use the **Chart Title** and **Axis Titles** selections within the **Labels** menu.

(continued)

[3]Source: McDonald's Corporation, *2008 Annual Report*, p. 19.

MINITAB

1. Open the Minitab data file CX02AIR. The labels, **Certificate Type** and **Number (thousands),** have already been entered at the top of columns **C1** and **C2,** respectively. The names of the categories have already been entered into **C1,** and the numerical values (thousands of pilots) have been entered into **C2.**
2. Click **Graph.** Click **Bar Chart.** In the **Bars represent** box, select **Values from a table.** Select **One column of values, Simple.** Click **OK.** Within the **Graph variables** menu, enter **C2.** Enter **C1** into the **Categorical variable** box. Click **Scale.** Select the **Axes and Ticks** tab and select **Transpose value and category scales.** Click **OK.** Click **OK.**

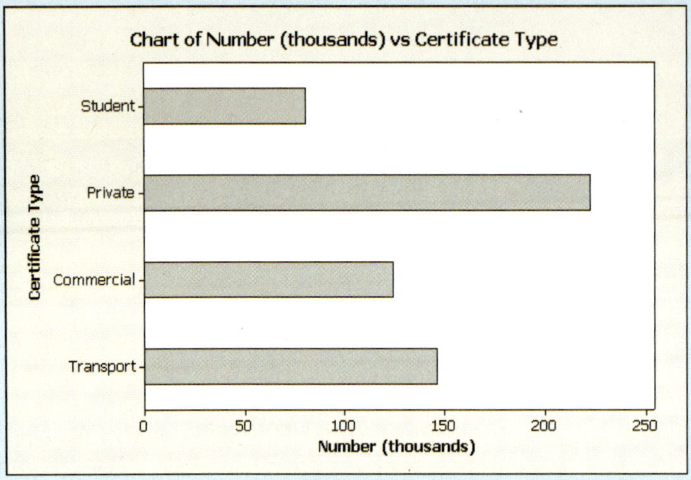

COMPUTER 2.5 SOLUTIONS

The Line Chart

EXCEL

	A	B
1	McDonald's Net Income	
2		
3	**Year**	**Net Income**
4	2003	1.471
5	2004	2.279
6	2005	2.602
7	2006	3.544
8	2007	2.395
9	2008	4.313
10		
11		
12		
13		

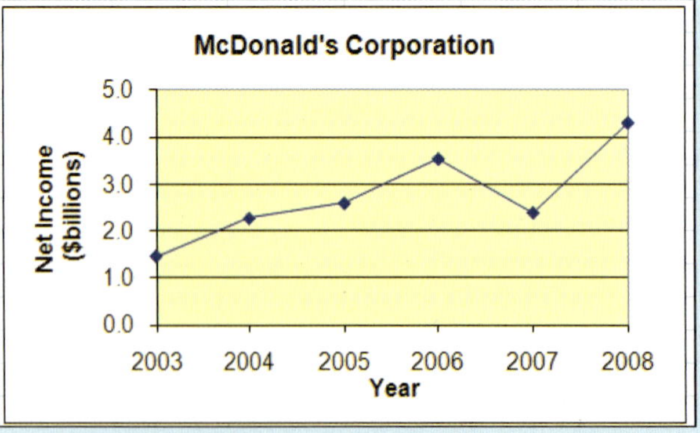

1. Open the Excel data file **CX02MCD**. The labels and data are already entered, as shown in the printout. Click on **A4** and drag to **B9** to select cells **A4:B9.** (*Note:* Within the two columns, the variable to be represented by the vertical axis should always be in the column at the right.)
2. From the **Insert** ribbon, click **Scatter** from the **Charts** menu. From the **Scatter** submenu, click the fifth option **(Scatter with Straight Lines).**
3. Further appearance improvements can be made by clicking on the chart and dragging the borders to expand it vertically and horizontally. If you don't want the little "Legend" box that originally appears, just right-click within it, then click **Delete.** Additional editing can include insertion of the desired labels: Double-click within the chart area. From the **Layout** ribbon, use the **Chart Title** and **Axis Titles** selections within the **Labels** menu.

MINITAB

1. Open the Minitab data file **CX02MCD**. The labels, **Year** and **Net Income,** have already been entered at the top of columns **C1** and **C2,** respectively. The years and the numerical values have been entered into **C1** and **C2,** respectively.
2. Click **Graph.** Click **Scatterplot.** Select **With Connect Line.** Click **OK.** Enter **C2** into the first line of the **Y variables** box. Enter **C1** into the first line of the **X variables** box. Click **OK.**
3. Double-click on the chart title and the vertical-axis labels and revise as shown.

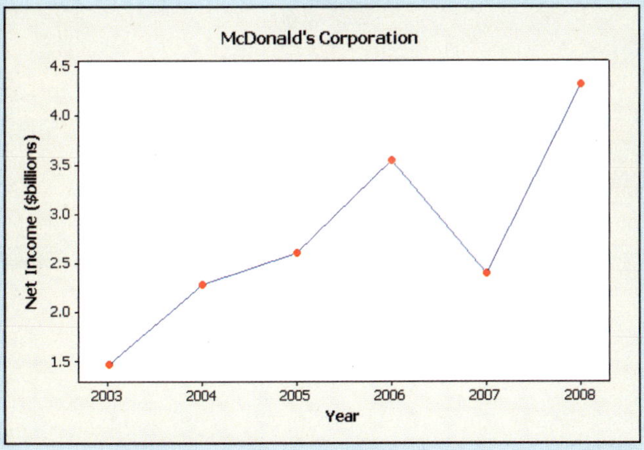

The Pie Chart

The **pie chart** is a circular display divided into sections based on either the number of observations within or the relative values of the segments. If the pie chart is not computer generated, it can be constructed by using the principle that a circle contains 360 degrees. The angle used for each piece of the pie can be calculated as follows:

$$\begin{matrix}\text{Number of degrees}\\ \text{for the category}\end{matrix} = \begin{matrix}\text{Relative value}\\ \text{of the category}\end{matrix} \times 360$$

For example, if 25% of the observations fall into a group, they would be represented by a section of the circle that includes (0.25×360), or 90 degrees. Computer Solutions 2.6 shows the Excel and Minitab procedures for generating a pie chart to show the relative importance of four major business segments in contributing to Home Depot Corporation's overall profit.[4] The underlying data file is **CX02HDEP**.

[4]Source: Home Depot Corporation, *2008 Annual Report*, pp. 3, 28.

The Pictogram

Using symbols instead of a bar, the **pictogram** can describe frequencies or other values of interest. Figure 2.2 is an example of this method; it was used by Panamco to describe soft drink sales in Central America over a 3-year period. In the diagram, each truck represents about 12.5 million cases of soft drink products. When setting up a pictogram, the choice of symbols is up to you. This is an important consideration because the right (or wrong) symbols can lend nonverbal or emotional content to the display. For example, a drawing of a sad child with her arm in a cast

FIGURE 2.2

In the pictogram, the symbols represent frequencies or other values of interest. This chart shows how soft drink sales (millions of cases) in Central America increased from 1996 through 1998.

Source: Panamco, *Annual Report 1998,* p. 24.

COMPUTER 2.6 SOLUTIONS

The Pie Chart

EXCEL

	A	B
1	Home Depot Sales ($billions)	
2		
3	Segment:	Sales
4	Plumbing, electrical and kitchen	22.099
5	Hardware and seasonal	19.961
6	Building materials, lumber, millwork	15.897
7	Paints and flooring	13.331
8		
9		
10		
11		
12		
13		

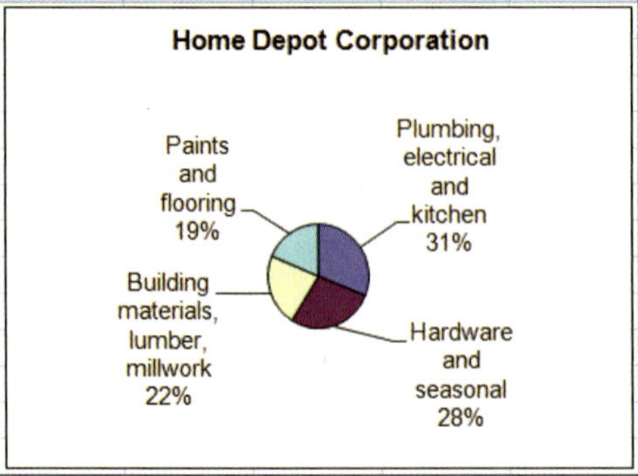

1. Open the Excel data file **CX02HDEP**. The segment names have already been entered as shown in the display, as have the sales for each segment.
2. Click on **A4** and drag to **B7** to select cells **A4:B7.** From the **Insert** ribbon and its **Charts** menu, click **Pie.** Click on the first option (**Pie**) in the **2-D Pie** menu.
3. Additional editing can include insertion of the desired chart title: Double-click within the chart area. From the **Layout** ribbon, use the **Chart Title** selection within the **Labels** menu.

MINITAB

1. Open the Minitab data file **CX02HDEP**. The labels, **Segment** and **Sales,** have already been entered at the top of columns **C1** and **C2,** respectively. The names of the segments have already been entered into **C1** and the numerical values (sales, billions of dollars) have been entered into **C2.**
2. Click **Graph.** Click **Pie Chart.** Select **Chart values from a table.** Enter **C1** into the **Categorical variable** box. Enter **C2** into the **Summary variables** box. Click **OK.**
3. Double-click on the chart title and revise as shown.

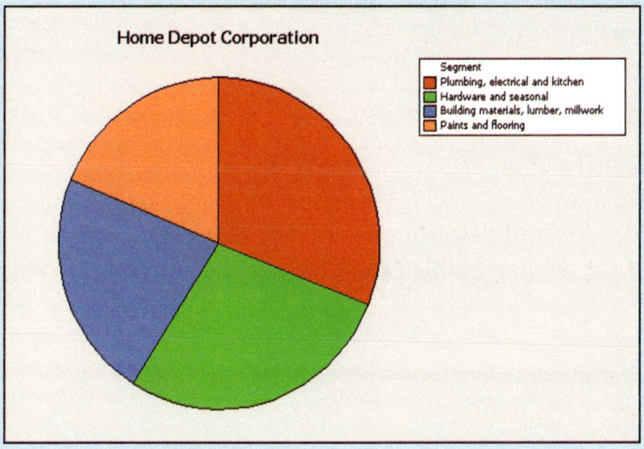

(each symbol representing 10,000 abused children) could help to emphasize the emotional and social costs of child abuse.

The Sketch

Varying in size depending on the frequency or other numerical value displayed, the **sketch** is a drawing or pictorial representation of some symbol relevant to the data. This approach will be demonstrated in part (c) of Figure 2.4.

Other Visuals

The preceding approaches are but a few of the many possibilities for the visual description of information. As noted in the chapter-opening vignette, *USA Today* readers are treated daily to a wide variety of interesting and informative displays, including the "Statistics That Shape Our ... " visuals in the lower-left corner of the front page of each section. One of these displays is shown in Figure 2.3.

FIGURE 2.3

One of the many creative ways *USA Today* and its "USA Snapshots" present informative statistics in an interesting way.

Source: From USA SNAPSHOTS®, "Workers Take Happiness Over Money", *USA Today*, September 20, 2001. Copyright 2001. Reprinted with permission.

The Abuse of Visual Displays

Remember that visuals can be designed to be either emotionally charged or purposely misleading to the unwary viewer. This capacity to mislead is shared by a great many statistical tests and descriptions, as well as visual displays. We will consider just a few of the many possible examples where graphical methods could be viewed as misleading.

FIGURE 2.4

The information presented may be distorted in several ways. Part (a) shows the effect of compressing the data by using a high endpoint for the vertical axis. In part (b), the change is exaggerated by taking a slice from the vertical axis. In part (c), although the sketch representing 2004 is 65.7% higher and 65.7% wider than the 2003 sketch, the increase is distorted because the area of the 2004 sketch is 274.6% as large as that of the 2003 sketch.

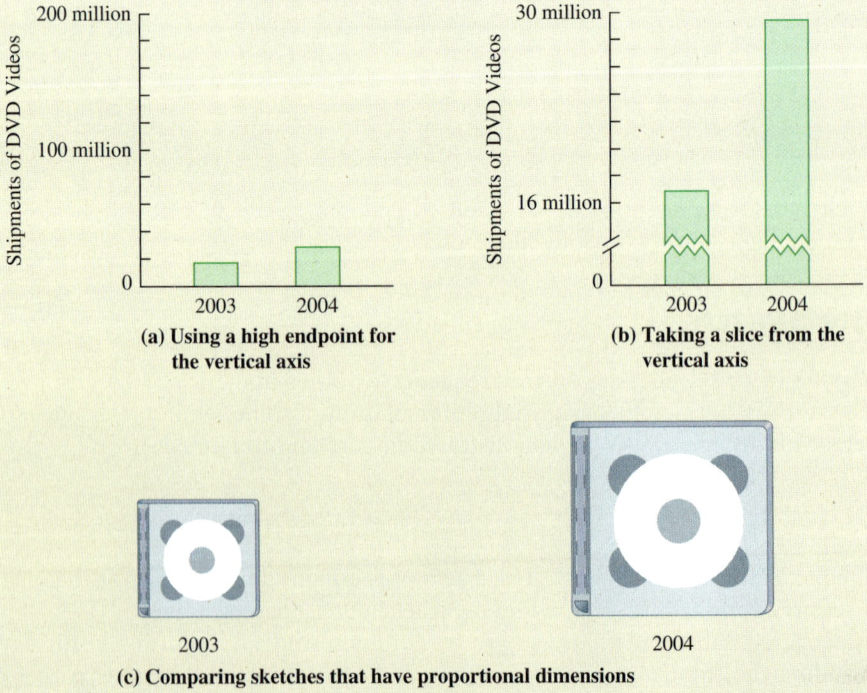

During 2004, 29.0 million DVD videos were shipped, compared to 17.5 million in 2003.[5] Using these data, we can construct the charts in Figure 2.4. In part (a), the vertical axis is given a very high value as its upper limit, and the data are "compressed" so that the increase in shipments isn't very impressive. In part (b) of Figure 2.4, taking a slice from the vertical axis causes the increase in shipments to be more pronounced. Another strategy for achieving the same effect is to begin the vertical axis with a value other than zero.

In part (c) of Figure 2.4, we see how shipments in the two years could be compared using a sketch of a DVD video case. Although the sketch on the right is 65.7% higher than the one on the left, it *looks more than* 65.7% larger. This is because area for each sketch is height times width, and both the height *and* the width are 65.7% greater for the one at the right. Because both height and width are increased, the sketch for 2004 shipments has an area 274.6% (based on 1.657 × 1.657) as large as its 2003 counterpart.

Figure 2.5 is another example of visual distortion by using a nonzero vertical axis. Over the years shown, sales increased by only 1.4%, and this dismal performance is evident from part (a) of the figure. However, in part (b), the starting

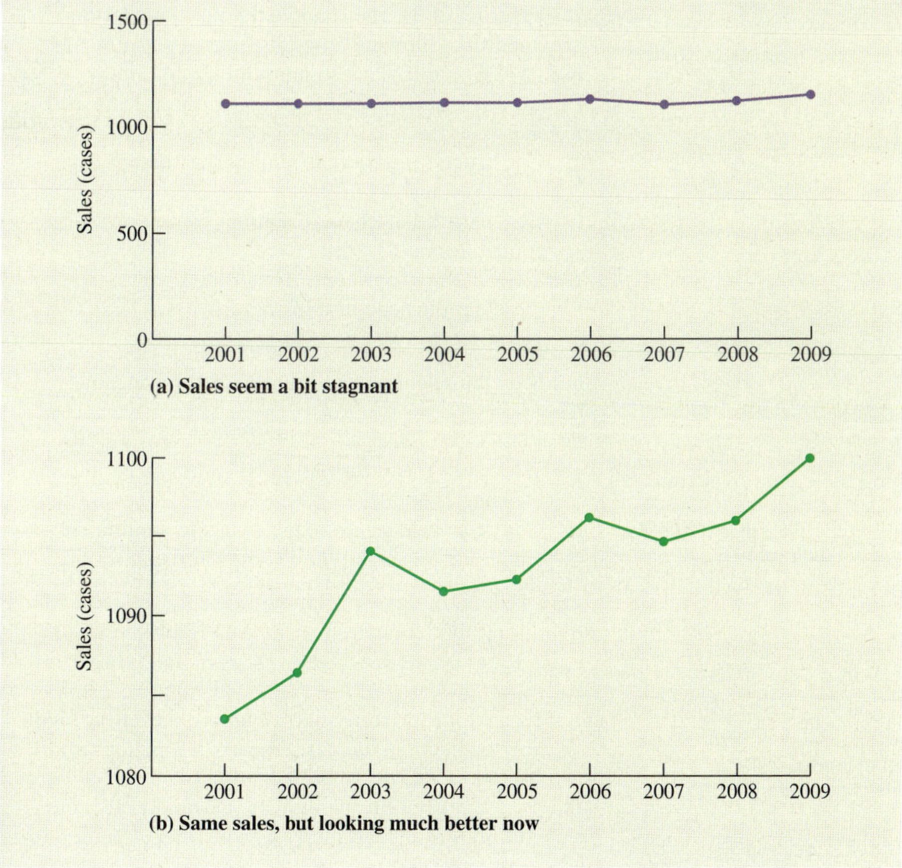

FIGURE 2.5

If the vertical scale is stretched and has a starting point that is greater than zero, the sales for this firm can look a lot better than the 1.4% growth that occurred over the 8-year period.

[5]Data source: Data (but not the visual displays) are from *The World Almanac and Book of Facts 2006*, p. 276.

point for the vertical axis is 1080 cases instead of 0, and the vertical axis is stretched considerably. As a result, it looks as though the company has done very well over the years shown.

EXERCISES

Note: These graphs and charts can be done by hand, but, if possible, use the computer and your statistical software.

2.29 What is the difference between a histogram and a bar chart? For what type of data would each be appropriate?

2.30 The following U.S. market shares for 2007 retail car sales have been reported. Source: *World Almanac and Book of Facts 2009*, p. 122.

Country or Region of Manufacture	2007 market share
North America	69.0%
Japan	15.5
Germany	7.4
Other	8.1

Express the data in the form of a bar chart.

2.31 The median income for a four-person family has been reported as shown here for 1993–2003. Source: *Time Almanac 2006*, p. 627.

1993	$45,161	1999	$59,981
1994	47,012	2000	62,228
1995	49,687	2001	63,278
1996	51,518	2002	62,732
1997	53,350	2003	65,093
1998	56,061		

Construct a line graph describing these data over time.

2.32 For McDonald's Corporation, net income per share of common stock was as follows for the period 1998–2008. Source: McDonald's Corporation, *2005 Annual Report*, p. 1; *2008 Annual Report*, p. 19.

1998	$1.10	2004	$1.79
1999	1.39	2005	2.04
2000	1.46	2006	2.83
2001	1.25	2007	1.98
2002	0.70	2008	3.76
2003	1.15		

Construct a line graph describing these data over time.

2.33 For the period 1998–2008, McDonald's Corporation paid the following dividends (rounded to nearest penny) per share of common stock. Source: McDonald's Corporation, *2005 Annual Report*, p. 1; *2008 Annual Report*, p. 19.

1998	$0.18	2004	$0.55
1999	0.20	2005	0.67
2000	0.22	2006	1.00
2001	0.23	2007	1.50
2002	0.24	2008	1.63
2003	0.40		

Using these data and the data in Exercise 2.32, draw a line graph that describes both net income and dividend per share of common stock over time. (Both lines will be plotted on the same graph.)

2.34 During 2001, the Campbell Soup Foundation provided the following amounts in grants: Camden, N.J., $1,336,700; plant communities, $341,500; Dollars for Doers, $179,600; other projects, $64,100. Source: The Campbell Soup Foundation, *2002 Annual Report*, p. 16.

a. Construct a pie chart to summarize these contributions.
b. Construct a bar chart to summarize these contributions.
c. Why is it appropriate to construct a bar chart for these data instead of a histogram?

2.35 It has been estimated that 92.9% of U.S. households own a telephone and that 68.5% have cable television. Source: Bureau of the Census, *Statistical Abstract of the United States 2009*, p. 696.

a. Why can't this information be summarized in one pie chart?
b. Construct two separate pie charts to summarize telephone ownership and cable television service.

2.36 Federal outlays for national defense totaled $553 billion in 2007 compared with just $134 billion in 1980. Source: Bureau of the Census, *Statistical Abstract of the United States 2009*, p. 324.

a. Using appropriate symbols that would reflect favorably on such an increase, construct a pictogram to compare 2007 with 1980.
b. Using appropriate symbols that would reflect unfavorably on such an increase, construct a pictogram to compare 2007 with 1980.

2.37 For the data in Exercise 2.36, use the sketch technique (and an appropriate symbol of your choice) to compare the 2007 and 1980 expenditures.

2.5 THE SCATTER DIAGRAM

There are times when we would like to find out whether there is a relationship between two quantitative variables—for example, whether sales is related to advertising, whether starting salary is related to undergraduate grade point average, or whether the price of a stock is related to the company's profit per share. To examine whether a relationship exists, we can begin with a graphical device known as the **scatter diagram,** or **scatterplot.**

Think of the scatter diagram as a sort of two-dimensional dotplot. Each point in the diagram represents a pair of known or observed values of two variables, generally referred to as y and x, with y represented along the vertical axis and x represented along the horizontal axis. The two variables are referred to as the **dependent** (y) and **independent** (x) variables, since a typical purpose for this type of analysis is to estimate or predict what y will be for a given value of x.

Once we have drawn a scatter diagram, we can "fit" a line to it in such a way that the line is a reasonable approximation to the points in the diagram. In viewing the "best-fit" line and the nature of the scatter diagram, we can tell more about whether the variables are related and, if so, in what way.

1. A *direct (positive) linear relationship* between the variables, as shown in part (a) of Figure 2.6. The best-fit line is linear and has a positive slope, with both y and x increasing together.
2. An *inverse (negative) linear relationship* between the variables, as shown in part (b) of Figure 2.6. The best-fit line is linear and has a negative slope, with y decreasing as x increases.

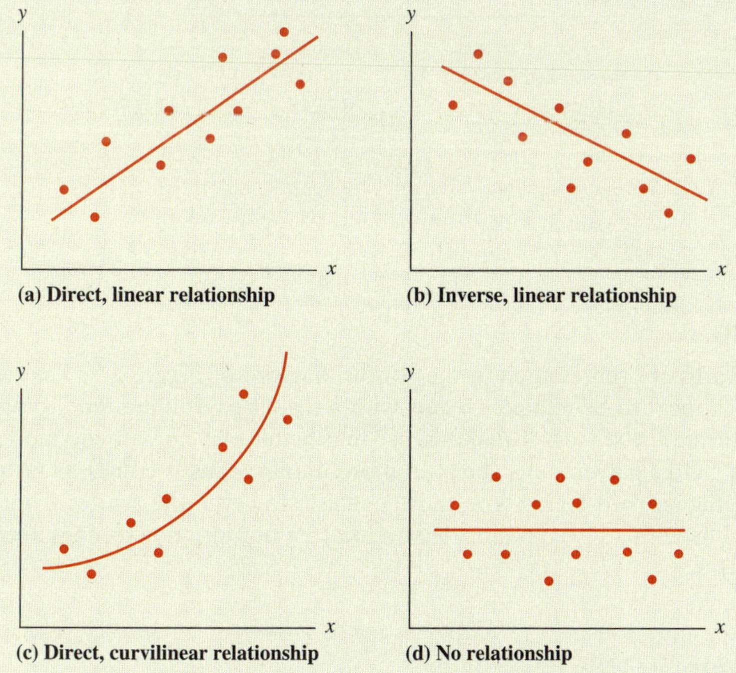

FIGURE 2.6

When two variables are related, as in parts (a) through (c), the relationship can be either direct (positive) or inverse (negative), and either linear or curvilinear. In part (d), there is no relationship at all between the variables.

3. A *curvilinear relationship* between the variables, as shown in part (c) of Figure 2.6. The best-fit line is a curve. As with a linear relationship, a curvilinear relationship can be either direct (positive) or inverse (negative).
4. *No relationship* between the variables, as shown in part (d) of Figure 2.6. The best-fit line is horizontal, with a slope of zero and, when we view the scatter diagram, knowing the value of x is of no help whatsoever in predicting the value of y.

In this chapter, we will consider only *linear* relationships between variables, and there will be two possibilities for fitting a straight line (i.e., a linear equation) to the data. The first possibility is the "eyeball" method, using personal judgment and a ruler applied to the scatter diagram. The second, more accurate, approach is to use the computer and your statistical software to fit a straight line that is mathematically optimum. This technique, known as the "least-squares" method, will be discussed in much more detail in Chapter 15, Simple Linear Regression and Correlation.

EXAMPLEEXAMPLEEXAMPLEEXAMPLEEXAMPLEEXAMPLEEX

EXAMPLE

Best-Fit Equation

In 2006, opening-day Major League Baseball team payrolls ranged from $194.633 million for the New York Yankees to just $14.999 million for the Florida Marlins.[6] During the 162-game season, the Yankees won 97 games; the Marlins, just 78.[7] The payroll and wins data for the Yankees, Marlins, and the other major league teams are contained in the data file CX02BASE. A partial listing of the data for the 30 teams is shown here:

Team	x = Payroll	y = Wins
New York Yankees	$194.633	97
Boston Red Sox	120.100	86
Toronto Blue Jays	71.915	87
Baltimore Orioles	72.586	70
⋮		
San Diego Padres	69.896	88
Colorado Rockies	41.233	76

SOLUTION

All 30 teams are represented in the scatter diagram of Figure 2.7. The two variables do appear to be related—teams with a higher payroll did tend to win more games during the season. It makes sense that better players will win more games, and that better players must be paid more money; thus it comes as no surprise that the higher-paid teams won more games. However, the scatter diagram has provided us with a visual picture that reinforces our intuitive feelings about wins and dollars.

[6]Source: content.usatoday.com, June 21, 2009.
[7]Source: mlb.com, June 21, 2009.

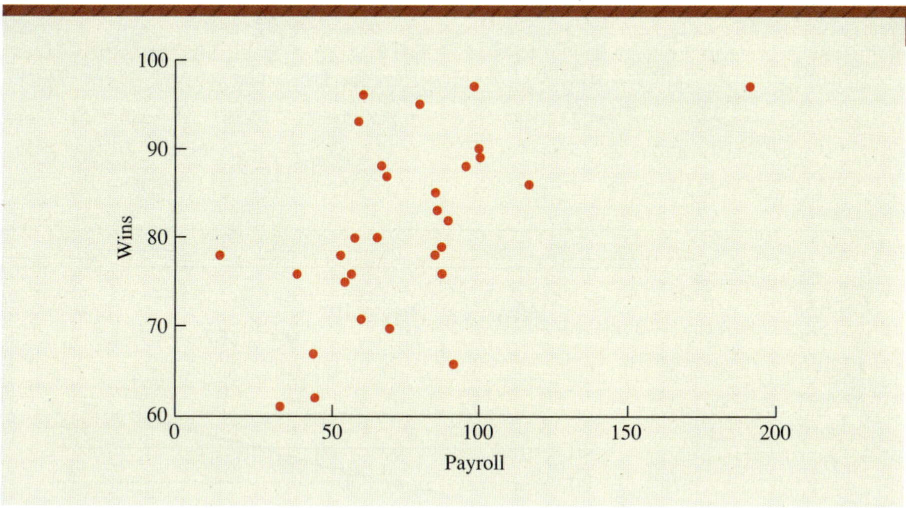

FIGURE 2.7

Scatter diagram of Wins versus Payroll (millions of dollars) for the 30 Major League Baseball teams during the 2006 season. Teams with a higher payroll tended to win more games.

Data sources: content.usatoday.com and mlb.com, June 21, 2009.

We can use Excel or Minitab to easily generate both a scatter diagram for the data and the linear equation that best fits the 30 data points. The procedures and results are shown in Computer Solutions 2.7, and the resulting equation can be written as

$$\text{Wins} = 67.980 + 0.167 \times \text{Payroll}$$

The following are among the important applications for this kind of analysis and best-fit equation:

1. Given a value for x, we can use the equation to estimate a value for y. For example, if a team had a payroll of \$50 million, we estimate that this team would have had 67.980 + 0.167(50), or 76.33 wins during this season. This is obtained simply by substituting x = \$50 million into the equation and calculating an estimated value for y = wins.
2. We can interpret the slope of the equation—in this case, 0.167. Keeping in mind that the slope of the equation is (change in y)/(change in x), we can say that, on average, raising the payroll by an extra \$1.0 million would tend to produce an extra 0.167 wins. Accordingly, an extra \$10 million would tend to produce an extra 1.67 wins.
3. We can interpret the slope in terms of the type of linear relationship between the variables. For the baseball data, the slope is positive (+0.167), reflecting a direct (positive) relationship in which payroll and wins increase and decrease together.
4. We can identify data points where the actual value of y is quite different from the value of y predicted by the equation. In this situation, we might be interested in teams that overperformed (won more games than their payroll would have predicted) or underperformed (won fewer games than their payroll would have predicted). For example, the Oakland Athletics played better than they were paid. Given their payroll of \$62.243 million, the equation would have predicted them to win just 67.980 + 0.167(62.243) = 78.375 games, but they actually won a lot more—93 games.

COMPUTER 2.7 SOLUTIONS

The Scatter Diagram

EXCEL

	A	B	C
1	Team	Payroll	Wins
2	New York Yankees	194.633	97
3	Boston Red Sox	120.100	86
4	Toronto Blue Jays	71.915	87
5	Baltimore Orioles	72.586	70
6	Tampa Bay Devil Rays	35.418	61
7	Cleveland Indians	56.032	78
8	Chicago White Sox	102.751	90
9	Detroit Tigers	82.613	95
10	Kansas City Royals	47.294	62
11	Minnesota Twins	63.396	96
12	Texas Rangers	68.229	80
13	Oakland Athletics	62.243	93
14	Seattle Mariners	87.960	78

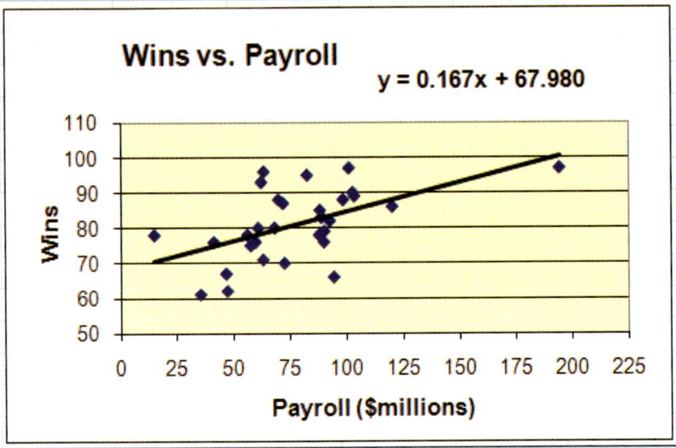

1. Open the Excel data file **CX02BASE**. The labels and data are already entered, as shown in the printout. Click on **B1** and drag to **C31** to select cells **B1:C31.** (*Note:* As with the line chart, of the two columns, the variable to be represented by the vertical axis should always be in the column to the right.)
2. From the **Insert** ribbon, click **Scatter** from the **Charts** menu. From the **Scatter** submenu, click the first option **(Scatter with only Markers).**
3. This optional step adds the best-fit straight line. Right-click on any one of the points in the scatter diagram. When the menu appears, click **Add Trendline.** In the **Trendline Options** menu, select the **Linear** and **Display Equation on chart** options. **Click Close.**
4. Further appearance improvements can be made by clicking on the chart and dragging the borders to expand it vertically and horizontally. If you don't want the little "Legend" box that originally appears, just right-click within it, then click **Delete.** Additional editing can include insertion of the desired labels: Double-click within the chart area. From the **Layout** ribbon, use the **Chart Title** and **Axis Titles** selections within the **Labels** menu.

MINITAB

The following steps produce the accompanying scatter diagram, complete with the best-fit equation. Along with the equation, the display includes information we will be covering later in the text.

1. Open the Minitab data file **CX02BASE**. The labels, **Payroll** and **Wins,** have already been entered at the top of columns **C2** and **C3,** respectively. The payroll values and the win totals have been entered into **C2** and **C3,** respectively.
2. Click **Stat.** Select **Regression.** Click **Fitted Line Plot.** Enter **C3** into the **Response (Y)** box. Enter **C2** into the **Predictor (X)** box. Click to select the **Linear** model.

3. Click **Options.** Enter **Wins vs. Payroll** into the **Title** box. Click **OK.** Click **OK.**

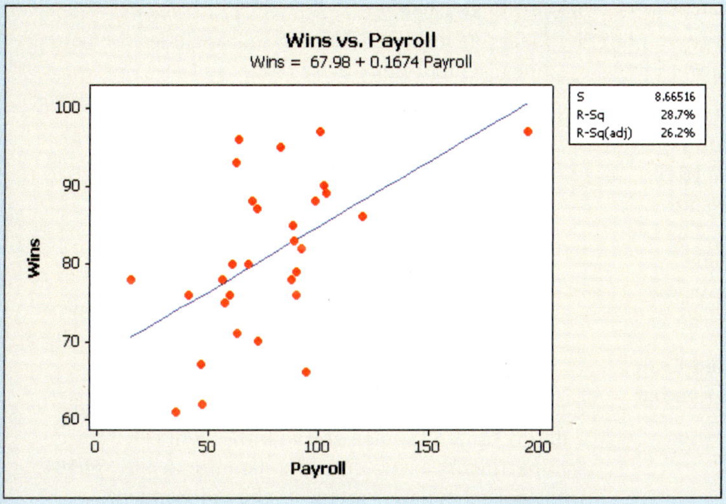

The nature of this chapter has allowed us to only briefly introduce the scatter diagram, the best-fit linear equation, and interpretation of the results. Later in the text, in Chapter 15, Simple Linear Regression and Correlation, we will cover this topic in much greater detail.

EXERCISES

2.38 What is a scatterplot, and for what kind of data is it a useful descriptive device?

2.39 Differentiate between a positive linear relationship and a negative linear relationship between variables.

2.40 When there is no relationship between two variables, what will tend to be the slope of the best-fit straight line through the scatter diagram?

2.41 *PC World* reports the following text and graphics speeds (pages per minute, or ppm) for its top-rated multifunction inkjet printers. Source: *PC World*, July 2009, p. 44.

x = ppm, Plain Text	y = ppm, Graphics	x = ppm, Plain Text	y = ppm, Graphics
30.0	20.0	26.0	17.0
33.0	31.0	7.5	4.5
28.0	23.0	27.0	19.0
32.0	24.0	34.0	33.0
8.4	5.6	33.0	32.0

a. Draw a scatter diagram representing these data.

b. Using the "eyeball" method, fit a straight line to the data.

c. Does there appear to be any relationship between the variables? If so, is the relationship direct or inverse?

2.42 For six local offices of a large tax preparation firm, the following data describe x = service revenues and y = expenses for supplies, freight, and postage during the previous tax preparation season:

x = Service Revenues (Thousands)	y = Supplies, Freight, Postage (Thousands)
$351.4	$18.4
291.3	15.8
325.0	19.3
422.7	22.5
238.1	16.0
514.5	24.6

a. Draw a scatter diagram representing these data.
b. Using the "eyeball" method, fit a straight line to the data.
c. Does there appear to be any relationship between the variables? If so, is the relationship direct or inverse?

2.43 In the 2009 stress tests applied to the nation's biggest banks, the U.S. government determined that 10 of them needed a greater cushion of capital in order to absorb losses in a worst-case economic scenario. The banks, the bailout they received, and the additional cushion they were deemed to need are as shown below.
Source: "Bank Stress-Test Results," *USA Today,* May 8, 2009, p. 3B.

Bank	x = Bailout Received	y = Added Cash Cushion Needed
Bank of America	$45.0 billion	$33.9 billion
Wells Fargo	25.0	13.7
GMAC	5.0	11.5
Citigroup	50.0	5.5
Regions	3.5	2.5
SunTrust	4.9	2.2
KeyCorp	2.5	1.8
Morgan Stanley	10.0	1.8
Fifth Third Bancorp	3.4	1.1
PNC	7.6	0.6

a. Draw a scatter diagram representing these data.
b. Using the "eyeball" method, fit a straight line to the data.
c. Does there appear to be any relationship between the variables? If so, is the relationship direct or inverse?

(DATA SET) *Note:* Exercises 2.44 and 2.45 require a computer and statistical software.

2.44 For a sample of 100 employees, a personnel director has collected data on x = the number of years an employee has been with the firm and y = the number of shares of company stock the employee owns. The data are in file **XR02044**.
a. Generate a scatter diagram that includes the best-fit linear equation for these data.
b. Does there appear to be any relationship between the variables? If so, is the relationship direct or inverse?
c. Interpret the slope of the equation generated in part (a).
d. Viewing the scatter diagram and equation in part (a), do you think there are any employees who appear to own an unusually high or low amount of stock compared to the ownership that the equation would have predicted?

2.45 A rental management executive has collected data on square footage and monthly rental fee for 80 apartments in her city. The data are in file **XR02045**. Considering square footage as the independent variable and monthly rental fee as the dependent variable:
a. Generate a scatter diagram that includes the best-fit linear equation for these data.
b. Does there appear to be any relationship between the variables? If so, is the relationship direct or inverse?
c. Interpret the slope of the equation generated in part (a).
d. Viewing the scatter diagram and equation in part (a), do you think there are any apartments that seem to have an unusually high or an unusually low rental fee compared to what would be predicted for an apartment of their size? Are there any variables that are not included in the data that might help explain such a differential? If so, what might they be?

2.6 TABULATION, CONTINGENCY TABLES, AND THE EXCEL PivotTable

When some of the variables represent categories, we can apply a popular and useful summarization method called **tabulation,** where we simply count how many people or items are in each category or combination of categories. If each person or item is also described by a quantitative variable (e.g., annual income), we can take the concept one step further and examine how members of one category (e.g., union employees) differ from members of another category (e.g., nonunion employees) in terms of average annual income.

EXAMPLE

Tabulation Methods

For 50 persons observed using a local automated teller machine (ATM), researchers have described the customers according to age category and gender category, and have used a stopwatch to measure how long the customer required to complete his or her transactions at the machine. In this situation, age category and gender are considered to be nominal-scale (category) variables, and time (measured in seconds) is a quantitative variable. The raw data are shown in part A of Table 2.3 and are in data file **CX02ATM**.

SOLUTIONS

Simple Tabulation

In **simple tabulation,** also known as **marginal** or **one-way tabulation,** we merely count how many people or items are in each category. The simple tabulation in part B of Table 2.3 shows us that the 50 ATM users consisted of 28 males and 22 females. An alternative would have been to express the counts in terms of percentages—e.g., 56% of the subjects were males and 44% were females.

Cross-Tabulation (Contingency Table)

The **cross-tabulation,** also known as the **crosstab** or the **contingency table,** shows how many people or items are in *combinations* of categories. The cross-tabulation in part C of Table 2.3 describes how many of the 50 ATM users were in each age–gender group; for example, 7 of the individuals were males in the youngest age group and 4 were females in the oldest age group. Because there are two category variables involved, part C of Table 2.3 could also be referred to as a **two-way crosstab.** Cross-tabulations help us to identify and examine possible relationships between the variables. For example, the crosstab in part C of Table 2.3 indicates that female ATM users outnumbered men in the youngest age category, but the reverse was true in the middle and older age categories. Given these results, bank management might want to interview customers to learn whether women in the middle and older age categories might have security concerns about the ATM location, lighting, or layout.

In a useful application of the concept of cross-tabulation, we can generate a tabular display that describes how a selected quantitative variable tends to differ from one category to another or from one combination of categories to another. For example, part D of Table 2.3 shows the average time (seconds) required for each of our six categories of ATM users. The average time for all 50 persons was 39.76 seconds, with males in the middle age group having the fastest average time (34.53 seconds) and females in the older age group having the slowest average time (49.95 seconds). From a gender perspective, the average time for males (38.36 seconds) was slightly faster than that for females (41.53 seconds).

EXAMPLEEXAMPLEEXAMPLEEXAMPLEEXAMPLEEXAMPLEEXAMPLEEXAMPLE

Cross-tabulations are very basic to many of the data analysis techniques in the remainder of the text, and this section includes two different Computer Solutions in which Excel and Minitab are applied. Computer Solutions 2.8 shows the

TABLE 2.3

Data representing age category, gender, and machine-usage time for 50 persons observed using an automated teller machine (ATM). Age category and gender are nominal-scale (category) variables, and time is a quantitative variable measured in seconds. The coding for the age and gender categories follows: AgeCat = 1 for age < 30, 2 for age 30–60, 3 for age > 60; Gender = 1 for male, 2 for female. The data are in file **CX02ATM**.

A. Raw Data

AgeCat	Gender	Seconds	AgeCat	Gender	Seconds	AgeCat	Gender	Seconds
1	2	50.1	2	2	46.4	2	1	33.2
1	1	53.0	2	1	31.0	2	2	29.1
2	2	43.2	2	1	44.8	2	1	37.7
1	2	34.9	2	1	30.3	2	2	54.9
3	1	37.5	2	1	33.6	1	2	29.4
2	1	37.8	3	2	56.3	3	1	37.1
3	1	49.4	2	2	37.8	3	2	42.9
1	1	50.5	1	2	34.4	1	2	44.3
3	1	48.1	2	1	32.9	3	2	45.0
1	2	27.6	1	1	26.3	2	2	47.6
3	2	55.6	2	1	30.8	1	1	58.1
2	2	50.8	1	1	24.1	1	2	32.0
3	1	43.6	1	2	43.6	2	1	36.9
1	1	35.4	2	2	29.9	1	1	27.5
1	2	37.7	2	1	26.3	3	1	42.3
3	1	44.7	3	1	47.7	2	1	34.1
2	1	39.5	1	2	40.1			

B. Simple Tabulation, by Gender

1 = Male	2 = Female	Total
28	22	50

C. Cross-Tabulation or Contingency Table, by Age Category and Gender

	Gender:		
Age Category:	1 = Male	2 = Female	Total
1 = under 30 yr.	7	10	17
2 = 30–60 yr.	13	8	21
3 = over 60 yr.	8	4	12
Total	28	22	50

D. Mean Usage Times (in Seconds), by Age Category and Gender

	Gender:		
Age Category:	1 = Male	2 = Female	Total
1 = under 30 yr.	39.27	37.41	38.18
2 = 30–60 yr.	34.53	42.46	37.55
3 = over 60 yr.	43.80	49.95	45.85
Total	38.36	41.53	39.76

procedures and the results when Excel and Minitab are used in generating a cross-tabulation, or contingency table, for the data in our ATM example. The underlying data are in **CX02ATM**, and the results correspond to the crosstab in part C of Table 2.3.

COMPUTER 2.8 SOLUTIONS

The Cross-Tabulation

EXCEL

	A	B	C	D	E	F	G
1	AgeCat	Gender	Seconds	Count of Gender	Gender		
2	1	2	50.1	AgeCat	1	2	Grand Total
3	1	1	53.0	1	7	10	17
4	2	2	43.2	2	13	8	21
5	1	2	34.9	3	8	4	12
6	3	1	37.5	Grand Total	28	22	50

1. Open the Excel data file CX02ATM. The labels and data are already entered, as shown in the printout. The variables are age category (1 = <30, 2 = 30–60, and 3 = >60), gender category (1 = male, 2 = female), and time (in seconds). Click on **A1** and drag to **C51** to select cells **A1:C51.**
2. From the **Insert** ribbon and its **Tables** menu, click **PivotTable.** In the **Create PivotTable** menu, click **Select a table or range.** The cells A1:C51 should already be represented in the **Table/Range** box; if not, enter this range into the box. Click to select **Existing Worksheet** and enter **D1** into the box. Click **OK.**
3. Click the **AgeCat** label at the right and drag it into the **Row Labels** rectangle. Click the **Gender** label at the right and drag it into the **Column Labels** rectangle. Click the **Gender** label again and drag it into the **Values** rectangle.
4. Right-click on any one of the data values within the table, then select **Value Field Settings.** In the **Summarize value field by** box, select **Count.** Click **OK.**
5. Click on the **Row Labels** cell and edit by entering **AgeCat.** Click on the **Column labels** cell and edit by entering **Gender.**

MINITAB

```
Tabulated statistics: AgeCat, Gender
 Rows: AgeCat   Columns: Gender

          1    2   All
 1        7   10    17
 2       13    8    21
 3        8    4    12
 All     28   22    50

 Cell Contents:        Count
```

1. Open the Minitab data file CX02ATM. The labels, **AgeCat, Gender,** and **Seconds,** have already been entered at the top of columns **C1, C2,** and **C3,** respectively. The corresponding age categories (1 = <30, 2 = 30–60, and 3 = >60), gender categories (1=male, 2=female), and times (in seconds) have been entered into **C1, C2,** and **C3,** respectively.
2. Click **Stat.** Select **Tables.** Click **Cross Tabulation and Chi-Square.** In the **Categorical variables** section, enter **C1** into the **For rows** box and **C2** into the **For columns** box. In the **Display** portion, click to place a check mark next to **Counts.** Click **OK.**

COMPUTER 2.9 SOLUTIONS

Cross-Tabulation with Cell Summary Information

EXCEL

	A	B	C	D	E	F	G
1	AgeCat	Gender	Seconds	Average of Seconds	Gender		
2	1	2	50.1	AgeCat	1	2	Grand Total
3	1	1	53.0	1	39.27	37.41	38.18
4	2	2	43.2	2	34.53	42.46	37.55
5	1	2	34.9	3	43.80	49.95	45.85
6	3	1	37.5	Grand Total	38.36	41.53	39.76
7	2	1	37.8				
8	3	1	49.4				
9	1	1	50.5				
10	3	1	48.1				
11	1	2	27.6				
12	3	2	55.6				
13	2	2	50.8				
14	3	1	43.6				
15	1	1	35.4				
16	1	2	37.7				
17	3	1	44.7				
18	2	1	39.5				
19	2	2	46.4				
20	2	1	31.0				
21	2	1	44.8				

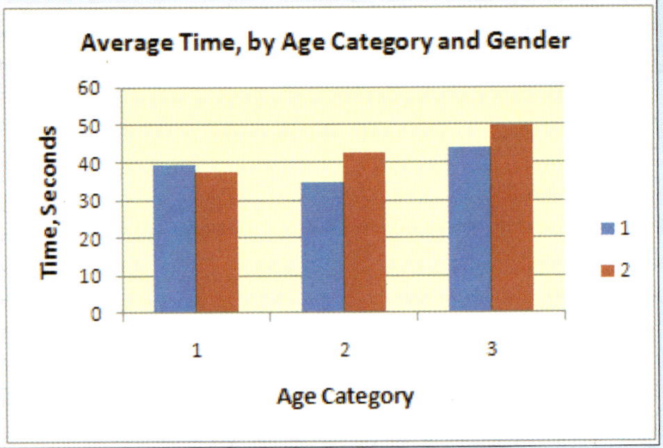

1. Open the Excel data file CX02ATM. Click on **A1** and drag to **C51** to select cells **A1:C51.**
2. From the **Insert** ribbon and its **Tables** menu, click **PivotTable.** In the **Create PivotTable** menu, click **Select a table or range.** The cells A1:C51 should already be represented in the **Table/Range** box; if not, enter this range into the box. Click to select **Existing Worksheet** and enter **D1** into the box. Click **OK.**
3. Click the **AgeCat** label at the right and drag it into the **Row Labels** rectangle. Click the **Gender** label at the right and drag it into the **Column Labels** rectangle. Click the **Seconds** label at the right and drag it into the **Values** rectangle.
4. Right-click on any one of the data values within the table, then select **Value Field Settings.** In the **Summarize value field by** box, select **Average.** Click **OK.** Click on **E3** and drag to **G6** to select cells **E3:G6.** Right-click within the field and select **Number Format.** Select **Number,** and specify **2** in the **Decimal places** box. Click **OK.** Click on the **Row Labels** cell and edit by entering **AgeCat.** Click on the **Column Labels** cell and edit by entering **Gender.**
5. Click on any cell within the table. From the **Insert** ribbon, select **Column** from the **Charts** menu. Select the first option **(Clustered Column)** from the **2-D Column** choices. The chart can now be edited and the desired labels inserted: Double-click within the chart area. From the **Layout** ribbon, use the **Chart Title** and **Axis Titles** selections within the **Labels** menu.

MINITAB

1. Open the Minitab data file CX02ATM. As in Computer Solutions 2.8, the labels, **AgeCat, Gender,** and **Seconds,** have already been entered at the top of columns **C1, C2,** and **C3,** respectively. The corresponding age categories (1 = <30, 2 = 30–60, and 3 = >60), gender categories (1 = male, 2 = female), and times (in seconds) have been entered into **C1, C2,** and **C3,** respectively.
2. Click **Stat.** Select **Tables.** Click **Descriptive Statistics.** In the **Categorical variables** section, enter **C1** into the **For rows** box and **C2** into the **For columns** box.

3. Click on **Display summaries for Associated Variables** and enter **C3** into the **Associated variables** box. Select **Means.** Click **OK.** Click **OK.**

```
Tabulated statistics: AgeCat, Gender
Rows: AgeCat    Columns: Gender

              1       2     All
1         39.27   37.41   38.18
              7      10      17

2         34.53   42.46   37.55
             13       8      21

3         43.80   49.95   45.85
              8       4      12

All       38.36   41.53   39.76
             28      22      50

Cell Contents:    Seconds   :  Mean
                               Count
```

EXERCISES

2.46 When variables are used as the basis for a contingency table, what scale of measurement must they represent?

2.47 Differentiate between simple tabulation and cross-tabulation, and give a real or hypothetical example of each.

Note: Exercises 2.48–2.50 can be done by hand, but computer statistical software is preferable, if available.

2.48 The 30 vehicles operated by a taxi service have been listed according to their engine (1 = diesel, 2 = gasoline), transmission (1 = manual, 2 = automatic), and whether they have air conditioning (1 = no air conditioning, 2 = air conditioning). These characteristics, along with the miles per gallon (mpg) achieved by each vehicle during the past month, are listed here. The data are also provided in file **XR02048**.

a. Construct a simple tabulation in which the counts are according to the type of engine.
b. Construct a cross-tabulation describing the fleet, using type of engine and type of transmission as the categorization variables.
c. Construct a display showing the average mpg according to type of engine and type of transmission. Do the categorization variables seem to be related to mpg? If so, how?

Engine	Trans	AC	mpg	Engine	Trans	AC	mpg
1	1	1	36.4	1	1	2	38.6
1	1	1	37.7	1	1	2	34.8
1	1	1	38.2	1	2	2	30.8
1	1	1	41.8	1	2	2	26.4
1	1	1	34.8	1	2	2	31.6
1	2	1	35.4	1	2	2	28.3
1	2	1	31.3	1	2	2	29.5
1	2	1	32.2	2	1	2	23.0
2	1	1	26.2	2	1	2	21.5
2	1	1	25.5	2	1	2	19.5
2	1	1	27.7	2	2	2	18.3
2	2	1	22.0	2	2	2	19.1
2	2	1	23.3	2	2	2	20.8
1	1	2	34.8	2	2	2	21.8
1	1	2	34.0	2	2	2	18.4

2.49 For the fleet described in Exercise 2.48,

a. Construct a simple tabulation in which the counts are according to the type of engine.
b. Construct a cross-tabulation describing the fleet, using type of engine and whether the vehicle has air conditioning as the categorization variables.
c. Construct a display showing the average mpg according to type of engine and whether the vehicle has air conditioning. Do the categorization variables seem to be related to mpg? If so, how?

2.50 For the fleet described in Exercise 2.48,

a. Construct a simple tabulation in which the counts are according to the type of transmission.
b. Construct a cross-tabulation describing the fleet, using type of transmission and whether the vehicle has air conditioning as the categorization variables.
c. Construct a display showing the average mpg according to type of transmission and whether the vehicle has air conditioning. Do the categorization variables seem to be related to mpg? If so, how?

(DATA SET) *Note:* Exercises 2.51–2.54 require a computer and statistical software.

2.51 A high school guidance counselor has examined the 60 colleges and universities that are on her "highly recommended" list for her school's graduating seniors. The data she has collected on each college or university include highest degree level offered (1 = bachelor's, 2 = master's, 3 = doctorate), primary support for the school (1 = public, 2 = private), type of campus setting (1 = urban, 2 = suburban, 3 = rural), and annual cost for tuition and fees (dollars). The data are in file **XR02051**.

a. Construct a simple tabulation in which the counts are according to the highest degree level offered.
b. Construct a cross-tabulation describing the schools, using the highest degree level offered and whether the school is public or private as the categorization variables.
c. Construct a display showing the average value for tuition and fees according to highest degree level offered and whether the school is public or private. Do the categorization variables seem to be related to the level of tuition and fees? If so, how?

2.52 For the schools described in Exercise 2.51,

a. Construct a simple tabulation in which the counts are according to the type of campus setting.
b. Construct a cross-tabulation describing the schools, using the type of campus setting and whether the school is public or private as the categorization variables.
c. Construct a display showing the average value for tuition and fees according to type of campus setting and whether the school is public or private. Do the categorization variables seem to be related to the level of tuition and fees? If so, how?

2.53 An interdisciplinary research team has examined 35 U.S. cities, reviewing and grading each according to the following performance areas: financial management, capital projects, personnel policies, information technology, and managing for results.[a] The grades awarded were letter grades such as students might receive in school. We have rounded off the "+" and "−" portions of the letter grades so that they are simply A, B, C, D, or F, and the data are coded so that A = 4, B = 3, C = 2, D = 1, and F = 0. We have also included one more variable for each city: its population according to the latest census statistics at the time.[b] The data for all 35 cities are in file **XR02053**. [a]Source: Richard Wolf, "Phoenix Is Managing To Get Everything Right," *USA Today*, January 31, 2000, p. 13A. [b]Source: *The New York Times Almanac 2000*, pp. 248–250.

a. Construct a simple tabulation in which the counts are according to the grade on financial management.
b. Construct a cross-tabulation describing the cities, using grade on financial management and grade on information technology as the categorization variables.
c. Construct a display showing the average population size according to grade on financial management and grade on information technology. Do the categorization variables seem to be related to the level of population? If so, how?

2.54 For the cities described in Exercise 2.53,

a. Construct a simple tabulation in which the counts are according to the grade on personnel policies.
b. Construct a cross-tabulation describing the cities, using the grade on personnel policies and the grade on managing for results as the categorization variables.
c. Construct a display showing the average population according to grade on personnel policies and the grade on managing for results. Do the categorization variables seem to be related to the level of population? If so, how?

2.7 SUMMARY

- **Frequency distributions and histograms**

To more concisely communicate the information contained, raw data can be visually represented and expressed in terms of statistical summary measures. When data are quantitative, they can be transformed to a frequency distribution or a histogram describing the number of observations occurring in each category.

The set of classes in the frequency distribution must include all possible values and should be selected so that any given value falls into just one category. Selecting the number of classes to use is a subjective process. In general, between 5 and 15 classes are employed. A frequency distribution may be converted to show either relative or cumulative frequencies for the data.

- **Stem-and-leaf displays, dotplots, and other graphical methods**

The stem-and-leaf display, a variant of the frequency distribution, uses a subset of the original digits in the raw data as class descriptors (stems) and class members (leaves). In the dotplot, values for a variable are shown as dots appearing along a single dimension. Frequency polygons, ogives, bar charts, line graphs, pie charts, pictograms, and sketches are among the more popular methods of visually summarizing data. As with many statistical methods, the possibility exists for the purposeful distortion of graphical information.

- **The scatter diagram**

The scatter diagram, or scatterplot, is a diagram in which each point represents a pair of known or observed values of two variables. These are typically referred to as the dependent variable (y) and the independent variable (x). This type of analysis is carried out to fit an equation to the data to estimate or predict the value of y for a given value of x.

- **Tabulation and contingency tables**

When some of the variables represent categories, simple tabulation and cross-tabulation are used to count how many people or items are in each category or combination of categories, respectively. These tabular methods can be extended to include the mean or other measures of a selected quantitative variable for persons or items within a category or combination of categories.

EQUATIONS

Width (Class Interval) of a Frequency Distribution Class

$$\text{Class interval} = \begin{array}{c}\text{Lower class limit of}\\ \text{next higher class}\end{array} - \begin{array}{c}\text{Lower class limit of}\\ \text{the class}\end{array}$$

Approximate Width of Classes When Constructing a Frequency Distribution

$$\text{Approximate class width} = \frac{\begin{array}{c}\text{Largest value}\\ \text{in raw data}\end{array} - \begin{array}{c}\text{Smallest value}\\ \text{in raw data}\end{array}}{\text{Number of classes desired}}$$

Midpoint of a Class in a Frequency Distribution

$$\text{Class mark} = \text{Lower class limit} + 0.5(\text{class interval})$$

Size of Each Piece of a Pie Chart

$$\begin{array}{c}\text{Number of degrees}\\ \text{for the category}\end{array} = \begin{array}{c}\text{Relative value of the}\\ \text{category}\end{array} \times 360$$

CHAPTER EXERCISES

2.55 The breakdown of U.S. cities having a population of at least 10,000 persons has been reported as follows. Source: Bureau of the Census, *Statistical Abstract of the United States 2009*, p. 34.

Population	Number of Cities
10,000–under 25,000	1510
25,000–under 50,000	684
50,000–under 100,000	430
100,000–under 250,000	191
250,000–under 500,000	37
500,000–under 1,000,000	25
1,000,000 or more	9

a. How many cities have a population of at least 25,000 but less than 500,000?
b. How many cities have a population less than 250,000?
c. How many cities have a population of at least 100,000 but less than 1,000,000? What percentage of cities are in this group?
d. What is the class mark for the 100,000–under 250,000 class?
e. Convert the table to a relative frequency distribution.

2.56 The National Oceanic and Atmospheric Administration reports the following record high temperatures (degrees Fahrenheit) in 50 U.S. cities. These data are also provided in the file **XR02056.** Source: Bureau of the Census, *Statistical Abstract of the United States 2009*, p. 229.

City	Temperature
Mobile, AL	105
Juneau, AK	90
Phoenix, AZ	122
Little Rock, AR	112
Los Angeles, CA	110
Denver, CO	105
Hartford, CT	102
Wilmington, DE	102
Miami, FL	98
Atlanta, GA	105
Honolulu, HI	95
Boise, ID	111
Chicago, IL	104
Indianapolis, IN	104
Des Moines, IA	108
Wichita, KS	113
Louisville, KY	106
New Orleans, LA	102
Portland, ME	103
Baltimore, MD	105
Boston, MA	102

(continued)

City	Temperature
Detroit, MI	104
Duluth, MN	97
Jackson, MS	107
St. Louis, MO	107
Great Falls, MT	106
Omaha, NE	114
Reno, NV	108
Concord, NH	102
Atlantic City, NJ	106
Albuquerque, NM	107
Buffalo, NY	99
Charlotte, NC	104
Bismarck, ND	111
Cleveland, OH	104
Oklahoma City, OK	110
Portland, OR	107
Pittsburgh, PA	103
Providence, RI	104
Columbia, SC	107
Sioux Falls, SD	110
Nashville, TN	107
Houston, TX	109
Salt Lake City, UT	107
Burlington, VT	101
Norfolk, VA	104
Seattle, WA	100
Charleston, WV	104
Milwaukee, WI	103
Cheyenne, WY	100

a. Construct a stem-and-leaf display for these data.
b. Construct a frequency distribution.
c. Determine the interval width and the class mark for each of the classes in your frequency distribution.
d. Based on the frequency distribution obtained in part (b), draw a histogram and a relative frequency polygon to describe the data.

2.57 The average daily cost to community hospitals for patient stays during 2006 for each of the 50 U.S. states was as follows:

AL	$1274	HI	1383	MA	1895
AK	2455	ID	1511	MI	1608
AZ	1930	IL	1676	MN	1379
AR	1304	IN	1699	MS	1100
CA	2056	IA	1093	MO	1687
CO	1864	KS	1155	MT	924
CT	1839	KY	1318	NE	1180
DE	1634	LA	1288	NV	1544
FL	1595	ME	1627	NH	1714
GA	1243	MD	2009	NJ	1910

(continued)

NM	1738	PA	1601	VT	1355
NY	1602	RI	1824	VA	1491
NC	1362	SC	1487	WA	2298
ND	966	SD	774	WV	1152
OH	1773	TN	1349	WI	1566
OK	1379	TX	1752	WY	906
OR	2202	UT	1914		

These data are also provided in the file **XR02057**.

Source: Bureau of the Census, *Statistical Abstract of the United States 2009*, p. 114.

a. Construct a stem-and-leaf display for these data.
b. Construct a frequency distribution.
c. Determine the interval width and the class mark for each of the classes in your frequency distribution.
d. Based on the frequency distribution obtained in part (b), draw a histogram and a relative frequency polygon.

2.58 In 2007, unemployment rates in the 50 U.S. states were reported as follows. These data are also provided in the file **XR02058**. Source: Bureau of the Census, *Statistical Abstract of the United States 2009*, p. 373.

	Percent		Percent		Percent
AL	4.0	LA	4.3	OH	5.6
AK	6.2	ME	4.7	OK	4.4
AZ	3.9	MD	3.6	OR	5.2
AR	5.6	MA	4.6	PA	4.3
CA	5.3	MI	7.1	RI	4.9
CO	3.7	MN	4.6	SC	5.6
CT	4.5	MS	6.1	SD	2.9
DE	3.5	MO	5.0	TN	4.6
FL	4.1	MT	3.6	TX	4.3
GA	4.3	NE	3.1	UT	2.6
HI	2.9	NV	4.6	VT	4.0
ID	3.0	NH	3.6	VA	3.1
IL	5.1	NJ	4.2	WA	4.6
IN	4.6	NM	3.7	WV	4.6
IA	3.7	NY	4.6	WI	5.0
KS	4.1	NC	4.5	WY	2.9
KY	5.4	ND	3.2		

a. Construct a stem-and-leaf display for these data.
b. Construct a frequency distribution for these data.
c. Determine the interval width and the class mark for each of the classes in your frequency distribution.
d. Based on the frequency distribution obtained in part (b), draw a histogram and a relative frequency polygon to describe the data.

2.59 During 2007, sales of new, privately owned homes in the total United States and in the western states were broken down into price categories as follows: Source: Bureau of the Census, *Statistical Abstract of the United States 2009*, p. 595.

	Sales (Thousands of Homes)	
Price of Home	**Total United States**	**Western States**
under $200,000	268	22
$200,000–under $300,000	227	54
$300,000–under $500,000	186	64
$500,000 or over	94	41

Convert these data to relative frequency distributions, one for the total United States, the other for the western states. Do the results appear to suggest any differences between total U.S. prices and those in the western states?

2.60 The following stem-and-leaf output has been generated by Minitab. The data consist of two-digit integers.

```
Stem-and-leaf of AGE    N=21
Leaf Unit=1.0

  1    3 3
  4    3 669
  4    4
  8    4 5667
 (6)   5 012233
  7    5 68
  5    6 0
  4    6 779
  1    7 0
```

a. From this display, is it possible to determine the exact values in the original data? If so, list the data values. If not, provide a set of data that could have led to this display.
b. Interpret the numbers in the leftmost column of the output.

2.61 The following stem-and-leaf output has been generated by Minitab. The data consist of three-digit integers.

```
Stem-and-leaf of SCORE    N=21
Leaf Unit=10

  2    1 89
  3    2 0
  3    2
  9    2 445555
  9    2
 (2)   2 89
 10    3 0011
  6    3 23
  4    3 455
  1    3
  1    3 8
```

a. From this display, is it possible to determine the exact values in the original data? If so, list the data values. If not, provide a set of data that could have led to this display.

b. Interpret the numbers in the leftmost column of the output.

2.62 For the period 2001–2008, the Bristol-Myers Squibb Company, Inc. reported the following amounts (in billions of dollars) for (1) net sales and (2) advertising and product promotion. The data are also in the file **XR02062**. Source: Bristol-Myers Squibb Company, *Annual Reports, 2005, 2008.*

Year	Net Sales	Advertising/Promotion
2001	$16.612	$1.201
2002	16.208	1.143
2003	18.653	1.416
2004	19.380	1.411
2005	19.207	1.476
2006	16.208	1.304
2007	18.193	1.415
2008	20.597	1.550

For these data, construct a line graph that shows both net sales and expenditures for advertising/product promotion over time. Some would suggest that increases in advertising should be accompanied by increases in sales. Does your line graph support this?

2.63 A social scientist would like to use a pictogram in comparing the number of violent crimes against persons in 2006 with the comparable figure for 1960. Considering that she can select from a wide variety of symbols to demonstrate these frequencies, suggest several possible symbols that would tend to underscore the violence and/or human tragedy of such crimes.

(DATA SET) *Note:* Exercises 2.64–2.72 require a computer and statistical software.

2.64 If possible with your statistical software, construct a dotplot for the data of Exercise 2.56 (use data file **XR02056**). Are there any cities for which the record high temperature is so high that it would appear to be an outlier? If so, which one(s)?

2.65 If possible with your statistical software, construct a dotplot for the data of Exercise 2.57 (use data file **XR02057**). Are there any states for which the average daily cost was so high that it would appear to be an outlier? If so, which one(s)?

2.66 If possible with your statistical software, use the dotplot method to compare the 50-state unemployment rates in 1980 with those in 2007. Use the same scale for each dotplot, then comment on whether unemployment appears to have changed in terms of its range (highest minus lowest) or in the general level of unemployment (perhaps a slight shift to the right or to the left) from 1980 to 2007. Data for both years are in the file **XR02066**. Source: Bureau of the Census, *Statistical Abstract of the United States 2009*, p. 373.

2.67 Among the information in its top-50 list of the nation's law schools, *U.S. News & World Report* provided two reputation scores (with maximum = 5.0) for each school: one from academicians, the other from lawyers/judges. The magazine also provided the most recent acceptance rate (%), student-faculty ratio (students/faculty), and the percentage of graduates who were employed immediately upon graduation. The data values are listed in file **XR02067**. Source: *U.S. News & World Report, Best Graduate Schools, 2010 edition*, p. 22.

a. Construct a scatter diagram where the variables are the two kinds of reputation scores. Do schools given higher scores by the academicians seem to also receive higher scores from the lawyers/judges?
b. Fit a linear equation to the scatter diagram. Is the slope positive or is it negative? Is the sign of the slope consistent with your intuitive observation in part (a)?

2.68 For the law schools discussed in Exercise 2.67, and using the data in file **XR02067**,

a. Construct a scatter diagram where the dependent variable is the percentage of graduates who were employed immediately upon graduation, and the independent variable is the reputation score provided by the lawyers/judges. Do schools given higher scores by the lawyers/judges also seem to have a higher percentage of their graduates employed immediately upon graduation?
b. Fit a linear equation to the scatter diagram. Is the slope positive or is it negative? Is the sign of the slope consistent with your intuitive observation in part (a)?

2.69 The household saving rates (percentage of disposable household income that goes into savings) have been reported for several countries—including the United States, Canada, France, Germany, and Japan—for the years 1980 through 1997. The percentages are listed in data file **XR02069**. Source: Security Industry Association, *1999 Securities Industry Fact Book*, p. 98.

a. Construct a scatter diagram using the household saving rates in the United States and Canada as the two variables. It doesn't really matter which one you consider to be the independent variable. In the years when U.S. saving rates were higher, did those in Canada seem to be higher as well?
b. Fit a linear equation to the scatter diagram. Is the slope positive or is it negative? Is the sign of the slope consistent with your intuitive observation in part (a)?

2.70 For the nations and household saving rates in Exercise 2.69, and using the data in file **XR02069**,

a. Construct a scatter diagram using the household saving rates in the United States and Germany as the two variables. It doesn't really matter which one you consider to be the independent variable. In the years when U.S. saving rates were higher, did those in Germany seem to be higher as well?
b. Fit a linear equation to the scatter diagram. Is the slope positive or is it negative? Is the sign of the slope consistent with your intuitive observation in part (a)?

2.71 An air-compressor manufacturer purchases mechanical components from two different suppliers (MechSup = 1 or 2) and electrical/motor components from three different suppliers (ElectSup = 1, 2, or 3), with final assembly being carried out by one of two technicians (Tech = 1 or 2). Following production, finished compressor units are tested to measure how much pressure they can exert (pounds per square inch, or psi). During the past few months, 100 compressors have been tested before shipment to customers, and the resulting data are listed in file **XR02071**.

a. Construct a simple tabulation in which the counts are according to which company supplied the mechanical components.
b. Construct a cross-tabulation describing the 100 compressors tested, using mechanical supplier and electrical/motor supplier as the categorization variables.
c. Construct a display showing the average pressure (psi) according to mechanical supplier and electrical/motor supplier. Which combination of mechanical and electrical/motor suppliers seems to result in the highest pressure exertion? The lowest pressure exertion?
d. Based on the crosstab and means in part (c), would it seem that any of the five suppliers should be examined further with regard to the effect their product might have on the final pressure capabilities of the finished product? Explain your answer.

2.72 For the situation and data described in Exercise 2.71, and using data file **XR02071**,

a. Construct a simple tabulation in which the counts are according to which company supplied the electrical/motor components.
b. Construct a cross-tabulation describing the 100 compressors tested, using electrical/motor supplier and final assembly technician as the categorization variables.
c. Construct a display showing the average pressure (psi) exerted according to electrical/motor supplier and final assembly technician. Which combination of electrical/motor supplier and final assembly technician seems to result in the highest average pressure exerted? The lowest pressure exerted?
d. Based on the crosstab and means in part (c), would it seem that the two technicians might not be equally adept at the final assembly task? Explain your answer.

INTEGRATED CASES

Thorndike Sports Equipment
(Meet the Thorndikes: See Video Unit One.)

Luke Thorndike, founder and current president of Thorndike Sports Equipment, had guided the business through 34 successful years and was now interested in bringing his favorite grandson Ted into the company.

The elder Thorndike, possessed of a sharp but fading wit, begins the meeting with, "Great to see you, Ted. You always were a high-strung kid. Thought you might like to join our tennis racquet division." Ted counters, "Not quite, but you're getting warm, Luke." The Thorndikes have always been a strange bunch.

"Seriously, Ted, I'm getting a little up in years, the microcomputer I bought myself for Christmas is collecting dust, and I think you and your business degree could bring some new blood to the company. I'd like you to be my executive vice president. I've been running this outfit by the seat of my pants for a lot of years now, and the world just seems to be getting too big and too complicated these days. I've got index cards and file folders just about piled up to the ceiling in that little room next door, and there's got to be a better way of making sense of all this information. Maybe I shouldn't have fought the fire department when they wanted to condemn the place."

"Besides all these records piling up, a lot of technical developments are affecting our business—things like composite bicycle wheels, aerodynamic golf balls, and oversize tennis racquets. Just yesterday one of our engineers came in and said she's come up with a new golf ball that will go farther than the conventional design. She seems trustworthy, but we might need to have some numbers to back up our claim if we decide to come out with the product."

After further discussion of the business and the position that Mr. Thorndike has proposed, Ted accepts the offer. As his first official duty, he sets up a test of the new golf ball that's supposed to travel farther than the conventional design. He decides to mix 25 of the new balls with 25 of the old type, have a golf pro hit all 50 of them at a driving range, then measure how far each goes. The results are provided here and are also listed in the data file **THORN02**.

(continued)

25 Drives with New Ball (Yards)

267.5	248.3	265.1	243.6	253.6	232.7	249.2	232.3	252.8
247.2	237.4	223.7	260.4	269.6	256.5	271.4	294.1	256.3
264.3	224.4	239.9	233.5	278.7	226.9	258.8		

25 Drives with Conventional Ball (Yards)

241.8	255.1	266.8	251.6	233.2	242.7	218.5	229.0	256.3
264.2	237.3	253.2	215.8	226.4	201.0	201.6	244.1	213.5
267.9	240.3	247.9	257.6	234.5	234.7	215.9		

1. Using 10-yard intervals beginning with 200.0–under 210.0, 210.0–under 220.0, on up to 290.0–under 300.0, construct a frequency distribution for the distances traveled by the new ball.
2. Using the same intervals as in part 1, construct a frequency distribution for the distances traveled by the conventional ball.
3. Place the frequency distribution for the new ball next to the one for the conventional ball. Does it appear that the new ball might be more "lively" than the conventional ball?

Springdale Shopping Survey*

The major shopping areas in the community of Springdale include Springdale Mall, West Mall, and the downtown area on Main Street. A telephone survey has been conducted to identify strengths and weaknesses of these areas and to find out how they fit into the shopping activities of local residents. The 150 respondents were also asked to provide information about themselves and their shopping habits. The data are provided in the computer file **SHOPPING**. The variables in the survey were as follows:

A. How Often Respondent Shops at Each Area (Variables 1–3)

	1. Springdale Mall	2. Downtown	3. West Mall
6 or more times/wk.	(1)	(1)	(1)
4–5 times/wk.	(2)	(2)	(2)
2–3 times/wk.	(3)	(3)	(3)
1 time/wk.	(4)	(4)	(4)
2–4 times/mo.	(5)	(5)	(5)
0–1 times/mo.	(6)	(6)	(6)

B. How Much the Respondent Spends during a Trip to Each Area (Variables 4–6)

	4. Springdale Mall	5. Downtown	6. West Mall
$200 or more	(1)	(1)	(1)
$150–under $200	(2)	(2)	(2)
$100–under $150	(3)	(3)	(3)
$ 50–under $100	(4)	(4)	(4)
$ 25–under $50	(5)	(5)	(5)
$ 15–under $25	(6)	(6)	(6)
less than $15	(7)	(7)	(7)

*Source: Materials for this case have been provided courtesy of The Archimedes Group, Indiana, PA. Data are based on actual responses obtained to this subset of the questions included in the survey; town and mall identities have been disguised.

C. General Attitude toward Each Shopping Area (Variables 7–9)

	7. Springdale Mall	8. Downtown	9. West Mall
Like very much	(5)	(5)	(5)
Like	(4)	(4)	(4)
Neutral	(3)	(3)	(3)
Dislike	(2)	(2)	(2)
Dislike very much	(1)	(1)	(1)

D. Which Shopping Area Best Fits Each Description (Variables 10–17)

	Springdale Mall	Downtown	West Mall	No Opinion
10. Easy to return/exchange goods	(1)	(2)	(3)	(4)
11. High quality of goods	(1)	(2)	(3)	(4)
12. Low prices	(1)	(2)	(3)	(4)
13. Good variety of sizes/styles	(1)	(2)	(3)	(4)
14. Sales staff helpful/friendly	(1)	(2)	(3)	(4)
15. Convenient shopping hours	(1)	(2)	(3)	(4)
16. Clean stores and surroundings	(1)	(2)	(3)	(4)
17. A lot of bargain sales	(1)	(2)	(3)	(4)

E. Importance of Each Item in Respondent's Choice of a Shopping Area (Variables 18–25)

	Not Important					Very Important	
18. Easy to return/exchange goods	(1)	(2)	(3)	(4)	(5)	(6)	(7)
19. High quality of goods	(1)	(2)	(3)	(4)	(5)	(6)	(7)
20. Low prices	(1)	(2)	(3)	(4)	(5)	(6)	(7)
21. Good variety of sizes/styles	(1)	(2)	(3)	(4)	(5)	(6)	(7)
22. Sales staff helpful/friendly	(1)	(2)	(3)	(4)	(5)	(6)	(7)
23. Convenient shopping hours	(1)	(2)	(3)	(4)	(5)	(6)	(7)
24. Clean stores and surroundings	(1)	(2)	(3)	(4)	(5)	(6)	(7)
25. A lot of bargain sales	(1)	(2)	(3)	(4)	(5)	(6)	(7)

F. Information about the Respondent (Variables 26–30)

26. Gender: (1) = Male (2) = Female
27. Number of years of school completed:
    (1) = less than 8 years (3) = 12–under 16 years
    (2) = 8–under 12 years (4) = 16 years or more
28. Marital status: (1) = Married (2) = Single or other
29. Number of people in household: ________ persons
30. Age: ________ years

Each respondent in this database is described by 30 variables. As an example of their interpretation, consider row number 1. This corresponds to respondent number 1 and contains the following information.

Variable number 6 = 7. The respondent usually spends less than $15 when she shops at West Mall.

Variable number 26 = 2. The respondent is a female.

5 5 5 6 7 7 4 4 4 1 1 3 1 1 1 1 1 6 7 7 6 5 6 6 7 2 3 2 4 2 1

Variable number 1 = 5. The respondent shops at Springdale Mall 2–4 times/month.

Variable number 19 = 7. Respondent believes "high quality of goods" is very important in choosing a shopping area.

The data from these 150 respondents will be the basis for further analyses later in the text. In applying some of the techniques from this chapter, the following questions could provide insights into the perceptions and behavior of Springdale residents regarding the three shopping areas.

1. Using the data file **SHOPPING**, construct three different frequency distributions, one each for variables 7, 8, and 9. How do the three areas compare in terms of residents' general attitudes toward each?
2. Do people tend to spend differently at the three areas? Construct and compare frequency distributions for variables 4, 5, and 6.
3. To find out more about specific strengths and weaknesses of the areas, set up a frequency distribution for variable 10 (i.e., how many "best-fits" votes did each area get for "easy to return/exchange goods"?). Repeat this for variables 11–17 and interpret the results. Which of the malls seems to be the "bargain mall"? Which area seems to have the most convenient shopping hours?
4. Generate a cross-tabulation in which the categorization variables are variable 26 (gender) and variable 28 (marital status). For each of the four subcategories in the cross-tabulation, have your statistical software include the average for variable 30 (respondent age). Interpret the results.

CHAPTER 3

Statistical Description of Data

EMPLOYEE ABSENTEEISM: SLICK TRICK OR TRULY SICK?

In a survey of 1100 British employers, researchers found quite a discrepancy in the annual sick days used by public- versus private-sector employees. Workers in the public sector used an average of 10.7 sick days per year, compared to just 7.8 sick days per year for those in the private sector. Does the British private sector attract younger and healthier workers, or is it possible that public-sector workers simply take advantage of the greater number of sick days they are generally allowed?

According to a survey of 574 U.S. human resource executives, the answer is the latter—allowing more paid sick days leads to more paid sick days being taken. Human resource directors suspect that only 28% of paid sick leave is actually used because of illness, but they have no way of knowing for sure which absent employees are really sick.

Rosemary Haefner, Senior Career Advisor for CareerBuilder.com, makes an interesting seasonal observation about the contrived absenteeism situation: "With the cold and flu season kicking in, it's a popular time of the year for employees to call in sick. However, the number of those who are actually feeling under the weather may not necessarily match up with unscheduled absences. Twenty percent of workers say they called into work because they just didn't feel like going to the office that day. One in four workers report they feel sick days are equivalent to taking extra vacation days and treat them as such."

The situation can also be viewed from the standpoint of union/management status, and using the median (the sick-days value where just as many workers are higher as lower) instead of the average, or mean. From this perspective, union hourly workers use a median of 6.1 sick days per year, nonunion hourly workers take 4.9, and management uses 4.0.

In the pages ahead, we'll talk more about the mean and the median, as well as several other important numerical measures that describe data. Try not to be absent.

Sources: "Public Sector Sick Days Cost Taxpayers Billions," management-issues.com, June 27, 2009. Del Jones, "Firms Take New Look at Sick Days," *USA Today,* October 8, 1996, p. 8B; and Kate Lorenz, "Why Are People Calling in Sick? What's Your Excuse?" Netscape.com, May 1, 2006.

"I won't eat the mean;
I won't eat the median;
I won't even eat the first decile."

LEARNING OBJECTIVES

After reading this chapter, you should be able to:

- Describe data by using measures of central tendency and dispersion.
- Convert data to standardized values.
- Use the computer to visually represent data with a box-and-whisker plot.
- Determine measures of central tendency and dispersion for grouped data.
- Use the coefficient of correlation to measure association between two quantitative variables.

(3.1) INTRODUCTION

In Chapter 2, we saw how raw data are converted into frequency distributions, histograms, and visual displays. We will now examine statistical methods for describing typical values in the data as well as the extent to which the data are spread out. Introduced in Chapter 1, these descriptors are known as **measures of central tendency** and **measures of dispersion:**

- **Measures of central tendency** Numbers describing *typical* data values. Chances are the typical subscriber to *The Wall Street Journal* earns more than the typical subscriber to *Mad* magazine. By using measures of central tendency, we can numerically describe the typical income of members of each group. The primary measures of central tendency discussed in this chapter are the arithmetic mean, weighted mean, median, and mode.
- **Measures of dispersion** Numbers that describe the *scatter* of the data. It's very likely that the variety of heights of centers in the National Basketball Association (most such players are between 6′8″ and 7′4″) is not as wide as the dispersion that exists within the general population, which includes individuals as tall as the NBA centers and persons as short as newborn babies. Measures of dispersion allow us to numerically describe the scatter, or *spread*, of measurements. Among the measures of dispersion discussed in this chapter are the range, quantiles, mean absolute deviation, variance, and standard deviation.

When we discuss measures of central tendency and dispersion, we'll be using two other key terms from Chapter 1: *population* and *sample*. Remember that (1) we often wish to use information about the sample to make inferences about the population from which the sample was drawn, and (2) characteristics of the population are referred to as *parameters,* while characteristics of the sample are referred to as *statistics*.

Chapter 2 also introduced the scatter diagram and best-fit linear equation as a graphical method of examining the possible linear relationship between two quantitative variables. In this chapter, we will extend this to two statistical measures of association: the coefficients of correlation and determination. As with the scatter diagram and best-fit equation, these will be discussed in greater detail in Chapter 15, Simple Linear Regression and Correlation.

STATISTICAL DESCRIPTION: MEASURES OF CENTRAL TENDENCY

3.2

We now discuss methods for representing the data with a single numerical value. This numerical descriptor has the purpose of describing the *typical* observation contained within the data.

The Arithmetic Mean

Defined as the sum of the data values divided by the number of observations, the **arithmetic mean** is one of the most common measures of central tendency. Also referred to as the *arithmetic average* or simply the *mean*, it can be expressed as μ (the population mean, pronounced "myew") or $\bar{x}$ (the sample mean, "x bar").

The population mean (μ) applies when our data represent *all* of the items within the population. The sample mean ($\bar{x}$) is applicable whenever data represent a *sample* taken from the population.

Population mean:

$$\mu = \frac{\sum_{i=1}^{N} x_i}{N} \quad \text{or simply} \quad \mu = \frac{\sum x_i}{N}$$

where μ = population mean
x_i = the ith data value in the population
$\sum$ = the sum of
N = number of data values in the population

NOTE

The leftmost version of the equation for the population mean (μ) includes complete summation notation. Notation such as the "$i = 1$" and "N" will be included in the remainder of the text only when the nature of the summation is not clear from the context in which it is being used.

Sample mean:

$$\bar{x} = \frac{\sum x_i}{n}$$

where $\bar{x}$ = sample mean
x_i = the ith data value in the sample
$\sum$ = the sum of
n = number of data values in the sample

In determining either a population mean (μ) or a sample mean ($\bar{x}$), the sum of the data values is divided by the number of observations. As an example of the calculation of a population mean, consider the following data for shipments of peanuts from a hypothetical U.S. exporter to five Canadian cities.

City	Peanuts (Thousands of Bags)
Montreal	64.0
Ottawa	15.0
Toronto	285.0
Vancouver	228.0
Winnipeg	45.0

Because these were the only Canadian cities identified as receiving U.S. peanuts from this supplier during this period, they can be considered a population. We can calculate the arithmetic mean (μ) for these data as follows:

$$\mu = \frac{\Sigma x_i}{N} = \frac{64.0 + 15.0 + 285.0 + 228.0 + 45.0}{5} = 127.4 \text{ thousand bags}$$

On the average, each Canadian destination for this U.S. company's peanuts received 127.4 thousand bags of peanuts during the time period involved.

To help understand what the arithmetic mean represents, picture a playground seesaw with markings ranging from 0 at one end to 300 at the other end, with five people of equal weight sitting at positions matching the shipments received by each Canadian city. Assuming a weightless seesaw, the individuals would be perfectly balanced when the pivot bar is located exactly at the 127.4 position, as shown in part (b) of Figure 3.1.

Figure 3.1 also reveals a potential weakness of the mean as a descriptor of typicalness. Notice in part (b) that three of the five cities received relatively moderate shipments, 15.0, 45.0, and 64.0 thousand bags, while the other two cities received 228.0 and 285.0 thousand bags of peanuts. Thus each of the latter received more than the other three cities combined. This results in three of the five cities having shipments that are far below the average, making the arithmetic mean not as typical as we might like. A solution to this difficulty is the median, to be discussed shortly.

FIGURE 3.1

The arithmetic mean, or average, is a mathematical counterpart to the center of gravity on a seesaw. Although the influence of the two values that are more than 200 thousand is not quite as great as that of Arnold, it causes the arithmetic mean to be greater than three of the five data values.

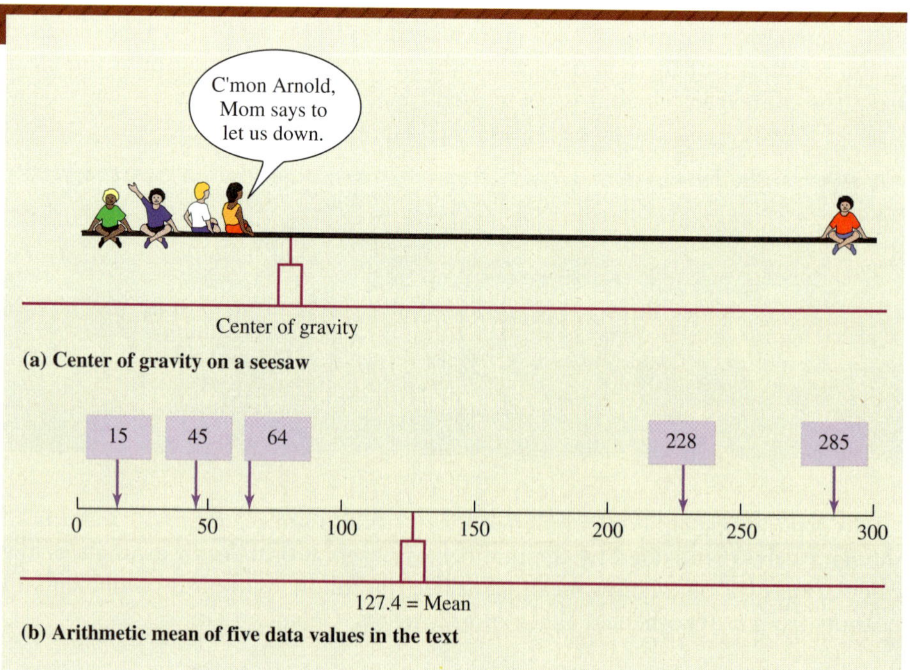

The Weighted Mean

When some values are more important than others, a **weighted mean** (sometimes referred to as a *weighted average*) may be calculated. In this case, each data value is weighted according to its relative importance. The formula for the weighted mean for a population or a sample will be as follows:

Weighted mean, μ_w (for a population) or $\bar{x}_w$ (for a sample):

$$\mu_w \text{ or } \bar{x}_w = \frac{\Sigma w_i x_i}{\Sigma w_i}$$

where w_i = weight assigned to the ith data value
x_i = the ith data value

Continuing with the peanut example, let's assume that shipments to the respective cities will be sold at the following profits per thousand bags: \$15.00, \$13.50, \$15.50, \$12.00, and \$14.00. (*Note:* In this example, we are trying to determine the weighted mean for the *profit;* accordingly, the preceding profit data will constitute the x_i values in our example and will be weighted by the shipment quantities to the five cities.)

The average profit per thousand bags will *not* be (15.00 + 13.50 + 15.50 + 12.00 + 14.00)/5, because the cities did not receive equal quantities of peanuts. A weighted mean must be calculated if we want to find the average profit per thousand bags for all shipments of peanuts from this U.S. exporter to Canada during this period. Using our assumed profit figures, this weighted mean is calculated as

$$\mu_w = \frac{\Sigma w_i x_i}{\Sigma w_i}$$

$$= \frac{64(\$15.00) + 15(\$13.50) + 285(\$15.50) + 228(\$12.00) + 45(\$14.00)}{64 + 15 + 285 + 228 + 45}$$

$$= \$14.04 \text{ per thousand bags}$$

The Median

In a set of data, the **median** is the value that has just as many values above it as below it. For example, the numbers of bags of peanuts (in thousands) shipped by the U.S. exporter to the five Canadian cities were

15.0 45.0 | 64.0 | 228.0 285.0

Two values below 64.0 | The median, 64.0 | Two values above 64.0

For these data, 64.0 is the value that has just as many values above it as below it. Montreal, which received 64.0 thousand bags of peanuts, is in the median position among the five Canadian cities.

The preceding example had an odd number of data values (5), and the median was the observation with two values above and two below. For some data, however, the number of values will be even. In such cases, the median will be halfway between the *two values* in the middle when the data are arranged in order of size.

For example, Ryder System, Inc. reported the following data for percentage return on average assets over an 8-year period.[1]

Raw data 2.8 7.0 1.6 0.4 1.9 2.6 3.8 3.8

Ordered data |0.4 1.6 1.9| |2.6 2.8| |3.8 3.8 7.0| % return on avg. assets

Three values

Three values

Three values are equal to or above the two in the middle, and three are equal to or below them. The median is halfway between 2.6 and 2.8, or 2.7.

For these data, the median, 2.7, is the value halfway between the two values in the middle of the ordered arrangement, 2.6 and 2.8. Unlike the mean, the median is not influenced by extreme high or low values in the data. For example, if the highest data value had been 600% instead of 7.0%, the median would still be 2.7%.

The Mode

In a set of data, the **mode** is a value that occurs with the greatest frequency. As such, it could be considered the single data value most typical of all the values. Consider again the 8 years of return-on-assets data reported by Ryder System, Inc.:

Ordered data 0.4 1.6 1.9 2.6 2.8 |3.8 3.8| 7.0 % return on avg. assets

Mode = 3.8

For these data, the mode is 3.8, since it occurs more frequently than any other value. In this case, the mode does not appear to be a very good descriptor of the data, as five of the other six values are all smaller than 3.8.

Depending on the data, there can be more than one mode. For example, if the leftmost data value had been 1.6 instead of 0.4, there would have been two modes, 1.6 and 3.8; if this were the case, the values 1.6 and 3.8 would each have occurred in two different years. When there are two modes, a distribution of values is referred to as **bimodal.**

Comparison of the Mean, Median, and Mode

As we have seen, the mean, median, and mode are alternative approaches to describing the central tendency of a set of values. In deciding which measure to use in a given circumstance, there are a number of considerations:

- The mean gives equal consideration to even very extreme values in the data, while the median tends to focus more closely on those in the middle of the data array. Thus, the mean is able to make more complete use of the data. However, as pointed out earlier, the mean can be strongly influenced by just

[1]Source: Ryder System, Inc., *2005, 2003,* and *1998 Annual Reports.*

one or two very low or high values. A demonstration of how a single observation can affect the mean and median is shown in Seeing Statistics Applet 1, at the end of the chapter.

- There will be just one value for the mean and one value for the median. However, as indicated previously, the data may have more than one mode.
- The mode tends to be less useful than the mean and median as a measure of central tendency. Under certain circumstances, however, the mode can be uniquely valuable. For example, when a television retailer decides how many of each screen size to stock, it would do him little good to know that the mean television set sold has a 38.53-inch screen—after all, there is no such thing as a 38.53-inch television. Knowledge that the mode is 30 inches would be much more useful.

Distribution Shape and Measures of Central Tendency

The relative values of the mean, median, and mode are very much dependent on the shape of the distribution for the data they are describing. As Figure 3.2 shows, distributions may be described in terms of *symmetry* and *skewness*. In a **symmetrical distribution,** such as that shown in part (a), the left and right sides of the distribution are mirror images of each other. The distribution in part (a) has a single mode, is bell shaped, and is known as the normal distribution. It will be discussed in Chapter 6—for now, just note that the values of the mean, median, and mode are equal.

Skewness refers to the tendency of the distribution to "tail off" to the right or left, as shown in parts (b) and (c) of Figure 3.2. In examining the three distribution

FIGURE 3.2

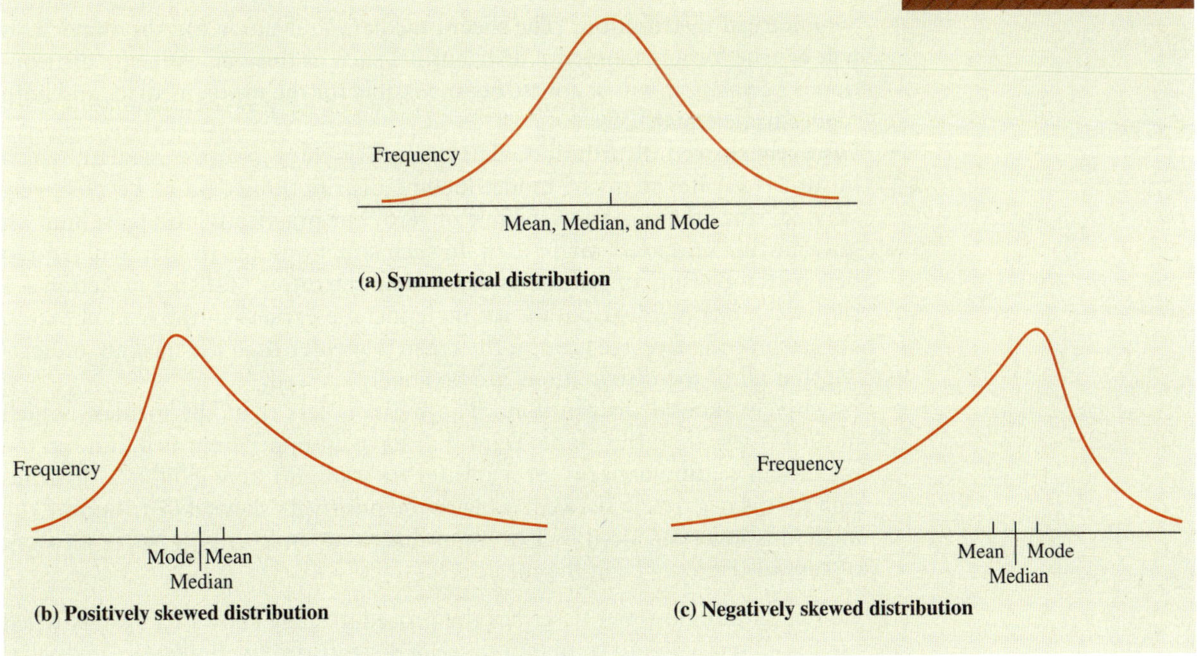

The shape of the distribution determines the relative positions of the mean, median, and mode for a set of data values.

STATISTICS IN ACTION 3.1

U.S. Males Post Skewed Incomes

Data values are skewed positively when the mean exceeds the median, and incomes are especially likely to exhibit this type of distribution. For example, the incomes of males who were 25 or older in 2000 were distributed as follows:

	Age Group				
Income	**25 to <35**	**35 to <45**	**45 to <55**	**55 to <65**	**65 or over**
under $10,000	8.4%	7.7%	8.4%	11.6%	18.7%
$10,000–under $25,000	29.9	21.0	17.7	24.4	45.2
$25,000–under $50,000	40.5	36.7	34.6	30.4	23.5
$50,000–under $75,000	13.8	19.6	20.3	16.7	6.4
$75,000 or over	7.4	15.0	19.0	16.9	6.3
	100.0	100.0	100.0	100.0	100.1*
Median (thousands)	$30.63	$37.09	$41.07	$34.41	$19.17
Mean (thousands)	$42.15	$53.48	$58.46	$61.05	$56.50

*Differs from 100.0 due to rounding.

For each of these distributions, the mean exceeds the median, reflecting that incomes are skewed positively (toward larger incomes) for all of these age groups. In addition, the gap between the median and the mean tends to grow as we proceed from the youngest age group to the oldest. The gap is especially pronounced for 65-or-over males—many of these people are on fixed incomes, but others may have incomes that are quite substantial.

Source: Bureau of the Census, U.S. Department of Commerce, *Statistical Abstract of the United States 2002,* pp. 439–440.

shapes shown in the figure, note the following relationships among the mean, median, and mode:

- **Symmetrical distribution** The mean, median, and mode are the same. This will be true for any *unimodal* distribution, such as this one. (When a distribution is *bimodal,* it will of course be impossible for the mean, median, and both modes to be equal.)
- **Positively skewed distribution** The mean is greater than the median, which in turn is greater than the mode. Income distributions tend to be positively skewed, since there is a lower limit of zero, but practically no upper limit on how much a select few might earn. In such situations, the median will tend to be a better measure of central tendency than the mean. Statistics in Action 3.1 shows the income distribution for male income-earners in the United States. For each of the five age groups, the mean is greater than the median, indicating that all of the distributions are positively skewed.
- **Negatively skewed distribution** The mean is less than the median, which in turn is less than the mode. Data that have an upper limit (e.g., due to the maximum seating capacity of a theater or stadium) may exhibit a distribution that is negatively skewed. As with the positively skewed distribution, the median is less influenced by extreme values and tends to be a better measure of central tendency than the mean.

Computer Solutions 3.1 describes the procedures and shows the results when we use Excel and Minitab to obtain descriptive statistics for the automated teller machine usage times for the 50 customers discussed in Chapter 2. The data are in

COMPUTER 3.1 SOLUTIONS

Descriptive Statistics

EXCEL

	A	B	C	D	E	F
1	AgeCat	Gender	Seconds		*Seconds*	
2	1	2	50.1			
3	1	1	53.0		Mean	39.76
4	2	2	43.2		Standard Error	1.26
5	1	2	34.9		Median	37.80
6	3	1	37.5		Mode	37.80
7	2	1	37.8		Standard Deviation	8.92
8	3	1	49.4		Sample Variance	79.49
9	1	1	50.5		Kurtosis	-0.87
10	3	1	48.1		Skewness	0.23
11	1	2	27.6		Range	34.0
12	3	2	55.6		Minimum	24.1
13	2	2	50.8		Maximum	58.1
14	3	1	43.6		Sum	1987.8
15	1	1	35.4		Count	50

1. Open the Excel data file CX02ATM. The labels and data values have already been entered as shown in the printout.
2. From the **Data** ribbon, click **Data Analysis** in the rightmost menu section. Within **Analysis Tools,** click **Descriptive Statistics.** Click **OK.**
3. In the **Input** section, enter **C1:C51** into the **Input Range** box. In the **Grouped By** section, click to select **Columns.** Because the input range includes the name of the variable in the first row, click to place a check mark in the **Labels in First Row** box.
4. In the **Output Options** section, click to select **Output Range,** then enter **E1** into the **Output Range** box. Click to place a check mark in the **Summary Statistics** box. Click **OK.**
5. Appearance improvements can be made by changing column widths and reducing the number of decimal places shown in the column of descriptive statistics.

MINITAB

Descriptive Statistics: Seconds

Variable	N	N*	Mean	SE Mean	StDev	Minimum	Q1	Median	Q3	Maximum
Seconds	50	0	39.76	1.26	8.92	24.10	32.67	37.80	46.70	58.10

1. Open the Minitab data file CX02ATM. The times (seconds) for the 50 ATM users are in column **C3.**
2. Click **Stat.** Select **Basic Statistics.** Click **Display Descriptive Statistics.** Enter **C3** into the **Variables** box. To get the printout shown here, click **OK.** [*Note:* As an option, we could easily obtain descriptive statistics for the males compared with those for the females. Just click to place a check mark next to **By variable,** then enter **C2** (the gender variable, where 1 = male, 2 = female) into the **By variable** box.]

file **CX02ATM**. The printouts are typical of what we would expect from most statistical software. Shown below are the printout items that pertain to the measures of central tendency discussed in this section and our interpretation of their values.

- **Sample size = 50** Excel refers to this as "count," while Minitab calls it "*N*". Computer printouts often include only capital letters, so "*N*" can still appear instead of the "*n*" that would be more appropriate for describing the size of a sample. (The *N** term in the Minitab printout shows how many data values were either blank or missing. There are no missing values in this database, so *N** is 0.)
- **Mean = 39.76 seconds and Median = 37.80 seconds** As the printout shows, the average customer required 39.76 seconds to complete his or her transaction, and just as many people had a time longer than 37.80 seconds as had a shorter time. Since the mean is greater than the median, the distribution is positively skewed.
- **Mode = 37.80 seconds** Excel reports the most frequent data value as 37.80 seconds. Minitab can also display the mode, but it has not been selected for this printout.
- **Skewness = 0.23** Excel has quantified the skewness and reported it as +0.23. The sign indicates a positively skewed distribution—that is, one in which the mean exceeds the median.

Not shown in either printout is a measure called the **trimmed mean,** in which statistical software ignores an equal number or percentage of data values at both the high end and the low end of the data before it is calculated. This descriptor can optionally be shown by Minitab, but not by Excel. Minitab's trimmed mean ignores the smallest 5% and the largest 5% of the data values. The purpose of the trimmed mean is to reduce the effect exerted by extreme values in the data.

EXERCISES

3.1 Determine the mean and the median wage rate for the data in Exercise 2.9.

3.2 Using the data in Exercise 2.11, determine the mean and the median number of goals per season that Mr. Gretzky scored during his 20 seasons in the National Hockey League.

3.3 Erika operates a website devoted to providing information and support for persons who are interested in organic gardening. According to the hit counter that records daily visitors to her site, the numbers of visits during the past 20 days have been as follows: 65, 36, 52, 70, 37, 55, 63, 59, 68, 56, 63, 63, 43, 46, 73, 41, 47, 75, 75, and 54. Determine the mean and the median for these data. Is there a mode? If so, what is its value?

3.4 A social scientist for a children's advocacy organization has randomly selected 10 Saturday-morning television cartoon shows and carried out a content analysis in which he counts the number of incidents of verbal or physical violence in each. For the 10 cartoons examined, the counts were as follows: 27, 12, 16, 22, 15, 30, 14, 30, 11, and 21. Determine the mean and the median for these data. Is there a mode? If so, what is its value?

3.5 For 1988 through 2007, the net rate of investment income for U.S. life insurance companies was as follows. What was the mean net rate of investment income for this period? The median? Source: American Council of Life Insurers, *Life Insurers Fact Book 2008*, p. 43.

Year	Percentage	Year	Percentage
1988	9.03	1998	6.95
1989	9.10	1999	6.71
1990	8.89	2000	7.05
1991	8.63	2001	6.31
1992	8.08	2002	5.38
1993	7.52	2003	5.03
1994	7.14	2004	4.80
1995	7.41	2005	4.90
1996	7.25	2006	5.35
1997	7.35	2007	5.71

3.6 The following is a list of closing prices for 15 of the stocks held by an institutional investor. What was the average closing price for this sample of stocks? The median closing price?

$31.69	56.69	65.50	83.50	56.88
72.06	121.44	97.00	42.25	71.88
70.63	35.81	83.19	43.63	40.06

3.7 A reference book lists the following maximum life spans (in years) for animals in captivity: Source: *The World Almanac and Book of Facts 2009,* p. 329.

Beaver	50	Grizzly bear	50
Bison	40	Horse (domestic)	50
Camel	50	Kangaroo	24
Cat (domestic)	37	Moose	27
Chimpanzee	60	Pig	27
Cow	30	Polar bear	45
Dog (domestic)	20	Rabbit	13
Asian elephant	77	Sea lion	34
Giraffe	36	Sheep	20
Goat	18	Tiger	26
Gorilla	54	Zebra	50

What is the mean of these maximum life spans? The median?

3.8 According to a utility company, utility plant expenditures per employee were approximately $50,845, $43,690, $47,098, $56,121, and $49,369 for the years 2005 through 2009. Employees at the end of each year numbered 4738, 4637, 4540, 4397, and 4026, respectively. Using the annual number of employees as weights, what is the weighted mean for annual utility plant investment per employee during this period?

3.9 A student scored 90 on the midterm exam, 78 on the final exam, and 83 on the term project. If these three components are weighted at 35%, 45%, and 20%, respectively, what is the weighted mean for her course performance?

3.10 An observer stands at an overpass and, for each motorcycle that passes below, records the value of x = the number of riders on the motorcycle.

a. What value would you anticipate for the mode of her data? Why?
b. Would you anticipate that the mean would be greater than or less than the mode? Why?
c. Would you anticipate that the mean would be greater than or less than the median? Why?

3.11 In preparation for upcoming contract negotiations, labor and management representatives of the Hoffer Concrete Company have been collecting data on the earnings of a sample of employees (ranging from the president to the lowest paid "go-fer") at similar firms. Assuming that such sample data are available, and that the mean and median have been identified:

a. Which of these measures of central tendency would be likely to best serve the purposes of the management representative? Why?
b. Which of these measures of central tendency would be likely to best serve the purposes of the union representative? Why?

(DATA SET) *Note:* Exercises 3.12–3.15 require a computer and statistical software.

3.12 For the *U.S. News & World Report* top-50 U.S. law schools in Exercise 2.67, find the mean and median scores that the schools received from the academic evaluators. Compare these with the mean and median scores awarded by the lawyer/judge evaluators. Does either set of evaluators seem to provide ratings that are positively skewed or negatively skewed? The data are in file **XR02067**.

3.13 Each of the 100 air compressors described in Exercise 2.71 was tested to measure the amount of pressure it could exert. Find the mean and the median pressure, and comment on possible skewness of the data. The data are in file **XR02071**.

3.14 A personnel administrator has collected data on 100 employees, including gender, age, and number of days absent during the previous year. Compare the mean and median number of days absent for females (coded as gender = 1) versus males (coded as gender = 2) within the company. The data are in file **XR03014**.

3.15 For the data in Exercise 3.14, compare the mean and median ages for female employees (gender code = 1) versus male employees (gender code = 2). The data are in file **XR03014**.

STATISTICAL DESCRIPTION: MEASURES OF DISPERSION (3.3)

Although the mean and other measures are useful in identifying the central tendency of values in a population or a set of sample data, it's also valuable to describe their dispersion, or scatter. As a brief introduction to the concept of dispersion, consider the three distributions in Figure 3.3. Although distributions B and C have the same mean (μ = 12.5 ounces), notice that B has a much wider dispersion

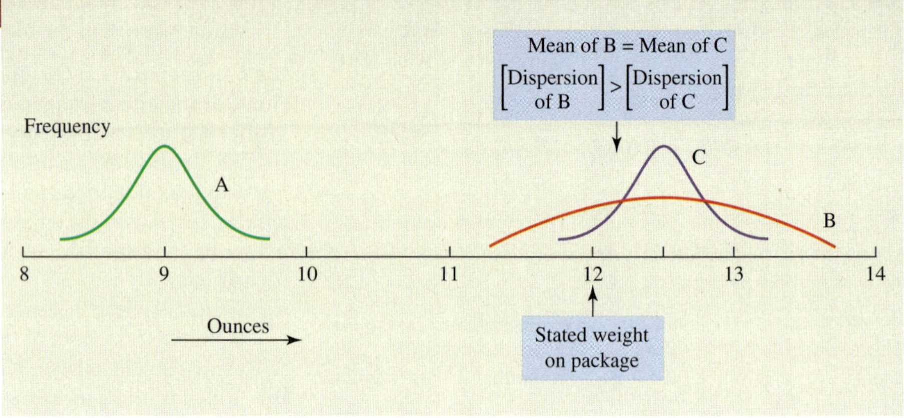

FIGURE 3.3

The dispersion of a distribution refers to the amount of scatter in the values. Distributions A, B, and C describe outputs for a machine that fills cereal boxes. Although B and C have the same mean, B has much greater dispersion and will result in many of the boxes being filled to less than the 12 ounces stated on the package.

and that both B and C have a much larger mean than A (μ = 9.0 ounces). Considering these three distributions as representing the output of a machine that fills boxes of breakfast cereal, we can make the following observations:

- Machine A is set too low. On the average, it fills boxes to just 9 ounces instead of the 12 ounces advertised on the package. The firm's legal counsel will soon be very busy.
- Machine B delivers an average of 12.5 ounces, but the dispersion is so wide that many of the boxes are being underfilled. The firm's legal counsel will soon be moderately busy.
- Machine C delivers an average of 12.5 ounces, and the dispersion is narrow enough that very few of the boxes are being underfilled. The firm's legal counsel can relax.

As this example shows, measures of dispersion can have great practical value in business. We'll now discuss several of the more common measures, including the range, quantiles, mean absolute deviation, variance, and standard deviation.

Range

The simplest measure of dispersion, the **range** is the difference between the highest and lowest values. Part (a) of Figure 3.4 lists new privately owned housing starts for each of the 50 U.S. states. When the data are arranged in numerical order in part (b) of Figure 3.4, we see a range of 270.2 thousand housing starts (271.4 thousand for California minus the 1.2 thousand for Wyoming).

Although the range is easy to use and understand, it has the weakness of being able to recognize only the extreme values in the data. For example, if we were to eliminate the three highest states, 94% of the states would still be represented, but the range would now be only 71.9 thousand starts (73.1 thousand for Georgia minus the 1.2 thousand for Wyoming). Thus the range would be cut to just one-fourth of its original value despite the unchanged relative positions of the other 47 states.

Ads such as "Earn from \$10 to \$10,000 a week by sharpening saws at home" take advantage of this aspect of the range. The vast majority of such entrepreneurs may well earn less than \$20 per week. However, it takes only one unusually successful person for the range to be so impressive. If there is even one very extreme value, the range can be a large number that is misleading to the unwary.

FIGURE 3.4

(a) Raw data (new housing starts, in thousands)

State	Starts	State	Starts	State	Starts	State	Starts	State	Starts
AL	17.2	HI	7.3	MA	39.2	NM	11.8	SD	2.5
AK	4.0	ID	4.3	MI	37.6	NY	61.9	TN	38.1
AZ	71.8	IL	38.7	MN	28.6	NC	70.7	TX	143.1
AR	9.9	IN	23.0	MS	8.8	ND	2.6	UT	16.5
CA	271.4	IA	5.2	MO	27.2	OH	33.0	VT	4.1
CO	32.8	KS	13.3	MT	2.0	OK	10.7	VA	64.1
CT	24.5	KY	13.8	NE	5.0	OR	11.3	WA	35.5
DE	4.6	LA	18.8	NV	14.0	PA	43.6	WV	1.5
FL	202.6	ME	8.1	NH	17.8	RI	5.4	WI	20.2
GA	73.1	MD	42.1	NJ	55.0	SC	32.8	WY	1.2

(b) Data arranged from smallest to largest

1.2	5.2	13.8	28.6	43.6
1.5	5.4	14.0	32.8	55.0
2.0	7.3	16.5	32.8	61.9
2.5	8.1	17.2	33.0	64.1
2.6	8.8	17.8	35.5	70.7
4.0	9.9	18.8	37.6	71.8
4.1	10.7	20.2	38.1	73.1
4.3	11.3	23.0	38.7	143.1
4.6	11.8	24.5	39.2	202.6
5.0	13.3	27.2	42.1	271.4

(c) Quartiles

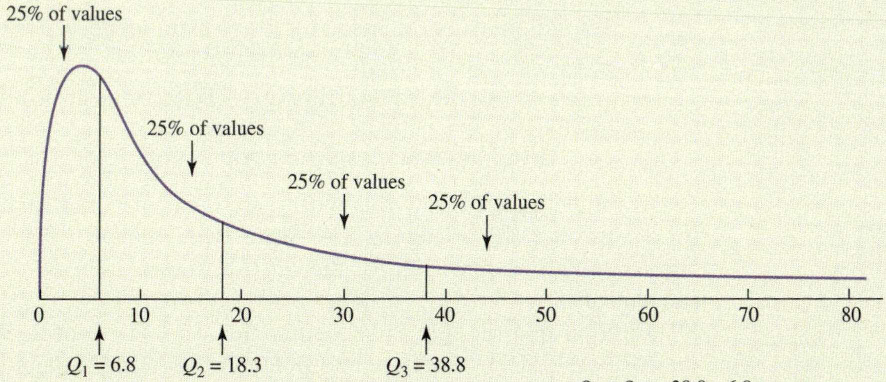

Interquartile range $= Q_3 - Q_1 = 38.8 - 6.8 = 32.0$

Quartile deviation $= \frac{Q_3 - Q_1}{2} = \frac{38.8 - 6.8}{2} = 16.0$

Raw data, ordered data, and quartiles for new privately owned housing starts in the 50 U.S. states
Source: Bureau of the Census, U.S. Department of Commerce, *Construction Reports*, series C20.

The **midrange,** a variant of the range, is the average of the lowest data value and the highest data value. Although it is vulnerable to the effects of extremely low or high outliers in the data, the midrange is sometimes used as a very approximate measure of central tendency. For the data in Figure 3.4, the midrange would be $(271.4 + 1.2)/2$, or 136.3 thousand housing starts.

Quantiles

We have already seen how the median divides data into two equal-size groups: one with values above the median, the other with values below the median. **Quantiles** also separate the data into equal-size groups in order of numerical value. There are several kinds of quantiles, of which **quartiles** will be the primary topic for our discussion.

PERCENTILES divide the values into 100 parts of equal size, each comprising 1% of the observations. The median describes the 50th percentile.

DECILES divide the values into 10 parts of equal size, each comprising 10% of the observations. The median is the 5th decile.

QUARTILES divide the values into four parts of equal size, each comprising 25% of the observations. The median describes the second quartile, below which 50% of the values fall. (*Note:* In some computer statistical packages, the first and third quartile values may be referred to as *hinges*.)

After values are arranged from smallest to largest, quartiles are calculated similarly to the median. It may be necessary to *interpolate* (calculate a position between) two values to identify the data position corresponding to the quartile.

For N values arranged from lowest to highest:

$$\text{First quartile, } Q_1 = \text{Data value at position } \frac{(N+1)}{4}$$

$$\text{Second quartile (the median), } Q_2 = \text{Data value at position } \frac{2(N+1)}{4}$$

$$\text{Third quartile, } Q_3 = \text{Data value at position } \frac{3(N+1)}{4}$$

(Use N if data represent a population, n for a sample.)

As an example of how this works, we'll calculate the quartiles for the data shown in part (b) of Figure 3.4. Since there are 50 data values, representing a population consisting of the 50 U.S. states, $N = 50$ will be used in the calculations.

The first quartile, Q_1, is located at the $(50 + 1)/4$, or 12.75 position. Since there is no "12.75th" state, we interpolate between the 12th and 13th, going 75% of the way from the 12th state to the 13th state:

Position	x_{12}	$x_{12.75}$	x_{13}
Data value	5.4	6.8	7.3 thousand housing starts

75% of the way from 5.4 to 7.3, $Q_1 = 6.8$
(6.825 rounded to one decimal place)

In calculating Q_2, which is also the median, we identify what would be the 2(50 + 1)/4, or 25.5th data value. This is halfway between the 25th state (17.8 thousand housing starts) and the 26th state (18.8), or $Q_2 = 18.3$. The third quartile corresponds to the position calculated as 3(50 + 1)/4, or the 38.25th data value. Interpolating 25% of the way from the 38th state (38.7 thousand housing starts) to the 39th state (39.2), we find $Q_3 = 38.8$ thousand housing starts (38.825 rounded to 38.8). Although the rounding is not strictly necessary, it keeps the quartiles to the same number of decimal places as the original data.

Quartiles offer a special feature that allows us to overcome the inability of the range to consider any but the most extreme values. In particular, there are two useful descriptors that involve quartiles:

The **interquartile range** is the difference between the third quartile and the first quartile, or $Q_3 - Q_1$. In percentile terms, this is the distance between the 75% and 25% values.

The **quartile deviation** is one-half the interquartile range. The quartile deviation is $(Q_3 - Q_1)/2$.

Because they reduce the importance of extreme high and low values, the interquartile range and quartile deviation can be more meaningful than the range as indicators of the dispersion of the data. In addition, they consider more than just the two extreme values, thus making more complete use of the data. Part (c) of Figure 3.4 shows a sketch of the approximate distribution of the 50-state housing-starts data, along with the quartiles, the interquartile range, and the quartile deviation. The distribution is positively skewed (i.e., it tapers off toward a few very high values at the right), and the mean (34.6) is much greater than the median (18.3).

Mean Absolute Deviation (MAD)

With this descriptor, sometimes called the *average deviation* or the *average absolute deviation,* we now consider the extent to which the data values tend to differ from the mean. In particular, the **mean absolute deviation** (*MAD*) is the average of the absolute values of differences from the mean and may be expressed as follows:

Mean absolute deviation (*MAD*) for a population:

$$MAD = \frac{\sum |x_i - \mu|}{N}$$

where μ = population mean
x_i = the *i*th data value
N = number of data values in the population

(To calculate *MAD* for a sample, substitute n for N and $\bar{x}$ for μ.)

In the preceding formula, we find the sum of the absolute values of the differences between the individual values and the mean, then divide by the number of individual values. The two vertical lines ("| |") tell us that we are concerned with *absolute values,* for which the algebraic sign is ignored. To illustrate how the mean absolute deviation is calculated, we'll examine the following figures, which represent annual research and development (R&D) expenditures by Microsoft Corporation.[2]

[2]Source: Microsoft Corporation, *2005 Annual Report.*

TABLE 3.1

Calculation of mean absolute deviation for annual research and development (R&D) expenditures for Microsoft Corporation. Data are in millions of dollars.

Year	R&D x_i	Deviation from Mean $(x_i - \mu)$	Absolute Value of Deviation from Mean $\lvert x_i - \mu \rvert$
2001	4379	−1868.2	1868.2
2002	6299	51.8	51.8
2003	6595	347.8	347.8
2004	7779	1531.8	1531.8
2005	6184	−63.2	63.2
	31,236 $= \Sigma x_i$	0.0	3862.8 $= \Sigma \lvert x_i - \mu \rvert$

Mean: $\mu = \dfrac{\Sigma x_i}{N} = \dfrac{31{,}236}{5} = \$6247.2$ million

Mean absolute deviation: $MAD = \dfrac{\Sigma \lvert x_i - \mu \rvert}{N} = \dfrac{3862.8}{5} = \$772.6$

Year	2001	2002	2003	2004	2005	
R&D	4379	6299	6595	7779	6184	(millions of dollars)

If we consider these data as representing the population of values for these 5 years, the mean will be μ. It is calculated as

$$\mu = \frac{4379 + 6299 + 6595 + 7779 + 6184}{5} = \$6247.2 \text{ million}$$

As shown in Table 3.1, the mean absolute deviation is the sum of the absolute deviations divided by the number of data values, or $3862.8/5 = \$772.6$ million. On the average, annual expenditures on research and development during these 5 years were $772.6 million away from the mean. In the third column of Table 3.1, note that the sum of the deviations is zero. This sum will always be zero, so it isn't of any use in describing dispersion, and we must consider only the *absolute* deviations in our procedure.

Variance and Standard Deviation

Variance

The **variance,** a common measure of dispersion, includes all data values and is calculated by a mathematical formula. For a population, the variance (σ^2, "sigma squared") is the average of squared differences between the N data values and the mean, μ. For a sample variance (s^2), the sum of the squared differences between the n data values and the mean, $\bar{x}$, is divided by $(n - 1)$. These calculations can be summarized as follows:

Variance for a population:

$$\sigma^2 = \frac{\Sigma (x_i - \mu)^2}{N}$$

where σ^2 = population variance
μ = population mean
x_i = the ith data value
N = number of data values in the population

Variance for a sample:

$$s^2 = \frac{\sum (x_i - \bar{x})^2}{n - 1}$$

where s^2 = sample variance
$\bar{x}$ = sample mean
x_i = the ith data value
n = number of data values in the sample

Using the divisor $(n - 1)$ when calculating the variance for a sample is a standard procedure that makes the resulting sample variance a better estimate of the variance of the population from which the sample was drawn. Actually, for large sample sizes (e.g., $n \geq 30$) the subtraction of 1 from n makes very little difference.

Table 3.2 shows the calculation of the variance for a sample of estimates of highway miles per gallon (mpg) for five sports-utility vehicles. In the table, notice the role played by **residuals.** These are the same as the deviations from the mean discussed in the preceding section. For each observation, the residual is the difference between the individual data value and the mean.

$$\text{Residual for a data value in a sample} = x_i - \bar{x}$$

where $\bar{x}$ = sample mean
x_i = value of the ith observation

For example, the residual for the third observation (21 mpg) is $21.0 - 20.0 = 1.0$ mpg. The residual is an important statistical concept that will be discussed further in later chapters.

TABLE 3.2

Calculation of variance (s^2) and standard deviation (s) for estimated highway fuel economy (mpg) for five 2009 sports-utility vehicles.

Model	Highway mpg (x_i)	x_i^2	Residual $(x_i - \bar{x})$	Residual² $(x_i - \bar{x})^2$
Saturn Outlook	23	529	3.0	9.0
Jeep Liberty	21	441	1.0	1.0
Subaru Tribeca	21	441	1.0	1.0
Land Rover LR3	17	289	−3.0	9.0
Porsche Cayenne GTS	18	324	−2.0	4.0
	100 $= \sum x_i$	2024 $= \sum x_i^2$		24.0 $= \sum (x_i - \bar{x})^2$

$$\bar{x} = \frac{\sum x_i}{5} = \frac{100}{5} = 20.0 \text{ miles per gallon}$$

$$s^2 = \frac{\sum (x_i - \bar{x})^2}{n - 1} = \frac{24.0}{5 - 1} = 6.0 \qquad s = \sqrt{6.0} = 2.45$$

Source: U.S. Environmental Protection Agency, *Fuel Economy Guide* 2009.

Standard Deviation

The positive square root of the variance of either a population or a sample is a quantity known as the **standard deviation.** The standard deviation is an especially important measure of dispersion because it is the basis for determining the proportion of data values within certain distances on either side of the mean for certain types of distributions (we will discuss these in a later chapter). The standard deviation may be expressed as

	For a Population	For a Sample
Standard Deviation	$\sigma = \sqrt{\sigma^2}$	$s = \sqrt{s^2}$

For the data in Table 3.2, the sample standard deviation is 2.45 miles per gallon. This is calculated by taking the square root of the sample variance, 6.0.

Variance and Standard Deviation: Additional Notes

- Unless all members of the population or sample have the same value (e.g., 33, 33, 33, 33, and so on), the variance and standard deviation cannot be equal to zero.
- If the same number is added to or subtracted from all the values, the variance and standard deviation are unchanged. For example, the data 24, 53, and 36 will have the same variance and standard deviation as the data 124, 153, and 136. Both the variance and standard deviation depend on the residuals (differences) between the observations and the mean, and adding 100 to each data value will also increase the mean by 100.

Computer Solutions 3.2 describes the procedures and shows the results when we use Excel and Minitab to obtain descriptive statistics for the housing-starts data shown in Figure 3.4. The data are in file **CX03HOUS**. The printout items pertaining to the measures of dispersion in this section and an interpretation of their values follow:

- **Standard Deviation (or StDev) = 49.78** The standard deviation is $s = 49.78$ thousand housing starts during the year. Excel also lists the variance, although this could be readily calculated as the square of the standard deviation.
- **Standard Error (SE Mean) = 7.04** This is the standard error for the mean, a term that will not be discussed until Chapter 8. Disregard it for now.
- **Minimum = 1.20 and Maximum = 271.40** These are the values for the minimum and maximum observations in the data. Excel also provides the range, which is simply 271.40 − 1.20, or 270.20 thousand housing starts.
- **Q1 = 6.82 and Q2 = 38.83** Reported by Minitab, the first and third quartiles are 6.82 and 38.83: 50% of the states had between 6.82 and 38.83 thousand housing starts. Excel does not include the first and third quartiles in its printout of summary statistics, but you can get at least approximate values by using the procedure described in Computer Solutions 3.2. If you are using the Data Analysis Plus add-in with Excel, the first and third quartiles are provided along with the Boxplot; this is demonstrated in Computer Solutions 3.3 in the next section.

Statistical software typically assumes the data are a sample instead of a population and uses (number of observations − 1) in calculating the standard deviation. This is usually not a problem because (1) much of the data in business statistics will

COMPUTER 3.2 SOLUTIONS

Descriptive Statistics

EXCEL

	A	B	C	D
1	**Hstarts**		*Hstarts*	
2	17.2			
3	4.0		Mean	34.65
4	71.8		Standard Error	7.04
5	9.9		Median	18.30
6	271.4		Mode	32.80
7	32.8		Standard Deviation	49.78
8	24.5		Sample Variance	2478.55
9	4.6		Kurtosis	12.20
10	202.6		Skewness	3.27
11	73.1		Range	270.20
12	7.3		Minimum	1.20
13	4.3		Maximum	271.40
14	38.7		Sum	1732.30
15	23.0		Count	50

Open the Excel data file CX03HOUS. The steps are the same as for Computer Solutions 3.1, except that we enter **A1:A51** into the **Input Range** box and **C1** into the **Output Range** box. Excel does not include the first and third quartiles in its printout of summary statistics. However, if you have a large number of data values, you can display at least approximate first and third quartiles by dividing the sample size by 4, rounding to the nearest integer if necessary, then entering this number into the **Kth Largest** box as well as the **Kth Smallest** box during step 3.

MINITAB

```
Descriptive Statistics: Hstarts

Variable   N  N*   Mean  SE Mean  StDev  Minimum    Q1  Median     Q3  Maximum
Hstarts   50   0  34.65     7.04  49.78     1.20  6.82   18.30  38.83   271.40
```

Open the Minitab data file CX03HOUS. The steps are the same as for Computer Solutions 3.1, except the 50 values for the housing starts are in column **C1.**

actually be sample data, and (2) there will not be an appreciable difference when the number of observations is large. However, if your data do represent a population, and you would like to know the value of σ instead of s, you can multiply s from the printout to get σ, as shown here:

$$\text{Standard deviation based on } (N) \text{ divisor} = \text{Standard deviation based on } (N-1) \text{ divisor} \times \sqrt{\frac{N-1}{N}}$$

EXERCISES

3.16 Provide a real or hypothetical example of a situation where the range could be misleading as a measure of dispersion.

3.17 For the sample of computer website hits described in Exercise 3.3, determine the range, the mean absolute deviation, the standard deviation, and the variance.

3.18 For the sample of Saturday-morning cartoon violence counts described in Exercise 3.4, determine the range, the mean absolute deviation, the standard deviation, and the variance.

3.19 During July 2005, the seven most-visited shopping websites (with number of visitors, in millions) were: eBay (61.7), Amazon (42.0), Wal-Mart (22.6), Shopping.com (22.4), Target (20.5), Apple Computer (17.4), and Overstock.com (17.2). Considering these as a population, determine: Source: *The World Almanac and Book of Facts 2006*, p. 390.

a. The mean, median, range, and midrange.
b. The mean absolute deviation.
c. The standard deviation and variance.

3.20 For a sample of 11 employers, the most recent hourly wage increases were 18, 30, 25, 5, 7, 2, 20, 12, 15, 55, and 40 cents per hour. For these sample data, determine:

a. The mean, median, range, and midrange.
b. The mean absolute deviation.
c. The standard deviation and variance.

3.21 According to the U.S. Environmental Protection Agency, a sample of 10 subcompact models shows the following estimated values for highway fuel economy (mpg): 40, 33, 32, 30, 27, 29, 27, 23, 21, and 10. For these sample data, determine:

a. The mean, median, range, and midrange.
b. The mean absolute deviation.
c. The standard deviation and variance.

3.22 For a sample of eight apartment buildings in Chicago, the percentage-occupancy rates during a recent month were as shown below:

50.0%	79.1	61.3	43.9
68.1	70.0	72.5	66.3

For these sample data, determine:

a. The mean, median, range, and midrange.
b. The mean absolute deviation.
c. The standard deviation and variance.

3.23 Determine the first, second, and third quartiles for the data in Exercise 3.20; then calculate the interquartile range and the quartile deviation.

3.24 Determine the first, second, and third quartiles for the data in Exercise 3.21; then calculate the interquartile range and the quartile deviation.

(DATA SET) *Note:* Exercises 3.25–3.27 require a computer and statistical software.

3.25 During the 1999 holiday shopping season, Keynote Systems, Inc. visited a number of e-commerce websites to see how long it took for the site to come up on the PC of a consumer using a 56K modem. The average time for Walmart.com was 23.35 seconds. Under the assumption that the company visited Wal-Mart's site on 50 occasions and obtained the sample times (in seconds) in data file **XR03025**, determine and interpret: Source: "How Key Web Sites Handle Holiday Shopping Rush," *USA Today*, November 24, 1999, p. 3B.

a. The mean, median, range, and midrange.
b. The mean absolute deviation.
c. The standard deviation and variance.

3.26 Data file **XR03014** includes the number of absences from work for a sample of 100 employees. For the variable representing number of days absent, determine and interpret the following:

a. The mean, median, range, and midrange.
b. The mean absolute deviation.
c. The standard deviation and variance.

3.27 Data file **XR02028** shows the depths (in meters) that 80 water-resistant watches were able to withstand just before leakage. For the watches in this sample, determine and interpret the following:

a. The mean, median, range, and midrange.
b. The mean absolute deviation.
c. The standard deviation and variance.

ADDITIONAL DISPERSION TOPICS (3.4)

The Box-and-Whisker Plot

Also referred to as a **box plot,** the **box-and-whisker plot** is a graphical device that simultaneously displays several of the measures of central tendency and dispersion discussed previously in the chapter. It highlights the first and third quartiles, the median, and the extreme values in the data, allowing us to easily identify these descriptors. We can also see whether the distribution is symmetrical or whether it is skewed either negatively or positively.

Computer Solutions 3.3 describes the procedures and shows the results when we use Excel and Minitab to generate a box plot for the housing-starts data

COMPUTER 3.3 SOLUTIONS

The Box Plot

EXCEL

	A	B	C	D	E	F	G
1	**Box Plot**						
2	*Hstarts*						
3	Smallest = 1.2						
4	Q1 = 6.825						
5	Median = 18.3						
6	Q3 = 38.825						
7	Largest = 271.4						
8	IQR = 32						
9	Outliers: 271.4, 202.6, 143.1,						

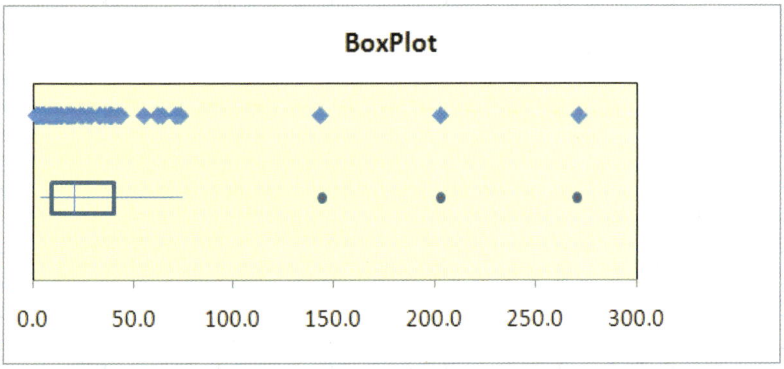

1. Open the Excel data file CX03HOUS. Click on **A1** and drag to **A51** to select the data values in cells **A1:A51.** From the **Add-Ins** ribbon, click **Data Analysis Plus** in the leftmost menu section.
2. In the **Data Analysis Plus** menu, click **Box Plot.** Click **OK.** Ensure that the **Input Range** box contains the range you specified in step 1. Click **Labels.** Click **OK.**
3. Improve the appearance by selecting and rearranging the components of the printout.

(continued)

MINITAB

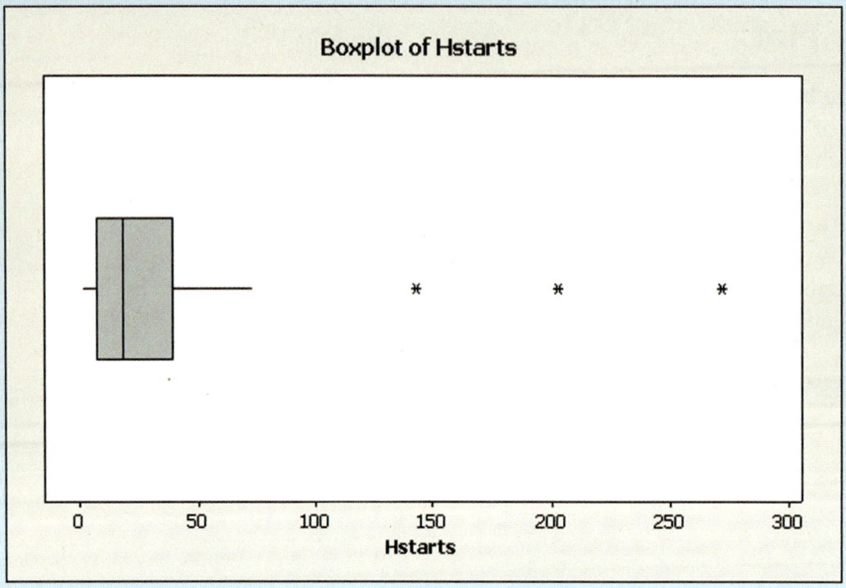

1. Open data file **CX03HOUS**. The 50 values for the housing starts are in column **C1.**
2. Click **Graph.** Click **Boxplot.** Select **One Y Simple** and click **OK.** In the **Graph variables** box, enter **C1.**
3. Click **Scale** and select the **Axes and Ticks** menu. Select **Transpose value and category scales.** (This causes the box plot to be oriented horizontally.) Click **OK.** Click **OK.**

shown in Figure 3.4. The data are in file **CX03HOUS**. The box plot has the following components:

- **Quartiles** The box extends horizontally from the first quartile (25th percentile) to the third quartile (75th percentile). Q_1 and Q_3 are shown at the left and right sides of the box, and the width of the box is the interquartile range.
- **Median** This is shown by the vertical line within the box. It is off-center within the box, being located slightly left of center. Because the median is closer to Q_1 than to Q_3, the distribution is skewed positively.
- **Extreme values** The whisker on the left side of the box extends to the smallest data value, Wyoming's 1.2 thousand housing starts. The whisker on the right side extends to cover all but the three most distant data values, which are considered outliers.

Chebyshev's Theorem

When either a population or a sample has a small standard deviation, individual observations will tend to be closer to the mean. Conversely, a larger standard deviation will result when individual observations are scattered widely about their mean. An early mathematician named Chebyshev set forth a theorem that quantifies this phenomenon. It specifies the minimum percentage of observations that will fall within a given number of standard deviations from the mean, regardless of the shape of the distribution.

CHEBYSHEV'S THEOREM For either a sample or a population, the percentage of observations that fall within k (for $k > 1$) standard deviations of the mean will be at least

$$\left(1 - \frac{1}{k^2}\right) \times 100$$

To demonstrate this rule, let's examine how it applies to the housing-starts data in Figure 3.4. For these data, values for the population mean and standard deviation are $\mu = 34.6$ and $\sigma = 49.3$, respectively (in thousands of housing starts). Using Chebyshev's theorem, we can find the percentage of values that should fall within $k = 2$ and $k = 3$ standard deviations of the mean. For example, if $k = 2$, this percentage should be at least $[1 - (1/2^2)](100)$, or 75%. Expressed another way, we should find that at least 75% of the states have housing starts that fall in the interval described by $34.6 \pm 2(49.3)$, or from -64.0 to 133.2. The actual percentage of data values falling within this interval is 94%, well above the minimum of 75% predicted by the theorem.

The Empirical Rule

Although Chebyshev's theorem applies to *any* distribution, regardless of shape, there is a rule of thumb that applies *only to distributions that are bell shaped and symmetrical,* like part (a) of Figure 3.2 and part (a) of Figure 3.3. This rule is known as the *empirical rule:*

THE EMPIRICAL RULE For distributions that are bell shaped and symmetrical:

- About 68% of the observations will fall within 1 standard deviation of the mean.
- About 95% of the observations will fall within 2 standard deviations of the mean.
- Practically all of the observations will fall within 3 standard deviations of the mean.

Let's consider part (a) of Figure 3.3 (page 68), which happens to be a bell-shaped, symmetrical distribution with a mean of 9.0 and a standard deviation of 0.3. If the filling machine inserts an average of 9.0 ounces of cereal into the boxes, with a standard deviation of 0.3 ounces, then approximately 68% of the boxes will contain $9.0 \pm 1(0.3)$ ounces of cereal, or between 8.7 and 9.3 ounces. Again applying the empirical rule, but using 2 standard deviations instead of 1, approximately 95% of the boxes will contain $9.0 \pm 2(0.3)$ ounces of cereal, or between 8.4 and 9.6 ounces. Practically all of the boxes will contain $9.0 \pm 3(0.3)$ ounces of cereal, or between 8.1 and 9.9 ounces.

Standardized Data

In our discussion of Chebyshev's theorem, we used the standard deviation as a unit of measurement, or measuring stick, for expressing distance from the mean to a data value. **Standardizing** the data is based on this concept and involves

TABLE 3.3

Standardization of estimated highway fuel economy (mpg) data for a sample of five 2009 sports-utility vehicles. Standardizing the data involves expressing each value in terms of its distance from the mean, in standard deviation units.

	Original Data		Standardized Data
Model	**Highway mpg (x_i)**	$x_i - \bar{x}$	$\left[z_i = \frac{(x_i - \bar{x})}{s}\right]$
Saturn Outlook	23.00	3.00	1.22
Jeep Liberty	21.00	1.00	0.41
Subaru Tribeca	21.00	1.00	0.41
Land Rover LR3	17.00	−3.00	−1.22
Porsche Cayenne GTS	18.00	−2.00	−0.82
			0.00
			$= \Sigma z_i$

Original Data	Standardized Data
$\bar{x} = 20.00$ mpg	$\bar{z} = 0.000$
$s_x = 2.45$ mpg	$s_z = 1.000$

expressing each data value in terms of its distance (in standard deviations) from the mean. For each observation in a sample:

$$z_i = \frac{x_i - \bar{x}}{s}$$

where z_i = standardized value for the ith observation
$\bar{x}$ = sample mean
x_i = the ith data value
s = sample standard deviation

(When data represent a population, μ replaces $\bar{x}$, and σ replaces s.)

What we're really doing is expressing the data in different units. It's like referring to a 6-footer as someone who is 2 yards tall. Instead of using feet, we're using yards. For example, the mpg estimates for the five sports-utility vehicles in Table 3.2 can be converted as shown in Table 3.3. After conversion, some standardized values are positive, others are negative. This is because we're using the mean as our reference point, and some data values are higher than the mean, others lower.

Standardized values have no units. This is true even though the original data, the standard deviation, and the mean may each represent units such as dollars, hours, or pounds. Because the data values and the standard deviation are in the same units, when one quantity is divided by the other, these units are in both the numerator and denominator, and the units cancel each other out.

A standardized value that is either a large positive number or a large negative number is a relatively unusual occurrence. For example, the estimated 23 mpg for the Saturn Outlook was easily the highest data value for the five vehicles in the sample. A standardized distribution will always have a mean of 0 and a standard deviation of 1. The standard deviation will be either s or σ, depending on whether the data represent a sample or a population. Standardizing the original data values makes it easier to compare two sets of data in which the original measurement units are not the same. Computer Solutions 3.4 describes the procedures and shows the results when we use Excel and Minitab to standardize the data for the five sports-utility vehicles.

COMPUTER 3.4 SOLUTIONS

Standardizing the Data

EXCEL

	A	B	C	D	E
1	MPG	STDMPG		Mean Is:	20.00
2	23	1.22		Std. Dev.Is:	2.45
3	21	0.41			
4	21	0.41			
5	17	-1.22			
6	18	-0.82			

1. Open the Excel data file CX03SUV. The data are entered in column **A,** as shown here. Type the labels **Mean is:** and **Std. Dev. is:** into cells **D1** and **D2,** respectively. Enter the equation **=AVERAGE(A2:A6)** into **E1.** Enter the equation **=STDEV(A2:A6)** into **E2.** (Alternatively, you can find the values for the mean and standard deviation by using a procedure like that of Computer Solutions 3.1 to obtain values for all the descriptive statistics.)
2. Enter the equation **=(A2-20.00)/2.45** into **B2.** Click on cell **B2.** Click on the lower right corner of cell **B2** and the cursor will turn to a "+" sign. Drag the "+" sign down to cell **B6** to fill in the remaining standardized values. Enter the label into cell **B1.**

MINITAB

```
Data Display

Row  MPG     STDMPG
  1   23    1.22474
  2   21    0.40825
  3   21    0.40825
  4   17   -1.22474
  5   18   -0.81650


Descriptive Statistics: MPG, STDMPG

Variable  N  N*    Mean  SE Mean  StDev  Minimum      Q1  Median     Q3
MPG       5   0   20.00     1.10   2.45    17.00   17.50   21.00  22.00
STDMPG    5   0  -0.000    0.447  1.000   -1.225  -1.021   0.408  0.816

Variable  Maximum
MPG         23.00
STDMPG      1.225
```

1. Open the Minitab data file CX03SUV. The five fuel-economy values are in column **C1.**
2. Click **Calc.** Click **Standardize.** Enter **C1** into the **Input column(s)** box. Enter **C2** into the **Store results in** box. Click to select **Subtract mean and divide by std. dev.** Click **OK.**
3. The standardized values are now in **C2,** which we can title by entering **STDMPG** into the label cell above the values.
4. Besides the original and standardized values, the printout shown here includes descriptive statistics for both. (*Note:* The mean and standard deviation of the standardized values are 0 and 1, respectively.) To get the printout: Click **Edit.** Click **Command Line Editor.** Enter **print C1 C2** into the first line. Enter **describe C1 C2** into the second line. Click **Submit Commands.**

The Coefficient of Variation

Expressing the standard deviation as a percentage of the mean, the **coefficient of variation (*CV*)** indicates the *relative amount of dispersion* in the data. This enables us to easily compare the dispersions of two sets of data that involve different measurement units (e.g., dollars versus tons) or differ substantially in magnitude. For example, the coefficient of variation can be used in comparing the variability in prices of diamonds versus prices of coal or in comparing the variability in sales, inventory, or investment of large firms versus small firms.

In comparing the variability of gold prices and slab zinc prices, for example, standard deviations would tell us that the price of gold varies a great deal more than the price of zinc. However, the dollar value of an ounce of gold is much higher than that of a pound of slab zinc, causing its variability to be exaggerated. Dividing each standard deviation by the mean of its data, we are better able to see how much each commodity varies relative to its own average price. The coefficient of variation is expressed as a percentage and is calculated as follows:

	For a Population	For a Sample
Coefficient of Variation	$CV = \frac{\sigma}{\mu} \times 100$	$CV = \frac{s}{\overline{x}} \times 100$

Suppose that, over a sample of 12 months, the mean price of gold had been $\overline{x}$ = \$810.6 per ounce, with a standard deviation of s = \$49.0 per ounce, while the mean and standard deviation for slab zinc during this same period had been \$1.04 per pound and \$0.08 per pound, respectively. Which commodity would have had the greater relative dispersion, or coefficient of variation?

We calculate the coefficient of variation for each commodity: The CV for gold was $s/\overline{x}$, or \$49.0/\$810.6 = 0.060, and the CV for slab zinc was \$0.08/\$1.04 = 0.077. Expressed as percentages, the coefficients of variation would be 6.0% for gold and 7.7% for slab zinc. Although the standard deviation of monthly gold prices was more than 612 times the standard deviation of zinc prices, from a *relative dispersion* perspective, the variability of gold prices was actually *less than* that for slab zinc.

EXERCISES

3.28 A manufacturing firm has collected information on the number of defects produced each day for the past 50 days. The data have been described by the accompanying Minitab box-and-whisker plot.

a. What is the approximate value of the median? The first and third quartiles?
b. What do the asterisks (*) at the right of the display indicate? What implications might they have for the production supervisor?
c. Does the distribution appear to be symmetric? If not, is it positively skewed or negatively skewed?

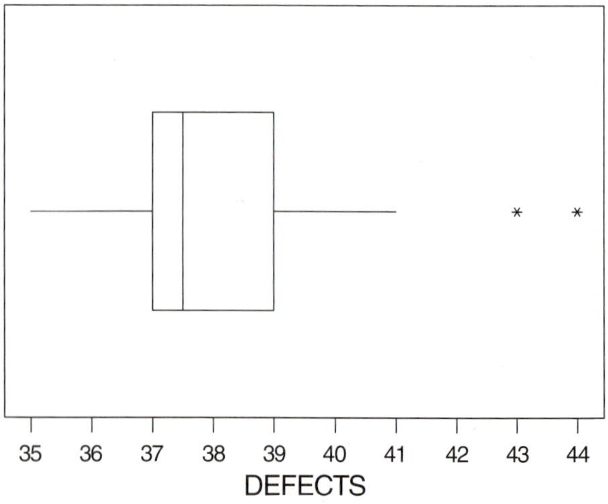

3.29 According to Chebyshev's theorem, what percentage of observations should fall

a. within 2.5 standard deviations of the mean?
b. within 3.0 standard deviations of the mean?
c. within 5.0 standard deviations of the mean?

3.30 For the data in Exercise 3.20, determine the percentage of observations that fall within $k = 1.5$ standard deviation units from the mean. Do the results support Chebyshev's theorem?

3.31 Standardize the data in Exercise 3.21, then identify the percentage of observations that fall within $k = 2.0$ standard deviation units from the mean. Do the results support Chebyshev's theorem?

3.32 The manufacturer of an extended-life lightbulb claims the bulb has an average life of 12,000 hours, with a standard deviation of 500 hours. If the distribution is bell shaped and symmetrical, what is the approximate percentage of these bulbs that will last

a. between 11,000 and 13,000 hours?
b. over 12,500 hours?
c. less than 11,000 hours?
d. between 11,500 and 13,000 hours?

3.33 For college-bound high school seniors from a certain midwestern city, math scores on the Scholastic Aptitude Test (SAT) averaged 480, with a standard deviation of 100. Assume that the distribution of math scores is bell shaped and symmetrical.

a. What is the approximate percentage of scores that were between 380 and 580?
b. What is the approximate percentage of scores that were above 680?
c. Charlie scored 580 on the math portion of the SAT. What is the approximate percentage of students who scored lower than Charlie?
d. Betty scored 680 on the math portion of the SAT. What is the approximate percentage of students who scored higher than Charlie but lower than Betty?

3.34 For data set A, the mean is $1235, with a standard deviation of $140. For data set B, the mean is 15.7 inches, with a standard deviation of 1.87 inches. Which of these two data sets has the greater relative dispersion?

3.35 A sample of 20 customers from Barnsboro National Bank reveals an average savings account balance of $315, with a standard deviation of $87. A sample of 25 customers from Wellington Savings and Loan reveals an average savings account balance of $8350, with a standard deviation of $1800. Which of these samples has the greater relative dispersion?

(DATA SET) *Note:* Exercises 3.36–3.38 require a computer and statistical software.

3.36 A resort catering to golfers has reported the average income of golfer households to be $95,000. Assume that this average had been based on the 100 incomes in file **XR03036.** Source: golfwashington.com, June 23, 2009.

a. Construct and interpret the box-and-whisker plot for the data.
b. Standardize the data; then determine the mean and standard deviation of the standardized values.

3.37 In Exercise 3.14, the personnel director has collected data describing 100 employees, with one of the variables being the number of absences during the past year. Using data file **XR03014**, describe the variable, number of absences.

a. Construct and interpret the box-and-whisker plot for the number of absences.
b. Standardize the data; then determine the mean and standard deviation of the standardized values.

3.38 Exercise 3.25 described a study in which e-commerce researchers visited walmart.com on multiple occasions during the holiday shopping season and measured the time (in seconds) it took for the site to come up on the computer screen. Data file **XR03025** contains results that we assumed could have been obtained during this study.

a. Construct and interpret the box-and-whisker plot for the times.
b. Standardize the data; then determine the mean and standard deviation of the standardized values.

DESCRIPTIVE STATISTICS FROM GROUPED DATA (3.5)

The frequency distribution, also referred to as **grouped data,** is a convenient summary of raw data, but it loses some of the information originally contained in the data. As a result, measures of central tendency and dispersion determined from the frequency distribution will be only approximations of the actual values. However, since the original data may not always be available to us, these approximations are sometimes the best we can do. Although other descriptors can be approximated from the frequency distribution, we will concentrate here on the mean and the standard deviation. Techniques for their approximation are

TABLE 3.4

The formulas in the text are used to calculate the approximate mean and standard deviation based only on a frequency distribution for the usage times of the 50 automated teller machine (ATM) customers in data file **CX02ATM**.

Time (seconds)	(1) Midpoint (m_i)	(2) Frequency (f_i)	(3) $f_i m_i$	(4) m_i^2	(5) $f_i m_i^2$
20–under 25	22.5	1	22.5	506.25	506.25
25–under 30	27.5	7	192.5	756.25	5,293.75
30–under 35	32.5	10	325.0	1056.25	10,562.50
35–under 40	37.5	9	337.5	1406.25	12,656.25
40–under 45	42.5	9	382.5	1806.25	16,256.25
45–under 50	47.5	6	285.0	2256.25	13,537.50
50–under 55	52.5	5	262.5	2756.25	13,781.25
55–under 60	57.5	3	172.5	3306.25	9,918.75
		50 $= \sum f_i = n$	1980.0 $= \sum f_i m_i$		82,512.50 $= \sum f_i m_i^2$

Approximate mean: $\bar{x} = \dfrac{\sum f_i m_i}{n} = \dfrac{1980.0}{50} = 39.6$

Approximate variance: $s^2 = \dfrac{\sum f_i m_i^2 - n\bar{x}^2}{n-1} = \dfrac{82{,}512.50 - 50(39.6)^2}{50-1}$

$= \dfrac{82{,}512.50 - 78{,}408}{50-1} = \dfrac{4104.5}{49} = 83.77$

Approximate standard deviation: $s = \sqrt{83.77} = 9.15$

summarized in Table 3.4 and explained subsequently. In this section, we'll assume that all we know about the usage times for the 50 automated teller machine (ATM) customers in data file **CX02ATM** is the frequency distribution shown in the first three columns of Table 3.4. We will approximate the mean and standard deviation, then compare them with the actual values (which happen to be $\bar{x} = 39.76$ seconds and $s = 8.92$ seconds).

Arithmetic Mean from Grouped Data

When approximating the mean from a frequency distribution, we consider each observation to be located at the midpoint of its class. As an example, for the 30–under 35 class, all 10 persons in this category are assumed to have a time of exactly 32.5 seconds, the midpoint of the class. Using this and other midpoints, we can find the approximate value of the mean.

Approximate value of the sample mean, from grouped data:

$$\bar{x} = \frac{\sum f_i m_i}{n}$$

where $\bar{x}$ = approximate sample mean
f_i = frequency for class i
m_i = midpoint for class i
n = number of data values in the sample

(If data are a population, μ replaces $\bar{x}$ and N replaces n.)

The formula assumes that each midpoint occurs f_i times—e.g., that 22.5 seconds occurs 1 time, 27.5 seconds occurs 7 times, and so on. Applying this formula as shown in Table 3.4, we calculate the mean as approximately 39.6 seconds. Again, this is an approximation—the actual value, computed from the original data, is 39.76 seconds. (*Note:* If each data value *really is* at the midpoint of its class, the calculated mean, variance, and standard deviation will be exact values, not approximations.)

Variance and Standard Deviation from Grouped Data

Variance

As in our previous discussion of the variance, its calculation here will depend on whether the data constitute a population or a sample. The previously determined approximate value for the mean (in this case, estimated as $\bar{x} = 39.6$) is also used. When computing the variance from grouped data, one of the following formulas will apply:

Approximate value of the variance from grouped data:

- **If the data represent a population:**

$$\sigma^2 = \frac{\sum f_i m_i^2 - N\mu^2}{N}$$

where μ = approximate population mean
f_i = frequency for class i
m_i = midpoint for class i
N = number of data values in the population

- **If the data represent a sample:**

$$s^2 = \frac{\sum f_i m_i^2 - n\bar{x}^2}{n-1}$$

where $\bar{x}$ = approximate sample mean
f_i = frequency for class i
m_i = midpoint for class i
n = number of data values in the sample

Calculation of the approximate population variance for the ATM customer data is shown in Table 3.4 and results in an estimate of 83.77 for the value of s^2. This differs from the $s^2 = 79.49$ reported by Excel in Computer Solutions 3.1, but remember that our calculation here is an approximation. If the original data are not available, such approximation may be our only choice.

Standard Deviation

The approximate standard deviation from grouped data is the positive square root of the variance. This is true for both a population and a sample. As Table 3.4 shows, the approximate value of s is the square root of 83.77, or 9.15 seconds. Like the variance, this compares favorably with the value that was based on the original data.

EXERCISES

3.39 Using the data in Exercise 3.5, construct a frequency distribution.

a. Determine the approximate values of the mean and standard deviation.
b. Compare the approximate values from part (a) with the actual values.
c. Construct a frequency distribution with twice as many classes as before, then repeat parts (a) and (b). Have the approximations improved?
d. If you were to construct a "frequency distribution" in which each data value was at the midpoint of its own class, what values do you think the "approximations" would have? Explain.

3.40 A sample consisting of 100 employees has been given a manual-dexterity test. Given the accompanying frequency distribution, determine the approximate mean and standard deviation for these data.

3.41 Eighty packages have been randomly selected from a frozen food warehouse, and the age (in weeks) of each package is identified. Given the frequency distribution shown, determine the approximate mean and standard deviation for the ages of the packages in the warehouse inventory.

Data for Exercise 3.40

Score	Number of Persons
5–under 15	7
15–under 25	9
25–under 35	12
35–under 45	14
45–under 55	13
55–under 65	9
65–under 75	8
75–under 85	11
85–under 95	10
95–under 105	7

Data for Exercise 3.41

Age (Weeks)	Number of Packages
0–under 10	25
10–under 20	17
20–under 30	15
30–under 40	9
40–under 50	10
50–under 60	4

3.6 STATISTICAL MEASURES OF ASSOCIATION

In Chapter 2, we introduced the scatter diagram as a method for visually representing the relationship between two quantitative variables, and we also showed how statistical software can be used to fit a linear equation to the points in the scatter diagram. This section introduces a way to numerically measure the strength of the linear relationship between two variables—the coefficient of correlation.

The **coefficient of correlation** (r, with $-1 \leq r \leq +1$) is a number that indicates both the direction and the strength of the linear relationship between the dependent variable (y) and the independent variable (x):

- **Direction of the relationship** If r is positive, y and x are directly related, as in parts (a) and (c) of Figure 3.5. If r is negative, the variables are inversely related, as in parts (b) and (d) of Figure 3.5.
- **Strength of the relationship** The larger the absolute value of r, the stronger the linear relationship between y and x. If $r = -1$ or $r = +1$, the best-fit linear equation will actually include all of the data points. This is the case in parts (a) and (b) of Figure 3.5. However, in parts (c) and (d), the respective equations are less than perfect in fitting the data, resulting in absolute values of r that are less than 1.

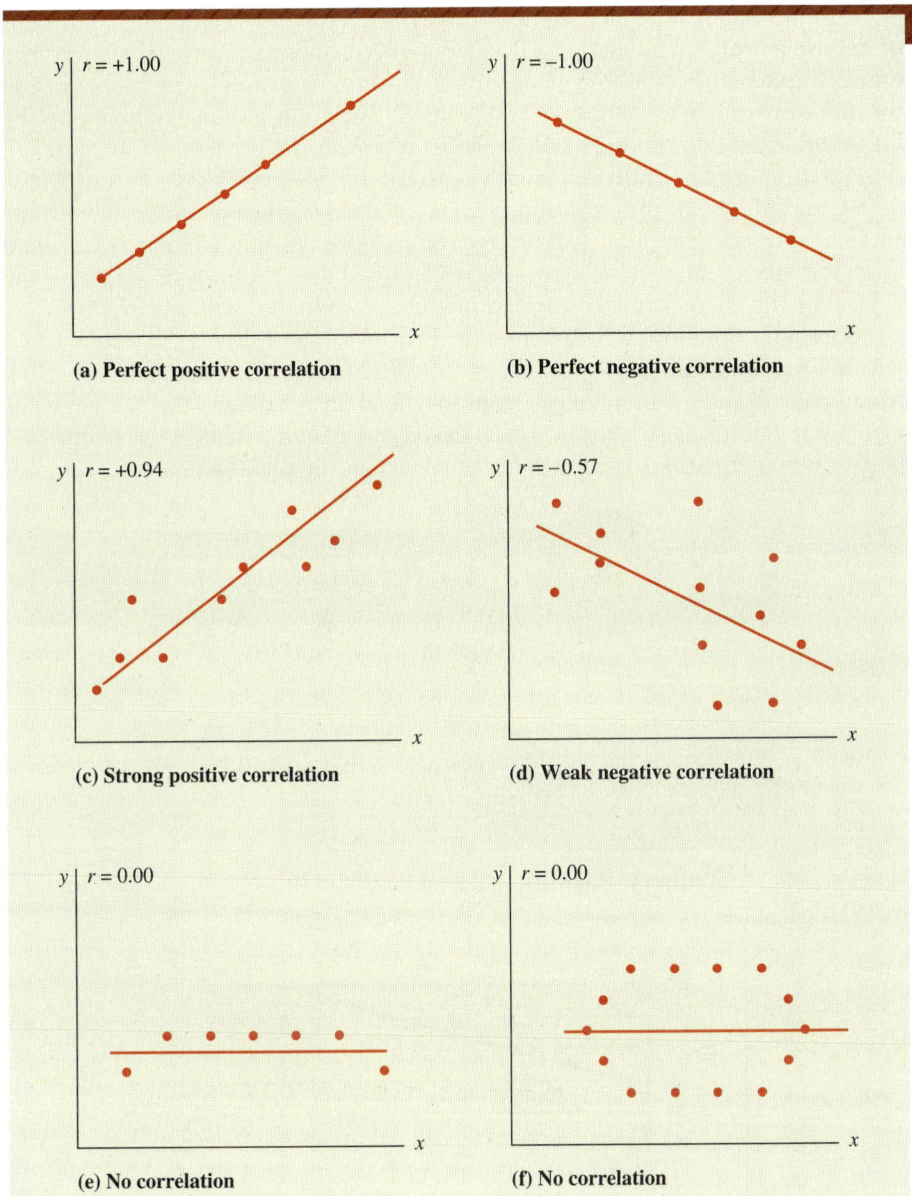

(a) Perfect positive correlation
(b) Perfect negative correlation
(c) Strong positive correlation
(d) Weak negative correlation
(e) No correlation
(f) No correlation

FIGURE 3.5

The stronger the linear relationship between the variables, the greater will be the absolute value of the correlation coefficient, *r*. When $r = 0$, the slope of the best-fit linear equation will also be 0.

In parts (e) and (f) of Figure 3.5, the slope is zero, *r* is zero, and there is no linear relationship whatsoever between *y* and *x*. In each case, the value of *y* doesn't depend on the value of *x*. At the end of the chapter, Seeing Statistics Applet 2 can be used to demonstrate how the value of *r* changes as the pattern of observations is changed.

In Chapter 15, Simple Linear Regression and Correlation, we will examine the coefficient of correlation and its method of calculation in greater detail. For now, it's important that you simply be aware of this important statistical measure of association. In this chapter, we'll rely on statistical software to do the calculations for us.

Another measure of the strength of the linear relationship is the **coefficient of determination, r^2.** Its numerical value is the proportion of the variation in y that is explained by the best-fit linear equation. For example, in parts (a) and (b) of Figure 3.5, r^2 is 1.00 and 100% of the variation in y is explained by the linear equation describing y as a function of x. On the other hand, in part (c), the equation does not fully explain the variation in y—in this case, $r^2 = 0.94^2 = 0.884$, and 88.4% of the variation in y is explained by the equation. As with the coefficient of correlation, it will be covered in more detail in Chapter 15. *A final note:* It doesn't matter which variable is identified as y and which one as x—the values for the coefficients of correlation and determination will not change.

Computer Solutions 3.5 describes the procedures and shows the results when we use Excel and Minitab to find the coefficient of correlation between a baseball team's payroll and the number of games they won during the season. In Chapter 2, Computer Solutions 2.7 showed the Excel and Minitab scatter diagrams and best-fit linear equations for the data, which are in file **CX02BASE**.

COMPUTER 3.5 SOLUTIONS

Coefficient of Correlation

EXCEL

The accompanying printout includes the scatter diagram and best-fit linear equation from Computer Solutions 2.7. It also shows the results for each of two options for determining the coefficient of correlation.

	A	B	C
1	**Team**	**Payroll**	**Wins**
2	New York Yankees	194.633	97
3	Boston Red Sox	120.100	86
4	Toronto Blue Jays	71.915	87
5	Baltimore Orioles	72.586	70
6	Tampa Bay Devil Rays	35.418	61
7	Cleveland Indians	56.032	78
8	Chicago White Sox	102.751	90
9	Detroit Tigers	82.613	95
10	Kansas City Royals	47.294	62
11	Minnesota Twins	63.396	96
12	Texas Rangers	68.229	80
13	Oakland Athletics	62.243	93
14	Seattle Mariners	87.960	78
15	Los Angeles Angels	103.472	89
16	Atlanta Braves	90.157	79
17	New York Mets	101.085	97
18	Philadelphia Phillies	88.273	85
19	Washington Nationals	63.143	71

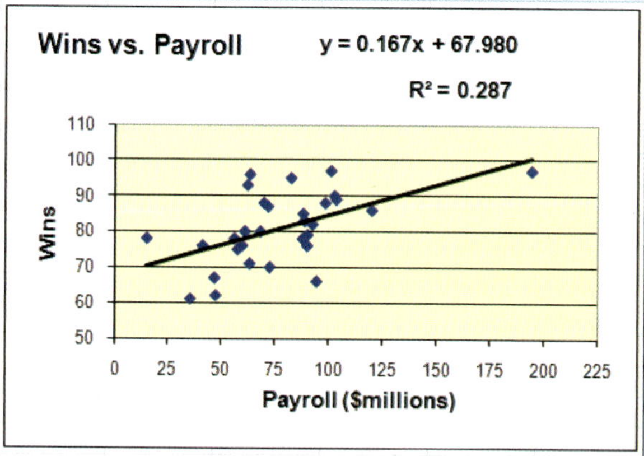

	Payroll	*Wins*
Payroll	1	
Wins	0.536	1

Excel Option 1, to simply obtain the coefficient of correlation:

1. Open the Excel data file **CX02BASE**. The labels and data are already entered, as shown in the printout. Click on **B1** and drag to **C31** to select cells **B1:C31.** From the **Data** ribbon, click **Data Analysis** in the rightmost menu.
2. In the **Analysis Tools** box, click **Correlation.** Click **OK.** Enter **B1:C31** into the **Input range** box. In the **Grouped By** section, click next to **Columns.** Click to place a check mark next to **Labels in First Row.** Enter **E17** into the **Output Range** box. Click **OK.**

(*continued*)

Excel Option 2, to display r^2 on an existing chart like the one from Computer Solutions 2.7:

1. Right-click anywhere on the best-fit line in the chart. Click **Format Trendline.**
2. In the **Trendline Options** menu, select **Display R-squared value on chart.** Click **Close.**

MINITAB

Minitab Option 1, to simply determine the coefficient of correlation:

```
Correlations: Payroll, Wins

Pearson correlation of Payroll and Wins = 0.536
```

1. Open the Minitab data file **CX02BASE**. The labels **Payroll** and **Wins** have already been entered at the top of columns **C2** and **C3,** respectively. The payroll values and the win totals have been entered into **C2** and **C3**, respectively.
2. Click **Stat.** Select **Basic Statistics.** Click **Correlation.** Enter **C2** and **C3** into the **Variables** box. Click **OK.**

Minitab Option 2, to display r^2 along with a scatter diagram and best-fit linear equation:

Generate the Minitab scatter diagram and best-fit linear equation as in the Minitab portion of Computer Solutions 2.7 (not shown here). The chart automatically includes "R-Sq = 28.7%" at the top.

Both software packages report the coefficient of correlation as $r = +0.536$. If we have also generated a scatter diagram and best-fit linear equation, as in Computer Solutions 2.7, the r^2 value (0.287, or 28.7%) is displayed either automatically (Minitab) or as an option (Excel). Interpreting the results, we conclude that there is a positive relationship between a team's payroll and the number of games it won during the season. Further, the size of the payroll explains 28.7% of the variation in the number of wins. From the opposite perspective, (100.0% − 28.7%), or 71.3% of the variation in wins is *not* explained by payroll. This makes sense, as there are probably a great many other variables—for example, level of fan support, quality of coaching, player injuries, strength of schedule—that also influence the number of games a team wins.

EXERCISES

3.42 What is the coefficient of determination and what does it tell us about the relationship between two variables?

3.43 For a set of data, r^2 is 0.64 and the variables x and y are inversely related. What is the numerical value of r?

(DATA SET) *Note:* Exercises 3.44–3.47 require a computer and statistical software.

3.44 In Exercise 3.14, two of the variables examined by the personnel director were age and number of absences during the past year. Using the data in file **XR03014**, generate a scatter diagram (with y = number of absences and x = age), fit a linear equation to the data, then determine and interpret the coefficients of correlation and determination.

3.45 In the *U.S. News & World Report* top-50 U.S. law schools in Exercise 2.67, the listing for each school included two rating scores: one from academic evaluators and another from lawyer/judge evaluators. Using the data in file **XR02067**, generate a scatter diagram with axes of your choice, fit a linear equation to the data, then determine and interpret the coefficients of correlation and determination.

3.46 The U.S. National Center for Health Statistics has compiled data on death rates (deaths per 100,000 people) from heart disease and cancer for the 50 states. The states and their respective death rates from these two diseases are provided in data file **XR03046**. Generate a scatter diagram with axes of your choice, fit a linear equation to the data, then determine and interpret the coefficients of correlation and determination. Source: Bureau of the Census, *Statistical Abstract of the United States 2009*, p. 87.

3.47 A report on the average price of a sample of brand-name prescription drugs with their generic equivalents has found the generics to be considerably cheaper. The drugs and the average prices for their brand-name versus generic versions are listed in data file **XR03047**. Generate a scatter diagram with axes of your choice, fit a linear equation to the data, then determine and interpret the coefficients of correlation and determination. Source: Julie Appleby, "Drug Makers Fight Back As Patents Near Expiration," *USA Today*, November 26, 1999, p. 4B.

3.7 SUMMARY

• Measures of central tendency and dispersion

In addition to summarization by frequency distributions and related visual techniques, raw data may be statistically described through the use of measures of central tendency and dispersion. As with the frequency distribution, this process loses some of the detail in the original data, but offers the benefit of replacing voluminous raw data with just a few numerical descriptors.

Measures of central tendency describe typical values in the data. Four of the most common are the arithmetic mean, weighted mean, median, and mode. Dispersion measures describe the scatter of the data, or how widely the values tend to be spread out. Common measures of dispersion include the range, quantiles, mean absolute deviation, variance, and standard deviation. The midrange is the average of the lowest and highest values and is sometimes used as a very rough measure of central tendency.

Chebyshev's theorem applies to any distribution, regardless of shape, and specifies a minimum percentage of observations that will fall within a given number of standard deviations from the mean. For distributions that are bell shaped and symmetrical, the empirical rule expresses the approximate percentage of observations that will fall within a given number of standard deviations from the mean.

• Conversion of data to standardized values

Standardization of the data involves expressing each value in terms of its distance from the mean, in standard deviation units. The resulting standardized distribution will have a mean of 0 and a standard deviation of 1. The coefficient of variation expresses the standard deviation as a percentage of the mean and is useful for comparing the relative dispersion for different distributions.

• Box-and-whisker plot and estimation from grouped data

Measures of central tendency and dispersion are part of the output whenever computer statistical packages have been used to analyze the data. In addition, the statistical package is likely to offer further description in the form of information such as the trimmed mean and the box-and-whisker plot. The approximate value of measures of central tendency and dispersion may be calculated from the grouped data of the frequency distribution.

The coefficient of correlation

The coefficient of correlation (r) is an important measure of the strength of the linear relationship between two quantitative variables. The coefficient of correlation is between -1 and $+1$, and the larger the absolute value of r, the stronger the linear relationship between the variables. The coefficient of determination (r^2) is the proportion of the variation in y that is explained by the best-fit linear equation, and r^2 is always ≤ 1.

EQUATIONS

Arithmetic Mean

- Population mean:

$$\mu = \frac{\Sigma x_i}{N}$$

where μ = population mean
x_i = the ith data value in the population
Σ = the sum of
N = number of data values in the population

- Sample mean:

$$\bar{x} = \frac{\Sigma x_i}{n}$$

where $\bar{x}$ = sample mean
x_i = the ith data value in the sample
Σ = the sum of
n = number of data values in the sample

Weighted Mean

Weighted mean, μ_w (for a population) or $\bar{x}_w$ (for a sample):

$$\mu_w \text{ or } \bar{x}_w = \frac{\Sigma w_i x_i}{\Sigma w_i}$$

where w_i = weight assigned to the ith data value
x_i = the ith data value

Range

- Range = Largest data value − Smallest data value
- Midrange = (Largest data value + Smallest data value)/2

Quartiles

In a set of N values arranged from lowest to highest:

- First quartile, Q_1 = Data value at position $\dfrac{(N+1)}{4}$
- Second quartile (the median), Q_2 = Data value at position $\dfrac{2(N+1)}{4}$
- Third quartile, Q_3 = Data value at position $\dfrac{3(N+1)}{4}$

  (Use N if data represent a population, n for a sample.)

Interquartile Range and Quartile Deviation

- Interquartile range: third quartile minus first quartile, or $Q_3 - Q_1$
- Quartile deviation: one-half the interquartile range, or $(Q_3 - Q_1)/2$

Mean Absolute Deviation (MAD)

Mean absolute deviation for a population:

$$MAD = \frac{\sum|x_i - \mu|}{N}$$

where μ = population mean
x_i = the ith data value
N = number of data values in the population

(To calculate MAD for a sample, substitute n for N and $\bar{x}$ for μ.)

Variance

- Variance for a population:

$$\sigma^2 = \frac{\sum(x_i - \mu)^2}{N}$$

where σ^2 = population variance
μ = population mean
x_i = the ith data value
N = number of data values in the population

- Variance for a sample:

$$s^2 = \frac{\sum(x_i - \bar{x})^2}{n - 1}$$

where s^2 = sample variance
$\bar{x}$ = sample mean
x_i = the ith data value
n = number of data values in the sample

Standard Deviation

For a Population	For a Sample
$\sigma = \sqrt{\sigma^2}$	$s = \sqrt{s^2}$

Residual

$$\text{Residual for a data value in a sample} = x_i - \bar{x}$$

where $\bar{x}$ = sample mean
x_i = value of the ith observation

Chebyshev's Theorem

For either a sample or a population, the percentage of observations that fall within k (for $k > 1$) standard deviations of the mean will be at least

$$\left(1 - \frac{1}{k^2}\right) \times 100$$

The Empirical Rule

For distributions that are bell shaped and symmetrical:

- About 68% of the observations will fall within 1 standard deviation of the mean.
- About 95% of the observations will fall within 2 standard deviations of the mean.
- Practically all of the observations will fall within 3 standard deviations of the mean.

Standardized Data

$$z_i = \frac{x_i - \bar{x}}{s}$$

where z_i = standardized value for the ith observation
$\bar{x}$ = sample mean
x_i = the ith data value
s = sample standard deviation

(When data represent a population, μ replaces $\bar{x}$ and σ replaces s.)

Coefficient of Variation (CV)

For a Population	For a Sample
$CV = \frac{\sigma}{\mu} \times 100$	$CV = \frac{s}{\bar{x}} \times 100$

Conversion of Computer-Output Standard Deviation Based on (*N* − 1) Divisor to a Population Standard Deviation Based on (*N*) Divisor

$$\text{Standard deviation based on } (N) \text{ divisor} = \text{Standard deviation based on } (N-1) \text{ divisor} \times \sqrt{\frac{N-1}{N}}$$

Mean and Variance from Grouped Data

- Approximate value of the mean from grouped data:

$$\mu = \frac{\Sigma f_i m_i}{N}$$

where μ = approximate population mean
f_i = frequency for class i
m_i = midpoint for class i
N = number of data values in the population

(If data are a sample, $\bar{x}$ replaces μ and n replaces N.)

- Approximate value of the variance from grouped data:
If data represent a population:

$$\sigma^2 = \frac{\Sigma f_i m_i^2 - N\mu^2}{N}$$

where μ = approximate population mean
f_i = frequency for class i
m_i = midpoint for class i
N = number of data values in the population

If data represent a sample:

$$s^2 = \frac{\Sigma f_i m_i^2 - n\bar{x}^2}{n-1}$$

where $\bar{x}$ = approximate sample mean
f_i = frequency for class i
m_i = midpoint for class i
n = number of data values in the sample

CHAPTER EXERCISES

Note: For many of the Exercises 3.48–3.64, a computer and statistical software will be desirable and useful. However, any necessary calculations can also be done with the aid of a pocket calculator. For readers using statistical software, keep in mind the file-naming key—for example, the data for Exercise 3.57 will be in data file **XR03057**.

3.48 The first seven customers of the day at a small donut shop have checks of \$1.25, \$2.36, \$2.50, \$2.15, \$4.55, \$1.10, and \$0.95, respectively. Based on the number of customers served each day, the manager of the shop claims that the shop needs an average check of \$1.75 per person to stay profitable. Given her contention, has the shop made a profit in serving the first seven customers?

3.49 A dental supplies distributor ships a customer 50 boxes of product A, 30 boxes of B, 60 boxes of C, and 20 boxes of D. The unit shipping costs (dollars per box) for the four products are \$5, \$2, \$4, and \$10, respectively. What is the weighted mean for shipping cost per unit?

3.50 The 2007 top-grossing feature film was *Spiderman 3*, a product that brought in \$336.5 million at the box office. The gross receipts for this and the rest of the top-20 feature films of that year are shown below. Determine the mean and median for these data. Is there a mode? If so, what is its numerical value? Source: *The New York Times Almanac 2009*, p. 413.

\$336.5	\$322.7	\$319.2	\$309.4	\$292.0
256.4	227.5	219.5	217.3	210.6
206.4	183.1	168.3	148.8	143.5
140.1	134.5	131.9	130.2	127.8

3.51 According to Honda Motor Co., Inc., the exchange rate (yen per U.S. dollar) from 1998 through 2005 was 123, 128, 112, 111, 125, 122, 113, and 108. Determine the mean and median for these data. Is there a mode? If so, what is its numerical value? Source: Honda Motor Co., Ltd., *2005 Annual Report*.

3.52 *Natural History* magazine has published a listing of the maximum speeds (in mph) for a wide variety of animals, including those shown in the table. Source: *The World Almanac and Book of Facts 2009*, p. 330.

Cheetah	70	Grizzly bear	30
Pronghorn antelope	61	Domestic cat	30
Wildebeest	50	Human	27.89
Lion	50	Elephant	25
Coyote	43	Black mamba snake	20
Mongolian wild ass	40	Squirrel	12
Giraffe	32	Spider	1.17
Wart hog	30	Giant tortoise	0.17
White-tailed deer	30	Garden snail	0.03

a. Determine the mean and the median for these data.
b. Is there a mode? If so, what is its numerical value?

3.53 The frequency distribution for population density (persons per square mile) for the 50 U.S. states is as follows: Source: Bureau of the Census, *Statistical Abstract of the United States 2009*, p. 18.

Population Density	Number of States
0–under 100	27
100–under 200	11
200–under 300	4
300–under 400	1
400–under 500	2
500–under 600	1
600–under 700	0
700–under 800	1
800–under 900	1
900–under 1000	0
1000–under 1100	1
1100–under 1200	1

Does this distribution appear to be symmetrical? If not, is it skewed positively or negatively?

3.54 The cafeteria manager at a manufacturing plant has kept track of the number of cups of coffee purchased during each of the past 90 working days.

a. Using the following descriptive statistics, construct a box-and-whisker plot for the data. Does the distribution appear to be symmetrical? If not, is it positively skewed or negatively skewed?

```
         N    Mean   Median  Tr Mean  StDev  SE Mean
CUPS    90  125.64   123.50   125.50  30.80     3.25

         Min     Max      Q1       Q3
CUPS   54.00  206.00  102.75   147.25
```

b. If this distribution of daily coffee totals is bell shaped and symmetrical, on approximately what percent of the days were 64 or fewer cups of coffee purchased?

3.55 A grocery store owner has found that a sample of customers purchased an average of 3.0 pounds of luncheon meats in the past week, with a sample standard deviation of 0.5 lb.

a. If the store's meat scale is off by 0.1 lb (e.g., a purchase listed as 1 lb actually weighs 1.1 lb), what will be the values of the sample mean and standard deviation after the owner corrects his data for the weighing error?
b. Using the mean and standard deviation you calculated in part (a), and assuming the distribution of purchase amounts was bell shaped and symmetrical, what luncheon meat purchase amount would have been exceeded by only 2.5% of the customers last week?

3.56 For a sample of five different years from the period 1960 through 2007, it is found that U.S. work stoppages involving at least 1000 workers occurred 268 times in one of these years, with 424, 235, 145, and 44 work stoppages in the other four. Source: Bureau of the Census, *Statistical Abstract of the United States 2009*, p. 418.

a. Determine the mean, median, range, and midrange.
b. Calculate the mean absolute deviation.
c. Calculate the standard deviation and variance.

3.57 A quality control supervisor has taken a sample of 16 bolts from the output of a thread-cutting machine and tested their tensile strengths. The results, in tons of force required for breakage, are as follows:

2.20	1.95	2.15	2.08	1.85	1.92
2.23	2.19	1.98	2.07	2.24	2.31
1.96	2.30	2.27	1.89		

a. Determine the mean, median, range, and midrange.
b. Calculate the mean absolute deviation.
c. Calculate the standard deviation and variance.

3.58 The accompanying box-and-whisker plot represents the number of gallons of water used by 80 households over a 1-day period. Determine the approximate values for the median, the first and third quartiles, and the range. Does the distribution appear to be skewed? If so, is it positively skewed or negatively skewed?

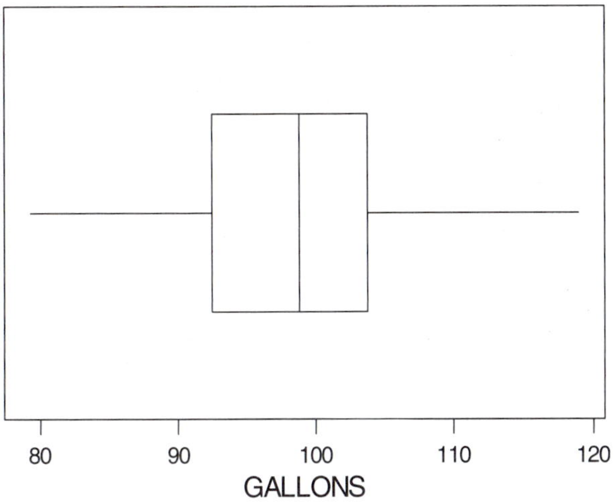

3.59 A testing firm has measured the power consumption of 40 dorm-size microwave ovens selected at random from the manufacturer's production line. Given the following box-and-whisker plot, determine the approximate values for the median, the first and third quartiles, and the range. Does the distribution appear to be skewed? If so, is it positively skewed or negatively skewed?

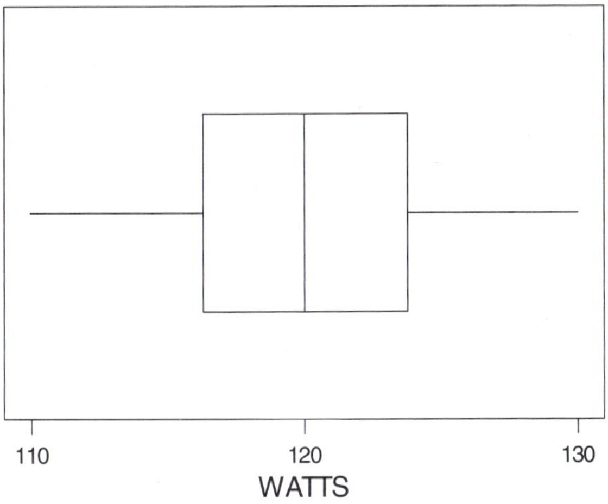

3.60 A law enforcement agency, administering breathalyzer tests to a sample of drivers stopped at a New Year's Eve roadblock, measured the following blood alcohol levels for the 25 drivers who were stopped:

0.00%	0.08%	0.15%	0.18%	0.02%
0.04	0.00	0.03	0.11	0.17
0.05	0.21	0.01	0.10	0.19
0.00	0.09	0.05	0.03	0.00
0.03	0.00	0.16	0.04	0.10

a. Calculate the mean and standard deviation for this sample.
b. Use Chebyshev's theorem to determine the minimum percentage of observations that should fall within $k = 1.50$ standard deviation units of the mean. Do the sample results support the theorem?
c. Calculate the coefficient of variation for these data.

3.61 Use the coefficient of variation to compare the variability of the data in Exercise 3.57 with the variability of the data in Exercise 3.60.

3.62 Using the frequency distribution in Exercise 3.53, determine the approximate mean and standard deviation for the underlying data.

3.63 The following dotplot describes the lengths, in pages, of a sample consisting of 20 reports generated by a consulting firm during the past 3 months. Based on the dotplot, what are the values for the median and the first and third quartiles?

```
                  .
      ..    . .  ::.:. ..     ..: .
-----+---------+---------+---------+------PAGES
    10        20        30        40        50
```

3.64 For the sample data presented in the dotplot of Exercise 3.63, construct a frequency distribution in which the classes are 10–under 20, 20–under 30, and 30–under 40. If this frequency distribution were all that we knew about the underlying data, what approximate values would be estimated for the mean and standard deviation of these data?

(DATA SET) *Note:* Exercises 3.65–3.69 require a computer and statistical software.

3.65 According to the Energy Information Administration, the average U.S. household spends $1196 per year for electricity. Assume that this finding could have been based on the electricity expenditures reported by 250 households, with the data in file **XR03065**. Source: eia.doe.gov, June 26, 2009.

a. Confirm the mean and determine the median and the standard deviation of the electricity expenditures.
b. Generate and interpret the box-and-whisker plot for the households from this region.
c. From the results in parts (a) and (b), do there appear to be any "outlier" households? If so, for what reason(s) might energy-conservation officials wish to examine the habits or characteristics of these households?

3.66 *Bride's* magazine reports the cost of the average honeymoon as $5111. Assume that their findings could have been based on the honeymoon expenditures of a sample consisting of 300 couples whose costs are listed in data file **XR03066**. Source: msnbc.msn.com, June 26, 2009.

a. Confirm the mean and determine the median and the standard deviation of the honeymoon expenditures for this sample.
b. Generate and interpret the box-and-whisker plot for the couples in the sample.
c. From the results in parts (a) and (b), do there appear to be any "outlier" couples who have spent an unusually large amount of money on their honeymoon? If so, what kinds of companies or organizations might have an interest in finding out how to identify such couples before they've made their honeymoon plans?

3.67 During 2005, 1,186,000 college-bound high school seniors took the ACT college admission test. The average score on the mathematics component was 20.7, with a standard deviation of 5.0. We are assuming that the math scores in data file **XR03067** could have been the math scores for a sample of 400 college-bound seniors who took the ACT that year. (*An important note:* The mean and the standard deviation for our sample are close, but not exactly the same as the mean and standard deviation of the population from which the sample was obtained—this is a key concept that is central to Chapter 8, Sampling Distributions.) Based on the assumed sample data in file **XR03067**, Source: nces.ed.gov, June 26, 2009.

a. Determine the mean, the median, and the standard deviation of the math scores in the sample.
b. Generate and interpret the box-and-whisker plot for the math scores in the sample.
c. What math score would a test-taker have to achieve to be higher than 75% of the sample members? To be higher than 25% of the sample members?

3.68 Data for a sample of college players eligible to enter the National Football League include weight (pounds) and time (seconds) required to sprint 40 yards. Using the data listed in file **XR03068**, and with weight as the x (independent) variable and time as the y (dependent) variable, generate a scatter diagram, fit a linear equation to the diagram, then determine and interpret the coefficients of correlation and determination. Source: "Rating the Prospects," *USA Today*, April 14, 2000, p. 19C.

3.69 During a five-year period, the Federal Aviation Administration took administrative actions (letters of correction or warning notices) and levied fines against U.S. airline companies for various safety-related reasons. For each airline involved, data file **XR03069** lists the number of administrative actions taken toward the airline along with the total amount of fines (in millions of dollars) levied against the airline during this time period. Using number of administrative actions as the x (independent) variable and fines as the y (dependent) variable, generate a scatter diagram, fit a linear equation to the diagram, then determine and interpret the coefficients of correlation and determination. Source: "Fines and Administrative Actions Against U.S. Airlines," *USA Today*, March 13, 2000, p. 3B.

INTEGRATED CASES

Thorndike Sports Equipment

When we last visited the Thorndikes, one of the company's engineers had just developed a new golf ball. According to the engineer, the new ball incorporates some changes that may enable it to travel farther than conventional golf balls. The engineer, an avid golfer herself, is very excited about the prospect that she may have developed an innovation that could revolutionize the game of golf. Old Luke says he will be happy only if the new ball revolutionizes his company's net profits.

Some testing has already been done. As reported in the Chapter 2 episode of the Thorndikes, Ted set up a test in which he mixed 25 of the new balls with 25 of the old type, then had a golf pro hit all 50 of them at a driving range. All 50 distances were measured and are listed in data file **THORN02**. For reader convenience, they are also repeated here. Ted has examined the visual descriptions of the data carried out in Chapter 2, but he would like to have some statistical summaries as well.

1. For the 25 drives made with the new balls, calculate statistics that you think would be appropriate in describing the central tendency and the dispersion of the data.
2. Repeat part 1 for the 25 drives made with the conventional golf balls.
3. Ted wants to write a one-paragraph memo to Luke describing the results as they are reflected in parts 1 and 2, but he had to fly to New York this morning. Help Ted by writing a first draft for him. Keep in mind that Luke appreciates plain English, so try to use your own words whenever possible.

25 Drives with New Ball (Yards):

267.5	248.3	265.1	243.6	253.6	232.7	249.2	232.3	252.8
247.2	237.4	223.7	260.4	269.6	256.5	271.4	294.1	256.3
264.3	224.4	239.9	233.5	278.7	226.9	258.8		

25 Drives with Conventional Ball (Yards):

241.8	255.1	266.8	251.6	233.2	242.7	218.5	229.0	256.3
264.2	237.3	253.2	215.8	226.4	201.0	201.6	244.1	213.5
267.9	240.3	247.9	257.6	234.5	234.7	215.9		

Springdale Shopping Survey

This exercise focuses on the Springdale Shopping Survey, the setting and data collection instrument for which were provided at the end of Chapter 2. The data are in the computer file **SHOPPING**.

We will concentrate on variables 18–25, which reflect how important each of eight different attributes is in the respondent's selection of a shopping area. Each of these variables has been measured on a scale of 1 (the attribute is not very important in choosing a shopping area) to 7 (the attribute is very important in choosing a shopping area). The attributes being rated for importance are listed below. Examining the relative importance customers place on these attributes can help a manager "fine-tune" his or her shopping area to make it a more attractive place to shop.

18 Easy to return/exchange goods
19 High quality of goods
20 Low prices
21 Good variety of sizes/styles
22 Sales staff helpful/friendly
23 Convenient shopping hours
24 Clean stores and surroundings
25 A lot of bargain sales

1. Perform the following operations for variables 18–25:
   a. Obtain descriptive statistics, including the mean and median.
   b. Generate a box-and-whisker plot for the variable. Does the distribution appear to be skewed? If so, is the skewness positive or negative?
2. Based on the results for question 1, which attributes seem to be the most important and the least important in respondents' choice of a shopping area?
3. Use their respective coefficients of variation to compare the relative amount of dispersion in variable 29 (number of persons in the household) with that of variable 30 (respondent's age).
4. Determine the coefficient of correlation between variable 29 (number of persons in the household) and variable 30 (age). What percentage of the variation in household size is explained by respondent age?

BUSINESS CASE

Baldwin Computer Sales (A)

Baldwin Computer Sales is a small company located in Oldenburg, Washington. The founder of the company, Jonathan Baldwin, began the business by selling computer systems through mail-order at discount prices. Baldwin was one of the first computer mail-order companies to offer a toll-free phone number to their customers for support and trouble-shooting. Although the company has grown over time, many new competitors have entered the computer mail-order marketplace so that Baldwin's share of this market has actually declined.

Five years ago, Bob Gravenstein, a marketing consultant, was contracted by Baldwin to develop long-term marketing plans and strategies for the company. After careful study, he recommended that Baldwin branch out to reach a new and growing segment of the computer sales market: college students. At that time, most universities were providing microcomputer labs and facilities for their students. However, many students were purchasing their own computers and printers so that they could have access to a computer at any time in their apartments or dorms. After graduation, students take the computers, with which they are already familiar, to their new jobs or use them at home after work hours. The percentage of college students owning a computer was increasing rapidly, and Bob Gravenstein recommended that Baldwin Computer Sales take advantage of this marketing opportunity.

The marketing plan developed by Bob Gravenstein worked as follows: Five universities were initially selected for the program with expansion to other universities planned over time. Any student in at least his or her sophomore year at one of these universities was eligible to purchase a discounted JCN-2001 microcomputer system with printer from Baldwin Computer Sales under the program. The JCN-2001 is a private-label, fully compatible system with all of the features of brand-name models. The student makes a small payment each semester at the same time regular tuition payments are due. When the student graduates and finds a job, the payments are increased so that the computer will be paid for within two years after graduation. If the student fails to make payments at any time, Baldwin could repossess the computer system.

The prospect of future sales of computer equipment to these students was also part of the marketing strategy. The JCN-2001 is an entry-level computer system that suffices for most academic and professional uses. Eventually, however, many students who purchased this system would outgrow it and require an upgrade to a more powerful machine with more features. Bob Gravenstein argued that after their good experience with the JCN-2001, these customers would make their future purchases from Baldwin Computer Company.

Today, five years later, Baldwin Computer Sales is still operating the student purchase program and has expanded it to several more universities. There are currently enough data available from the early days of the program for Baldwin to determine whether the program has been successful and whether it should be continued. To discuss the future of the student purchase program, Jonathan Baldwin has called a meeting with Ben Davis, who has been in charge of the program since it began, and Teresa Grimes, the new vice president of marketing.

Baldwin: "I called you both in here today to discuss the student purchase program. As you know, the program has been in place for approximately five years and we need to make decisions about where to go from here. Ben, weren't you telling me last week that we now have enough data to evaluate the program?"

Davis: "Yes, sir. Any student who began the program as a sophomore five years ago should be out of school and employed for at least two years."

Baldwin: "Well, based on your information, would you say the program has been a success or a failure?"

Davis: "That's a tough call, sir. While most of the participants in the program eventually pay their account in full, we have had a high rate of defaults, some while the students were still in school, but mostly after they graduated."

Baldwin: "How much are we losing on those who default?"

Davis: "Each case is different. As I said, some default early and others after they have graduated. There are also the costs of repossession and repair to bring the product back up to resale quality. In many instances we were not able to retrieve the computer systems. Our data suggest that our average loss on each student-customer who defaults is about $1200. On the other hand, our average profit from participants who pay their accounts in full is approximately $750. Overall, we are close to just breaking even."

Grimes: "Ben, have you considered 'qualifying' students for the program, much like a loan officer would qualify someone for a loan?"

Davis: "We had initially thought about doing that, but we didn't believe that there was much information, if any, in the way of a credit history on most college students. Applicants are still requested to provide as much information as possible on their application, including their class, grade point average, work experience, scholarships, and how much of their college expenses were earned through work. However, we weren't sure that this information was particularly useful for screening applicants."

Grimes: "Knowing that we were going to have this discussion today, I had one of my assistants who is well-versed in statistics look over some of those data last week. She has come up with a 'screening test' based only on the information you have been collecting from student applicants. By being more selective about whom we allow to participate in the student purchase program, it may be possible to increase our profit from the program."

Davis: "It would be easy enough to check out her screening test by trying it out on our early data from the program. In those cases, we know whether or not the student actually defaulted."

Baldwin: "Why don't the two of you spend some time on this idea and get back to me next week? At that time I want a recommendation to either discontinue the program, continue it as is, or continue it using this 'screening test' idea. Make sure you have the evidence to back up your recommendation."

Assignment

Ben Davis and Teresa Grimes must analyze the data from the student purchase program and make a recommendation to Mr. Baldwin about the future of the program. The necessary data are contained in the file **BALDWIN**. A description of this data set is given in the Data Description section that follows. Using this data set and other information given in the case, help Ben Davis and Teresa Grimes evaluate the student purchase program and the potential usefulness of the screening test developed by the assistant to Teresa Grimes. The case questions will assist you in your analysis of the data. Use important details from your analysis to support your recommendations.

Data Description

The **BALDWIN** file contains data on all participants in the student purchase program who by now should have either paid in full or defaulted (i.e., those participants who should have graduated and held a job for at least two years). A partial listing of the data is shown here.

Student	School	Default	When	Score
6547	1	1	1	64
4503	2	0	0	58
1219	2	0	0	52
9843	4	1	0	56
6807	1	0	0	47
6386	4	0	0	58
⋮	⋮	⋮	⋮	⋮

SEEING STATISTICS: APPLET 1

Influence of a Single Observation on the Median

The dots in this applet can be viewed as showing the speeds of a sample of nine cars being driven through a residential area. Eight of the values are fixed, but we can change the "speed" of one of them to see how this change affects the median for the sample data. By clicking on the data value for the ninth car (represented by the green dot at the far right), we can drag the value upward or downward and change this car's speed. The median for the data is shown at the top of the applet.

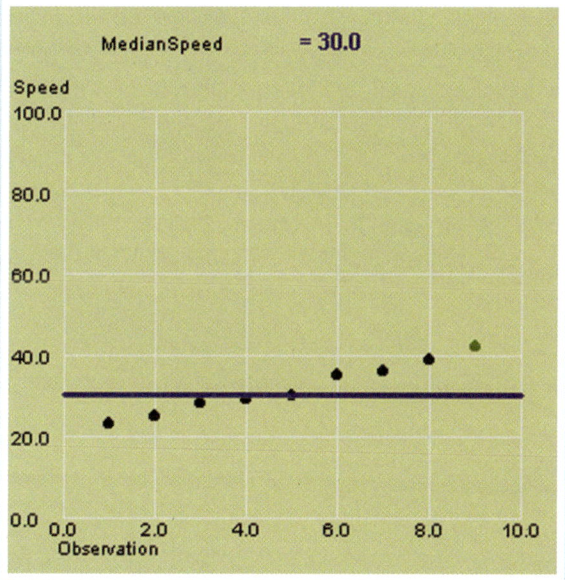

Applet Exercises

1.1 Click on the rightmost green dot and drag it so that the speed of the ninth car is 40 mph. What is the value for the median?

1.2 Click on the rightmost green dot and drag it so that the speed of the ninth car increases from 40 mph to 100 mph. What effect has this had on the value of the median?

1.3 By adjusting the speed of the ninth car, identify the highest and lowest possible values for the median speed for the nine cars in the applet.

1.4 Suppose the ninth car were a jet-powered supersonic vehicle and traveled through the neighborhood at 1000 mph. Although such a speed lies far beyond the scale in the applet, what would be the value of the median? What would be the approximate value for the mean? Which of these two sample descriptors has been more greatly affected by this extreme value in the data?

Try out the Applets: http://www.cengage.com/bstatistics/weiers

These data are coded as follows:

STUDENT: Student transaction number for identification purposes

SCHOOL: University where the student was enrolled

DEFAULT: 1, in the event of default,
0, if account was paid in full on time

WHEN: 1, if default occurred (or paid in full) before graduation,
0, if default occurred (or paid in full) after graduation

SCORE: Score on screening test based on student applicant information such as his or her class, grade point average, work experience, scholarships, and how much of their college expenses were earned through work

1. Generate appropriate descriptive statistics of the screening test scores for those who did not default on their computer loan. Be sure to include the mean and the third quartile. Do the same for those who did default, then compare the results. Is the mean score on the screening test higher for those who did not default?
2. In the descriptive statistics of the screening test scores for the students who did not default, identify the value of the third quartile and interpret its meaning. If this numerical value had been established as a cutoff for

SEEING STATISTICS: APPLET 2

Scatter Diagrams and Correlation

This applet displays a scatter diagram that we can rearrange by using the slider at the bottom. By clicking on the slider and adjusting its position, we cause the points to gravitate to positions that make the linear relationship between the variables either stronger or weaker. For any given configuration, the best-fit straight line is shown, the coefficient of correlation is displayed at the top of the diagram, and we can reverse the nature of the relationship by clicking on the "Switch Sign" button.

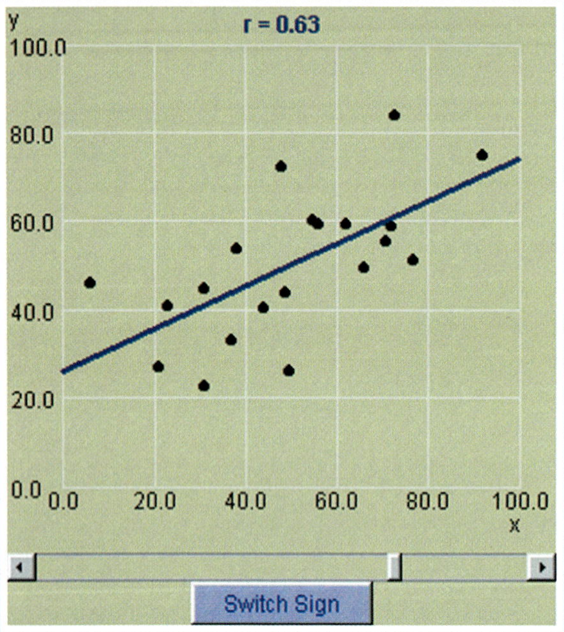

Applet Exercises

2.1 Click on the slider and adjust it so that the best-fit straight line is as close as possible to horizontal. What is the value of the coefficient of correlation?

2.2 Click on the slider and adjust it so that the coefficient of correlation is +1.0. Describe the best-fit line and its relationship to the pattern of points.

2.3 Adjust the slider so that the coefficient of correlation is as close as possible to +0.60. Describe how the best-fit line and its relationship to the pattern of points have changed.

2.4 Click the "Switch Sign" button. Describe how the best-fit line and its relationship to the pattern of points have changed.

2.5 With the slider positioned at the far left, gradually move it to the far right. Describe the changes taking place in the best-fit line and the pattern of points along the way.

Try out the Applets: http://www.cengage.com/bstatistics/weiers

receiving a computer loan, what percentage of those who repaid would have been denied a loan in the first place? Comment on the advisability of setting a cutoff score that is this high.

3. Construct adjacent dotplots (use box plots if the dotplot feature is not available on your computer statistical package) and visually compare the screening test scores of students who did not default on their computer loan to the scores of those who defaulted. Does the distribution of screening test scores for those who did not default appear to be shifted to the right of the distribution of scores for those who did default?
4. Based on your results and responses to the previous questions, does the screening test appear to be potentially useful as one of the factors in helping Baldwin predict whether a given applicant will end up defaulting on his or her computer loan?

CHAPTER 4

Data Collection and Sampling Methods

U.S. CENSUS: EXTENSIVE, EXPENSIVE, BUT UNDERCOUNTING IS A PROBLEM

Since 1790, at 10-year intervals, the United States has taken a count of the nation's populace. Required by the Constitution, this enumeration is used to allocate seats in the U.S. House of Representatives, to determine election districts, and to distribute more than $1.3 trillion to local and state governments over the next 10 years. In addition, the census shapes important decisions about public health, education, senior services, and transportation. The 2010 census is at least as expensive as it is important. In carrying out this enumeration, the Bureau of the Census will spend more than $14 billion and employ about 1.4 million field operators, more than the entire population of New Hampshire.

Despite such Herculean efforts, undercounting has been a persistent problem. It has been estimated that approximately 5 million people were not included in the 2000 enumeration and that those not counted were especially likely to include blacks, Hispanics, illegal immigrants, and the poor. As a result of undercounting, such communities tend to receive less than proportionate shares of political representation and federal funds.

In adjusting its methodology to better reflect the population and its geographical distribution, the Bureau of the Census is relying partly on sampling approaches similar to those described in this chapter. It has also proposed the controversial technique of allowing citizens to be counted by going online. Changes in enumeration methods can cause great controversy in the political arena, partly because some states (e.g., Texas and California) would benefit from including more of the "uncounteds," while others (e.g., Alabama and Pennsylvania) would tend to suffer.

Thus, although charged with the seemingly straightforward job of determining how many people live in the United States, the Bureau of the Census finds itself caught in the middle of a controversy with significant political overtones. To the extent that federal representation and funding are allocated according to the population distribution, the way in which the official count is determined could greatly influence the fortunes of the individual states affected.

Source: Haya el Nasser, "For 2010 Census, Counting Gets Tougher," usatoday.com, June 27, 2009.

Survey research is a popular method of collecting data.

LEARNING OBJECTIVES

After reading this chapter, you should be able to:

- Describe the types of business research studies and their purposes.
- Explain the difference between primary and secondary data.
- Describe how survey research is conducted.
- Discuss the major sources of error in survey research.
- Describe the experimentation and observational approaches to research.
- Differentiate between internal and external secondary data, and identify sources for each.
- Understand the relevance of the data warehouse and data mining.
- Explain why a sample is generally more practical than a census.
- Differentiate between the probability and nonprobability approaches to sampling, and understand when each approach is appropriate.
- Differentiate between sampling error and nonsampling error.
- Understand that probability samples are necessary to estimate the amount of sampling error.

4.1 INTRODUCTION

Statistics can be used as (1) a *descriptive* tool for expressing information relevant to business and (2) an *inferential* device for using sample findings to draw conclusions about a population. To use statistics for either of these applications, *data* are required, and such data may be gathered from a wide variety of sources. The techniques presented in subsequent chapters are only as good as the data on which they are based.

In this chapter, we'll examine some of the popular approaches used in collecting statistical data. Our discussion of business and survey research will be on a strictly introductory level, but it should make you more familiar with the sources and procedures used in obtaining the kinds of data seen in this textbook, as well as in the business and general media. We will also discuss the importance of sampling and describe some of the more popular sampling methods used in business research today.

4.2 RESEARCH BASICS

Types of Studies

Business research studies can be categorized according to the objective involved. Accordingly, we can identify four types of studies: exploratory, descriptive, causal, and predictive.

Exploratory

Often the initial step, **exploratory research** helps us become familiar with the problem situation, identify important variables, and use these variables to form hypotheses that can be tested in subsequent research. The hypothesis will be a statement about a variable or the relationship between variables—for example, "Production will increase if we switch the line locations of operators A and B." The hypothesis may not be true, but it is useful as an assertion that can be examined through the collection of sample data for a test period during which the operators have traded line positions.

Exploratory research can also be of a qualitative nature. One such method is the *focus group interview*, in which a moderator leads a small-group discussion about a topic and the client watches and listens from behind a one-way mirror. The U.S. Environmental Protection Agency has used focus groups to learn what people think and know about their drinking water. Among the results: People expressed distrust in water companies and knew that some contaminants could be in the water, but most people were unable to specify any of the specific contaminants that could be there.[1]

Descriptive

As might be expected, **descriptive research** has the goal of describing something. For example, a survey by the National Association of Home Builders found that the average size of single-family homes ballooned from 1500 square feet in 1970 to nearly 2400 square feet in 2004.[2]

Causal

In **causal research,** the objective is to determine whether one variable has an effect on another. In Suginami City, Japan, there were a record 1710 break-ins during 2002. A few years later, a neighborhood watch group noticed that there seemed to be fewer burglaries on streets where houses had flower boxes in their street-facing windows. Thus began Operation Flower, and the number of burglaries had fallen 80% by 2008. According to city official Kiyotaka Ohyagi, "By planting flowers facing the street, more people will be keeping an eye out while taking care of the flowers or watering them." In Suginami City, Flower Power took on a whole new meaning.[3]

Regarding causal studies, it should be pointed out that statistical techniques alone cannot *prove* causality. Proof must be established on the basis of quantitative findings along with logic. In the case of Suginami City's Operation Flower and the reduction in burglaries, it would seem obvious that causation was not in the *reverse* direction (i.e., lower burglary statistics were not causing flowers to grow). However, we must consider the possibility that one or more other variables might have contributed to the reduced number of break-ins—some of these possibilities could be greater participation in neighborhood watch groups, increased police presence, or the widespread installation of security cameras.

Predictive

Predictive research attempts to forecast some situation or value that will occur in the future. A common variable for such studies is the expected level of future sales. As might be expected, forecasts are not always accurate. For example,

[1]"Consumer Feedback Focus Group Results," U.S. Environmental Protection Agency, epa.gov, July 25, 2006.
[2]David Struckey and Robert W. Ahrens, "The Swelling Dwelling," *USA Today*, April 10, 2006, p. 1A.
[3]"Tokyo Residents Fight Burglars with Flower Power," reuters.com, June 11, 2009.

in 1984, the Semiconductor Industry Association predicted a 22% increase for 1985, a year in which sales actually fell by about 17%.[4] With any forecast, there will tend to be an error between the amount of the forecast and the amount that actually occurs. However, for a good forecasting model, the amount of this error will be consistently smaller than if the model were not being used. Forecasting will be one of the topics in Chapter 18.

The Research Process

A research study begins by defining the problem or question in specific terms that can be answered by research. Relevant variables must also be identified and defined. The steps in the research process include (1) defining the problem, (2) deciding on the type of data required, (3) determining through what means the data will be obtained, (4) planning for the collection of data and, if necessary, selection of a sample, (5) collecting and analyzing the data, (6) drawing conclusions and reporting the findings, and (7) following through with decisions that take the findings into consideration.

Primary versus Secondary Data

Statistical data can be categorized in a number of ways, including primary versus secondary. **Primary data** refer to those generated by a researcher for the specific problem or decision at hand. Survey research (Section 4.3), experimentation, and observational research (both discussed in Section 4.4) are among the most popular methods for collecting primary data.

Secondary data have been gathered by someone else for some other purpose. Secondary data can be **internal** or **external** depending on whether the data were generated from within your firm or organization or by an outside person or group. Some of the many possible sources of secondary data will be described in Section 4.5.

EXERCISES

4.1 What is the difference between primary data and secondary data? Between internal secondary data and external secondary data?

4.2 A published article reports that 60% of U.S. households have cable television. The president of an electronics firm clips the item from the paper and files it under "Cable TV data." Would this material be considered primary data or secondary data?

4.3 A pharmaceutical firm's annual report states that company sales of over-the-counter medicines increased by 48.1%.

a. To the firm itself, would this information be primary data or secondary data?
b. To a competitor, what type of information would this represent?

4.4 Bertram Pearlbinder, owner of an apartment building near the campus of Hightower University, is considering the possibility of installing a large-screen color television in the building's recreation room. For each of the following, indicate whether the associated data are primary or secondary:

a. Bertram interviews the current residents of the building to determine whether they would rather have the color television or a pool table.
b. Bertram, while reading *Newsweek*, notices an article describing the property damage in college dormitories and apartments.
c. One of the tenants tells Bertram that many of the residents are "very excited" about the prospect of getting the big TV.

[4]Source: Otis Port, "A Rosy Forecast for Chips Has Skeptics Hooting," *BusinessWeek*, October 14, 1985, p. 104.

SURVEY RESEARCH (4.3)

In **survey research,** we communicate with a sample of individuals in order to generalize on the characteristics of the population from which they were drawn. It has been estimated that approximately $20 billion per year is spent on survey research worldwide.[5]

Types of Surveys

The Mail Survey

In a **mail survey,** a mailed questionnaire is typically accompanied by a cover letter and a postage-paid return envelope for the respondent's convenience. A good cover letter will be brief, readable, and will explain who is doing the study, why it's being done, and why it's important for the reader to fill out and return the enclosed questionnaire.

The Personal Interview

In the **personal interview,** an interviewer personally secures the respondent's cooperation and carries out what could be described as a "purposeful conversation" in which the respondent replies to the questions asked of her. The personal interview tends to be relatively expensive compared to the other approaches, but offers a lot of flexibility in allowing the interviewer to explain questions, to probe more deeply into the answers provided, and even to record measurements that are not actually asked of the respondent. For example, along with obtaining responses, the interviewer may record such things as the respondent's gender, approximate age, physical characteristics, and mode of dress.

The Telephone Interview

The **telephone interview** is similar to the personal interview, but uses the telephone instead of personal interaction. Telephone interviews are especially useful for obtaining information on what the respondent is doing at the time of the call (e.g., the television program, if any, being watched). Also, use of the telephone makes it possible to complete a study in a relatively short span of time.

Questionnaire Design

Also referred to as the *data collection instrument*, the **questionnaire** is either filled out personally by the respondent or administered and completed by an interviewer. The questionnaire may contain any of three types of questions: (1) **multiple choice,** in which there are several alternatives from which to choose; (2) **dichotomous,** having only two alternatives available (with "don't know" or "no opinion" sometimes present as a third choice); and (3) **open-ended,** where the respondent is free to formulate his or her own answer and expand on the subject of the question.

In general, multiple choice and dichotomous questions can be difficult to formulate, but data entry and analysis are relatively easily accomplished. The reverse tends to be true for open-ended questions, where a respondent may state or

[5]Source: "Top 25 Research Companies Maintain Position in 2003," esomar.org, July 25, 2006.

write several paragraphs in response to even a very short question. Because they give the respondent an opportunity to more fully express his feelings or describe his behaviors, open-ended questions are especially useful in exploratory research.

Proper wording of each question is important, but often difficult to achieve. A number of problems can arise, including the following: (1) the vocabulary level may be inappropriate for the type of person being surveyed; (2) the respondent may assume a frame of reference other than the one the researcher intended; (3) the question may contain "leading" words or phrases that unduly influence the response; and (4) the respondent may hesitate to answer a question that involves a sensitive topic. Examples of each situation follow:

- **Inappropriate vocabulary level**

  Poor wording "Have you patronized a commercial source of cinematic entertainment within the past month?"

  Problem Vocabulary level will be too high for many respondents.

  Better wording "Have you gone to a movie within the past month?"

- **Confusing frame of reference**

  Poor wording "Are you in better shape than you were a year ago?"

  Problem To what does "shape" refer—physical, financial, emotional?

  Better wording (if the desired frame of reference is physical) "Are you in better physical condition than you were a year ago?"

- **"Leading" words/phrases**

  Poor wording "To help maintain the quality of our schools, hospitals, and public services, do you agree that taxes should be increased next year?"

  Problems "Schools, hospitals, and public services" has an emotional impact. Also, "do you agree" suggests that you *should* agree.

  Better wording "Do you support an increase in taxes next year?"

- **Sensitive topic**

  Poor wording "How much money did you make last year?"

  Problem This question requests detailed, personal information that respondents are hesitant to provide. The question is also very blunt.

  Better wording "Which of the following categories best describes your income last year?"

[ ] under $20,000	[ ] $60,000–$79,999
[ ] $20,000–$39,999	[ ] $80,000–$99,999
[ ] $40,000–$59,999	[ ] $100,000 or more

The preceding represent only a few of the pitfalls that can be encountered when designing a questionnaire. In general, it's a good idea to pretest the questionnaire by personally administering it to a small number of persons who are similar to the eventual sample members.

Sampling Considerations in Survey Research

The sampling techniques presented later in the chapter are readily applicable to survey research. As we will discuss further in Section 4.7, we can make statistical inferences from the sample to the population only if we employ a *probability sample*—one in which every person in the population has either a known or a calculable chance of being included in the sample.

Two of the problems unique to telephone surveys are cell-only households and unlisted numbers. About 16% of all households have wireless or cell-phone service only. Of those who have "landline" telephones, nearly one-third do not have their number listed in the directory. In some locales, the percentage is much higher, with some metropolitan areas having an unlisted percentage up to 60%. Households that are cell-only or unlisted may have different characteristics than those that are listed. For example, studies have found unlisted households are more likely to be urban, young, poor, and mobile compared to households whose number is in the telephone directory.[6]

Random-digit dialing is a technique that has been developed to overcome the problem of unlisted numbers. In this approach, at least the final four digits are randomly selected, either by a computer or from a random number table. The area code and the three-digit exchange number can also be randomly selected if the survey is national in scope. The use of randomly selected numbers results in both listed and unlisted households being included in the sample.

If we are to use information from a few thousand sample members to make inferences about a population numbering in the millions, the sample selection process must be both precise and rigorous. As a result, a number of companies specialize in the generation of samples for survey research clients. Survey Sampling, Inc., a leading supplier of statistical samples for the research industry, is such a company. Each year, Survey Sampling "marks" millions of telephone numbers that have been selected from its Random Digit Sampling Database. Any number that is called is automatically eliminated from consideration for a period of 6 months. One of the purposes of this practice is to help ensure that respondents will be "fresher" and more likely to cooperate.[7]

Mailing lists are readily available for the identification of practically any population a business researcher would ever want to consider. For example, infoUSA, Inc. has a database of 14 million businesses across the United States, and these are coded according to such criteria as geographical location and standard industrial classification (SIC). If you wish to mail a questionnaire to a simple random sample from populations like the 20,338 sporting goods retailers, the 22,114 tanning salons, or the 13,509 siding contractors, you can even obtain the mailing labels directly from the list company.[8]

Either an entire list or a sample from it can be generated on the basis of geographical location, at levels down to individual states and counties. Of particular interest is the ability to select samples of persons from what are known as *compiled* versus *response* lists. The distinction between these two kinds of samples is described here:

COMPILED LIST **A sample consisting of firms or persons who are alike in some way, such as being (1) physicians, (2) homeowners, (3) personal computer owners, or (4) residents of Westmoreland County, Pennsylvania. A compiled list is passive in that you can be placed on one without having done anything.**

(*continued*)

[6]Source: surveysampling.com, June 28, 2009.
[7]Ibid.
[8]Source: infousa.com, June 28, 2009.

RESPONSE LIST A sample consisting of firms or persons who have engaged in a specified behavior or activity, such as (1) subscribing to *PC Magazine*, (2) contributing to Greenpeace, (3) buying from a mail-order catalog, or (4) applying for an American Express credit card.

Like questionnaire design, survey sampling can itself be the topic of an entire textbook. This is especially true when the design involves either proportionate or disproportionate stratified sampling, two of the more advanced designs discussed later in the chapter.

Errors in Survey Research

Survey research may lead to several different kinds of errors, and it's recommended that you be a skeptical "consumer" of statistical findings generated by this mode of research. These errors may be described as sampling error, response error, and nonresponse error. **Sampling error,** discussed below, is a random error. It can also be described as nondirectional or **nonsystematic,** because measurements exhibiting random error are just as likely to be too high as they are to be too low. On the other hand, **response** and **nonresponse errors** are both of the directional, or **systematic,** type.

Sampling Error

Sampling error occurs because a sample has been taken instead of a complete census of the population. If we have taken a simple random sample, the methods described in Chapter 9 can be used to estimate the likely amount of error between the sample statistic and the population parameter. One of the procedures discussed in Chapter 9 is the determination of the sample size necessary to have a given level of confidence that the sample proportion will not be in error by more than a specified amount.

Response Error

Some respondents may "distort" the truth (to put it kindly) when answering a question. They may exaggerate their income, understate their age, or provide answers that they think are "acceptable." Biased questions can even encourage such response errors, for example, "Shoplifting is not only illegal, but it makes prices higher for everyone. Have you ever shoplifted?"

Nonresponse Error

Not everyone in the sample will cooperate in returning the questionnaire or in answering an interviewer's questions. This would not be a problem, except that those who respond may be different from those who don't. For example, if we're using a mail questionnaire to find out the extent to which people are familiar with the works of William Shakespeare, those who are less literate or less interested in this classic author may also be less likely to complete and return our questionnaires. As a result, we could "measure" a much higher level of interest than actually exists.

EXERCISES

4.5 What are the major approaches to carrying out survey research?

4.6 Provide an example of a survey question that would tend to exceed the vocabulary level of the typical adult.

4.7 Comment on the appropriateness of the question, "Have you ever broken the law and endangered the lives of others by running a red light?"

4.8 What is random-digit dialing and why is it used?

4.9 How does a compiled mailing list differ from a response mailing list?

4.10 Explain what is meant by sampling error, response error, and nonresponse error in survey research.

4.11 A research firm finds that only 33% of those who reported buying Kellogg's Frosted Flakes had actually purchased the product during the period monitored. What type of survey research error does this represent?

4.12 In order to increase the response rate to mail questionnaires, researchers sometimes include a dollar bill or other monetary incentive to reward the respondent for his or her cooperation. Could there be occasions where a relatively large reward—e.g., the inclusion of a $20 bill with a short questionnaire—might end up exerting a biasing influence on the answers the respondent provides? If so, give a real or hypothetical example of such an occasion.

4.13 A company has constructed two prototypes of a new personal digital assistant (PDA) and would like to find out which one of the two looks easier to use. Of the three types of surveys discussed in this section, which one would the company *not* wish to employ?

EXPERIMENTATION AND OBSERVATIONAL RESEARCH (4.4)

Experimentation

In **experiments,** the purpose is to identify cause-and-effect relationships between variables. There are two key variables in an experiment: (1) the *independent* variable, or treatment, and (2) the *dependent* variable, or measurement.

We must also consider what are known as *extraneous* variables—outside variables that are not part of the experiment, but can influence the results. Persons or objects receiving a treatment are said to be in an *experimental* group, while those not exposed are in the *control* group. In symbolic terms, we will refer to the independent variable as "*T*" and each measurement of the dependent variable as "*O*":

O_1, 1st measurement:		*T*, treatment:		O_2, 2nd measurement:
In 1974, 53.0% of Norwegian males smoked daily.	→	In 1975, Norway banned tobacco advertising.	→	In 1984, 41.6% of Norwegian males smoked daily.

In evaluating experiments, two kinds of validity must be considered: internal and external. We will examine each in terms of the Norwegian ban on tobacco advertising.[9]

Internal Validity

Internal validity refers to whether *T* really made the difference in the measurements obtained. In the preceding experiment, was it really the advertising ban

[9]Source: Jack Burton, "Norway Up in Smoke," *Advertising Age,* April 14, 1986, p. 82.

that made the difference? For example, extraneous variables included a high cigarette tax ($1 to $2 a pack), public service announcements, and extensive educational efforts on the hazards of smoking.

External Validity

Even if *T did* make the difference, **external validity** asks whether the results can be generalized to other people or settings. For example, would similar results have been obtained in the United States or Great Britain, or if the tobacco ad ban had been enacted 10 or 20 years earlier instead of in 1975?

Health groups have long studied such phenomena as the effect of cigarette smoking on longevity. However, it is difficult to remove the influence of other variables on this relationship. For example, the heavy smoker is more apt to be a heavy drinker or a compulsive workaholic, while the nonsmoker is more likely to be a nondrinker or light drinker and to be more attentive to other factors (such as exercise and weight control) that tend to increase life span.

It can be useful to design an experiment in which people or objects are randomly placed into either the experimental or control group, then measurements are made on the dependent variable. Consider the following example in which 200 persons are randomly assigned, 100 to each group:

- Experimental group ($n_1 = 100$):

  T_1: [Members sample soft drink in bottle with a *new* label design.] → O_1: [65% say "tastes great."]

- Control group ($n_2 = 100$):

  T_2: [Members sample the same soft drink, but in bottle with the *old* label design.] → O_2: [39% say "tastes great."]

These results suggest that the new label is effective in causing testers to evaluate the soft drink as tasting good. Although this is just one of many possible experimental designs, it shows how an experimental group and a control group can be used in carrying out an experiment. To facilitate internal validity, it's best if persons or objects are randomly assigned to the experimental and control groups. The significance of the difference between the 65% for the experimental group and the 39% for the control group can be obtained using the hypothesis testing methods presented in Chapter 11.

Observational Research

Observation relies on watching or listening, then counting or measuring. For example, observational research allows us to passively watch highway travelers and monitor variables such as speed and adherence to other traffic laws, or the extent to which drivers from one age group, gender group, or vehicle group behave differently from those of other groups. One of the more interesting applications of observational research involves actually contriving a situation that is to be observed—for example, hiring a "mystery shopper" to present retail store personnel with a given problem, product return, or complex request, then have the hired

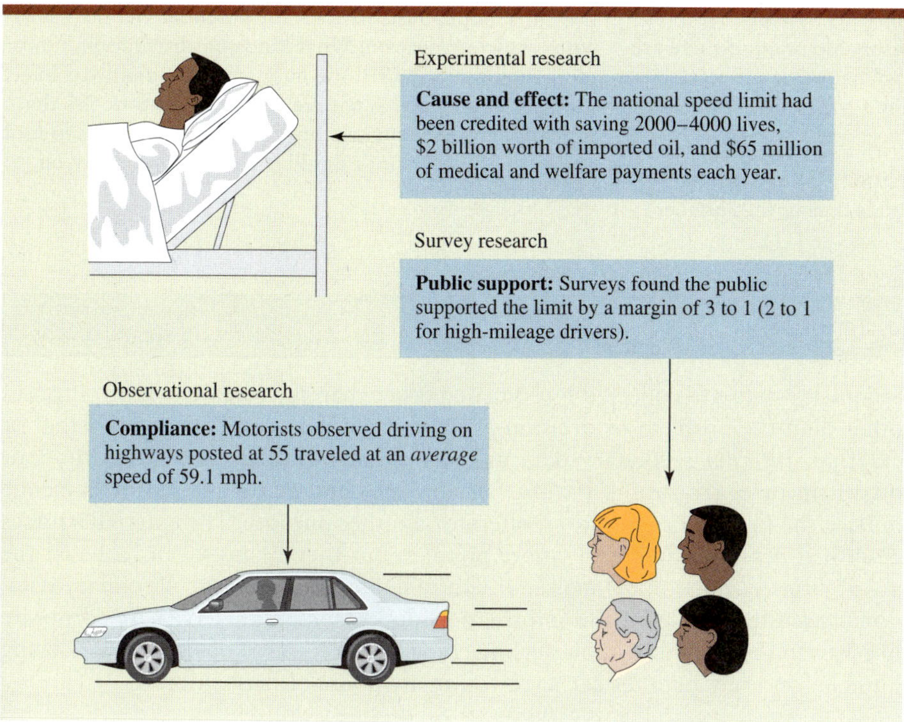

FIGURE 4.1

News media often report research results in the context of an article or story. This figure dramatizes the unusual case of a single article that included the findings of three different types of investigations centered on the same topic: the 55-mph national speed limit that was in effect between 1974 and 1987. Note the contradiction between the support indicated by surveys and the lack of compliance measured in observational studies.

Data source: Geoff Sundstrom, "Speed Limit Retention Is Favored," *Automotive News,* December 3, 1984, p. 6.

shopper submit a report on how he or she was treated, and the extent to which her contrived situation was handled according to corporate policy.

The results of surveys, experiments, and observational studies are often reported in news media. Figure 4.1 shows data from a single article containing elements of all three research methods. The research dealt with the former 55-mph national speed limit and involved a 2-year study by the Transportation Research Board of the National Research Council. The study seems to have included some rather contradictory findings.

EXERCISES

4.14 In studying the effectiveness of monetary rewards on seat belt usage, a firm has been giving out lottery tickets to employees wearing their belts when entering the company parking area. At the end of each week, the holder of the winning ticket gets a free dinner at a local restaurant. Before the experiment, only 36% of employees wore seat belts. After 1 month of the lottery, the usage rate is up to 57%.

a. Identify the dependent and independent variables in this experiment.
b. Is a control group involved in this experiment? If so, who?

4.15 A direct-mail firm tries two different versions of a mailing, one of which includes a 50-cent monetary incentive for the respondent. Responses to the mailing with the incentive were 15% higher than for a similar mailing without it.

a. Identify the dependent and independent variables in this experiment.
b. Is a control group involved in this experiment? If so, who?

4.16 In a health study, the dietary habits of 420,000 men were examined over a 3-year period. For those who ate no fried foods, the death rate was 1208 per 100,000 man-years. For men who consumed fried foods more than 15 times a week, the corresponding rate was only 702. Considering that the relative ages of members of these two groups were not mentioned, critique the internal validity of the experiment. Source: "Science Finds Fried Foods Not Harmful," *Moneysworth* magazine, November 24, 1975, p. 1.

4.17 When traveling, some motorists assume that a restaurant has good food and reasonable prices if there are a lot of trucks parked outside. What other observed characteristics of a restaurant might a traveler use in deciding whether to stop and eat?

4.18 In doing "observational studies" while driving through suburban neighborhoods during the summer months, home burglars look for potential victims who are away on vacation. What kinds of observation indicators would tend to tell the burglars that a family is likely away on vacation? What are some of the strategies that vacationing homeowners might use in attempting to fool potential burglars into thinking they are *not* away on vacation?

4.5 SECONDARY DATA

Secondary data are collected by someone other than the researcher, for purposes other than the problem or decision at hand; such data can be either internal or external. *Internal* secondary data are those that have been generated by your own firm or organization. *External* secondary data were gathered by someone outside the firm or organization, often for another purpose. The external sources described here are among the more popular, but they by no means exhaust the possibilities. As a practical matter, if you're in search of relevant secondary data, don't hesitate to consult the information desk of your university or corporate library. Don't even think about feeling embarrassed—this is one case where being humble for a few minutes can save you many hours of frustration.

Internal Sources

Internal secondary data have traditionally existed in the form of accounting or financial information. Among the many other possibilities are annual reports, as well as more specific reports that may be available from departments such as production, quality control, engineering, marketing, sales, and long-range planning. In short, anything in written form that has ever been generated within the company falls within the realm of internal secondary data. In addition, some information may not even exist in written form, but may be found in the collective memories and observations that executives, managers, and engineers have recorded over the years.

External Sources

Government Agencies

Federal, state, and local governments are extremely prolific generators and purveyors of secondary data, with the federal government, through the U.S. Department of Commerce and its Bureau of the Census, being an especially important source. The following are among the many U.S. Department of Commerce publications that are highly relevant to business research:

- *Statistical Abstract of the United States* Published annually, this volume is an especially extensive collection of social, economic, and political data from a wide variety of sources, both governmental and private.
- *Census of Population* Conducted every 10 years, this is the research for which the Bureau of the Census is probably best known. It provides detailed information on the U.S. population, including age, education, employment, and a broad range of other demographic characteristics.
- *Census of Housing* Also published every 10 years, this report presents voluminous data on such variables as home ownership or rental, type of

structure, number of residents, market value, heating source, major appliances, and much more.

- ***Economic Censuses*** Conducted in years ending in 2 and 7, these studies provide data on retail and wholesale trade, the construction and transportation industries, and the manufacturing and minerals industries. Data are aggregated by national, regional, and local levels.
- ***Survey of Current Business*** A monthly publication containing information on a number of statistical topics, including population, trade, and transportation. It includes over 2500 updated statistical series.

Other key publications of the U.S. Department of Commerce include *Historical Statistics of the United States from Colonial Times to 1970, County Business Patterns, State and Metropolitan Area Data Book,* and the *U.S. Industrial Outlook*. The Department of Labor provides employment and wage information in the *Monthly Labor Review,* while the Federal Reserve System publishes the *Federal Reserve Bulletin,* which includes statistics on banking, industrial production, and international trade. In addition to the staggering quantity of information available from the federal government, published data are also provided by governmental agencies at the state, county, and local levels.

Other Published Sources

This category includes a variety of periodicals, handbooks, and other publications. The following list is a sample of the information available:

- ***Business Periodicals Index*** Published monthly, this is a bibliography of articles in selected business publications, all indexed by subject area.
- ***Reader's Guide to Periodical Literature*** Similar to the *Business Periodicals Index,* but indexes popular, rather than business, publications.
- ***Rand McNally Commercial Atlas and Marketing Guide*** Includes statistical data for each county in the United States and for more than 120,000 cities and towns.

Other published sources include (1) indexes, such as the *New York Times Index* and the *Wall Street Journal Index*, and (2) popular business publications, including *American Demographics, BusinessWeek, Fortune, Sales and Marketing Management, Dun's Business Month, Business Horizons,* and *Industry Week.*

Commercial Suppliers

Information may also be purchased from outside suppliers who specialize in business research. One of the best-known commercial suppliers is the A. C. Nielsen Company, which provides a wide variety of data services for client firms. Although some data are supplied to a number of subscribing firms, studies may also be tailored to the specific needs of a single research customer.

Trade Associations

Trade associations collect and disseminate data that are related to their industry; these groups may be identified in the *Encyclopedia of Associations*. This publication is also an excellent source of private organizations having special interests that may be related to your informational needs. Most such groups generate regular publications that are available upon request or special order.

Data Warehousing and Data Mining

Advancements in computing power, mass data storage, and data analysis techniques have led to a relatively recent and exciting data source—the **data warehouse.** This is a large collection of data from inside and outside the company, put together for the express purpose of using sophisticated analytical techniques to uncover patterns and discover interrelationships within the data. The procedure by which these analytical techniques are applied is known as **data mining**—it is analogous to physically sifting through great quantities of ore in the quest for gold.

The data warehouse typically contains massive information from a variety of sources that pertains to customers, vendors, products, or any other organizational or environmental subjects of interest to the company's decision makers. It is estimated that over 90% of large corporations worldwide either have or are building data warehouses, and that the market for data warehousing hardware and software grew from practically nil in 1994 to more than $15 billion by 2000.[10]

The tools for data mining comprise many of the techniques covered in this text, such as descriptive statistics, cross-tabulations, regression, and correlation, but also include many methods that are far beyond the scope of our text, such as those briefly described here.

Factor analysis A set of techniques for studying interrelationships among variables. A very large set of variables can be reduced to a smaller set of new variables (factors) that are more basic in meaning but contain most of the information in the original set.

Cluster analysis When each of a great many individuals or objects is represented by a set of measurements or variables, cluster analysis categorizes the individuals or objects into groups where the members tend to be similar. In marketing, one application of cluster analysis is to categorize consumers into groups, a procedure associated with what is referred to as *market segmentation.*

Discriminant analysis This technique identifies the variables that best separate members of two or more groups. It can also be used to predict group membership on the basis of variables that have been measured or observed.

Multidimensional scaling Based on a ranking that reflects the extent to which various pairs of objects are similar or dissimilar, multidimensional scaling is a procedure that maps the location of objects in a perceptual space according to dimensions that represent important attributes of the objects. In this approach to multidimensional scaling, one of the biggest challenges is the interpretation of exactly what these dimensions mean.

Data warehouses and data mining can be useful to many kinds of businesses. For example, the descriptive and predictive models developed through data mining can help telecommunications companies answer questions like these: (1) What kinds of customers tend to be profitable or unprofitable to the company? (2) How can we predict whether customers will buy additional products or services, like call waiting? (3) Why do customers tend to call more at certain times? (4) Which products or services are the most profitable?[11] Likewise, companies in the insurance industry can benefit from models that help them identify

[10]Source: Davis, Duane, *Business Research for Decision Making* (Pacific Grove, CA: Duxbury Press, 2000), p. 355.

[11]Source: Groth, Robert, *Data Mining* (Upper Saddle River, NJ: Prentice Hall, 2000), p. 209.

which insurance claims are most likely to be fraudulent, and investment firms can profit from models that help them better differentiate between initial public offerings (IPOs) that should be pursued and those that are better avoided.

SAS, SPSS, and IBM are among those offering data mining software and assistance to a variety of firms and industries. For example, IBM's Intelligent Miner™ software was selected by Bass Export, the United Kingdom's largest beer exporter. As stated by Bass executive Mike Fisher, "Each week, we deliver orders to 23,000 customers. It's vitally important that we understand the buying habits of each of them, the profitability of each individual account, and the profitability of each brand we deliver there."[12] Reuters, a leading provider of financial news and information, has used the SPSS-developed Clementine software to develop models for reducing error rates in its compilation and reporting of the massive amount of data it receives from worldwide sources.[13]

Internet Data Sources

Thousands of database services—some are commercial, but many are free—offer bibliographical and statistical data for personal computer (PC) users. The specific set of available data will depend on your Internet provider, your Web-browsing software, and on the search engines at your disposal for specifying World Wide Web sites for further perusal. The PC and access to the vast electronic "ether" make it very convenient to locate great quantities of information relevant to a particular business situation or decision. The author's seventh-grade teacher had a favorite saying, "The more you know, the less you know you know, because you know there's so much to know." This certainly applies to our familiarity with the Internet and its capabilities, and this section will barely scratch the surface of what's out there.

Corporate News and Annual Reports Online

If you want to read a company's latest annual report, in many cases all you need to do is visit the firm's website and look for the "information for investors" or similar section. There's a good chance that your browser's home page includes a sponsored section where stock quotes (usually with a 20-minute or more delay) are available for companies on the major stock exchanges; if so, current news about a company or its industry is available just by pursuing the news items that accompany the quote.

Secondary Data Online

Many of the sources of published secondary data discussed previously offer information online as well as in hard-copy form. For example, the U.S. Bureau of the Census website (census.gov) can be the starting point for data from sources such as *USA Statistics in Brief* or the *Statistical Abstract of the United States*. Its "State & County QuickFacts" section is particularly interesting and can provide detailed economic and demographic data about the county in which you live. *As with any useful site that you visit, be sure to refer to the "Related Sites" section.* In this case, the Bureau provides links to international statistical agencies, state data centers, more than 100 federal statistical agencies, and to its parent agency, the U.S. Department of Commerce.

[12]Source: Groth, Robert, *Data Mining* (Upper Saddle River, NJ: Prentice Hall, 2000), p. 11.
[13]Ibid.

STATISTICS IN ACTION 4.1

Get That Cat off the Poll!

When a public opinion poll is conducted professionally, respondents are carefully and scientifically selected to ensure that they are representative. However, as with many other human endeavors, public opinion polls are sometimes carried out using methods that are not quite so professional. One unfortunate polling firm was the target of special scrutiny by a federal grand jury in New Haven, Connecticut, for techniques it allegedly employed during some of its studies between 2002 and 2004. According to prosecutors, the polling firm's owners and managers had instructed fieldworkers to fabricate survey results, to complete surveys even if they had to make up the answers themselves, and to meet their response quotas "even if they had to talk to dogs and cats."

Source: Noreen Gillespie, "Polling Firm Accused of Falsifying Results," *The Indiana Gazette,* March 10, 2005, p. 7.

Finding Just About Anything—Those Incredible Search Engines

Tonight Show host Johnny Carson had a segment that he called "Stump the band," where audience members would state a fictitious song title and the band would try to play music that seemed to fit the title. Regardless of whether the band was "stumped," the audience member always received a free dinner or other prize.

If there were ever a contest called "Stump the Internet search engine," it would have few winners. Regardless of how esoteric the information or topic we are seeking, we can use a search engine like Google to obtain a wealth of guidance in leading us toward our goal. A prerequisite to using the search engines effectively is the strategic selection of the key words to be used in the search. It can also be helpful to use the "search within a search" or "power search" features to enter more specific key words to be found within a search that has already been conducted.

Tapping Others' Expertise—Searching Message Boards and Interest Groups

Message boards and interest groups can be viewed as the Internet counterpart to the "brick and mortar" organizations listed in the *Encyclopedia of Associations*. For practically any human interest or pursuit, you'll find some forum with participants who are enthusiastic about the topic and possess high levels of knowledge and experience. Select the "Groups" option at google.com, enter key words that describe your problem or interest, and you'll get a list of postings that contain one or more of the key words. *Note:* There are some privacy issues here—for example, if you've posted to a discussion group, a present or potential employer can search the groups for your name to find out more about *you,* your interests, and the footprints you've made during your electronic travels.

Evaluating Secondary Data

Secondary data should be selected, evaluated, and used with caution, since the information may be irrelevant to your needs, out of date, not expressed in units or levels of aggregation that are compatible with your informational requirements, or may not have been collected or reported objectively. When considering the use of secondary information, be sure to keep in mind both who collected the information and whether the individual or group may have had some stake in the results of their own research. Honesty and reputation are also very important. Statistics in Action 4.1 describes a situation where a company may have been more interested in obtaining respondents than in obtaining facts.

EXERCISES

4.19 What are secondary data? Differentiate between internal and external secondary data.

4.20 Besides accounting and financial information, what are some of the other sources of internal secondary data?

4.21 What kinds of external secondary data are readily available online?

4.22 Briefly, what kinds of information does the U.S. Department of Commerce provide in its *Census of Population?* In its *Census of Housing?*

4.23 What is the *Encyclopedia of Associations,* and how can this publication be useful to the business researcher?

4.24 What is data mining, and how can it be useful to a business or other organization? What are some of the analytical tools it employs?

Note: Exercises 4.25–4.31 require a computer and access to the Internet. Web addresses were accurate at this writing. If a site address is no longer applicable, use a search engine (e.g., google.com) to identify the company or organization's new home page.

4.25 Visit the U.S. Bureau of the Census (census.gov), and find the following descriptive statistics for the county where your college or university is located:

a. Median household income
b. Number of Social Security recipients
c. Per-capita retail sales

4.26 Repeat Exercise 4.25, but obtain the requested descriptive statistics for Dare County, North Carolina.

4.27 Referring to each company's online annual report, find the number of employees and net income in the most recent fiscal year for Coca-Cola, McDonald's, and Microsoft.

4.28 Visit google.com and search message boards for postings that pertain to (a) a professional sports team and (b) a hobby or recreational activity of your choice.

4.29 Using whitepages.com, find out how many telephone listings are in your state for people with the same last name as yours. How many listings are there in your hometown for people named Smith?

4.30 Using one or more search engines and key words of your choice, find (a) a website that reports the average life span of Saint Bernard dogs and (b) a website that advises business professionals on strategies for writing an effective memo.

4.31 Using a search engine and the key words "spy equipment," identify some of the high-tech gadgets that companies or individuals can use in carrying out industrial espionage on other companies or individuals.

THE BASICS OF SAMPLING (4.6)

Selected Terms

The following terms—some of which we have discussed previously in other contexts—are of special relevance to our discussion of samples, the importance of sampling, and the methods by which samples are selected:

POPULATION The set of all possible elements that could theoretically be observed or measured; this is sometimes referred to as the *universe*.

SAMPLE A selected portion from the elements within the population, with these elements actually being measured or observed.

CENSUS The actual measurement or observation of all possible elements from the population; this can be viewed as a "sample" that includes the entire population.

PARAMETER A characteristic of the population, such as the population mean (μ), standard deviation (σ), or population proportion (π).

STATISTIC A characteristic of the sample, such as the sample mean ($\bar{x}$), standard deviation (s), or sample proportion (p). In practice, this is used as an estimate of the (typically unknown) value of the corresponding population parameter.

Why a Sample Instead of a Census?

A Census Can Be Extremely Expensive and Time-Consuming

A complete census of the population is not a problem whenever the population happens to consist of a small number of elements that are easily identified. For example, there are only 29 nuclear electric power generation firms that have 500 or more employees.[14] In this setting, a study involving a complete census would present little difficulty. On the other hand, infoUSA, Inc., lists 77,240 pizza parlors in the United States.[15] Contacting every member of such a large population would require great expenditures of time and money, and a sampling from the list can provide satisfactory results more quickly and at much lower cost.

A Census Can Be Destructive

The Gallo Wine Company, like every other winery, employs wine tasters to ensure the consistency of product quality. Naturally, it would be counterproductive if the tasters consumed *all* of the wine, since none would be left to sell to thirsty customers. Likewise, a firm wishing to ensure that its steel cable meets tensile-strength requirements could not test the breaking strength of its entire output. As in the Gallo situation, the product "sampled" would be lost during the sampling process, so a complete census is out of the question.

In Practice, a Sample Can Be More Accurate Than a Census

Because of the tremendous amounts of time and money required for a complete census of a large population, a sample can actually be more accurate than a census in conducting a real-world study. For example, if we hired 10 people to work 16 hours a day telephoning each of the more than 100 million television households in the United States, it would take years to ask all of them whether they watched the CNN program on political campaigns that aired last night. By the time the last batch of calls was made, nobody would be able to remember whether they had watched a given program that was aired 2 or 3 years earlier.

Sampling can also contribute to accuracy by reducing *nonsampling error*. This occurs because we can now afford to spend more time and attention on the measuring instrument and related procedures. For example, if a survey question is biased ("Most studies find that capital punishment does not deter crime. Do you agree that capital punishment should be abolished?"), asking that same question of everyone in the entire population will only ensure that you get the same biased answers from everyone instead of from just a sample. With some of the money saved by conducting a sample instead of a census, we can fine-tune the measurement procedure through such strategies as pretesting the questionnaire, hiring more-experienced interviewers, or following up on those who did not initially respond. This accuracy advantage of a sample over a census is especially valuable to survey research. As was observed in the vignette at the beginning of this chapter, even the U.S. Bureau of the Census has proposed that its census efforts be supplemented with survey information based on samples.

[14]Source: Bureau of the Census, *Statistical Abstract of the United States 2009*, p. 565.
[15]Source: infousa.com, June 28, 2009.

Although a census is *theoretically* more accurate than a sample, *in practice* we are limited by funding constraints such that a fixed amount of money must be divided between the sampling (or census) process and the measurement process. If we allocate more funds to the sampling or census process, we have less to spend on the assurance of accuracy in the measurement process. Here is a brief review of the concepts of sampling and nonsampling error:

SAMPLING ERROR This is the error that occurs because a sample has been taken instead of a census. For example, the sample mean may differ from the true population mean simply by chance, but is just as likely to be too high as too low. When a probability sample (discussed in the next section) is used, the likely amount of sampling error can be statistically estimated using techniques explained in Chapter 9 and can be reduced by using a larger sample size. For a visual demonstration of sampling error, see Seeing Statistics Applet 3 at the end of the chapter.

NONSAMPLING ERROR This type of error is also referred to as *bias,* because it is a directional error. For example, survey questions may be worded in such a way as to invite responses that are biased in one direction or another. Likewise, a weighing scale may consistently report weights that are too high, or a worn micrometer may tend to report rod diameters larger than those that actually exist. *Nonsampling errors cannot be reduced by simply increasing the size of a sample.* To reduce nonsampling error, it is necessary to take some action that will eliminate the underlying cause of the error. For example, we may need to adjust a machine or replace parts that have become worn through repeated use.

EXERCISES

4.32 What is the difference between a sample and a census? Why can it be advantageous to select a sample instead of carrying out a census?

4.33 Differentiate between sampling error and nonsampling error.

4.34 Provide a real or hypothetical example of a situation in which the sampling process is destructive.

4.35 Differentiate between the terms *parameter* and *statistic*. Which one will be the result of taking a sample?

4.36 To impress the interviewer, many respondents in a survey of 400 persons exaggerated their annual incomes by anywhere from 2% to 50%. Does such exaggeration represent sampling error or nonsampling error? Explain.

4.37 In survey research, what are some of the methods by which we can reduce nonsampling error?

SAMPLING METHODS (4.7)

Sampling methods can be categorized as *probability* or *nonprobability*. The distinction is that with **probability sampling,** each person or element in the population has a known (or calculable) chance of being included in the sample. Such samples also allow us to estimate the maximum amount of sampling error between our sample statistic and the true value of the population parameter being estimated. **Nonprobability sampling** is primarily used in exploratory research studies where there is no intention of making statistical inferences from the sample to the population. Our emphasis in this and the chapters to follow will be on samples of the probability type.

Probability Sampling

In probability sampling, each person or element in the population has some (nonzero) known or calculable chance of being included in the sample. However, every person or element may not have an *equal* chance for inclusion.

The Simple Random Sample

In the **simple random sample,** every person or element in the population has an equal chance of being included in the sample. This type of sample is the equivalent of "drawing names from a hat" but is usually generated using procedures that are more workable than the familiar "hat" analogy.[16]

A practical alternative to placing names in a hat or box is to identify each person or element in the population with a number, then use a random number table to select those who will make up the sample. A portion of the random number table in Appendix A is reproduced here as Table 4.1 and will be used in the example that follows.

EXAMPLEEXAMPLEEXAMPLEEXAMPLEEXAMPLEE

EXAMPLE

Simple Random Sampling

A firm has 750 production employees and wishes to select a simple random sample of 40 workers to participate in a quality-assurance training program.

SOLUTION

Using the random digits listed in Table 4.1, the following steps can be taken to determine the 40 individuals to be included in the sample:

1. Identify the employees with numbers ranging from 001 to 750.
2. Arbitrarily select a starting point in the random number table. This can be done by closing your eyes and pointing to a position on the table or by some other relatively random method of your choice. As our starting point for the example, we will begin with digits 11–13 in row 2 of Table 4.1. Three-digit numbers are necessary because the largest employee identification number is three digits.
3. Work your way through the table and identify the first 40 unique (to avoid duplicates, since we're sampling without replacement) numbers that are between 001 and 750.

[16]For practicality, we are using the *operational definition* of the simple random sample. The *theoretical definition* says that, for a given sample size, each possible sample constituency has the same chance of being selected as any other sample constituency. For more details, the interested reader may wish to refer to a classic work on sampling: Leslie Kish, *Survey Sampling,* New York: John Wiley & Sons, 1965, pp. 36–39.

Row	Columns 1–10		Columns 11–20		
1	37220	84861	59998	77311	
2	31618	06840	**451**67	13747	Employee #451 is in sample.
3	53778	71813	**503**06	47415	Employee #503 is in sample.
4	82652	58162	**903**52	10751	903 > 750. 903 is ignored.
5	40372	53109	**769**95	24681	769 > 750. 769 is ignored.
6	24952	30711	**225**35	51397	Employee #225 is in sample.
7	94953	96367	**871**43	71467	871 > 750. 871 is ignored.
8	98972	12203	**907**59	56157	907 > 750. 907 is ignored.
9	61759	32900	**052**99	56687	Employee #052 is in sample.
10	14338	44401	**226**30	06918	Employee #226 is in sample.
⋮					⋮
					The process continues, until all 40 sample members have been selected.

TABLE 4.1

A portion of the table of random digits from Appendix A. In the example, a simple random sample of 40 is being selected from a population of 750 employees. The digits in row 2, columns 11–13, serve as an arbitrary starting point.

The procedure is summarized in Table 4.1 and begins with the selection of employee #451, who also happens to be at the randomly selected starting point. The next number, 503, also falls within the limit of 750 employees; at this point we have placed individuals #451 and #503 into the sample. Because the next number (903) exceeds 750, it is ignored and we continue down the column.

Computer Solutions 4.1 shows an alternative to using the random number table. Here we show the procedures and results when Excel and Minitab are used to select a simple random sample *from an existing list of data values*. In this case, we are selecting a simple random sample of 10 usage times (seconds) from the 50 ATM customers whose times were examined in Chapters 2 and 3. The data file is **CX02ATM**.

Although the simple random sampling approach just described is a straightforward procedure, it does require that we have a listing of the population members. This is not a problem if we are dealing with telephone directories or members of the *Fortune 500*, but it may not be practical in other situations. For example, it would be both expensive and awkward to take down the names and telephone numbers of all the shoppers at the local mall tomorrow so that we can later narrow them down to a simple random sample of 100 persons.

In some cases, such as when products are coming off an assembly line, we can use a random selection process to determine which ones to sample. For example, if we would like to subject 10% of the units to a detailed inspection, a random number sequence could be followed. As each product comes off the line, we advance to the next random digit in a table of random numbers. If the digit associated with a particular product happens to be a "9" (an arbitrary selection), it and all other units that are associated with the digit 9 receive the detailed inspection. Although it takes a slightly different form, this is a variation of the simple random sampling method, and each unit has the same probability (in this case, 0.10) of being sampled.

COMPUTER 4.1 SOLUTIONS

Simple Random Sampling

These procedures select a simple random sample from a list of existing data values.

EXCEL

	A	B	C	D	E
1	AgeCat	Gender	Seconds		40.1
2	1	2	50.1		50.8
3	1	1	53.0		37.8
4	2	2	43.2		42.9
5	1	2	34.9		40.1
6	3	1	37.5		32.9
7	2	1	37.8		29.4
8	3	1	49.4		49.4
9	1	1	50.5		37.8
10	3	1	48.1		39.5

1. Open the Excel data file **CX02ATM**. The 50 ATM usage times are in column C, and the variable name (Seconds) is in C1. From the **Data** ribbon, click **Data Analysis** in the rightmost menu. Click **Sampling.** Click **OK.**
2. Enter **C1:C51** into the **Input Range** box. Click to place a check mark into the **Labels** box (because the input range has the name of the variable at the top). Select **Random** and enter **10** into the **Number of Samples** box. Enter **E1** into the **Output Range** box. Click **OK.** The simple random sample of 10 values from the 50 ATM usage times are listed in column E.

MINITAB

```
Data Display

C5
   37.1   46.4   31.0   29.9   47.6   44.7   33.6   42.3   34.9   47.7
```

1. Open the Minitab file **CX02ATM**. The data are already listed in columns C1 (AgeCat), C2 (Gender), and C3 (Seconds).
2. Click **Calc.** Select **Random Data.** Click **Sample From Columns.** Enter **10** into the **Number of rows to sample:** box. Enter **C3** into the box below the **From columns:** label. Enter **C5** into the **Store samples in** box. (Do not select "Sample with replacement.") Click **OK.** The 10 data values randomly selected from the 50 usage times will be in column C5.

The Systematic Sample

Although similar in concept to the simple random sample, the **systematic sample** is easier to apply in practice. In this approach, we randomly select a starting point between 1 and k, then sample every kth element from the population. For example, in selecting a sample of 100 from a population of 6000, we would select a random number between 1 and 60, then include that person and every 60th person until we have reached the desired sample size. The choice of $k = 60$ is made on the basis that the population is 60 times as large as the desired sample size.

FIGURE 4.2

The systematic sample is a probability sample in which every kth person or element in the population is included in the sample. In this example, $k = 3$.

In the context of a shopping mall, we might simply elect to survey every 10th person passing a given point in the mall during the entire day. Such a process is illustrated in Figure 4.2, except with $k = 3$ as our selection rule. As in the simple random sample, we have a rule to follow, and personal judgment is not allowed to enter the picture. The use of such rules, and their elimination of personal judgment or bias in the selection process, is basic to all probability sampling procedures. If such rules are not followed, we could find ourselves avoiding interviews with mall shoppers who are carrying heavy bags, who appear to be in a hurry, or who in some other way cause us to be either less likely or more likely to interview them.

One potential problem with the systematic sample is **periodicity,** a phenomenon where the order in which the population appears happens to include a cyclical variation in which the length of the cycle is the same as the value of k that we are using in selecting the sample. For example, if daily sales receipts are arranged in order from January 1 to December 31, it would not be wise to select $k = 7$ in choosing a systematic sample. Under these conditions, we could end up with a sample consisting entirely of Mondays, Wednesdays, or some other given day of the week. Likewise, a production machine may have some quirk by which every 14th unit produced happens to be defective. If we were to use $k = 7$ or 14 in selecting our systematic sample, we would end up with a sample that is not representative of production quality. Although such periodicity is not a common problem, the possibility of its existence should be considered when undertaking a systematic sample.

Other Probability Sampling Techniques

A number of more advanced probability sampling methods are also available. Among them are the stratified and cluster sampling approaches.

In the **stratified sample,** the population is divided into layers, or *strata*; then a simple random sample of members from each stratum is selected. Strata members have the same percentage representation in the sample as they do in the population, so this approach is sometimes referred to as the *proportionate* stratified sample. Part A of Table 4.2 is an example of the stratified sampling approach in which the stratification has been done on the basis of education level. The selection of education as the variable on which to stratify could be important if our purpose is to measure the extent to which individuals provide charitable contributions to the performing arts.

A sample can also be stratified on the basis of two or more variables at the same time. Part B of Table 4.2 is similar to part A, except that the stratification is based on two variables. This is referred to as a *two-way* stratified sample since

TABLE 4.2

Examples of stratified sampling using one variable (part A) and two variables (part B) for stratification. In each case, strata members have the same percentage representation in the sample as they do in the population, and sample members from each stratum are selected through the use of probability sampling.

A. Stratified Sample Based on One Variable: Highest Education Level

	Population: U.S. Employed Civilians	**Stratified Sample of n = 1000**
<high school graduate	10,466,000	95
high school graduate	34,011,000	307
some college	31,298,000	283
college graduate	34,874,000	315
	Total = 110,649,000	Total = 1000

Persons with a college diploma comprise 31.5% of the population. They will also comprise 31.5% of the sample.

B. Stratified Sample Based on Two Variables: Education and Gender

	Population: U.S. Employed Civilians		**Stratified Sample of n = 1000**	
	Male	**Female**	**Male**	**Female**
<high school graduate	6,397,000	4,069,000	58	37
high school graduate	18,188,000	15,823,000	164	143
some college	15,613,000	15,685,000	141	142
college graduate	19,033,000	15,841,000	172	143
	Total = 110,649,000		Total = 1000	

Males with some years of college comprise 14.1% of the population. They will also comprise 14.1% of the sample.

Source: Based on data in Bureau of the Census, U.S. Department of Commerce, *Statistical Abstract of the United States 2002*, p. 385.

the sample has been forced to take on the exact percentage breakdown as the population in terms of *two* different measurements or characteristics.

The stratified sample forces the composition of the sample to be the same as that of the population, at least in terms of the stratification variable(s) selected. This can be important if some strata are likely to differ greatly from others with regard to the variable(s) of interest. For example, if you were doing a study to determine the degree of campus support for increased funding for women's sports programs, you might be wise to use a sample that has been stratified according to male/female so that your sample has the same percentage male/female breakdown as the campus population.

The **cluster sample** involves dividing the population into groups (e.g., based on geographic area), then randomly selecting some of the groups and taking either a sample or a census of the members of the groups selected. Such an approach is demonstrated in Figure 4.3. As in the stratified sample, members of the population may not have the *same* probability of inclusion, but these probabilities could be determined if we wished to exert the time and effort.

Nonprobability Sampling

In nonprobability sampling, not every unit in the population has a chance of being included in the sample, and the process involves at least some degree of

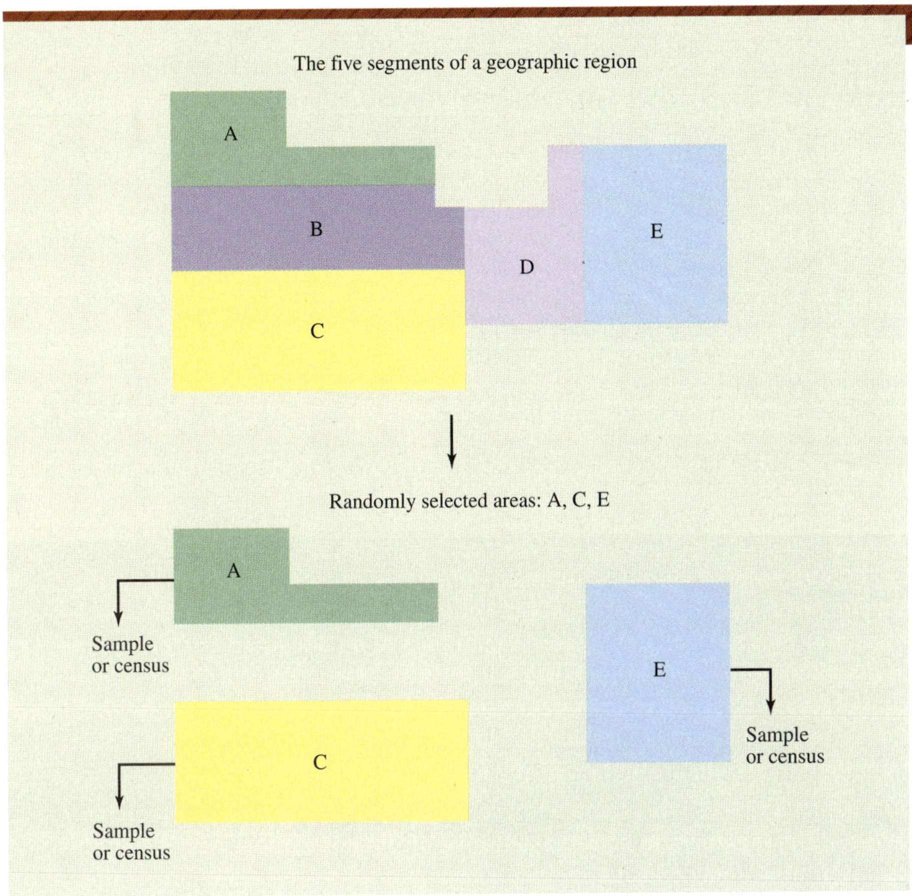

FIGURE 4.3

The cluster sample involves dividing the population into groups or areas, then randomly selecting some of the groups or areas and taking either a sample or a census of each. Shown here is an example of how this approach would apply to a city that has been divided into areas A through E.

personal subjectivity instead of following predetermined, probabilistic rules for selection. Although less important for the purposes of this text, such sampling can be useful in small-scale, exploratory studies where one wishes simply to gain greater familiarity with the population rather than to reach statistical conclusions about its characteristics. We will briefly examine four types of nonprobability samples:

CONVENIENCE SAMPLE Members of such samples are chosen primarily because they are both readily available and willing to participate. For example, a fellow student may ask members of your statistics class to fill out a questionnaire for a study she is doing as an assignment in her sociology course. So-called person-on-the-street interviews by television reporters are often identified as "random," although they are usually conducted at shopping areas near the station or at other locations selected primarily because they happen to be convenient for the interviewer and camera crew.

QUOTA SAMPLE This is similar to the stratified probability sample described previously, except that members of the various strata are *not* chosen through the use of a probability sampling technique. For example, if the sample is to include 30 undergraduates, all 30 could arbitrarily be selected from the marching band. Although it looks good on paper (e.g., the male/female percentage breakdowns are the same for the population and the sample), this type of sample is far inferior to the stratified sample in terms of representativeness.

(*continued*)

STATISTICS IN ACTION 4.2

A Sample of Sampling by Giving Away Samples

A sampling approach used by a drug company ended up giving a federal judge a headache. The company had claimed its brand of pain reliever to be recommended by 70,000 doctors.

It seemed that the firm had provided free samples of its product to 250,000 physicians, with about half of them returning a postcard that entitled them to receive still more free samples. The company then surveyed 404 of those who had returned the postcard, and 65% of them said they recommended use of this brand of pain reliever. Applying the 65% figure to all of those who had returned the postcard, the company came up with its claim that the product was recommended by 70,000 doctors.

Although some might view the company's claim as technically correct, the judge felt it was misleading and forbade the company from engaging in further deceptive advertising.

Source: Based on William Power, "A Judge Prescribes a Dose of Truth To Ease the Pain of Analgesic Ads," *The Wall Street Journal,* May 13, 1987, p. 33.

PURPOSIVE SAMPLE In this sample, members are chosen specifically because they're *not* typical of the population. For example, the maker of a new pocket calculator may submit prototypes to members of an advanced doctoral seminar in mathematics at the Massachusetts Institute of Technology. If these individuals cannot figure out how to operate it, the calculator should obviously be simplified. Conversely, if a sample of individuals known to have IQs below 80 are able to operate it, it should present no problem to the general population.

JUDGMENT SAMPLE This sample is selected on the basis that the researcher *believes* the members to be representative of the population. As a result, the representativeness of such a sample is only as good as the judgment of the person who has selected the sample. The president of a manufacturing firm may really believe that products produced by the 11:00 P.M.–7:00 A.M. shift are typical of those produced by all three work shifts. This belief is all that is necessary for a sample of output from the 11–7 shift to fall into the category of a judgment sample.

As this chapter has pointed out, statistically generalizing to the population on the basis of a nonprobability sample is not valid. Statistics in Action 4.2 describes how a company used a rather bizarre nonprobability sampling approach, then tried to sell the results as accurately describing the population.

EXERCISES

4.38 Attendees at an industrial trade show are given the opportunity to fill out a card that entitles them to receive a free software package from the sponsoring firm. The cards are placed in a bin, and 30 winners will be drawn. Considering the 30 winners as a sample from all of the persons filling out such cards, is this a probability sample or a nonprobability sample? What specific type of probability or nonprobability sample does this procedure represent?

4.39 To obtain a sample of the invoices over the past year, a clerk uses the computer to select all invoice numbers that are a multiple of 13. If the invoice numbers are sequential, beginning with 1, what type of sample does this represent?

4.40 Differentiate between a judgment sample and a convenience sample.

4.41 In what way are the quota sample and the stratified sample similar? In what way are they different? What effect does this difference have on the applications for which such samples might be appropriate?

4.42 What is the difference between a probability sample and a nonprobability sample? Which type is necessary if we wish to statistically generalize from the sample to the population?

4.43 For each of the following situations, would you recommend a sample or a census? Explain your reasoning in each case.

a. A purchasing agent has just received a shipment of shock-resistant watches and wants to find out approximately how far they can be dropped onto a concrete surface without breaking the crystal.
b. General Mills wants to learn the age, gender, and income characteristics of persons who consume Cheerios breakfast cereal.
c. The producers of NBC's *Tonight Show* want to find out what percentage of TV viewers recognize a photo of host Jay Leno.
d. A small company wants to determine whether U.S. companies that manufacture nuclear submarines might be interested in a new technique for purifying air when such craft are submerged.

4.44 Using the random number table in Appendix A, select a simple random sample of 5 from the 50 states of the United States.

4.45 What is meant by *periodicity*, and how can this present a problem when taking a systematic sample? Provide an example of a systematic sampling situation where periodicity could be a factor.

4.46 The manufacturer of a fast-drying, highly effective glue is concerned about the possibility that some users might accidentally glue their fingers together, raising potential legal problems for the firm. Could purposive sampling help determine the extent to which people might use the glue improperly? If so, describe the composition of at least one purposive sample that could be employed for this purpose.

(DATA SET) *Note:* Exercises 4.47–4.49 require a computer and statistical software capable of selecting random or systematic samples.

4.47 Given the 250 household electricity expenditures in Exercise 3.65 (file **XR03065**), generate a simple random sample consisting of 30 data values.

4.48 Given the honeymoon expenditures for the 300 couples in Exercise 3.66 (file **XR03066**), generate a simple random sample consisting of 30 data values.

4.49 A study found that the average American buys 4.6 movie tickets per year, compared to an average of 3.2 in Ireland. Assuming that the data values in file **XR04049** represent the number of movie ticket purchases reported by 100 American respondents. Source: spartatheater.com, June 28, 2009.

a. Select a simple random sample of 20 data values. Compute the sample mean.
b. Select another simple random sample of 20 data values, and compute the mean.
c. To two decimal places, compare the mean of the sample in part (a) with the mean of the sample in part (b). Are the values identical? Would you expect them to be? How do they compare to the mean of the 100 data values from which the samples were drawn? (We will learn more about this topic in Chapter 8, Sampling Distributions.)

SUMMARY (4.8)

Types of business research studies

Business research studies can be categorized according to the objective involved and may be exploratory, descriptive, causal, or predictive. Prior to beginning the study, the problem or question should be defined in specific terms that can be answered by research.

Survey research methods, data, and error sources

In business research, the data collected can be either primary or secondary. Primary data are generated by the researcher for the problem or situation at hand, while secondary data have been collected by someone else for some other purpose. Secondary data can be classified as internal or external, depending on whether the data have been generated from within or outside the researcher's firm or organization. Primary data tend to require more time and expense, but have the advantage of being more applicable to the research problem or situation.

Survey, experimental, and observational research are among the most popular methods for collecting primary data. Surveys may be carried out by mail questionnaires, personal interviews, and telephone interviews. The questionnaire, or data collection instrument, will contain any of three types of questions: multiple choice, dichotomous, and open-ended. In survey research, there are three major sources of error: sampling error, response error, and nonresponse error.

• Experimental and observational studies

Experimental studies attempt to identify cause-and-effect relationships between variables. Evaluation of the results of an experimental study involves the assessment of both internal and external validity. Observational studies emphasize what individuals do rather than what they say, with behavior observed and recorded on a data collection instrument resembling the questionnaire used in survey research.

• Internal secondary data and data mining

Internal sources of secondary data may take a variety of forms within the researcher's firm or organization, with accounting and financial records being especially useful. The many sources of secondary data include government agencies, publications, commercial suppliers, and trade associations. The data warehouse is a large collection of data put together to apply the sophisticated analytical tools of data mining to identify patterns and interrelationships in the data. More than ever, the Internet is an indispensable source of information and expertise.

• Sampling can be better than a census

When a population contains only a small number of elements, a complete census of its members is easily done. In practice, however, most studies involve relatively large populations for which the taking of a sample can provide satisfactory results much more quickly and at considerably lower cost. In some cases, the measurement process can be destructive, necessitating the use of sampling instead of a census. Because of the time and money it saves, a sample can even be more accurate than a census. This occurs because some of the limited funds available for the study can be used to improve the measurement process itself, thereby reducing nonsampling error.

• Sampling and nonsampling errors

Error is the difference between the actual value of the population parameter (e.g., μ or π) and the value of the sample statistic (e.g., $\bar{x}$ or p) that is being used to estimate it. Sampling error is a random, nondirectional error that is inevitable whenever a sample is taken instead of a census, but the amount of this error can be estimated whenever a sample is of the probability type. For probability samples, sampling error can be reduced by increasing the sample size. Nonsampling error is a tendency toward error in one direction or the other and can be present even if a complete census is taken.

• Probability and nonprobability sampling

In probability sampling, each person or element in the population has some (nonzero) known or calculable chance of being included in the sample. In the simple random sample, each person or element has the same chance for inclusion. Other probability samples discussed in the chapter are the systematic sample, the stratified sample, and the cluster sample.

In nonprobability sampling, not everyone in the population has a chance of being included in the sample, and the process involves at least some degree of personal subjectivity instead of predetermined, probabilistic selection rules.

Such samples can be useful in small-scale, exploratory studies where the researcher does not intend to make statistical generalizations from the sample to the population. Nonprobability techniques discussed in the chapter include convenience, quota, purposive, and judgment sampling.

CHAPTER EXERCISES

4.50 The Sonic Travel Agency has not kept pace with the current boom in family vacation air travel between the agency's northern city and sunny Florida. For each of the following, indicate whether the associated data are primary or secondary. If the data are secondary, further indicate whether they are internal or external.

a. Sonic's research department conducts a survey of past customers to determine their level of satisfaction with the agency's service.
b. The Florida Bureau of Tourism sends travel agencies across the nation the results of a study describing the characteristics of winter vacationers to the state, including where and how long people tended to stay.
c. To get a better grasp of the problem, the president of Sonic examines the company's annual reports for the past 10 years.

4.51 A labor union official has considered three possible negotiating stances that the union might take during upcoming contract negotiations. Members of the rank and file, on the other hand, could differ in terms of the approach they would most like to see. The union official is concerned with properly representing his membership in talks with company representatives. For this situation, formulate a specific question that research might be able to answer.

4.52 For each of the following report titles, indicate whether the study involved was exploratory, descriptive, causal, or predictive, and explain your reasoning.

a. "The Popularity of Bowling as a Participant Sport in Ohio"
b. "The Effects of TV Violence on Crime in Small Towns"
c. "A Preliminary Investigation of the Market for Pet Foods"
d. "Family Vacationing in the 21st Century"

4.53 In an attempt to measure the availability of component parts for a new product line, a manufacturer mails a questionnaire to 15 of the 25 companies that supply these components. Past experience with similar surveys has shown that such companies tend to exaggerate their ability to supply desired components on time and in the quantities desired. Of the companies surveyed, only 8 complete and return the questionnaire. Given this scenario, identify possible sources of sampling error, response error, and nonresponse error.

4.54 To collect information on how well it is serving its customers, a restaurant places questionnaire cards on the tables, with card deposit boxes located near the exit.

a. Do you think that nonresponse error might influence the results of this effort? In what way(s)?
b. In addition to being a source of information to management, are the cards providing any other benefits? In what way(s)?

4.55 A mail survey of junior executives is designed to measure how much time they spend reading *The Wall Street Journal* each day and how they feel about the newspaper's content and layout. Formulate one question of each of the following types that might be included in the questionnaire: (a) multiple-choice; (b) dichotomous; (c) open-ended.

4.56 The curator of a fossil museum finds there were 1450 visitors in May and 1890 visitors in June. At the beginning of June, he had hired Bert McGruff, a popular local athlete, to serve as a tour guide. Mr. McGruff, pointing out the increase in attendance, has demanded a raise in pay. Considering this situation as an experiment:

a. Identify the dependent and independent variables in this experiment.
b. Are there any extraneous variables that might have influenced this experiment?
c. Given your response to part (b), evaluate the internal validity of the experiment. Does Bert's request for a raise seem to be justified?

4.57 Timetech, Inc., has been experimenting with different approaches to improving the performance of its field sales force. In a test involving salespersons in Maine, the company found that giving sales leaders a free snowmobile caused a sales increase of 25% in that state. The experiment in Maine was so successful that company executives are now planning to increase sales nationwide by awarding free snowmobiles to sales leaders in all 50 states. Critique the external validity of this experiment.

4.58 A security officer uses a one-way mirror to watch department store shoppers and identify potential shoplifters. In addition to observing whether a person is attempting to steal merchandise, what other personal or behavioral characteristics can be observed in this setting?

4.59 Using either the Internet or the *Encyclopedia of Associations*, identify a group or organization that might be able to supply information on
a. the popularity of bowling.
b. child abuse in the United States.
c. industrial accidents.
d. antique and classic cars.

4.60 When customers purchase goods at a Radio Shack store, they are typically asked for their name and phone number so their purchase information can be stored. This would be one portion of the vast internal and external data available to the company. How might a company like Radio Shack utilize data warehousing and data mining?

4.61 In general, what are some of the considerations in evaluating the suitability of data that have been generated by someone else? For what types of sources should secondary data be regarded with special skepticism?

4.62 Shirley will be graduating soon and is anticipating job offers from several national-level corporations. Depending on the company she joins, she could end up living in any of four different locations in the United States.
a. What are some of the ways that Shirley could utilize the Internet for information to help her decide which company she would most prefer to join?
b. What are some of the ways that company personnel could use the Internet to find out more about Shirley and her suitability for their company? Would you consider any of these methods to be unethical? If so, which one(s) and why?

Note: Exercises 4.63–4.66 require a computer and access to the Internet. Web addresses were accurate at this writing. If a site address is no longer applicable, use a search engine (e.g., google.com) to identify the company or organization's new home page.

4.63 Visit the U.S. Bureau of the Census (census.gov), and find the following descriptive statistics for Kalamazoo County, Michigan:
a. Of those who are 25 or older, the percentage who are high school graduates
b. The number of persons who are 65 or older
c. The median household income

4.64 Using the U.S. Bureau of the Census (census.gov) as your starting point, use the links it provides and find the most recent population total for Canada.

4.65 One of the capabilities on the Internet is called "reverse lookup." Enter the residential telephone number for someone you know. Evaluate the accuracy of the name and address provided, along with the validity of the map that can be generated as additional information. Was the reverse lookup successful in identifying the person's e-mail address as well?

4.66 Visit google.com and search message boards for postings that pertain to Sony digital cameras. Given the kinds of discussions that take place in such settings, what benefits could Sony obtain by having someone within the company "lurk" (i.e., observe without participating) in one or more groups where digital cameras are being discussed?

4.67 A researcher wants to find out to what extent local residents participate in the recreation and entertainment activities available in their community. She selects a simple random sample of numbers from the telephone directory and proceeds to call these numbers between 6 P.M. and 9 P.M. If the line is busy or if there is no answer, she uses the next number on her list as a replacement. Comment on both the randomness and the representativeness of the sample of persons who end up being interviewed.

4.68 Researchers at a university with an enrollment of 12,000 take a census of the entire student population, asking, "Shoplifting is not only illegal, it raises the prices that we all have to pay. Have you ever shoplifted from the university bookstore?" Based on the preceding, discuss the possible presence of (a) sampling error and (b) nonsampling error.

4.69 According to the U.S. Bureau of the Census, 27% of U.S. adults are college graduates. In a stratified sample of 200 U.S. adults, approximately how many persons in each of the following categories should be questioned: (a) college graduate and (b) not a college graduate?
Source: *Statistical Abstract of the United States 2009*, p. 147.

4.70 To test the effectiveness of a new type of plastic handcuffs, a law enforcement agency puts the cuffs on a sample of 30 persons who belong to the local weightlifting club. None is able to break free from the cuffs. What type of sampling technique does this represent? Explain your reasoning.

4.71 Of a company's 1000 employees, 200 are managers, 700 are factory employees, and 100 are clerical. In selecting a sample of 100 employees, a researcher starts at the beginning of the alphabet and continues until she has collected the names of 20 managers, 70 factory employees, and 10 clerical workers. What type of sampling technique does this represent? Explain your reasoning.

4.72 A college of business has 20 faculty members in Accounting, 30 in Marketing, 20 in Management, 15 in Finance, and 15 in Information Systems. The dean of the college asks her secretary to arbitrarily select 4, 6, 4, 3, and 3 persons from the respective departments to serve on a faculty committee. Is this a probability sample or a nonprobability sample? What specific type of probability or nonprobability sample does this procedure represent?

(**DATA SET**) *Note:* Exercises 4.73 and 4.74 require a computer and statistical software capable of selecting simple random samples.

4.73 In 2006, the average U.S. production employee worked 38.9 hours per week. An assistant in the human resources department of Acme Eyebolts, Inc., is curious as to how the workers in her company compared to this figure during the first week of this month. The assistant will be reaching her conclusions based on a simple random sample of 30 data values out of the 300 Acme workers' weekly hours listed in file **XR04073**. Help the assistant by generating the simple random sample and computing its mean. Based on the sample, compare the Acme workweek with that of production workers nationally.
Source: *Statistical Abstract of the United States 2009*, p. 409.

4.74 Unknown to a quality assurance technician, the tensile strengths (in pounds per square inch, psi) for all 500 heavy-duty construction bolts in a recent shipment are as listed in file **XR04074**. Because a bolt must be broken to measure its strength, the testing process is destructive. The technician plans to collect a simple random sample of 20 bolts, then measure how much tension each one withstands before it breaks. Generate the simple random sample and compute its mean breaking strength. Assuming that the bolt manufacturer has advertised that, on average, such bolts will withstand 10,000 psi, refer to your sample result in commenting on the manufacturer's claim. (We will discuss the use of sample data in evaluating claims in much greater detail in Chapter 10, Hypothesis Testing.)

INTEGRATED CASE

Thorndike Sports Equipment—Video Unit Two

The first Thorndike video segment takes us back in time, to when Ted got his business degree and first joined his grandfather's company. At the time of this video, Luke isn't much in need of data mining—to reach the things at the bottom of his document heaps, he needs good old-fashioned *shovel* mining.

To get a feel for the industry in which his grandfather operates, Ted wants to find out more about the sporting goods manufacturing business and collect some data on the competition that Thorndike Sports Equipment faces with some of its products, like racquetball and tennis racquets. Please give Ted some help by getting onto the Internet and providing him with the information requested here. Internet search engines and strategically selected key words will come in handy.

1. Approximately how much money do Americans spend on sports equipment annually? If possible, identify which types of sports equipment are the biggest sellers.
2. Who are some of the leading manufacturers of racquetball and tennis racquets? If possible, find out their relative importance in terms of manufacturing volume or retail sales of these products.
3. For two of the companies that manufacture equipment for racquet sports, visit the companies' websites and, if possible, read their most recent annual report. For each company, what have been the trends in overall company sales and profitability over the most recent years reported?
4. Select a company that makes racquetball racquets, then search message boards for comments that group participants may have made about the quality or reputation of racquets carrying that company's brand name.

SEEING STATISTICS: APPLET 3

Sampling

When this applet loads, it contains 100 blue circles. As you click on any 10 of them, their blue covers disappear and each circle changes to either red or green. After you have taken a sample of 10, click on **Show All** and the display at the top will show both the proportion of your sample that were red and the actual proportion of the population that were red. Of course, because of sampling error, it is very likely that your sample proportion will be different from the actual population value. Click the **Reset** button to return the circles to blue so you can try another sample.

Applet Exercises

3.1 Take 20 different samples and for each one, record both the sample proportion and the actual population proportion. How many times was your sample proportion less than the actual population value? Equal to the actual population value? Greater than the actual population value?

3.2 For each of the 20 different samples from Applet Exercise 3.1, calculate the difference between your sample proportion and the actual population proportion. (Remember to retain the positive or negative sign associated with each difference.) What was the average difference for the samples you took?

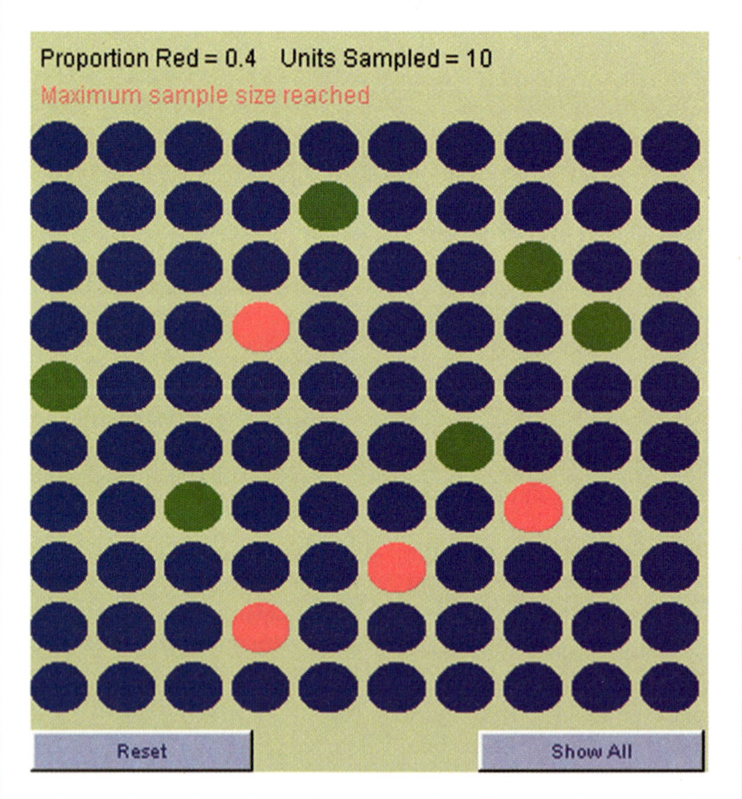

CHAPTER 5

Probability: Review of Basic Concepts

BUMPING UP YOUR CHANCES OF GETTING "BUMPED"

Among air travelers, it's common knowledge that airlines overbook flights to make up for no-shows. For the more clever air travelers, it becomes a challenge to get "bumped" from the flight in return for a free ticket, a voucher, or some other monetary goody. During 2007, more than 622,000 passengers agreed to give up their seats on overbooked flights.

There are fewer "no-shows" than in years past, and improved forecasting methods have helped the airline companies more closely predict how many people will need to be "bumped" from a given flight. However, to improve your chances of being bumped and rewarded with free travel or other compensation, experts advise to travel during peak travel periods, book flights that are already nearly full, and favor heavily booked flights on relatively small-bodied aircraft, such as the 727s and 737s. It also helps to arrive early, ask the agent if the flight is fully booked, and tell her you are willing to consider being bumped.

In 2007, about one of every 830 passengers was "bumped" from a flight. To boost your likelihood of becoming a "bumpee," follow these tips and pick up some more from the sources cited below.

Sources: Bestfares.com, July 26, 2006; BudgetTravel.com, July 26, 2006; bts.gov, June 28, 2009.

Facing a low-probability event

LEARNING OBJECTIVES

After reading this chapter, you should be able to:

- Understand the basic concept of probability, including the classical, relative frequency, and subjective approaches.
- Interpret a contingency table in terms of the probabilities associated with the two variables.
- Determine the probability that an event will occur.
- Construct and interpret a tree diagram with sequential events.
- Use Bayes' theorem and additional information to revise a probability.
- Determine the number of combinations and the number of possible permutations of *n* objects *r* at a time.

5.1 INTRODUCTION

Uncertainty plays an important role in our daily lives and activities as well as in business. In some cases, uncertainty even helps heighten our interest in these activities. For example, some sports fans refuse to watch the evening news whenever they're going to view a taped replay of a sports event that occurred earlier in the day—the replay is more exciting if they don't know who won the game.

In a business context, investment counselors cannot be sure which of two stocks will deliver the better growth over the coming year, and engineers try to reduce the likelihood that a machine will break down. Likewise, marketers may be uncertain as to the effectiveness of an ad campaign or the eventual success of a new product. The following situations provide other examples of the role of uncertainty in our lives:

- In the last game of the World Series, the home team is down one run with the bases loaded and two out in the bottom of the ninth. The pitcher, whose batting average is only 0.105, is due up. The manager sends in a pinch hitter with a 0.320 batting average to hit for the pitcher.
- Eddie is standing at an intersection, facing the "Don't Walk" light in a city where jaywalking carries a $20 fine. After looking around to be sure there are neither cars nor police, he crosses the street.
- Arthur is 75 years old, has had three heart attacks, and works as a bomb disposal expert with his local police department. When he applies for life insurance, he is turned down as an unacceptable risk.

In each of these cases, chance was involved. The baseball manager selected the batter who was more likely to get a hit. Eddie was trying to reduce the chance that something undesirable would happen to him. Arthur's life insurance company felt that his age, health, and occupation greatly reduced the likelihood of his surviving for many more years.

We'll soon offer several more precise definitions, but for now let's just consider probability as "the chance of something happening." Although humans have been dealing with probabilities ever since cave dwellers discovered their chances for survival would be better if they carried spears, only within the past couple of centuries have probabilities really been a focal point of mathematical attention. Gamblers were among the first to take an interest in probabilities, followed by the insurance industry and its mortality tables that expressed the likelihood that a person of a given age would live an additional year. More recently, probabilities have proven vital in many ways to both business and the social sciences.

The next section considers the three basic approaches to probability. This will be followed by a discussion of unions and intersections of events, concepts that are fundamental to describing situations of uncertainty. Sections 5.4 and 5.5 will examine the rules for calculating probability under different circumstances.

Sometimes, it is useful to *revise* a probability on the basis of additional information that we didn't have before. This will be the subject of Section 5.6. Finally, Section 5.7 will explain how permutations and combinations can help us calculate the number of ways in which things can happen.

PROBABILITY: TERMS AND APPROACHES (5.2)

Basic Terms

The discussion in this and later sections relies heavily on four important, interrelated definitions:

EXPERIMENT An activity or measurement that results in an outcome

SAMPLE SPACE All possible outcomes of an experiment

EVENT One or more of the possible outcomes of an experiment; a subset of the sample space

PROBABILITY A number between 0 and 1 that expresses the chance that an event will occur

The *event* in the preceding definitions could be almost anything of interest to the observer—e.g., rain tomorrow, a 30-point increase in the Dow Jones Industrial Average (DJIA) next week, or even that you'll take time out for a soft drink before finishing this page. As an example application, consider the possibility that the DJIA will increase by at least 30 points during the next week:

- **Experiment** Next week's trading activities on the New York Stock Exchange
- **Sample space** The sample space consists of two possible outcomes: (1) the DJIA goes up by at least 30 points, and (2) it does not.
- **Event** A = the DJIA goes up by at least 30 points next week.
- **Probability** The chance that the DJIA will increase by at least 30 points next week. Represented by $P(A)$, the probability for this and any other event will have the following properties:

$0 \leq P(A) \leq 1$	For any event, the probability will be no less than 0 and no greater than 1.
$P(A) + P(A') = 1$	Either the event will occur (A) or it will not occur (A'). A' is called the **complement** of A.

As indicated above, the probability for any event will be a number between 0 and 1. These extreme values represent events that are either impossible or certain. For example:

$P(B) = 0$ if B = you awake tomorrow morning and find yourself on Jupiter.

$P(C) = 1$ if C = it will rain someplace in the world tomorrow afternoon.

The Classical Approach

The classical approach describes a probability in terms of *the proportion of times that an event can be theoretically expected to occur:*

Classical probability:

For outcomes that are equally likely,

$$\text{Probability} = \frac{\text{Number of possible outcomes in which the event occurs}}{\text{Total number of possible outcomes}}$$

For example, in rolling a 6-sided die, the probability of getting a "three" is

$$\frac{1}{1 + 1 + 1 + 1 + 1 + 1}$$

Out of 6 possible (and equally likely) outcomes, there is one in which a "three" occurs.

Likewise, the probability of tossing a coin and having it land "heads" is $\frac{1}{2}$. In this case there are two possible (and equally likely) outcomes, only one of which includes the event "heads." Because we can determine the probability without actually engaging in any experiments, the classical approach is also referred to as the *a priori* ("before the fact") approach. Classical probabilities find their greatest application in games of chance, but they are not so useful when we are engaged in real-world activities where either (1) the possible outcomes are not equally likely or (2) the underlying processes are less well known.

The Relative Frequency Approach

In the relative frequency approach, probability is *the proportion of times an event is observed to occur in a very large number of trials:*

Relative frequency probability:

For a very large number of trials,

$$\text{Probability} = \frac{\text{Number of trials in which the event occurs}}{\text{Total number of trials}}$$

The relative frequency approach to probability depends on what is known as the **law of large numbers:** Over a large number of trials, the relative frequency with which an event occurs will approach the probability of its occurrence for a single trial. For example, if we toss a coin one time, the result will be either heads or tails. With *just one toss,* our relative frequency for heads will be either 0 or 1. If we toss it a great many times, however, the relative frequency will approach the 0.5 that it is theoretically "supposed to be."

On a more serious note, the life insurance industry relies heavily on relative frequencies in determining both candidate acceptability and the premiums that must be charged. Life insurance premiums are based largely on relative frequencies that reflect overall death rates for people of various ages. Table 5.1 is a portion of a mortality table of the type used by the life insurance industry. Of particular interest:

1. The death rate starts out relatively high for newborns, decreases slightly through the younger years, then steadily increases until (at the highest figure

	Deaths per 1000			Deaths per 1000			Deaths per 1000	
Age	Male	Female	Age	Male	Female	Age	Male	Female
0	4.18	2.89	35	2.11	1.65	70	39.51	22.11
5	0.90	0.76	40	3.02	2.42	75	64.19	38.24
10	0.73	0.68	45	4.55	3.56	80	98.84	65.99
15	1.33	0.85	50	6.71	4.96	85	152.95	116.10
20	1.90	1.05	55	10.47	7.09	90	221.77	190.75
25	1.77	1.16	60	16.08	9.47	95	329.96	317.32
30	1.73	1.35	65	25.42	14.59			

Source: American Council of Life Insurance, *1999 Life Insurance Fact Book*, pp. 182–186.

TABLE 5.1

Life insurance companies use mortality tables to predict how many policy holders in a given age group will die each year. When the death rate is divided by 1000, the result is the probability that a typical policyholder in the age group will pass away during the year.

shown, 95 years) the likelihood of death within the coming year is about one chance out of three.

2. Each death rate can be converted to a probability by dividing by 1000. The resulting probability isn't necessarily the exact probability of death for a specific individual but rather the probability for a "typical" individual in the age group. For example, a typical male policyholder of age 95 would have a 329.96/1000, or a 0.32996, probability of dying within the coming year.
3. Examination of male versus female death rates shows the rate for males is higher *for every age group*. Especially noteworthy is the drastic increase in male death rates between the ages of 10 and 20, a period during which female death rates show a much smaller increase.

The Subjective Approach

The subjective approach to probability is judgmental, representing *the degree to which one happens to believe that an event will or will not happen.* **Subjective probabilities** can also be described as hunches or educated guesses. It has been estimated that wagers in 2000 amounted to \$866 billion for all forms of gambling, ranging from casinos to bingo.[1]

Individuals place such bets only if they think they have some chance of winning. In a more negative direction, insurance customers paid \$9.4 billion in fire insurance premiums during 2006.[2] They would not have paid such sums if they felt their buildings had a zero probability of catching fire.

As the terms "hunch" and "educated guess" imply, subjective probabilities are neither based on mathematical theory nor developed from numerical analyses of the frequencies of past events. This doesn't make them meaningless, however. For example, the following subjective probabilities might be considered reasonable with regard to your activities today:

Event	Subjective Probability
You will be mugged while reading this page.	Low (Prob. $\leq$ 0.001)
You'll talk to someone on the phone today.	Medium (Prob. $\geq$ 0.600)
Your shoes are on the right feet.	High (Prob. $\geq$ 0.999)

[1]Source: OnPointRadio.org, July 26, 2006.
[2]Source: Bureau of the Census, *Statistical Abstract of the United States 2009*, p. 739.

STATISTICS IN ACTION 5.1

Probabilities, Stolen Lawn Mowers, and the Chance of Rain

If you're a thief in need of money, one solution is to steal a lawn mower and sell it on e-Bay. That's exactly what a German perpetrator did in 2009. However, if you've just had your lawn mower stolen, you're going to be shopping for another one. That's exactly what the victim did. In a highly unlikely scenario, the two crossed paths when the victim perused e-Bay, went to examine the lawn mower he intended to purchase, and noticed it was the very same one he used to own. He called the police, got his mower back, and the thief was promptly arrested. The probability of such a rare turn of events is about as tall as grass in the Sahara.

Although precipitation probabilities have been used since the 1960s, people still don't quite get it when the TV meteorologist says "30% chance of rain today." Some think it will rain all day in just 30% of the local area, while others believe it will rain for 30% of the day. Susan Joslyn, of the Psychology Department at the University of Washington in Seattle, found so much misunderstanding in her sampling of 450 undergraduates in the rainy Pacific Northwest that she fears confusion would be even greater in other parts of the country. In terms of probability, what's the meteorologist really saying? The answer: If there is a 30% chance of rain today, there is a 30% probability that it will rain somewhere within the local area sometime during the day. Keep the umbrella handy.

Sources: "Thief Nabbed Selling Stolen Mower to Owner," reuters.com, June 25, 2009; Doyle Rice, "With a Chance of Misapprehension," *USA Today,* June 24, 2009, p. 7D.

As intelligent human beings, we are the sum total of our experiences and memories, and these enable us to make intuitive estimates of the probabilities of events with which we are somewhat familiar. Such probabilities, though subjective, can be important in drawing conclusions and making decisions. Although you may not be able to precisely quantify the probability that the fast-talking salesperson in the plaid suit is really offering you a sound used car, this shouldn't prevent you from assigning a low probability to such an event. Statistics in Action 5.1 provides two interesting examples of subjective probabilities—one slim, the other often misunderstood.

Probabilities and "Odds"

The term **odds** is sometimes used as a way of expressing the likelihood that something will happen. When someone says "the odds for this outcome are three to one," she is really saying that the chance of the event occurring is three times the chance that it will *not* occur. In this case, the odds might be expressed as "three to one" or "3:1." When you see odds listed in reference to an event, you can convert them to a probability as follows:

Conversion from odds to probability, and vice versa:

1. **If the odds in favor of an event happening are *A* to *B*, or *A:B*, the probability being expressed is**

$$\frac{A}{A + B}$$

2. **If the probability of an event is x (with $0 \le x \le 1$), the odds in favor of the event are "x to $(1 - x)$," or "x:$(1 - x)$."**

Odds are typically expressed in terms of the lowest applicable integers. For example, "1.5 to 4" would become "3 to 8." Also, the odds *against* the occurrence of an event are the reverse of the odds in favor. For example, if the odds in favor of an event are 2 to 3, the odds against would be 3 to 2.

Like the other probabilities we've discussed, odds can be classical, based on relative frequencies, or even subjective. For example:

- **Classical** In flipping a fair coin, the probability of heads is 0.5. The resulting odds are "0.5 to 0.5," or "1 to 1."
- **Relative frequency** For example, 16.1% of U.S. households have a cell phone with Internet access, a probability of 0.161 for a randomly selected household.[3] If the event is "has cell phone with Internet access," the odds in favor of the household having this device would be 0.161 to 0.839, or approximately 1 to 5.
- **Subjective** In speculation about the upcoming 2000 Major League Baseball season, an oddsmaker assigned odds that each team would end up winning the World Series. The New York Yankees were given odds of 1 to 3 of winning, which translates into a subjective probability of 1/(1 + 3), or 0.25. The odds given the Milwaukee Brewers were much less favorable—just 1 to 10,000,000.[4]

NOTE

1. As discussed earlier, when presenting the odds *against* an event happening, the odds are listed in the opposite order. For example, the published odds *against* the Yankees were "3:1." For purposes of consistency with our previous examples, we've converted the published version to the "odds for" order.
2. In sports events, points are often added to or subtracted from the score of one of the teams (creating a "point spread") in an attempt to convert the odds to 1:1.

EXERCISES

5.1 The president of a computer manufacturing firm states that there is a 70% chance that industry shipments of notebook computers will double in the next 5 years. Is this a classical probability, a relative frequency probability, or a subjective probability?

5.2 If A = "Air France will begin making daily flights to Pocatello, Idaho, next year," identify the sample space and possible events that could occur. Use your judgment in arriving at an approximate probability for the event(s) and complement(s) you've identified.

5.3 If B = "IBM's net profit will increase next year," identify the sample space and possible events that could occur. Use your judgment in arriving at an approximate probability for the event(s) and complement(s) you've identified.

5.4 Electric meters on private homes in a community register the cumulative consumption of kilowatt-hours with a five-digit reading. If a home in the community were randomly selected, it's possible that the right-hand digit (which spins the fastest) would be a "7." Using the concept of classical probability, what is the probability that the digit is a "7"?

5.5 Regarding Exercise 5.4, how might an observer employ the law of large numbers to verify the classical probability that you identified?

5.6 If a die is rolled one time, classical probability would indicate that the probability of a "two" should be $\frac{1}{6}$. If the die is rolled 60 times and comes up "two" only 9 times, does this suggest that the die is "loaded"? Why or why not?

5.7 A newspaper article reported that Massachusetts had become the second state (after Montana) to prohibit sex to be used as a factor in determining insurance premiums. Considering the information in Table 5.1, would such unisex insurance tend to increase or decrease the life insurance premiums previously paid (a) by men? (b) by women? Source: John R. Dorfman, "Proposals for Equal Insurance Fees for Men and Women Spark Battle," *The Wall Street Journal*, August 27, 1987, p. 23.

5.8 It has been reported that about 35% of U.S. adults attended a sports event during the previous year. What are the odds that a randomly selected U.S. adult attended a sports event during the year? Source: Bureau of the Census, *Statistical Abstract of the United States 2009*, p. 747.

5.9 If the odds are 4:7 that an event will occur, what is the corresponding probability?

[3]Source: Adrienne Lewis, "Early Adopter Technologies," *USA Today*, October 17, 2005, p. 2B.
[4]Source: "World Series Odds," *USA Today*, March 14, 2000, p. 11C.

5.3 UNIONS AND INTERSECTIONS OF EVENTS

As discussed previously, an event is one or more of the possible outcomes in an experiment, and the sample space is the entire set of possible outcomes. Sample spaces and events may be portrayed in either a table or a visual display, and these representations can contribute to our discussion of the probability concepts in sections to follow.

The tabular method involves the contingency table, or cross-tabulation, introduced in Chapter 2. This shows either frequencies or relative frequencies for two variables at the same time. The visual mechanism, called a **Venn diagram,** simultaneously displays both the sample space and the possible events within. We will use the contingency table and the Venn diagram in demonstrating several key terms that are important to the consideration of probability:

MUTUALLY EXCLUSIVE EVENTS If one event occurs, the other cannot occur. An event (e.g., A) and its complement (A') are always mutually exclusive.

EXHAUSTIVE EVENTS A set of events is *exhaustive* if it includes all the possible outcomes of an experiment. The mutually exclusive events A and A' are exhaustive because one of them *must* occur. When the events within a set are both mutually exclusive and exhaustive, the sum of their probabilities is 1.0. This is because one of them *must* happen, and they include *all of the possibilities.* The entries within a relative frequency contingency table are mutually exclusive and exhaustive, and their sum will always be 1.0.

INTERSECTION OF EVENTS Two or more events occur at the same time. Such an intersection can be represented by "A and B," or "A and B and C," depending on the number of possible events involved.

UNION OF EVENTS At least one of a number of possible events occurs. A union is represented by "A or B," or "A or B or C," depending on the number of events.

The contingency tables in parts (1) and (2) of Table 5.2 show frequencies and relative frequencies describing the sex and age of persons injured by fireworks in 1995. The relative frequencies in part (2) are obtained by dividing the corresponding entries in part (1) by 11,449, the total number of persons injured by fireworks during that year. For example, part (1) shows that 3477 of those injured were males under the age of 15. The relative frequency for persons in this sex/age category is 3477/11,449, or 0.304, as shown in the contingency table in part (2).

Referring to the frequencies in part (1) of Table 5.2, we can identify several that illustrate the terms defined earlier in the section:

- **Mutually exclusive events** The victims were either male (event A, 8913 persons) or female (event A', 2536 persons).
- **Exhaustive events** The four mutually exclusive events (A and B), (A and B'), (A' and B), and (A' and B') are also exhaustive since a victim *must* be in one of these sex/age categories.
- **Intersection of events** There were 3477 victims who were males under the age of 15, the "A and B" category.
- **Union of events** There were 10,162 victims who were either male, or under age 15, or both, the "A or B" category. This category includes everyone except females who were 15 or older and its frequency is calculated as $5436 + 3477 + 1249 = 10,162$.

TABLE 5.2

Frequency and relative frequency contingency tables for fireworks victims.

(1) *Frequencies*

			Age		
			B Under 15	*B′* 15 or Older	
Sex	*A*	Male	3477	5436	8913
	A′	Female	1249	1287	2536
			4726	6723	11,449

(2) *Relative Frequencies*

			Age		
			B Under 15	*B′* 15 or Older	
Sex	*A*	Male	0.304	0.475	0.779
	A′	Female	0.109	0.112	0.221
			0.413	0.587	1.000

Source: U.S. Consumer Product Safety Commission, National Electronic Injury Surveillance System.

Figure 5.1 presents a series of Venn diagrams that also describe the age/sex breakdown of the fireworks victims. The information is the same as that contained in the contingency table of Table 5.2, part (1), but in a more visual form. Examples of calculations for finding the number of persons in the union or the intersection of events are shown in the figure, and the results are identical to those based on the contingency table of frequencies.

FIGURE 5.1

Venn diagrams describing persons injured by fireworks.

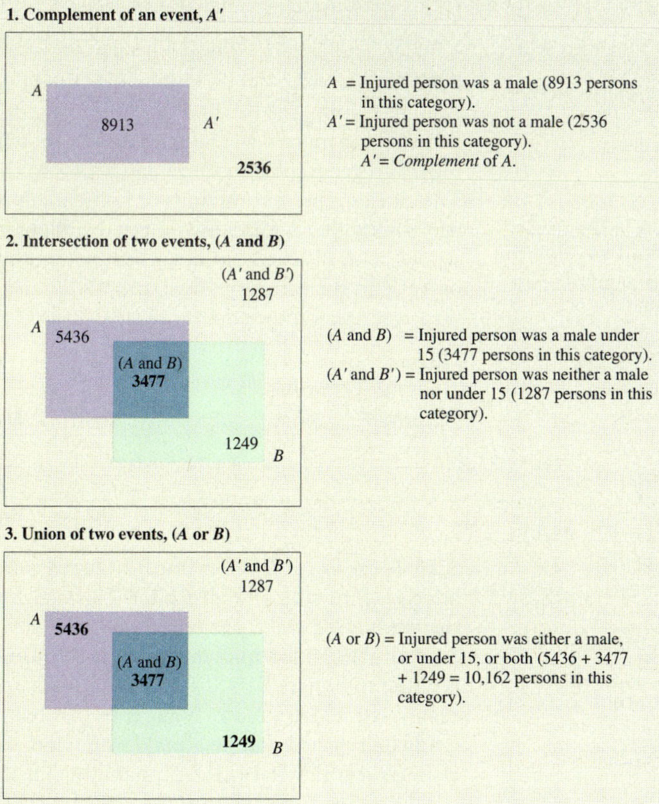

EXERCISES

5.10 A sample space includes the events A, B, and C. Draw a Venn diagram in which the three events are mutually exclusive.

5.11 A sample space includes the events A, B, and C. Draw a Venn diagram in which events A and B are mutually exclusive but events B and C are not mutually exclusive.

5.12 The following contingency table of frequencies is based on a 5-year study of fire fatalities in Maryland. For purposes of clarity, columns and rows are identified by the letters A–C and D–G, respectively. Source: National Fire Protection Association, *The 1984 Fire Almanac*, p. 151.

		Blood Alcohol Level of Victim			
		A	B	C	
Age		0.00%	0.01–0.09%	≥0.10%	
D	0–19	142	7	6	155
E	20–39	47	8	41	96
F	40–59	29	8	77	114
G	60 or over	47	7	35	89
		265	30	159	454

a. For this table, identify any two events that are mutually exclusive.
b. For this table, identify any two events that intersect.

5.13 Using the table in Exercise 5.12, how many victims were in the category described by:
a. $(A \text{ and } A')$?
b. $(C \text{ or } F)$?
c. $(A' \text{ and } G')$?
d. $(B \text{ or } G')$?

5.14 Using the table in Exercise 5.12, what is the probability that a randomly selected victim would have been at least 60 years old and have had a blood alcohol level of at least 0.10%?

5.15 The following table represents gas well completions during 1986 in North and South America. Source: American Gas Association, *1987 Gas Facts*, p. 50.

		D Dry	*D'* Not Dry	
N	North America	14,131	31,575	45,706
N'	South America	404	2,563	2,967
		14,535	34,138	48,673

a. Draw a Venn diagram that summarizes the information in the table.
b. Identify the region of the Venn diagram that represents $(N \text{ and } D)$.
c. Identify the region of the Venn diagram that represents $(N' \text{ and } D')$.
d. Identify the region of the Venn diagram that represents $(N \text{ or } D)$.
e. Identify the region of the Venn diagram that represents $(N' \text{ or } D')$.

5.16 Using the table in Exercise 5.15, assume that one well has been selected at random from the 48,673.
a. What is the probability that the well was drilled in North America and was dry?
b. What is the probability that the well was drilled in South America and was not dry?

5.17 The owner of a McDonald's restaurant in France is considering the possibility of opening up a new franchise at the other end of her town. At the same time, the manager of the Pennsylvania Turnpike is deciding whether to recommend raising the speed limit by 10 miles per hour. Draw a single Venn diagram representing their possible decisions. Explain any assumptions you have made regarding mutual exclusivity of events.

5.18 A shopping mall developer and a historical society are the only two bidders for a local historical landmark. The sealed bids are to be opened at next week's city council meeting and the winner announced. Draw a single Venn diagram describing purchase/nonpurchase of the landmark by these two parties. Explain any assumptions you have made regarding mutual exclusivity of events.

5.19 According to data from the U.S. Energy Information Administration, the 60.0 million U.S. households with personal computers were distributed as shown here with regard to geographic location and access to the Internet. The entries represent millions of households. What is the probability that a randomly selected computer household would be in the category described by (South or Midwest or Yes)? In the category (West and No)? Source: Bureau of the Census, *Statistical Abstract of the United States 2006*, p. 638.

Access to Internet?	Region				
	Northeast	Midwest	South	West	
Yes	9.7	11.8	16.9	12.2	50.6
No	1.2	2.3	3.8	2.1	9.4
	10.9	14.1	20.7	14.3	60.0

5.20 There are 100 males and 120 females in the graduating class of a local high school. Thirty-five percent of the graduating males are on school sports teams, as are 30% of the graduating females. A local businessperson is going to donate $1000 to the favorite charity of a randomly selected graduate. What is the probability that the graduate selected will be a male who is not a member of a sports team? What is the probability that the graduate will be either a female or a member of a sports team?

ADDITION RULES FOR PROBABILITY (5.4)

There are occasions where we wish to determine the probability that one or more of several events will occur in an experiment. The determination of such probabilities involves the use of **addition rules,** and the choice of a given rule will depend on whether the events are mutually exclusive. To illustrate these rules, we will use the two Venn diagrams in Figure 5.2.

When Events Are Mutually Exclusive

When events are *mutually exclusive,* the occurrence of one means that none of the others can occur. In this case, the probability that one of the events will occur is the sum of their individual probabilities. In terms of two events, the rule can be stated as follows:

Rule of addition when events are mutually exclusive:

$$P(A \text{ or } B) = P(A) + P(B)$$

Part (1) of Figure 5.2 is a relative frequency breakdown of colors of sport/compact cars made in North America during the 2004 model year. As the Venn diagram indicates, the possible events are mutually exclusive. For a given vehicle, only one color was applied. Using the letters A and B to identify the colors silver

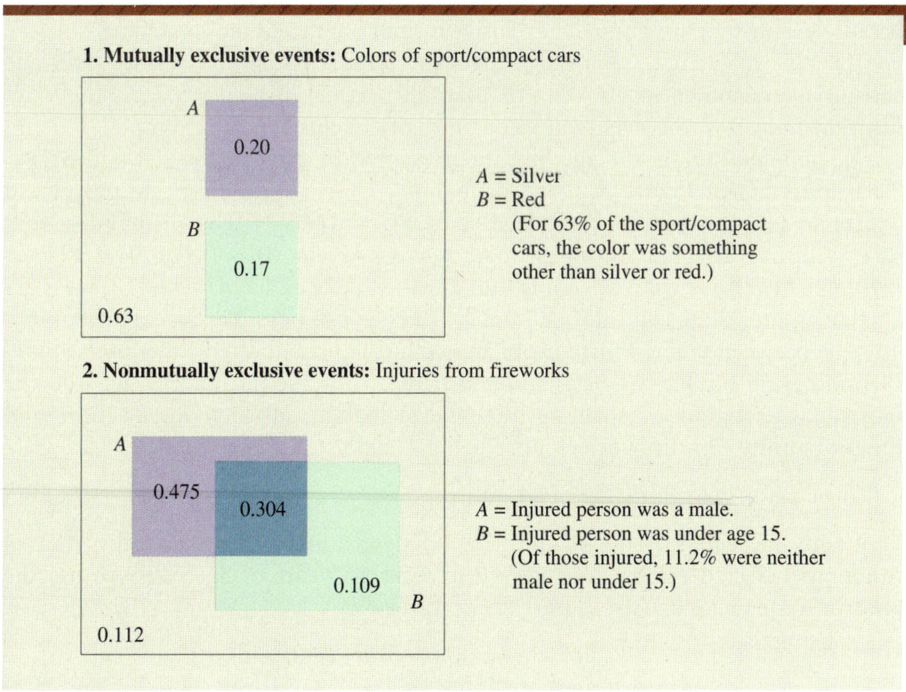

FIGURE 5.2
Venn diagrams showing relative frequencies for (1) colors of sport/compact cars made in North America during the 2004 model year, and (2) persons injured by fireworks.

Sources: Part (1) *Time Almanac 2006*, p. 609. Part (2) U.S. Consumer Product Safety Commission, National Electronic Injury Surveillance System.

and red, respectively, and considering a sport/compact car selected at random from that model year, we can arrive at the following probabilities:

- Probability that the car is either silver or red:

$$P(A \text{ or } B) = P(A) + P(B) = 0.20 + 0.17 = 0.37$$

- Probability that the car is either silver or some color other than silver or red:

  Although we have not specifically identified "neither silver nor red" with a letter designation, it can be expressed as (A' and B'). Using this identification, we can determine the probability that the randomly selected car is either silver or some color other than silver or red:

$$P(A \text{ or } [A' \text{ and } B']) = P(A) + P(A' \text{ and } B') = 0.20 + 0.63 = 0.83$$

When Events Are Not Mutually Exclusive

When events are *not mutually exclusive,* two or more of them can happen at the same time. In this case, the *general rule of addition* can be used in calculating probabilities. In terms of two events, it can be stated as follows:

General rule of addition when events are not mutually exclusive:

$$P(A \text{ or } B) = P(A) + P(B) - P(A \text{ and } B)$$

As befits its name, the general rule of addition can also be applied when the events *are* mutually exclusive. In this case, the final term in the expression will become zero because events that are mutually exclusive cannot happen at the same time.

Relative frequencies for the fireworks victims in the various sex and age categories are shown in the Venn diagram in part (2) of Figure 5.2. As the Venn diagram indicates, the events are not mutually exclusive. For example, a victim can be male *and* under the age of 15, represented by the category (A and B).

In part (2) of Figure 5.2, the probability that a randomly selected victim is a male (event A) is 0.475 + 0.304, or $P(A) = 0.779$. The probability that this randomly selected victim is under age 15 (event B) is 0.109 + 0.304, or $P(B) = 0.413$. If we were to simply add $P(A) = 0.779$ and $P(B) = 0.413$, we'd get a value greater than 1 (a total of 1.192), which is not possible since a probability must be between 0 and 1. The problem is that the probability of the intersection, $P(A \text{ and } B) = 0.304$, *has been counted twice.* To make up for this, we add the individual probabilities, then subtract the probability of their intersection, and the probability of a fireworks victim being male or under age 15 is:

$$P(A \text{ or } B) = P(A) + P(B) - P(A \text{ and } B) = 0.779 + 0.413 - 0.304 = 0.888$$

Using the general rule of addition in determining the probability that a randomly selected victim is female or at least 15 years of age, we will use the relative frequencies shown in part (2) of Table 5.2:

$$P(\text{female}) = P(A') = 0.221$$
$$P(15 \text{ or older}) = P(B') = 0.587$$
$$P(\text{female and [15 or older]}) = P(A' \text{ and } B') = 0.112$$

and

$$P(A' \text{ or } B') = P(A') + P(B') - P(A' \text{ and } B')$$
$$= 0.221 + 0.587 - 0.112 = 0.696$$

EXERCISES

5.21 A financial advisor frequently holds investment counseling workshops for persons who have responded to his direct mailings. The typical workshop has 10 attendees. In the past, the advisor has found that in 35% of the workshops, nobody signs up for the advanced class that is offered; in 30% of the workshops, one person signs up; in 25% of the workshops, two people sign up; and in 10% of the workshops, three or more people sign up. The advisor is holding a workshop tomorrow. What is the probability that at least two people will sign up for the advanced class? What is the probability that no more than one person will sign up? Draw a Venn diagram that includes the possible events and their relative frequencies for this situation.

5.22 A survey of employees at a large company found the following relative frequencies for the one-way distances they had to travel to arrive at work:

Number of Miles (One-Way)

	A ≤5	B 6–10	C 11–15	D 16–20	E 21–30	F ≥31
Relative Frequency	0.38	0.25	0.16	0.09	0.07	0.05

a. What is the probability that a randomly selected individual will have to travel 11 or more miles to work?
b. What is the probability that a randomly selected individual will have to travel between 6 and 15 miles to work?
c. Draw a Venn diagram that includes the relative frequency probabilities in the table.
d. Using the letter identifications provided, calculate the following probabilities: $P(A \text{ or } B \text{ or } E)$; $P(A \text{ or } F)$; $P(A' \text{ or } B)$; $P(A \text{ or } B \text{ or } C')$.

5.23 In 2008, McDonald's had 31,967 restaurants systemwide. Of these, 21,328 were operated by franchisees, 6502 by the company, and 4137 by affiliates. What is the probability that a randomly selected McDonald's restaurant is operated by either a franchisee or an affiliate? Source: McDonald's Corporation, *2008 Annual Report*, p. 44.

5.24 For three mutually exclusive events, $P(A) = 0.3$, $P(B) = 0.6$, and $P(A \text{ or } B \text{ or } C) = 1.0$. What is the value of $P(A \text{ or } C)$?

5.25 It has been reported that the 49,600 employees of United Airlines are distributed among the following corporate functions: Source: United Airlines, *Fact Sheet 2008*.

Pilots	13%	(*A*)
Management	17	(*B*)
Flight attendants	27	(*C*)
Mechanics	11	(*D*)
Other	32	(*E*)
	100%	

For the sample space consisting of United employees:
a. Draw a Venn diagram representing events *A*, *B*, *C*, *D*, and *E*.
b. What is the value of $P(A)$?
c. What is the value of $P(A \text{ or } B)$?
d. What is the value of $P(A \text{ or } D')$?

5.26 In 2003, Entergy Corporation had 2,631,752 electricity customers. Of these, 86.0% were in the residential category (*R*); 11.8% were commercial (*C*); 1.6% were industrial (*I*); and 0.6% were government and municipal (*G*). Source: Entergy Corporation, *2003 Annual Report*.
a. Draw a Venn diagram representing events *R*, *C*, *I*, and *G*.
b. What is the value of $P(R)$?
c. What is the value of $P(C \text{ or } G)$?
d. What is the value of $P(R \text{ or } C')$?

5.27 According to the Bureau of Labor Statistics, there are 777,000 men and 588,000 women in the life, physical, and social science occupations category; and 786,000 men and 204,000 women in the farming, fishing, and forestry occupations category. For these 2,355,000 individuals, what is the probability that a randomly selected individual will either be a male or be in the life, physical, and social sciences category? Source: *Encyclopaedia Britannica Almanac 2006*, p. 835.

5.28 Using the information presented in the table in Exercise 5.12, calculate the following probabilities:
a. $P(A \text{ or } D)$ b. $P(B \text{ or } F)$
c. $P(C \text{ or } G)$ d. $P(B \text{ or } C \text{ or } G)$

5.29 Using the information presented in the table in Exercise 5.15, calculate the following probabilities:
a. $P(D \text{ or } N)$ b. $P(D' \text{ or } N')$
c. $P(D \text{ or } N')$ d. $P(D' \text{ or } N)$

5.5 MULTIPLICATION RULES FOR PROBABILITY

While addition rules are used to calculate the probability that *at least one* of several events will occur, in this section we'll consider rules for determining the probability that two or more of the events will *all* occur. These rules involve several important terms:

MARGINAL PROBABILITY The probability that a given event will occur. No other events are taken into consideration. A typical expression is $P(A)$.

JOINT PROBABILITY The probability that two or more events will all occur. A typical expression is $P(A \text{ and } B)$.

CONDITIONAL PROBABILITY The probability that an event will occur, given that another event has already happened. A typical expression is $P(A|B)$, with the verbal description, "the probability of A, given B." A conditional probability may be determined as follows:

Conditional probability of event A, given that event B has occurred:

$$P(A|B) = \frac{P(A \text{ and } B)}{P(B)}$$ (*Note:* This applies only if $P(B) > 0$.)

The rules that follow are called **multiplication rules;** they determine the probability that two events will *both* happen or that three or more events will *all* happen. There are two multiplication rules, and the one that is applicable will depend on whether the events are *independent* or *dependent:*

INDEPENDENT EVENTS Events are independent when the occurrence of one event has no effect on the probability that another will occur.

DEPENDENT EVENTS Events are dependent when the occurrence of one event changes the probability that another will occur.

When Events Are Independent

When events are independent, their joint probability is the product of their individual probabilities. In the case of two events, the multiplication rule is as follows:

Multiplication rule when events are independent:

$$P(A \text{ and } B) = P(A) \times P(B)$$

To illustrate this rule, we'll use a device called a **tree diagram,** which visually summarizes the occurrence or nonoccurrence of two or more events. The tree diagram is especially useful in visualizing events that occur in sequence. Figure 5.3 illustrates two consecutive tosses of a fair coin. Regardless of the result of the first toss, the probability of heads on the next toss is going to be 0.5; the tree diagram in Figure 5.3 shows all the possibilities, along with their joint probabilities.

For example, the probability of getting heads twice in a row is 0.5×0.5, or 0.25. Likewise, the joint probability of any other given sequence (e.g., $H1 - T2$, $T1 - T2$, or $T1 - H2$) will also be 0.25. Since the diagram includes all four of the possible sequences, the joint probabilities add up to 1.00.

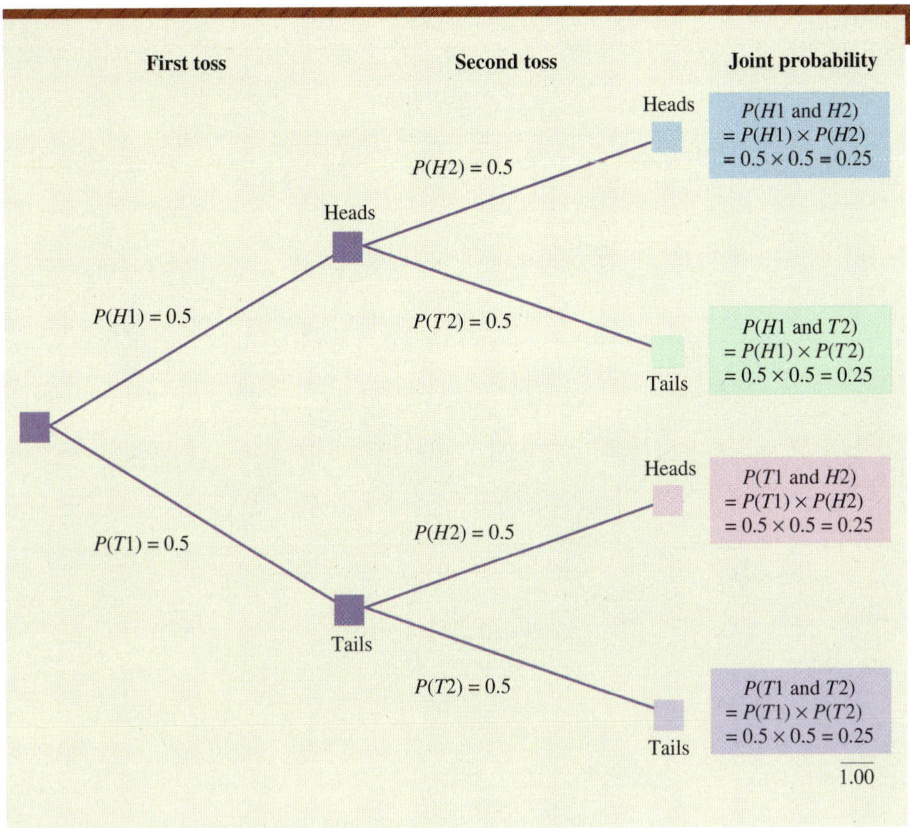

FIGURE 5.3

This tree diagram illustrates two coin tosses, a process involving independent events. The probability of heads on one toss has no effect on the probability of obtaining heads on the next.

When Events Are Not Independent

When events are *not* independent, the occurrence of one will influence the probability that another will take place. Under these conditions, a more general multiplication rule applies:

Multiplication rule when events are not independent:

$$P(A \text{ and } B) = P(A) \times P(B \mid A)$$

Table 5.3 shows marginal, joint, and conditional probabilities for the fireworks victims. It indicates that a fireworks victim is more likely to be male than female. The respective marginal probabilities are $P(A) = 0.779$ and $P(A') = 0.221$. Of the victims, 30.4% were males under the age of 15, and the joint probability for this category is $P(A \text{ and } B) = 0.304$.

The contingency table can be complemented by a tree diagram like the one in Figure 5.4. Using the tree diagram, we can visually describe the two variables and probabilities related to them. The tree diagram in the figure includes the marginal, joint, and conditional probabilities shown in Table 5.3.

The first set of branches in the tree diagram shows the marginal probabilities. These probabilities are based on relative frequencies and have the following values:

$$\text{Probability that the victim is a male} = P(A) = 0.779$$
$$\text{Probability that the victim is a female} = P(A') = 0.221$$

The second set of branches in the tree diagram includes conditional probabilities dependent on the result of each initial branch. For example, given that

TABLE 5.3

Relative frequency contingency table and marginal, joint, and conditional probabilities for a randomly selected fireworks victim.

		Age		
		B Under 15	*B′* 15 or Older	
Sex	*A* Male	0.304	0.475	0.779
	A′ Female	0.109	0.112	0.221
		0.413	0.587	1.000

Joint probabilities

$P(A \text{ and } B) = 0.304$

$P(A \text{ and } B') = 0.475$

$P(A' \text{ and } B) = 0.109$

$P(A' \text{ and } B') = 0.112$

Marginal probabilities

$P(A) = 0.779$

$P(A') = 0.221$

$P(B) = 0.413$

$P(B') = 0.587$

Conditional probabilities, e.g.,

$$P(A|B) = \frac{P(A \text{ and } B)}{P(B)} = \frac{0.304}{0.413} = 0.736$$

$$P(A'|B') = \frac{P(A' \text{ and } B')}{P(B')} = \frac{0.112}{0.587} = 0.191$$

FIGURE 5.4

This tree diagram illustrates the sex and age of fireworks accident victims. The events are not independent, so the probabilities in the second set of branches depend on what has happened in the first set.

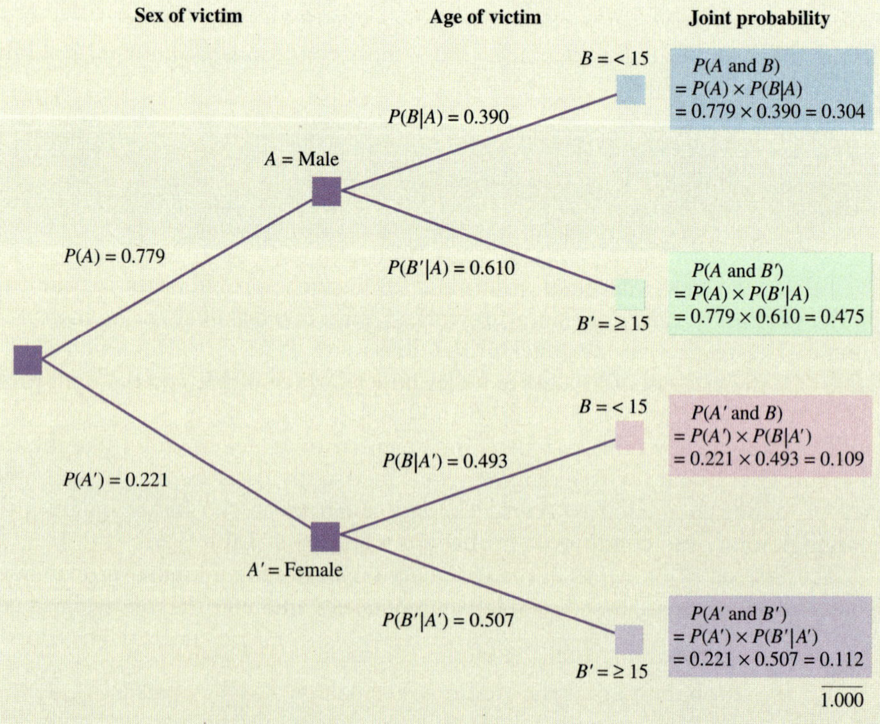

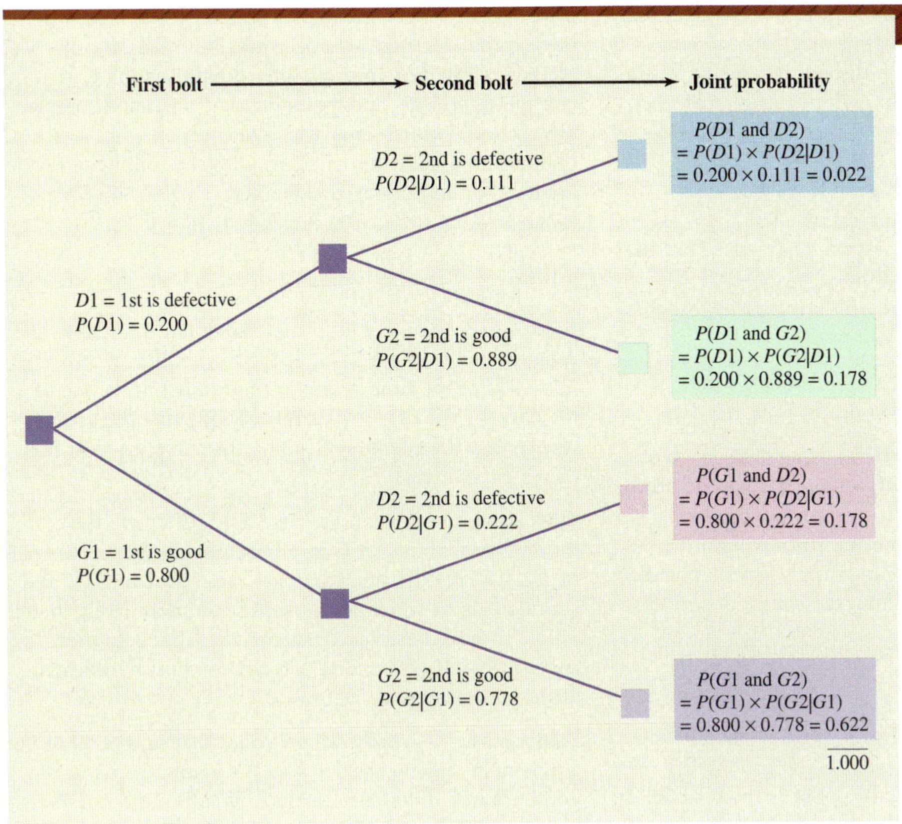

FIGURE 5.5

When sampling without replacement from a bin containing 8 good bolts and 2 defective bolts, the probability that the second bolt is defective will depend on whether or not the first one selected was defective. If the first bolt is defective, the probability that the second one will be defective is $P(D2|D1) = 0.111$.

the victim is a male, the conditional probability that he is under 15 years old is

$$P(B|A) = \frac{P(B \text{ and } A)}{P(A)} = \frac{0.304}{0.779} = 0.390$$

The other conditional probabilities are calculated in a similar manner; then the four different joint probabilities shown in the figure can also be computed. As in Figure 5.3, these are the only possibilities that can occur, so the sum of the joint probabilities is 1.

Events can be dependent when we *sample without replacement*. Figure 5.5 represents the selection of two bolts, without replacement, from a bin of 10 bolts in which 2 are defective. The probability that the first bolt is defective will be $P(D1) = \frac{2}{10}$, or 0.200. If the first bolt *is* defective, then the conditional probability that the second one will be defective is $P(D2|D1) = \frac{1}{9}$, or 0.111. This is because there will be just one defective in the 9 bolts that remain. If the first bolt is *not* defective, the probability that the second one will be defective is $P(D2|G1) = \frac{2}{9}$, or 0.222.

EXERCISES

5.30 What is the difference between a marginal probability and a joint probability?

5.31 It is possible to have a sample space in which $P(A) = 0.7$, $P(B) = 0.6$, and $P(A \text{ and } B) = 0.35$. Given this information, would events A and B be mutually exclusive? Would they be independent?

5.32 If events A and B are independent, will $P(A|B)$ be greater than, less than, or equal to $P(A)$? Explain.

5.33 It has been reported that 57% of U.S. households that rent do not have a dishwasher, while only 28% of homeowner households do not have a dishwasher. If one household is randomly selected from each ownership category, determine the probability that Source: Bureau of the Census, *Statistical Abstract of the United States 2009*, p. 601.

a. neither household will have a dishwasher.
b. both households will have a dishwasher.
c. the renter household will have a dishwasher, but the homeowner household will not.
d. the homeowner household will have a dishwasher, but the renter household will not.

5.34 A fair coin is tossed three times. What is the probability that the sequence will be heads, tails, heads?

5.35 A kitchen appliance has 16 working parts, each of which has a 0.99 probability of lasting through the product's warranty period. The parts operate independently, but if one or more malfunctions, the appliance will not work. What is the probability that a randomly selected appliance will work satisfactorily throughout the warranty period?

5.36 An optometry practitioner group has found that half of those who need vision correction are patients who require bifocal lenses.

a. For a randomly selected group of three people who require vision correction, what is the probability that all three will require bifocals? What is the probability that none of the three will require bifocals?
b. If the three individuals had all been selected from an organization consisting of retirees over age 65, do you believe the joint probabilities calculated in part (a) would still be accurate? If not, why not?

5.37 Through April 30 of the 2009 filing season, 15.8% of all individual U.S. tax returns were prepared by H&R Block. Source: H&R Block, Inc., *Fast Facts 2009*.

a. If two individuals are randomly selected from those filing tax returns during this period, what is the probability that both of their tax returns were prepared by H&R Block?
b. In part (a), what is the probability that neither return was prepared by H&R Block?
c. In part (a), what is the probability that exactly one of the two returns was prepared by H&R Block?

5.38 Of employed U.S. adults age 25 or older, 90.4% have completed high school, while 34.0% have completed college. For H = completed high school, C = completed college, and assuming that one must complete high school before completing college, construct a tree diagram to assist your calculation of the following probabilities for an employed U.S. adult: Source: Bureau of the Census, *Statistical Abstract of the United States 2009*, p. 146.

a. $P(H)$
b. $P(C|H)$
c. $P(H \text{ and } C)$
d. $P(H \text{ and } C')$

5.39 A taxi company in a small town has two cabs. Cab A stalls at a red light 25% of the time, while cab B stalls just 10% of the time. A driver randomly selects one of the cars for the first trip of the day. What is the probability that the engine will stall at the first red light the driver encounters?

5.40 Using the table in Exercise 5.12, calculate the conditional probability of C given each of the age groups, or $P(C|D)$, $P(C|E)$, etc. Compare these probabilities and speculate as to which age groups seem more likely than others to have been (according to the legal definition at that time, 0.10% blood alcohol content) intoxicated at the time they were victims.

5.41 Charlie has read a survey result that says 60% of the adults in his town consider Wendy's hamburgers to taste good. Charlie drives into the local Wendy's and questions a young couple about to enter the restaurant. According to Charlie, there's only a 0.36 (i.e., 0.6×0.6) probability that both of these people will say Wendy's hamburgers taste good. Do you think Charlie is correct? Explain.

5.42 Based on the information in Exercise 5.20, if the student chosen is known to be on a sports team, what is the probability that the student is a female?

5.43 Based on the information in Exercise 5.21, if the advisor has at least one person sign up for the advanced class, what is the probability that at least three people have signed up?

5.44 Based on the information in Exercise 5.22, if a person is known to travel a one-way distance of at least 11 miles to work, determine the probability that he or she drives at least 31 miles to work.

5.6 BAYES' THEOREM AND THE REVISION OF PROBABILITIES

In the 1700s, Thomas Bayes developed a theorem that is an extension of the concept of conditional probability discussed in the previous section. In Bayes' application of conditional probability, the emphasis is on sequential events; in particular, information obtained from a second event is used to revise the probability that a first event has occurred. The theorem will be demonstrated by means of an example, and key terms will be introduced along the way.

EXAMPLE

Bayes' Theorem

Present Information

A dryer manufacturer purchases heating elements from three different suppliers: Argostat, Bermrock, and Thermtek. Thirty percent of the heating elements are supplied by Argostat, 50% by Bermrock, and 20% by Thermtek. The elements are mixed in a supply bin prior to inspection and installation. Based on past experience, 10% of the Argostat elements are defective, compared to only 5% of those supplied by Bermrock, and just 4% of those from Thermtek. An assembly worker randomly selects an element for installation. What is the probability that the element was supplied by Argostat?

SOLUTION

- Events

A_1 = The element was produced by Argostat
A_2 = The element was produced by Bermrock
A_3 = The element was produced by Thermtek

- **Prior probability** This is an initial probability based on the present level of information. On this basis, $P(A_1) = 0.300$ since Argostat supplies 30% of the heating elements.

Additional Information

Upon testing the element before installation, an inspector finds it to be defective.

- Events

B = A tested element is defective
B' = A tested element is not defective

- **Posterior probability** This is a revised probability that has the benefit of additional information. It is a conditional probability and can be expressed as $P(A_1|B)$.

Argostat's quality record is the worst of the three suppliers, so the finding of a defective element would tend to suggest that $P(A_1|B)$ is higher than $P(A_1)$. From an intuitive standpoint, the posterior (revised) probability should be higher than 0.300. But how much higher? As a first step in finding out, consider the tree diagram and joint probabilities in Figure 5.6.

Because the events are dependent, the marginal probabilities in the first branches of the tree diagram are multiplied by the conditional probabilities in the second set of branches to obtain the joint probabilities of the final column. However, there is a problem: In order to construct the tree diagram, we had to use a time sequence that proceeded from the supplier source to the eventual testing of the element.

What we really need is the *reverse* of this tree diagram, that is, proceeding from the present (a known defective) to the past (the probability that it was supplied by Argostat). The joint probabilities of Figure 5.6 can be used in reversing the time sequence. In doing so, it is useful to look at the joint probabilities as relative frequencies out of 1000 cases. Taking this perspective, we observe that in 63 (i.e., 30 + 25 + 8) of the 1000 cases, the element was defective, and in 30 of these the element came from Argostat. In other words, we have just determined that

FIGURE 5.6

Heating-element tree diagram, shown in the order in which the elements are received and found to be either defective or good.

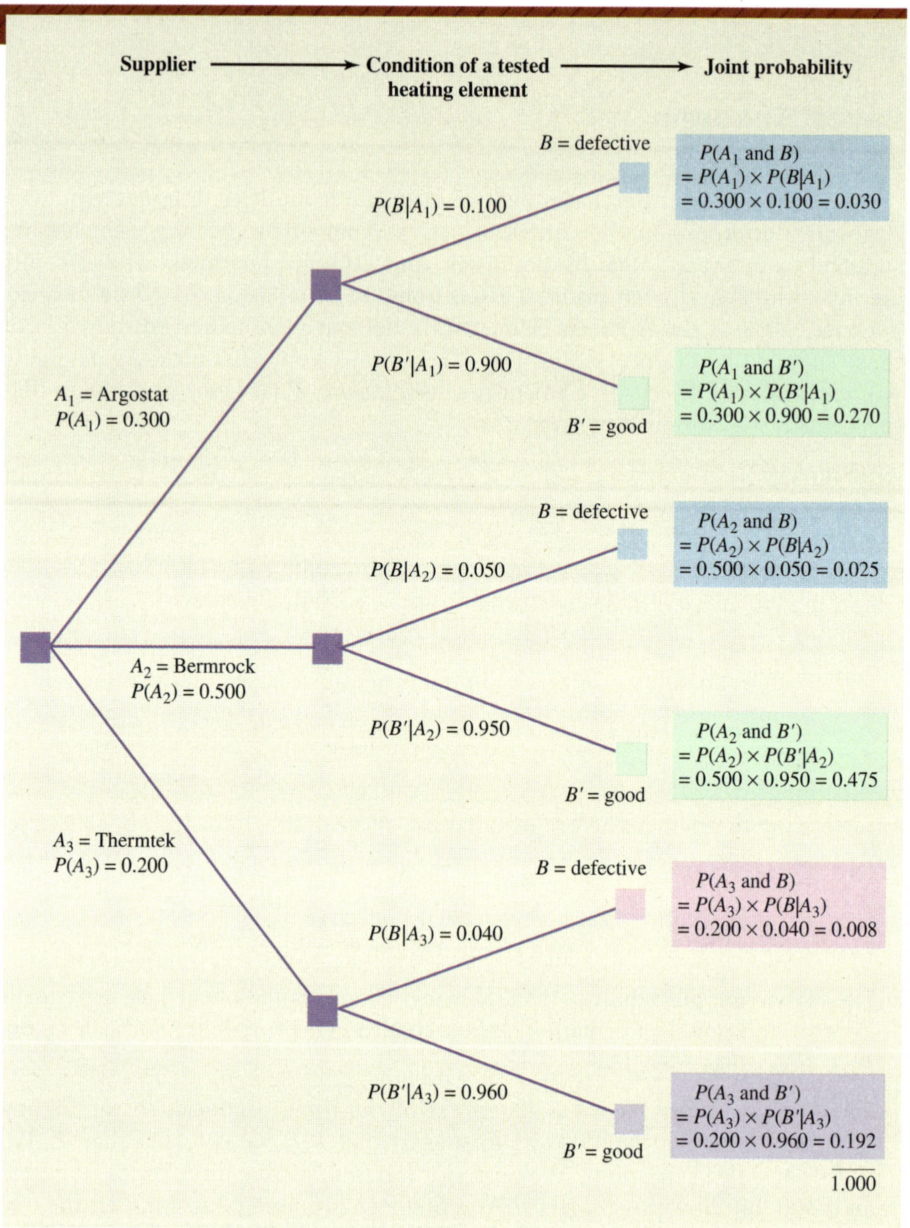

the revised probability, $P(A_1|B)$, is 30/63, or 0.476. This revised probability is among those shown in the *reversed* tree diagram in Figure 5.7.

Probabilities can also be revised by using **Bayes' theorem.** The theorem deals with sequential events, using information obtained about a second event to revise the probability that a first event has occurred.

Bayes' theorem for the revision of probability:

- **Events A and B** Probability of A, given that event B has occurred:

$$P(A|B) = \frac{P(A \text{ and } B)}{P(B)} = \frac{P(A) \times P(B|A)}{[P(A) \times P(B|A)] + [P(A') \times P(B|A')]}$$

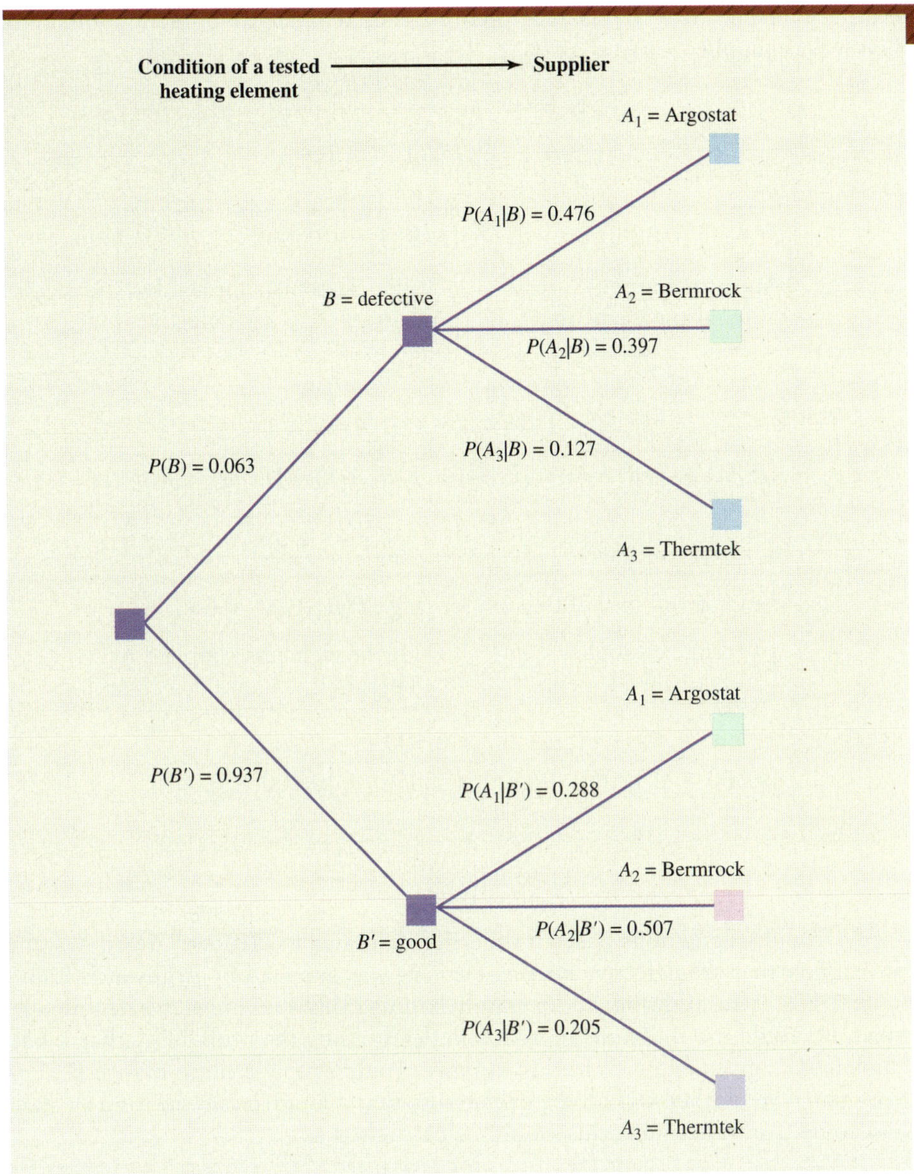

FIGURE 5.7

This reversal of the tree diagram in Figure 5.6 shows that the probability that the defective heating element came from Argostat is 0.476. The prior probability, $P(A_1) = 0.300$, has been converted to a revised probability, $P(A_1|B) = 0.476$.

- **General form** Probability of event A_i, given that event B has occurred:

$$P(A_i|B) = \frac{P(A_i \text{ and } B)}{P(B)} = \frac{P(A_i) \times P(B|A_i)}{\sum_{j=1}^{k} P(A_j) \times P(B|A_j)}$$

where A_i = the ith of k mutually exclusive and exhaustive events

Applying the general form of Bayes' theorem, the heating element had to come from one of $k = 3$ suppliers: Argostat (event A_1), Bermrock (event A_2), or Thermtek (event A_3). Additional information is available in the form of knowledge that the heating element is defective (event B). We must use the general form of the theorem because the number of suppliers is more than two. Given that the heating element is defective, Bayes' theorem is used in determining the probability that it was supplied by Argostat:

$$P(A_i | B) = \frac{P(A_i) \times P(B | A_i)}{\sum_{j=1}^{k} P(A_j) \times P(B | A_j)}$$

$P(A_1) = 0.300$

Argostat supplies 30% of the elements.

$P(B | A_1) = 0.100$

Of the elements supplied by Argostat, 10% are defective.

$$P(A_1 | B) = \frac{P(A_1) \times P(B | A_1)}{[P(A_1) \times P(B | A_1)] + [P(A_2) \times P(B | A_2)] + [P(A_3) \times P(B | A_3)]}$$

e.g., $P(A_3) = 0.200$

Thermtek supplies 20% of the elements.

e.g., $P(B | A_3) = 0.040$

Of the elements supplied by Thermtek, 4% are defective.

$$= \frac{0.300 \times 0.100}{[0.300 \times 0.100] + [0.500 \times 0.050] + [0.200 \times 0.040]}$$

and

$$P(A_1 | B) = \frac{0.030}{0.063} = 0.476$$

As in our calculations based on the tree diagrams in Figures 5.6 and 5.7, the probability that the defective heating element was supplied by Argostat is found to be 0.476. If the additional information had revealed a *nondefective* heating element, we could use the same approach in determining the probability that it had come from a given supplier. In this case, we would simply replace B with B'. For example, if we wished to find the probability that a nondefective heating element had come from Thermtek, this would be calculated as

$$P(A_3 | B') = \frac{P(A_3) \times P(B' | A_3)}{[P(A_1) \times P(B' | A_1)] + [P(A_2) \times P(B' | A_2)] + [P(A_3) \times P(B' | A_3)]}$$

$$= \frac{0.200 \times 0.960}{[0.300 \times 0.900] + [0.500 \times 0.950] + [0.200 \times 0.960]}$$

and

$$P(A_3 | B') = \frac{0.192}{0.937} = 0.205$$

For the heating-element example, Table 5.4 shows a contingency table with relative frequencies for suppliers and defectives. Like Table 5.3, Table 5.4 shows marginal, joint, and conditional probabilities. Given that a randomly selected element is defective, we can determine the conditional probability that it was supplied by an individual supplier. This is exactly what we did earlier by using tree diagrams (Figures 5.6 and 5.7) and by using Bayes' theorem for the revision of a probability. In each case, we have taken advantage of the fact that a revised probability can be viewed as a conditional probability.

TABLE 5.4

A contingency table showing relative frequencies and marginal, joint, and conditional probabilities for the heating-element example of Figures 5.6 and 5.7.

		Supplier			
		A_1	A_2	A_3	
The element is		Argostat	Bermrock	Thermtek	
B	Defective	0.030	0.025	0.008	0.063
B'	Good	0.270	0.475	0.192	0.937
		0.300	0.500	0.200	1.000

Joint probabilities, e.g.,
$P(A_1 \text{ and } B) = 0.030$
$P(A_2 \text{ and } B') = 0.475$
$P(A_3 \text{ and } B) = 0.008$

Marginal probabilities, e.g.,
$P(A_1) = 0.300$
$P(A_2) = 0.500$
$P(B) = 0.063$

Conditional probabilities, e.g.,

$$P(A_1|B) = \frac{P(A_1 \text{ and } B)}{P(B)} = \frac{0.030}{0.063} = 0.476$$

$$P(A_3|B') = \frac{P(A_3 \text{ and } B')}{P(B')} = \frac{0.192}{0.937} = 0.205$$

The concept of additional information and revised probabilities is an important part of decision making, an activity in which both statistics and additional information are important. This topic will be covered in greater detail in Chapter 19 and its online appendix.

EXERCISES

5.45 For U.S. live births, $P(\text{boy})$ and $P(\text{girl})$ are approximately 0.51 and 0.49, respectively.[a] According to a newspaper article, a medical process could alter the probabilities that a boy or a girl will be born. Researchers using the process claim that couples who wanted a boy were successful 85% of the time, while couples who wanted a girl were successful 77% of the time.[b] Assuming that the medical process does have an effect on the sex of the child: [a]Source: Bureau of the Census, *Statistical Abstract of the United States 2009*, p. 64. [b]Source: Steve Twedt, "Boy-Girl Selection Available Soon Here," *Pittsburgh Press*, December 31, 1987, pp. 1, 10.

a. Without medical intervention, what is the probability of having a boy?
b. With medical intervention, what is the conditional probability that a couple who wants a boy will have a boy?
c. With medical intervention, what is the conditional probability that a couple who wants a girl will have a girl?

5.46 Using the information in Exercise 5.45, assume that a couple who wanted a girl was randomly placed into either the treatment group (receiving the medical process described) or the control group (no treatment administered) in a test involving the medical procedure.

a. What is the couple's prior probability for being in the treatment group?
b. The couple's newborn child is a girl. Given this additional information, what is the revised probability that the couple was in the treatment group?

5.47 A magician has two coins: One is unbalanced and lands heads 60% of the time; the other is fair and lands heads 50% of the time. A member of the audience randomly selects one of the coins and flips it. The result is heads.

a. What is the prior probability that the fair coin was selected?
b. Given additional information in the form of the single flip that came up heads, what is the revised probability that the coin is the fair one?

5.48 For the information provided in Exercise 5.39, the cabbie finds that his car stalls at the first red light he encounters. Given this additional information, what is the probability that he has selected cab A?

5.49 Machine A produces 3% defectives, machine B produces 5% defectives, and machine C produces 10% defectives. Of the total output from these machines, 60% of the items are from machine A, 30% from B, and 10% from C. One item is selected at random from a day's production.

a. What is the prior probability that the item came from machine C?
b. If inspection finds the item to be defective, what is the revised probability that the item came from machine C?

5.7 COUNTING: PERMUTATIONS AND COMBINATIONS

Closely related to the concept of probability is the topic of *counting*. In this section, we'll see that counting can be much more than adding up football scores or keeping track of how many people enter a theater.

Fundamentals of Counting

If there are m ways in which a first event can happen, and n ways in which a second event can then occur, the total number of possibilities is $m \times n$. For example, if you have two different cars and four different routes to travel to work, you can arrive in 2×4, or 8, different ways. This can be described as the **principle of multiplication:**

The principle of multiplication:

- **If, following a first event that can happen in n_1 ways, a second event can then happen in n_2 ways, the total number of ways both can happen is $n_1 n_2$.**
- **Alternatively, if each of k independent events can occur in n different ways, the total number of possibilities is n^k.**

As an example of the second approach, assume that a state's auto license plates include 6 different numbers or letters, and that all 26 letters and 10 digits are available. As a result, there can be $(36)^6$, or 2,176,782,336 different license plates. If we include the possibility of blank spaces, this becomes $(37)^6$, or 2,565,726,409, more than enough to satisfy even the most populous state. As a matter of fact, this number of unique plates is more than three times as large as the number of motor vehicles in the entire world.[5]

In an application similar to the selection of letters and numbers for license plates, consonants and vowels can be randomly selected to either create words or fill in blank spaces that are remaining in a word. Statistics in Action 5.2 shows an example in which the computer randomly fills in the remaining three letters of the word "_ _ _ TECH."

In some situations, several events all have different possibilities, but the choice you make for one limits the number of choices for subsequent events. This might occur, for example, if you have five different errands to run and

[5]Source: As of 2005, 862 million motor vehicles were registered throughout the world; *New York Times Almanac 2009*, p. 423.

STATISTICS IN ACTION 5.2

Computer Assistance in the Name Game

When selecting a name for a company or a product, one alternative is to randomly generate combinations of letters, then scan the list for "words" that either look attractive or inspire further ideas. A computer might be programmed to randomly list five-letter words that begin with "k" and end with "x," have a desired order of consonants or vowels, or even include desired sequences like "corp," "tronics," or "electro."

For example, we might generate seven-letter words of the form *cvc*TECH, where c = any of 21 consonants and v = any of 5 vowels. In so forming "_ _ _ TECH," the number of possibilities is 21 × 5 × 21, or 2205. The computer becomes a high-speed variant of the proverbial typewriting monkey who will eventually generate a sentence. As this alphabetized partial listing shows, most of the "words" are not suitable for naming a company, a product, or even a fish. A few of them, however, are fairly reasonable and illustrate the possibilities of this method.

BUGTECH	CAVTECH	DUFTECH	FOZTECH	GAGTECH	HOWTECH
JARTECH	JETTECH	KAPTECH	KUBTECH	LOJTECH	MAZTECH
METTECH	MIPTECH	NESTECH	NOGTECH	PASTECH	QUKTECH
RELTECH	ROSTECH	SAXTECH	SOTTECH	TAGTECH	TOYTECH
VUZTECH	WIJTECH	XOGTECH	YESTECH	ZETTECH	ZIQTECH

must choose the order in which you will do them. In this case, the number of possibilities becomes

$$5 \times 4 \times 3 \times 2 \times 1 = 120$$

There are 5 choices available for the errand to be done first.
Once the 1st errand is chosen, 4 choices are left.
Once the 2nd errand is chosen, 3 choices are left.
Once the 3rd errand is chosen, 2 choices are left.
Once the 4th errand is chosen, only 1 choice remains.

The product ($5 \times 4 \times 3 \times 2 \times 1$) is called a **factorial** and can be described by the symbol "5!." For example, 3! is the same as $3 \times 2 \times 1$. The exclamation point is just a mathematical way of saving space. This approach to counting can be described as the *factorial rule of counting:*

Factorial rule of counting:

$$n! = n \times (n - 1) \times (n - 2) \times \cdots \times 1 \quad (\textit{Note: } 0! \text{ is defined as } 1.)$$

Permutations

Permutations refer to the number of different ways in which objects can be arranged in order. In a permutation, each item can appear only once, and each order of the items' arrangement constitutes a separate permutation. The number of possible arrangements can be determined as follows:

Number of permutations of n objects taken r at a time:

$$\frac{n!}{(n - r)!}$$

EXAMPLE

Permutations

Bill has 6 textbooks, but can fit only 4 of them onto a small shelf. If the other 2 books have to sit on the desk, in how many ways can Bill arrange the shelf?

SOLUTION

In this example, there are $n = 6$ objects, taken $r = 4$ at a time, and the number of choices available for the shelf arrangement would be

$$\frac{n!}{(n-r)!} = \frac{6!}{(6-4)!} = \frac{6 \times 5 \times 4 \times 3 \times 2 \times 1}{2 \times 1} = 360$$

The situation can be viewed as follows: (1) the first shelf position can be filled in any of 6 ways, (2) the second shelf position can then be filled in any of 5 ways, (3) the third shelf position can then be filled in any of 4 ways, and (4) the fourth shelf position can then be filled in any of 3 ways. The total number of ways of filling these four shelf positions is thus $6 \times 5 \times 4 \times 3$, or 360. In the equation above, note that the product (2×1) appears in both the numerator and the denominator and is therefore canceled out, reducing the equation to $6 \times 5 \times 4 \times 3 = 360$.

This might seem like a rather large number of possibilities for just 6 books on a 4-book shelf, but each different order of arrangement is regarded as a different permutation. In permutations, the exact order in which the objects are arranged is paramount.

EXAMPLE

Permutations

A tax auditor has 9 different returns to audit, but will have time to examine only 5 of them tomorrow. In how many different orders can tomorrow's task be carried out?

SOLUTION

In this case, there will be $n = 9$ objects, taken $r = 5$ at a time, and the number of different orders for doing tomorrow's work will be

$$\frac{n!}{(n-r)!} = \frac{9!}{(9-5)!} = \frac{9 \times 8 \times 7 \times 6 \times 5 \times 4 \times 3 \times 2 \times 1}{4 \times 3 \times 2 \times 1} = 15{,}120$$

Combinations

Unlike permutations, **combinations** consider only the possible sets of objects, *regardless of the order in which the members of the set are arranged*. The number of possible combinations of n objects taken r at a time will be as follows:

Number of combinations of n objects taken r at a time:

$$\binom{n}{r} = \frac{n!}{r!(n-r)!}$$

In the bookshelf example, there would be 15 different combinations by which Bill could select 4 of the 6 books for storage on the shelf. This is calculated by

$$\binom{n}{r} = \frac{n!}{r!(n-r)!} = \frac{6!}{4!(6-4)!} = \frac{6 \times 5 \times 4 \times 3 \times 2 \times 1}{(4 \times 3 \times 2 \times 1)(2 \times 1)} = 15$$

For greater ease of computation, this calculation can be simplified by taking advantage of the fact that 4! is actually a component of 6!. By making this substitution, the calculation becomes simpler because 4! is canceled out by appearing in both the numerator and denominator:

$$\binom{n}{r} = \frac{n!}{r!(n-r)!} = \frac{6!}{4!(6-4)!} = \frac{6 \times 5 \times 4!}{4!(2 \times 1)} = \frac{6 \times 5}{2 \times 1} = 15$$

In the tax return example, if the tax auditor does not consider the order in which tomorrow's audits are carried out, there will be 126 different compositions for the group of persons whose returns are audited that day. This is calculated as

$$\binom{n}{r} = \frac{n!}{r!(n-r)!} = \frac{9!}{5!(9-5)!} = \frac{9 \times 8 \times 7 \times 6 \times 5!}{5! \times 4!} = \frac{3024}{24} = 126$$

As these examples suggest, for given values of n and r, with $n > r > 1$, the number of permutations will always be greater than the number of combinations. This is because there will be $r!$ permutations for every possible combination. For example, if the auditor selects a given set of 5 returns to audit tomorrow, she can conduct the audits in 5! different orders. Thus, the number of combinations will be just 1/(5!) as large as the number of permutations. This can also be seen by comparing the respective formulas: They are identical except that the formula for combinations has $r!$ in the denominator, while that for permutations does not.

EXERCISES

5.50 A tax accountant has three choices for the method of treating one of a client's deductions. After this choice has been made, there are only two choices for how a second deduction can be treated. What is the total number of possibilities for the treatment of the two deductions?

5.51 A committee consists of eight members, each of whom may or may not show up for the next meeting. Assuming that the members will be making independent decisions on whether or not to attend, how many different possibilities exist for the composition of the meeting?

5.52 Ten prominent citizens have been nominated for a community's "Citizen of the Year" award. First- and second-place trophies are awarded to the two persons receiving the highest and second-highest number of votes. In how many different ways can the trophies be awarded?

5.53 In advertising her oceanfront cottage for summer rental, a property owner can specify (1) whether or not pets are permitted, (2) whether the rent will be \$700, \$900, or \$1200 per week, (3) whether or not children under 10 are allowed, and (4) whether the maximum stay is 1, 2, or 3 weeks. How many different versions of the ad can be generated?

5.54 A state's license plate has 6 positions, each of which has 37 possibilities (letter, integer, or blank). If the buyer of a personalized plate wants his first three positions to be BMW, in how many ways can his request be satisfied?

5.55 When we are considering n objects taken r at a time, with $r > 1$, why will the number of combinations be less than the number of permutations?

5.56 An investment counselor would like to meet with 12 of his clients on Monday, but he has time for only 8 appointments. How many different combinations of the clients could be considered for inclusion into his limited schedule for that day?

5.57 How many different combinations are possible if 6 substitute workers are available to fill 3 openings created by employees planning to take vacation leave next week?

5.58 A roadside museum has 25 exhibits but enough space to display only 10 at a time. If the order of arrangement is considered, how many possibilities exist for the eventual display?

5.59 A sales representative has 35 customers throughout the state and is planning a trip during which 20 will be visited. In how many orders can the visits be made?

5.8 SUMMARY

• The concept of probability

Uncertainty, an inherent property of both business and life, can be expressed in terms of probability. For any event, the probability of occurrence will be no less than 0 and no more than 1. All possible outcomes of an experiment are included in the sample space.

• Classical, subjective, and relative frequency probabilities

The classical approach describes probability in terms of the proportion of times that an event can be theoretically expected to occur. Because the probability can be determined without actually engaging in any experiments, this is known as the *a priori* approach. In the relative frequency approach, the probability of an event is expressed as the proportion of times it is observed to occur over a large number of trials.

The subjective approach to probability represents the degree to which one believes that an event will happen. Such probabilities can also be described as hunches or educated guesses. Another way to express the probability of an event is to indicate the odds that it will or will not occur.

• Venn diagrams and contingency tables

Events are mutually exclusive if, when one occurs, the other(s) cannot. They are exhaustive if they include all of the possible outcomes. An intersection of events represents the occurrence of two or more at the same time, while their union means that at least one of them will happen. The Venn diagram displays both the sample space and the events within. Frequencies or relative frequencies can also be summarized in a contingency table.

• Determining probabilities

The addition rules describe the probability that at least one of several events will occur, e.g., $P(A \text{ or } B)$. If the events are mutually exclusive, this probability will be the sum of their individual probabilities. When the events are not mutually exclusive, the general rule of addition applies. This rule takes into account the probabilities that two or more of the events will happen at the same time.

The multiplication rules describe the joint probability that two or more events will occur, e.g., $P(A \text{ and } B)$. When the events are independent, this probability is the product of their individual (marginal) probabilities. When the events are not independent, the general rule of multiplication applies. This rule considers the conditional probability that an event will occur, given that another one has already happened.

• Bayes' theorem and tree diagrams

Bayes' theorem, an application of the conditional probability concept, applies to two or more events that occur in sequence. It begins with an initial (prior) probability for an event, then uses knowledge about a second event to calculate a revised (posterior) probability for the original event. The application of Bayes' theorem can be visually described by a tree diagram that shows the possible sequences of events along with their marginal, conditional, and joint probabilities.

Permutations and combinations

Closely related to the concept of probability are the principles of counting. Two such rules are the principle of multiplication and the factorial rule of counting. The concept of permutations is concerned with the number of possible arrangements of n objects taken r at a time. Unlike permutations, combinations consider only the particular set of objects included, regardless of the order in which they happen to be arranged.

EQUATIONS

Classical Approach to Probability

For outcomes that are equally likely,

$$\text{Probability} = \frac{\text{Number of possible outcomes in which the event occurs}}{\text{Total number of possible outcomes}}$$

Relative Frequency Approach to Probability

For a very large number of trials,

$$\text{Probability} = \frac{\text{Number of trials in which the event occurs}}{\text{Total number of trials}}$$

Probabilities and "Odds"

1. If the odds in favor of an event happening are A to B, or A:B, the probability being expressed is

$$\frac{A}{A + B}$$

2. If the probability of an event is x (with $0 \le x \le 1$), the odds in favor of the event are x to $(1 - x)$, or x:$(1 - x)$. (*Note:* In practice, the ratio is converted to a more visually pleasing form in which the numbers are the lowest-possible integers.)

Addition Rule When Events Are Mutually Exclusive

$$P(A \text{ or } B) = P(A) + P(B)$$

Addition Rule When Events Are Not Mutually Exclusive

$$P(A \text{ or } B) = P(A) + P(B) - P(A \text{ and } B)$$

Conditional Probability

$$P(A|B) = \frac{P(A \text{ and } B)}{P(B)}$$ *Note:* This applies only if $P(B) > 0$.

Multiplication Rule When Events Are Independent

$$P(A \text{ and } B) = P(A) \times P(B)$$

Multiplication Rule When Events Are Not Independent

$$P(A \text{ and } B) = P(A) \times P(B|A)$$

Bayes' Theorem for the Revision of a Probability

- **Events A and B** Probability of A, given that event B has occurred:

$$P(A|B) = \frac{P(A \text{ and } B)}{P(B)} = \frac{P(A) \times P(B|A)}{[P(A) \times P(B|A)] + [P(A') \times P(B|A')]}$$

- **General form** Probability of event A_i, given that event B has occurred:

$$P(A_i|B) = \frac{P(A_i \text{ and } B)}{P(B)} = \frac{P(A_i) \times P(B|A_i)}{\sum_{j=1}^{k} P(A_j) \times P(B|A_j)}$$

where A_i = the ith of k mutually exclusive and exhaustive events

The Principle of Multiplication

If, following a first event that can happen in n_1 ways, a second event can then happen in n_2 ways, the total number of ways both can happen is $n_1 n_2$. If each of k independent events can occur in n different ways, the total number of possibilities is n^k.

The Factorial Rule of Counting

$$n! = n \times (n - 1) \times (n - 2) \times \cdots \times 1 \qquad \textit{Note: } 0! \text{ is defined as } 1.$$

Number of Permutations of n Objects Taken r at a Time

$$\frac{n!}{(n - r)!}$$

Number of Combinations of n Objects Taken r at a Time

$$\binom{n}{r} = \frac{n!}{r!(n - r)!}$$

CHAPTER EXERCISES

5.60 It has been estimated that the odds of being struck by lightning in a given year are 1 to 240,000. Using this information: Source: lightningtalks.com, July 27, 2006.

a. Express the odds in terms of "odds against."
b. What is the probability that a randomly selected individual will be struck by lightning this year?
c. What is the probability that the person in part (b) will be struck by lightning at least once during the next 20 years?

5.61 A fast-food chain gives each customer a coupon, the surface of which can be scratched to reveal whether a prize will be received. The odds for winning $1000 per week for life are listed as 1 to 200,000,000, while the odds for winning a free hamburger are 1 to 15. Sheila is going to have lunch at her local franchise of the chain and will receive one coupon during her visit. Determine the probability that she will win

a. $1000 per week for life.
b. a free hamburger.
c. either the lifetime $1000/week or a free hamburger.
d. both the lifetime $1000/week and a free hamburger.

5.62 For the situation in Exercise 5.61, what is the probability that Sheila will visit the franchise

a. 5 times without winning a free hamburger?
b. 10 times without winning a free hamburger?
c. 20 times without winning a free hamburger?
d. 40 times without winning a free hamburger?

5.63 The "daily number" of a state lottery is a 3-digit integer from 000 to 999.

a. Sam buys a ticket for the number 333. What is the probability that he will win?
b. Is the probability found in part (a) a classical, relative frequency, or subjective probability? Explain your answer.
c. Shirley buys a ticket for the number 418 since it came up yesterday and she thinks it's "hot." Is there a flaw in her logic? Explain.

5.64 The U.S. Bureau of Justice released the following probabilities for those arrested for committing various felony crimes in the United States:

	Probability of Being		
Crime	**Prosecuted**	**Convicted**	**Jailed for >1 Year**
Homicide	0.91	0.75	0.73
Assault	0.79	0.64	0.15
Burglary	0.88	0.81	0.28
Arson	0.88	0.72	0.28
Drug offenses	0.78	0.69	0.19
Weapons	0.83	0.70	0.13
Public disorder	0.92	0.85	0.12

Allen has been arrested for burglary, Bill has been arrested for a weapons offense, and Charlie has been arrested on a public-disorder charge. Assuming these individuals are typical perpetrators and the decisions regarding their respective fates are unrelated, determine the probability that Source: U.S. Bureau of Justice, as reported in Sam Meddis, "Felony Arrests: Short Terms," *USA Today*, January 18, 1988, p. 9A.

a. Allen will be jailed for more than a year.
b. either Allen or Bill (or both) will be convicted.
c. none of the three will be jailed for more than a year.
d. Allen and Bill will be convicted, but Charlie will be found innocent.
e. none will be prosecuted.

5.65 For the three perpetrators in Exercise 5.64, determine the number of possibilities in which
a. just one person is convicted.
b. exactly two of the three persons are convicted.
c. all three persons are convicted.

5.66 Dave has been arrested for arson and has been convicted. Edgar is currently being prosecuted for a drug offense. Applying the U.S. Bureau of Justice data in Exercise 5.64, determine the probability that
a. Dave will spend more than a year in jail.
b. Edgar will be convicted.
c. Edgar will spend more than a year in jail.

5.67 A firm has two computer systems available for processing telephone orders. At any given time, system A has a 10% chance of being "down," while system B has just a 5% chance of being "down." The computer systems operate independently. For a typical telephone order, determine the probability that
a. neither computer system will be operational.
b. both computer systems will be operational.
c. exactly one of the computer systems will be operational.
d. the order can be processed without delay.

5.68 An industrial hoist is being used in an emergency job where the weight exceeds the design limits of two of its components. For the amount of weight being lifted, the probability that the upper attachment hook will fail is 0.20. The probability that the lower hook will fail is 0.10. What is the probability that the hoisting job will be successfully completed?

5.69 The following relative frequency distribution describes the household incomes for families living in a suburban community:

Income	Relative Frequency
Less than \$20,000	0.10
\$20,000–under \$40,000	0.20
\$40,000–under \$60,000	0.30
\$60,000–under \$80,000	0.25
\$80,000 or more	0.15

a. For a randomly selected household, what is the probability that its annual income is less than \$80,000?
b. If a household is known to have an income of at least \$20,000, what is the probability that its annual income is in the \$60,000–under \$80,000 category?
c. Two households are randomly selected from those whose income is at least \$40,000 per year. What is the probability that both of these households are in the \$80,000 or more category?

5.70 Collecting data on traffic accident fatalities, the National Highway Traffic Safety Administration has found that 58.7% of the victims have 0.0% blood alcohol content (BAC), 5.8% of the victims have from 0.01 to 0.07% BAC, and 35.5% of the victims have at least 0.08% BAC. For a randomly selected victim: Source: *Statistical Abstract of the United States 2009*, p. 681.
a. What is the probability that the victim's BAC was at least 0.01%?
b. Given that the victim had been drinking prior to the accident, what is the probability that this victim's BAC was at least 0.08%?

5.71 During fiscal 1995, Ashland Exploration drilled 7 wells in the United States and 3 in Nigeria. One of the wells in Nigeria was productive and 2 of the wells in the United States were productive. The remaining wells were dry. For a well selected at random from those drilled during fiscal 1995, determine the following probabilities: Source: Ashland Oil, Inc., 1995 Fiscal Year Operational Supplement, p. 20.

a. P(drilled in the United States)
b. P(drilled in the United States or dry)
c. P(dry | drilled in Nigeria)

5.72 Avis, Inc., has reported that its fleet consists of 200,000 vehicles. If the vehicles are independent in terms of accident incidence and each has a 0.99999 chance of making it through the year without being in a major accident, what is the probability that the entire fleet will avoid a major accident during the year? Source: Avis.com, July 27, 2006.

5.73 Data from the Federal Bureau of Investigation show that 1 of every 184 motor vehicles was stolen during 2003. Applying this statistic to 5 motor vehicles randomly selected from the nation's vehicle population: Source: Bureau of the Census, *Statistical Abstract of the United States 2009*, pp. 194, 711.
a. What is the probability that none of the 5 motor vehicles will be stolen?
b. What is the probability that all 5 motor vehicles will be stolen?
c. How many possibilities exist in which 2 of the 5 motor vehicles are stolen?

5.74 Of the participants in a corporate meeting, 20% have had Red Cross training in cardiopulmonary resuscitation (CPR). During a break in the session, one participant is randomly selected to go across the street and pick up some hoagies from the local deli. A patron at the deli is stricken by a severe heart attack. His survival probability is just 0.40 if he does not receive immediate CPR, but will be 0.90 if CPR can be administered right away. The meeting participant and the stricken person are the only customers in the deli, and none of the employees knows CPR.
a. Without additional information, what is the probability that the stricken patron will survive?
b. Given that the stricken patron survives, what is the probability that the meeting participant knew how to administer CPR?
c. Given that the stricken patron does not survive, what is the probability that the meeting participant knew how to administer CPR?

5.75 Of the adults in Jefferson County, 10% have had CPR training. For a randomly selected group of adults from this county, how large would the group have to be in order for the probability that at least one group member has had CPR training to be at least 0.90?

5.76 Over the years, a realtor estimates that 50% of those who contact her about a listing end up allowing her to provide a guided tour of the property, 20% of those who take a guided tour follow up with a request for more information about the property, and 2% of those who make the follow-up informational request actually end up making an offer on the property. The realtor has just taken a young couple on a guided tour of one of her listed properties. Based on the probability estimates provided above, what is the probability that the young couple will actually make an offer on the property?

5.77 According to Sears, two-thirds of U.S. homeowners have an appliance from Sears. In a randomly selected group of three homeowners, what is the probability that all three would have an appliance from Sears? That at least one of the three would have an appliance from Sears? Source: Sears advertisement insert, *Indiana Gazette*, May 24, 2003.

5.78 A computer specialty manufacturer has a two-person technical support staff who work independently of each other. In the past, Tom has been able to solve 75% of the problems he has handled, and Adam has been able to solve 95% of the problems he has handled. Incoming problems are randomly assigned to either Tom or Adam. If a technical problem has just been assigned to the support department, what is the probability that it will be handled by Adam? If it turns out that the problem was solved, what is the probability that it was handled by Adam?

5.79 When a machine is properly calibrated, 0.5% of its output is defective, but when it is out of adjustment, 6% of the output is defective. The technician in charge of calibration can successfully calibrate the machine in 90% of his attempts. The technician has just made an attempt to calibrate the machine.
a. What is the probability that the machine is in adjustment?
b. The first unit produced after the calibration effort is found to be defective. Given this information, what is the probability that the machine is in adjustment?

5.80 In examining borrower characteristics versus loan delinquency, a bank has collected the following information: (1) 15% of the borrowers who have been employed at their present job for less than 3 years are behind in their payments, (2) 5% of the borrowers who have been employed at their present job for at least 3 years are behind in their payments, and (3) 80% of the borrowers have been employed at their present job for at least 3 years. Given this information:
a. What is the probability that a randomly selected loan account will be for a person in the same job for at least 3 years who is behind in making payments?
b. What is the probability that a randomly selected loan account will be for a person in the same job for less than 3 years or who is behind in making payments?
c. If a loan account is behind, what is the probability that the loan is for a person who has been in the same job for less than 3 years?

5.81 A security service employing 10 officers has been asked to provide 3 persons for crowd control at a local carnival. In how many different ways can the firm staff this event?

5.82 A corporate board of directors consisting of 15 persons is to form a subcommittee of 5 persons to examine an environmental issue currently facing the firm. How many different subcommittees are possible?

5.83 The chairperson of the accounting department has three summer courses available: Accounting 201, Accounting 202, and Accounting 305. Twelve faculty members are available for assignment to these courses, and no faculty member can be assigned to more than one course. In how many ways can the chairperson assign faculty to these courses?

5.84 A test to compare the taste of 6 soft drinks is being arranged. Each participant will be given 3 of the soft drinks and asked to indicate which one tastes best.

a. How many different possibilities exist for the set of soft drinks that a given participant will taste?
b. If we consider the order in which the participant in part (a) tastes the 3 soft drinks, how many possibilities exist?

5.85 A certain brand of bicycle lock requires that each of four adjacent wheels be rotated to exactly the correct position in order to open the lock and separate the chain to which it is attached. If each wheel is labeled with the digits 0 through 9, how many different possibilities are available to the manufacturer in "programming" the sequence of numbers for opening any given lock?

INTEGRATED CASES

Thorndike Sports Equipment

Ted Thorndike has been with his grandfather's firm for 5 weeks and is enjoying his job immensely. The elder Thorndike is also pleased—so pleased that he took the morning off to do a little fishing. However, while old Luke is catching some carp, young Ted is catching some flak.

At the other end of this morning's telephone call is an irate customer whose aluminum tennis racquet has cracked in half after just 14 months. The caller is an avid player who gets out on the court about two or three times a week, and he claims the racquet has never been abused in any way. He not only wants Thorndike Sports Equipment to send him another racquet, he also demands that the company reimburse him the $30 he lost after his racquet broke and he was no longer competitive with his opponent.

Ted assures the indignant caller that the company will replace his racquet at no cost, since it is only slightly over the 12-month warranty and the company values his loyalty to its products. Unfortunately, the $30 wager cannot be covered, even though it was a triple-or-nothing bet from the caller's losses in the two previous matches.

On Mr. Thorndike's return to the office, Ted mentions the interesting phone call, and the elder Thorndike is not a bit surprised. He says the firm has been getting a lot of complaints in recent months, mainly from longtime customers who claim their latest Thorndike aluminum racquet didn't hold up as well as the ones they had used in the past.

Speculating, Mr. Thorndike goes on to point out that the company has had two aluminum suppliers for many years, but added a third supplier just a year and a half ago. He suspects that the most recent supplier may be shipping an aluminum alloy that is more brittle and prone to failure. He's not sure, but he thinks the racquets that are failing are more likely than not to be constructed from the aluminum purchased from the newest of the three suppliers.

When the company sends out a replacement racquet, it does so only after it has received the defective one. All of the racquets that have been returned over the past 5 years or so are in a wooden crate in the basement. Mr. Thorndike isn't sure why, but 6 years ago he mandated that each racquet produced have a serial number that identifies when it was made and who supplied the aluminum for its construction. If someone were to ferret through the wooden crate, maybe he could shed some light on this business of customers calling up with broken racquets.

Mr. Thorndike goes on, "Say, Ted, why don't you come in over the weekend and take a look at those broken racquets downstairs, and see if maybe there's something going on here that I should know about? Maybe I should have stayed with the two suppliers I've been using. The prices the new folks were offering seemed too good to pass up, but maybe it was all too good to be true. Net profits from our tennis business aren't going to be very strong if we have to keep on sending people all these free racquets."

As Ted is rummaging through the broken racquets in the basement, what kind of information should he be seeking and how might he structure it in the form of a contingency table? What kinds of probabilities might be useful regarding racquet failure, and what implications could Ted's findings have for helping his grandfather better satisfy the company's racquet customers?

Springdale Shopping Survey

The contingency tables and relative frequency probabilities in this exercise are based on the Springdale Shopping Survey database. Data are in the computer file **SHOPPING**. Information like that gained from the two parts of this exercise could provide helpful insights into the nature of the respondents, their perceptions, and their spending behaviors. In particular, part 2 examines how conditional probabilities related to spending behavior might vary, depending on the gender of the respondent.

1. Based on the relative frequencies for responses to each variable, determine the probability that a randomly selected respondent
   a. [variable 4] spends at least $15 during a trip to Springdale Mall.
   b. [variable 5] spends at least $15 during a trip to Downtown.
   c. [variable 6] spends at least $15 during a trip to West Mall.

   Comparing the preceding probabilities, which areas seem strongest and weakest in terms of the amount of money a shopper spends during a typical shopping visit?
   d. [variable 11] feels that Springdale Mall has the highest-quality goods.
   e. [variable 11] feels that Downtown has the highest-quality goods.
   f. [variable 11] feels that West Mall has the highest-quality goods.

   Comparing the preceding probabilities, which areas are strongest and weakest in terms of the quality of goods offered?
2. Set up a contingency table for the appropriate variables, then determine the following probabilities:
   a. [variables 4 and 26] Given that the random respondent is a female, what is the probability that she spends at least $15 during a trip to Springdale Mall? Is a male more likely or less likely than a female to spend at least $15 during a visit to this area?
   b. [variables 5 and 26] Given that the random respondent is a female, what is the probability that she spends at least $15 during a trip to Downtown? Is a male more likely or less likely than a female to spend at least $15 during a visit to this area?
   c. [variables 6 and 26] Given that the random respondent is a female, what is the probability that she spends at least $15 during a trip to West Mall? Is a male more likely or less likely than a female to spend at least $15 during a visit to this area?

Based on the preceding probabilities, at which shopping areas are males and females most likely and least likely to spend $15 or more during a shopping visit?

BUSINESS CASE

Baldwin Computer Sales (B)

In the Baldwin Computer Sales case, visited previously in Chapter 3, one of the areas of concern regarding the student purchase program was whether the probability of defaulting on the computer payments might be related to the university attended by the student. Using the **BALDWIN** data file, first construct an appropriate contingency table of frequencies and determine the overall percentage of students who defaulted on their computer payments. Then address the following:

1. Given that a student attended university number 1, determine the conditional probability that he or she defaulted on the computer payments.
2. Repeat question 1 for each of the other universities. Do any of the conditional probabilities seem especially high or low compared to the others? In general, which school is associated with those most likely to default, and which school is associated with those least likely to default?
3. If a student is randomly chosen from those who have defaulted on their computer payments, determine the revised probability that the student is from the "most likely to default" university identified in question 2.

CHAPTER 6

Discrete Probability Distributions

ATM USERS ARE DISCRETE AS WELL AS DISCREET

Yes, we know automatic teller machine (ATM) customers use discretion as to when and how they use the ATM. For example, many zealously shield the keyboard when entering their personal ID number into the machine.

But ATM users are also discrete: They are members of a discrete probability distribution. For x = the number of persons arriving at the machine over a period of time, x will be a discrete variable—0, 1, 2, 3, and so on.

The next time you're in the vicinity of your local ATM, relax and enjoy a soft drink for 5 minutes while you count how many people arrive at the machine during that time. Do this on 9 more occasions. Now figure out what percent of the time x was equal to 0, what percent of the time x was equal to 1, and so on. Congratulations, you've just created a discrete probability distribution, a type of distribution that is the topic of this chapter.

More specifically, the phenomenon you've been observing tends to be well represented by a theoretical discrete probability distribution called the Poisson distribution, discussed later in the chapter. This distribution is useful in both describing and predicting the number of people arriving at an ATM, a tollbooth, an emergency room, and many other kinds of service facilities over a given period of time.

Stuck in the Friday afternoon discrete probability distribution . . .

LEARNING OBJECTIVES

After reading this chapter, you should be able to:

- Understand both the concept and the applications of a probability distribution for a random variable.
- Determine whether the random variable in a probability distribution is of the discrete type or the continuous type.
- Differentiate among the binomial, hypergeometric, and Poisson discrete probability distributions and their applications.
- Understand what is meant by a Bernoulli process and how this applies to the consecutive trials associated with the binomial distribution.
- Use the appropriate probability distribution in determining the probability that a discrete random variable will have a given value or a value in a given range.

6.1 INTRODUCTION

Chapter 2 introduced the relative frequency distribution as a way of describing the proportion of actual observations that either had one of several possible values or fell into one of a number of different ranges of values. In Chapter 5, we discussed classical probability as a probability that could be theoretically determined without engaging in any experiments [e.g., $P(\text{heads}) = 0.5$ for a coin flip]. In this chapter, we'll see how a probability distribution is really a very special combination of these two concepts.

In general, we can define a **probability distribution** as *the relative frequency distribution that should theoretically occur for observations from a given population.* In business and other contexts, it can be helpful to proceed from (1) a basic understanding of how a natural process seems to operate in generating events to (2) identifying the probability that a given event may occur. By using a probability distribution as a model that represents the possible events and their respective likelihoods of occurrence, we can make more effective decisions and preparations in dealing with the events that the process is generating.

Random Variables: Discrete versus Continuous

A **random variable** is a variable that can take on different values according to the outcome of an experiment. Random variables can be either *discrete* or *continuous*, a distinction discussed briefly in Chapter 1:

DISCRETE RANDOM VARIABLE A random variable that can take on only certain values along an interval, with the possible values having gaps between them. For example, in a given group of five children, the number who got at least one electronic toy for Christmas would be 0, 1, 2, 3, 4, or 5. It could not be a number between any of these values, such as 2.338.

CONTINUOUS RANDOM VARIABLE A random variable that can take on a value at *any* point along an interval. The exact temperature outside as you're reading this book could be 13.568, 78.352, or 83.815 degrees Fahrenheit, or any of an infinity of other values in the range of temperatures where colleges and universities are located. (The temperature examples have been expressed to just three decimal places here, but there would be countless possibilities along the temperature continuum.)

A random variable is described as *random* because we don't know ahead of time exactly what value it will have following the experiment. For example, when we toss a coin, we don't know for sure whether it will land heads or tails. Likewise, when we measure the diameter of a roller bearing, we don't know in advance what the exact measurement will be. In this chapter, the emphasis will be on discrete random variables and their probability distributions. In the next, we'll cover variables of the continuous type.

The Nature of a Discrete Probability Distribution

A **discrete probability distribution** is a listing of all possible outcomes of an experiment, along with their respective probabilities of occurrence. We'll begin our discussion of such distributions with a small-scale example.

EXAMPLE

Discrete Probability Distribution

An experiment is conducted in which a fair coin is flipped twice. The result of the experiment will be the random variable x = the number of times that heads comes up.

SOLUTION

Since the two tosses are independent, the probability of heads remains constant at $P(\text{heads}) = 0.5$ from one toss to the next. Each of four possible sequences is equally likely, and each has a probability of 0.25:

Two tosses of a coin: Possible sequences and their probabilities

Sequence	x = Number of Heads	Probability
HH	2	0.25 = (0.5 × 0.5)
HT	1	0.25
TH	1	0.25
TT	0	0.25
		1.00

Because the possible sequences are mutually exclusive (only one of them can happen for a given series of two tosses), the sum of the joint probabilities is 1.00. Two of the sequences (*HT* and *TH*) involve one head, and we can use the addition rule from Chapter 5 to find the probability that $x = 1$. This can be expressed as $P(HT \text{ or } TH) = P(HT) + P(TH) = 0.25 + 0.25 = 0.50$. We have just generated the following *probability distribution* for the discrete random variable x = number of heads in two coin tosses:

Two tosses of a coin: Discrete probability distribution for the random variable x = number of heads

x	$P(x)$
0	0.25
1	0.50
2	0.25
	1.00

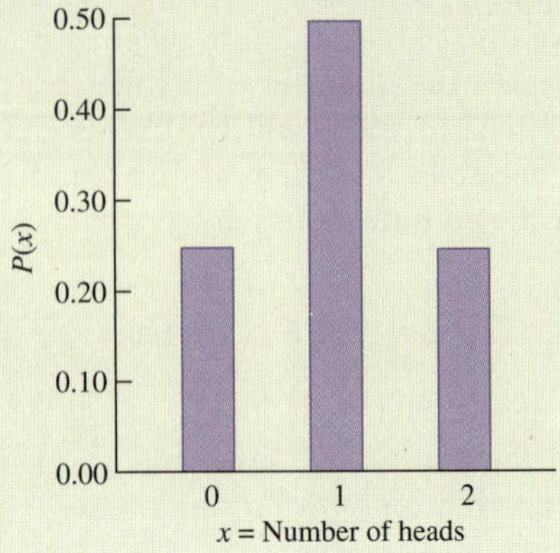

FIGURE 6.1

Discrete probability distribution for the random variable x = number of heads in two consecutive tosses of a coin.

This distribution can also be described graphically, as in Figure 6.1. This discrete probability distribution reflects several important characteristics shared by all such distributions:

Characteristics of a discrete probability distribution:

1. For any value of x, $0 \leq P(x) \leq 1$.
2. The values of x are *exhaustive:* The probability distribution includes all possible values.
3. The values of x are *mutually exclusive:* Only one value can occur for a given experiment.
4. The sum of their probabilities is one, or $\sum P(x_i) = 1.0$.

Although discrete probability distributions are often based on classical probabilities (as in Figure 6.1), they can be the result of relative frequencies observed from past experience or even from subjective judgments made about future events and their likelihood of occurrence. As an example of a discrete probability distribution based on relative frequencies, consider the following situation.

EXAMPLEEXAMPLEEXAMPLE

EXAMPLE

Discrete Probability Distribution

A financial counselor conducts investment seminars with each seminar limited to 6 attendees. Because of the small size of the seminar group and the personal attention each person receives, some of the attendees become clients following the seminar. For the past 20 seminars she has conducted, x (for x = the number of attendees who become clients) has had the relative frequency distribution shown in Figure 6.2.

SOLUTION

The probability distribution for x = the number of attendees who become clients can also be described in tabular form. The events and their respective

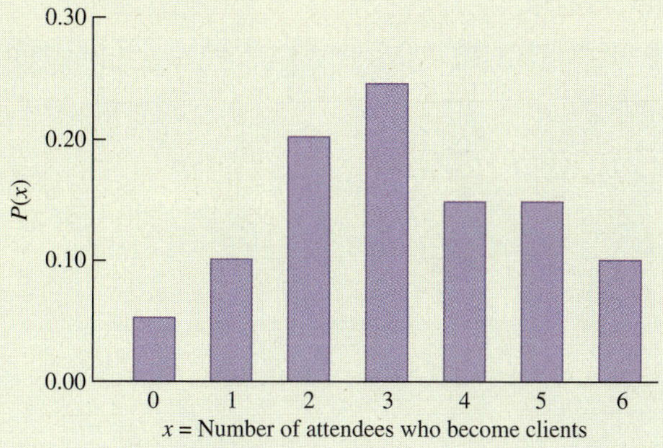

FIGURE 6.2

Relative frequency distribution for the number of attendees in the seminar group who become clients of the seminar leader.

probabilities are as follows:

Twenty investment seminars: Discrete probability distribution for the random variable x = number of attendees who become clients →

x	$P(x)$
0	0.05
1	0.10
2	0.20
3	0.25
4	0.15
5	0.15
6	0.10
	1.00

As in the classical relative frequency distribution of Figure 6.1, the events in Figure 6.2 are mutually exclusive and exhaustive. Interpreting their relative frequencies as probabilities, as shown in Figure 6.2, the sum of these probabilities will be 1.0. In a given seminar, the probability that nobody will become a client is $P(0) = 0.05$, while the probability that at least 4 will become clients is $P(x \geq 4) = P(x = 4) + P(x = 5) + P(x = 6)$, or $0.15 + 0.15 + 0.10 = 0.40$.

AMPLEEXAMPLEEXAMPLEEXAMPLE

The Mean and Variance of a Discrete Probability Distribution

For the observed data in Chapter 3, we used measures of central tendency and dispersion to describe *typical* observations and *scatter* in the data values, respectively. Similarly, it can be useful to describe a discrete probability distribution in terms of its central tendency and dispersion. For example, the investment counselor of Figure 6.2 may be interested both in how many attendees from a seminar *typically* become clients and in the extent to which the number who become clients *varies*.

The *mean* of a discrete probability distribution for a discrete random variable is called its *expected value*, and this is referred to as $E(x)$, or μ. In the following formula, it can be seen to be a weighted average of all the possible outcomes, with each outcome weighted according to its probability of occurrence. The *variance* is the expected value of the squared difference between the random variable and its mean, or $E[(x - \mu)^2]$:

General formulas for the mean and variance of a discrete probability distribution:

- **Mean:**

$$\mu = E(x) \quad \text{or} \quad \mu = \sum x_i P(x_i) \text{ for all possible values of } x$$

- **Variance:**

$$\sigma^2 = E[(x - \mu)^2] \quad \text{or} \quad \sigma^2 = \sum (x_i - \mu)^2 P(x_i)$$

for all possible values of x, and the standard deviation is

$$\sigma = \sqrt{\sigma^2}$$

NOTE

An alternative formula for σ^2 is $\sigma^2 = \left[\sum x_i^2 P(x_i)\right] - \mu^2$. It provides greater ease of calculation when the data values are large or when there are a great many possible values of x.

To demonstrate how the mean and variance are determined, we will compute these descriptors for the discrete probability distribution in Figure 6.2.

Mean and Variance for x = the Number of Attendees Who Become Clients (Represented by the Discrete Probability Distribution of Figure 6.2)

- Mean:

$$\begin{aligned}\mu &= E(x) = \sum x_i P(x_i)\\ &= 0(0.05) + 1(0.10) + 2(0.20) + 3(0.25) + 4(0.15) + 5(0.15) + 6(0.10)\\ &= 3.2 \text{ attendees who become clients}\end{aligned}$$

- Variance:

$$\begin{aligned}\sigma^2 &= E[(x - \mu)^2] = \sum (x_i - \mu)^2 P(x_i)\\ &= (0 - 3.2)^2(0.05) + (1 - 3.2)^2(0.10) + (2 - 3.2)^2(0.20)\\ &\quad + (3 - 3.2)^2(0.25) + (4 - 3.2)^2(0.15) + (5 - 3.2)^2(0.15)\\ &\quad + (6 - 3.2)^2(0.10)\\ &= 2.66\end{aligned}$$

and

$$\sigma = \sqrt{2.66} = 1.63 \text{ attendees who become clients}$$

In a given seminar, there will never be exactly $x = 3.2$ attendees who become clients. However, this is the *expected value,* or *mean* of x, and it will tend to be the average result if the probability distribution remains unchanged and a great many such seminars are conducted.

The formulas in this section can be used to determine the mean and variance for *any* discrete probability distribution. For some distributions, however, the mean and variance can be described and computed by using expressions that are both more practical and more easily calculated. As we discuss the distributions to follow, such expressions for the mean and variance will be presented as they become relevant.

Before we move on to examine three particular discrete probability distributions, the binomial, the hypergeometric, and the Poisson, consider the probability distribution represented by a state's three-digit daily lottery number. There are

STATISTICS IN ACTION 6.1

The Discrete Uniform Distribution That Took a Day Off

A feature of the Pennsylvania State Lottery is the "daily number," a three-digit number selected at 7 P.M. each weekday evening. The number is obtained from the first ball that pops from each of three machines containing balls numbered from 0 to 9. The big event is televised, with a senior-citizen witness and a lottery official present to verify the number of each ball.

Since the daily number can range from 000 to 999, the chance of any specific result is just 0.001. Lottery officials rigorously weigh the balls and conduct extensive testing to make sure some numbers don't have a higher probability than others. A number of years ago, an individual was convicted for injecting a latexlike substance that made some of the balls heavier. (The author and many others watching television noticed that some of the bouncing balls seemed a little "tired.") As a result of this one-day nonuniformity of the probability distribution, security was heightened considerably, and officials are now more scrupulous than ever in their efforts to ensure that the probability distribution is indeed uniform.

Despite the uniformity of this discrete probability distribution, some players still insist on playing a "lucky" number based on their birth date, wedding date, or digits from an auto license number. Some even avoid playing yesterday's number because they think it's been "used up" for awhile. Many seem to prefer 333 or 777, although 923, 147, or any other choice would have the same probability of being selected.

1000 possible outcomes (000 to 999) and, because all of the outcomes are equally likely, this is an example of what is called a *discrete uniform distribution.* Attempts to tamper with this uniformity are not well received by state officials. See Statistics in Action 6.1.

EXERCISES

6.1 Why is a random variable said to be "random"?

6.2 Indicate whether each of the following random variables is discrete or continuous.

a. The diameter of aluminum rods coming off a production line
b. The number of years of schooling employees have completed
c. The Dow Jones Industrial Average
d. The volume of milk purchased during a supermarket visit

6.3 Indicate whether each of the following random variables is discrete or continuous.

a. The number of drive-through customers at your local McDonald's today
b. The amount spent by all customers at the local McDonald's today
c. The time, as reported by your digital watch
d. The length of a pedestrian's stride

6.4 The values of x in a discrete probability distribution must be both exhaustive and mutually exclusive. What is meant by these two terms?

6.5 A discrete random variable can have the values $x = 3$, $x = 8$, or $x = 10$, and the respective probabilities are 0.2, 0.7, and 0.1. Determine the mean, variance, and standard deviation of x.

6.6 Determine the mean, variance, and standard deviation of the following discrete probability distribution:

x	0	1	2
$P(x)$	0.60	0.30	0.10

6.7 Determine the mean, variance, and standard deviation of the following discrete probability distribution:

x	0	1	2	3	4
$P(x)$	0.10	0.30	0.30	0.20	0.10

6.8 Ed Tompkins, the assistant dean of a business school, has applied for the position of dean of the school of business at a much larger university. The salary at the new university has been advertised as $200,000, and Ed is very excited about this possibility for making a big career move. He has been told by friends within the administration of the larger university that his resume is impressive and his chances of getting the position he seeks are "about 60%." If Ed stays at his current position, his salary next year will be $120,000. Assuming that his friends have accurately assessed his chances of success, what is Ed's expected salary for next year?

6.9 Scam artists sometimes "stage" accidents by purposely walking in front of slow-moving luxury cars whose drivers are presumed to carry large amounts of liability insurance. The "victim" then offers to accept a small monetary cash settlement so the driver can avoid an insurance claim, a blot on his driving record, and increased insurance premiums. Marlin has just tried to stage such a scheme by walking into a slow-moving Mercedes and faking a minor injury. Assume that the possible outcomes for Marlin are −$500 (i.e., he is arrested and fined for his scam), $0 (the driver recognizes the scam and refuses to pay), and $100 (the driver simply pays up to get rid of Marlin). If the probabilities for these outcomes are 0.1, 0.3, and 0.6, respectively, what is Marlin's expected monetary outcome for trying to scam the Mercedes driver?

6.10 A regional manager visits local fast-food franchises and evaluates the speed of the service. If the manager receives her meal within 45 seconds, the server is given a free movie-admission coupon. If it takes more than 45 seconds, the server receives nothing. Throughout the company's franchises, the probability is 0.60 that a meal will be served within 45 seconds. What is the expected number of coupons a counter employee will receive when serving the regional manager?

6.11 A consultant has presented his client with three alternative research studies that could be conducted. If the client chooses alternative *a*, the consultant will have to hire 1 additional employee. Two new employees will have to be hired if the client chooses *b*, and 5 if the client chooses *c*. The probability that the client will select alternative *a* is 0.1, with $P(b) = 0.5$ and $P(c) = 0.4$. Identify the discrete random variable in this situation and determine its expected value.

6.12 A contractor must pay a $40,000 penalty if construction of an expensive home requires more than 16 weeks. He will receive a bonus of $10,000 if the home is completed within 8 weeks. Based on experience with this type of project, the contractor feels there is a 0.2 chance the home will require more than 16 weeks for completion, and there is a 0.3 chance it will be finished within 8 weeks. If the price of the home is $550,000 before any penalty or bonus adjustment, how much can the buyer expect to pay for her new home when it is completed?

6.13 A music shop is promoting a sale in which the purchaser of a compact disk can roll a die, then deduct a dollar from the retail price for each dot that shows on the rolled die. It is equally likely that the die will come up any integer from 1 through 6. The owner of the music shop pays $5.00 for each compact disk, then prices them at $9.00. During this special promotion, what will be the shop's average profit per compact disk sold?

6.14 Laura McCarthy, the owner of Riverside Bakery, has been approached by insurance underwriters trying to convince her to purchase flood insurance. According to local meteorologists, there is a 0.01 probability that the river will flood next year. Riverside's profits for the coming year depend on whether Laura buys the flood insurance and whether the river floods. The profits (which take into consideration the $10,000 premium for the flood insurance) for the four possible combinations of Laura's choice and river conditions are:

		The River	
		Does Not Flood	Floods
Insurance Decision	No Flood Insurance	$200,000	−$1,000,000
	Get Flood Insurance	$190,000	$200,000

a. If Laura decides not to purchase flood insurance, use the appropriate discrete probability distribution to determine Riverside's expected profit next year.
b. If Laura purchases the flood insurance, what will be Riverside's expected profit next year?
c. Given the results in parts (a) and (b), provide Laura with a recommendation.

6.15 In 2007, nearly 24 million tons of steel mill products went to construction and contracting companies. Transco Steel, a hypothetical manufacturer specializing in the production of steel for this market, is faced with aging production facilities and is considering the possible renovation of its plant. The company has reduced its alternatives to just three: (1) no renovation at all, (2) minor renovation, and (3) major renovation. Whichever strategy it chooses, the company's profits will depend on the size of the construction and contracting industry in future years. Company executives have a subjective probability of 0.3 that the industry will not grow at all, a probability of 0.5 that it will grow moderately, and a probability of 0.2 that it will show a high rate of growth. The company's estimated profits (in millions of dollars) for each combination of renovation strategy and industry growth possibility are:

Source: *Statistical Abstract of the United States 2009*, p. 630.

		Industry Growth		
		None	Moderate	High
Level of Plant Renovation	None	$35	$38	$42
	Minor	$28	$45	$55
	Major	$21	$40	$70

Given the executives' subjective probabilities for the three possible states of industry growth, what will be the company's expected profit if it chooses not to renovate the plant? If it chooses minor renovations to the plant? If it chooses major renovations to the plant? Based on these expected values, provide the company's management with a recommendation as to which level of renovation they should choose.

THE BINOMIAL DISTRIBUTION 6.2

One of the most widely used discrete distributions, the **binomial distribution** deals with consecutive trials, each of which has two possible outcomes. This latter characteristic is the reason for the "bi" part of the descriptor *binomial*. The outcomes can be identified in general terms, such as "success" versus "failure" or "yes" versus "no," or by more specific labels, such as "milk in the refrigerator" versus "no milk in the refrigerator." The binomial distribution relies on what is known as the **Bernoulli process:**

Characteristics of a Bernoulli process:

1. There are two or more consecutive trials.
2. In each trial, there are just two possible outcomes—usually denoted as *success* or *failure*.
3. The trials are statistically independent; that is, the outcome in any trial is not affected by the outcomes of earlier trials, and it does not affect the outcomes of later trials.
4. The probability of a success remains the same from one trial to the next.

Description and Applications

In the binomial distribution, the discrete random variable, x, is the number of successes that occur in n consecutive trials of the Bernoulli process. The distribution, which has a wide variety of applications that will be demonstrated through examples and chapter exercises, can be summarized as follows:

The binomial probability distribution:

The probability of exactly x successes in n trials is

$$P(x) = \frac{n!}{x!(n-x)!}\pi^x(1-\pi)^{n-x}$$

where n = number of trials
x = number of successes
π = the probability of success in any given trial
$(1-\pi)$ = the probability of failure in any given trial

- Mean:

$$\mu = E(x) = n\pi$$

- Variance:

$$\sigma^2 = E[(x-\mu)^2] = n\pi(1-\pi)$$

In the binomial distribution, we can view the n trials as comprising (1) the x trials that are successes and (2) the $(n - x)$ trials that are failures. The second component in the formula, π^x, is the probability that x of x trials will all be successful. The third component, $(1 - \pi)^{n-x}$, is the probability that $(n - x)$ of $(n - x)$

trials will all be failures. The first component is the number of possible combinations for n items taken x at a time; in the context of the binomial distribution, this is the number of possible sequences where there are exactly x successes in n trials.

The binomial distribution is really a *family* of distributions, and the exact *member* of the family is determined by the values of n and π. The following observations may be made regarding the Bernoulli process and the requirement that the probability of success (π) remain unchanged:

1. If **sampling** is done **with replacement** (the person or other item selected from the population is observed, then put back into the population), π will be constant from one trial to the next.
2. If, when **sampling without replacement,** the number of trials (n) is very small compared to the population (N) of such trials from which the sample is taken, as a practical matter π can be considered to be constant from one trial to the next. This would occur, for example, if $n = 5$ interview participants were randomly selected from a population of $N = 1000$ persons. As a rule of thumb, if the population is at least 20 times as large as the number of trials, we will assume the "constant π" assumption has been satisfactorily met.

EXAMPLEEXAMPLEEXAMPLEEXAMPLEEXAMPLEEXAMPLEEXAMPLEEXAMPLEEXAMPLEEXAMPLEEXAMP

EXAMPLE

Binomial Distribution

Of the 18,000,000 part-time workers in the United States, 20% participate in retirement benefits.[1] A group of 5 people is to be randomly selected from this group, and the discrete random variable is x = the number of persons in the group who participate in retirement benefits.

SOLUTION

This example can be considered a Bernoulli process, with $n = 5$ trials and $\pi = 0.2$. The size of the population ($N = 18{,}000{,}000$) is extremely large compared to the number of trials, so π can be assumed to be constant from one trial to the next. A worker either participates in retirement benefits or does not participate, so there are only two possible outcomes for each of the 5 trials.

What Is the Expected Value of *x*?

The expected number of people in the group who participate can be calculated as $E(x) = n\pi = 5(0.2)$, or $E(x) = 1.0$. The expectation would be that 1.0 persons in the group would participate in retirement benefits. (In other words, if we were to randomly select a great many such groups, we would expect the average value of x to be 1.0.)

What Is the Probability That the Group Will Include Exactly Two Participants?

Using the binomial formula, with $n = 5$, $\pi = 0.2$, and $(1 - \pi) = (1.0 - 0.2) = 0.8$, the probability of x = exactly 2 participants is

$$P(x) = \frac{n!}{x!(n-x)!}\pi^x(1-\pi)^{n-x}$$

[1]Source: *Statistical Abstract of the United States 2006*, pp. 367, 635.

and

$$P(2) = \frac{5!}{2!(5-2)!}(0.2)^2(0.8)^{5-2} = \boxed{0.205}$$

$$\boxed{\text{i.e., } \frac{5 \cdot 4 \cdot 3 \cdot 2 \cdot 1}{(2 \cdot 1)(3 \cdot 2 \cdot 1)} = \frac{5 \cdot 4}{2 \cdot 1}}$$

For Each of the Possible Values of *x*, What Is the Probability That Exactly This Number of People Out of the Five Will Be Participants?

The values $x = 0$ through $x = 5$ include all of the possibilities for the number of persons in the sample who are participants; their probabilities would be

$$P(0) = \frac{5!}{0!(5-0)!}(0.2)^0(0.8)^5 = \boxed{0.328}$$

(*Note:* Remember that $0! = 1$ and that any nonzero number raised to the zero power is also equal to 1.)

$$P(1) = \frac{5!}{1!(5-1)!}(0.2)^1(0.8)^4 = \boxed{0.410}$$

$$P(2) = \frac{5!}{2!(5-2)!}(0.2)^2(0.8)^3 = \boxed{0.205} \quad \text{(as calculated earlier)}$$

$$P(3) = \frac{5!}{3!(5-3)!}(0.2)^3(0.8)^2 = \boxed{0.051}$$

$$P(4) = \frac{5!}{4!(5-4)!}(0.2)^4(0.8)^1 = \boxed{0.006}$$

$$P(5) = \frac{5!}{5!(5-5)!}(0.2)^5(0.8)^0 = \boxed{0.000} \quad \text{(rounded)}$$

We can verify that $\mu = E(x) = n\pi$, or $5(0.20) = 1.0$, by calculating $E(x) = \Sigma x_i P(x_i)$. Because we have rounded the individual probabilities, the result differs slightly from the actual value, $\mu = 1.000$ persons:

$$\begin{aligned} E(x) &= \Sigma x_i P(x_i) \\ &= 0(0.328) + 1(0.410) + 2(0.205) + 3(0.051) + 4(0.006) + 5(0.000) \\ &= 0.997 \text{ persons (differs from } \mu = 1.000 \text{ due to rounding of the individual probabilities)} \end{aligned}$$

What Is the Probability That the Group Will Include at Least Three Persons Who Are Participants?

Because x can take on only one of the possible values between 0 and 5, the outcomes are mutually exclusive. (Notice that the sum of the six probabilities just calculated will be 1.000.) Thus $P(x \geq 3)$ will be

$$\begin{aligned} P(x \geq 3) &= [P(x = 3) + P(x = 4) + P(x = 5)] \\ &= 0.051 + 0.006 + 0.000 = 0.057 \end{aligned}$$

The binomial distribution for the discrete random variable in this example is shown in Figure 6.3. The distribution is positively skewed. Binomial distributions can have different shapes, including being positively skewed, negatively skewed, or even symmetrical.

FIGURE 6.3

Binomial probability distribution for text example with $n = 5$ and $\pi = 0.20$. Of 18 million part-time workers, only 20% participate in retirement benefits. Because the population is very large compared to the number in the sample, π can be considered as remaining constant from one trial to the next.

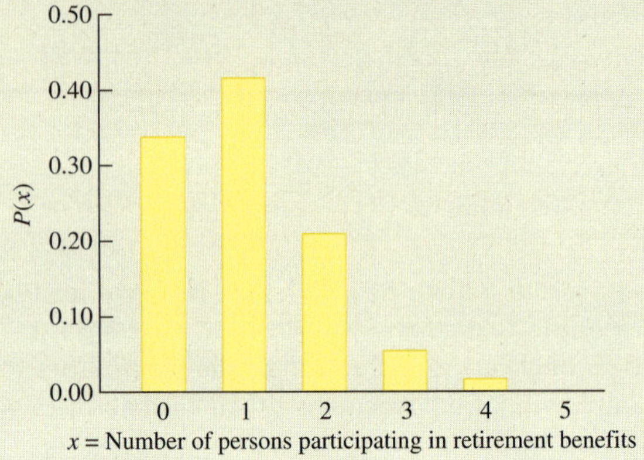

Using the Binomial Tables

As the preceding example indicates, calculating binomial probabilities can be tedious. Fortunately, binomial distribution tables, such as those in Appendix A, make the task a lot easier. We will continue with our example in demonstrating their use.

A given binomial probability distribution depends on the values of π and n, so a separate listing is necessary for various combinations of their values. For the five randomly selected part-time workers, $n = 5$ trials, and we will rely on the individual binomial probabilities table (Table A.1), a portion of which is reproduced here as Table 6.1. The "π" columns in the table range from 0.1 to 0.9 in increments of 0.1. The value of π in our example was $\pi = 0.2$, so this will be the

TABLE 6.1

A portion of the table of *individual* binomial probabilities from Appendix A. For $n = 5$ trials, each entry is $P(x = k)$ for the specified value of the population proportion, π.

Binomial Distribution, Individual Probabilities for x = number of successes in n trials, prob($x = k$)

$n = 5$

	π	0.1	0.2	0.3	0.4	0.5	0.6	0.7	0.8	0.9
	0	0.5905	0.3277	0.1681	0.0778	0.0313	0.0102	0.0024	0.0003	0.0000
	1	0.3281	0.4096	0.3601	0.2592	0.1562	0.0768	0.0284	0.0064	0.0005
k	**2**	0.0729	0.2048	0.3087	0.3456	0.3125	0.2304	0.1323	0.0512	0.0081
	3	0.0081	0.0512	0.1323	0.2304	0.3125	0.3456	0.3087	0.2048	0.0729
	4	0.0005	0.0064	0.0284	0.0768	0.1562	0.2592	0.3601	0.4096	0.3281
	5	0.0000	0.0003	0.0024	0.0102	0.0313	0.0778	0.1681	0.3277	0.5905

For $n = 5$ and $\pi = 0.2$, $P(x = 2)$ is 0.2048.

appropriate column for determining the probabilities that we've just calculated the hard way. Briefly summarizing our earlier treatment of this example, $n = 5$, $\pi = 0.2$, and the probabilities calculated for the possible values of x were

x	0	1	2	3	4	5
$P(x)$	0.328	0.410	0.205	0.051	0.006	0.000

Using the individual binomial probabilities in Table 6.1, we refer to the "$\pi = 0.2$" column within the "$n = 5$" section. For example, Table 6.1 shows $P(x = 0)$ to be 0.3277. Likewise, $P(x = 1)$ and $P(x = 2)$ are 0.4096 and 0.2048, respectively.

Appendix A also includes a table for *cumulative* binomial probabilities (Table A.2), a portion of which is reproduced as Table 6.2. Using this table, we find that $P(x \leq 3) = 0.9933$ for x = the number of participants among the five randomly selected workers.

Using the table of cumulative binomial probabilities is the best way to determine "greater than or equal to" probabilities, such as $P(x \geq 2)$. For example, to find $P(x \geq 2)$,

$$P(x \leq 1) + P(x \geq 2) = 1.000$$

and

From the table of cumulative probabilities, $P(x \leq 1) = 0.7373$.

$$P(x \geq 2) = 1.0000 - P(x \leq 1)$$

so

$$P(x \geq 2) = 1.0000 - 0.7373 = 0.2627$$

As an alternative, the individual binomial probabilities in Table 6.1 could also be used in finding $P(x \geq 2)$. Using this table would be more cumbersome,

TABLE 6.2

A portion of the table of *cumulative* binomial probabilities from Appendix A. For $n = 5$ trials, each entry is $P(x \leq k)$ for the specified value of the population proportion, π.

Binomial Distribution, Cumulative Probabilities for x = number of successes in n trials, prob $(x \leq k)$

$n = 5$

k \ π	0.1	0.2	0.3	0.4	0.5	0.6	0.7	0.8	0.9
0	0.5905	0.3277	0.1681	0.0778	0.0313	0.0102	0.0024	0.0003	0.0000
1	0.9185	0.7373	0.5282	0.3370	0.1875	0.0870	0.0308	0.0067	0.0005
2	0.9914	0.9421	0.8369	0.6826	0.5000	0.3174	0.1631	0.0579	0.0086
3	0.9995	0.9933	0.9692	0.9130	0.8125	0.6630	0.4718	0.2627	0.0815
4	1.0000	0.9997	0.9976	0.9898	0.9688	0.9222	0.8319	0.6723	0.4095
5		1.0000	1.0000	1.0000	1.0000	1.0000	1.0000	1.0000	1.0000

For $n = 5$ and $\pi = 0.2$, $P(x \leq 3)$ is 0.9933.

however, since we would have to add up each of the probabilities for $x = 2$ through $x = 5$, or

$$P(x \geq 2) = P(x = 2) + P(x = 3) + P(x = 4) + P(x = 5)$$
$$= 0.2048 + 0.0512 + 0.0064 + 0.0003$$
$$= 0.2627$$

Excel and Minitab can provide either individual or cumulative binomial probabilities. The procedures and the results are shown in Computer Solutions 6.1.

COMPUTER 6.1 SOLUTIONS

Binomial Probabilities

These procedures show how to get binomial probabilities associated with $n = 5$ trials when the probability of a success on any given trial is $\pi = 0.20$.

For each software package:

Procedure I: Individual or cumulative probabilities for $x = 2$ successes.

Procedure II: Complete set of individual or cumulative probabilities for $x = 0$ successes through $x = 5$ successes.

EXCEL

	A	B	C
1	x	P(X = x)	P(X <= x)
2	0	0.3277	0.3277
3	1	0.4096	0.7373
4	2	0.2048	0.9421
5	3	0.0512	0.9933
6	4	0.0064	0.9997
7	5	0.0003	1.0000

I. Excel procedure for just one individual or cumulative binomial probability

1. Click on the cell into which Excel will place the probability. From the **Formulas** ribbon, click on the **Insert Function** symbol at the extreme left. Click **Statistical** in the **Category** box. Click **BINOMDIST** in the **Function** box. Click **OK.**
2. Enter the number of successes **(2)** into the **Number_s** box. Enter the number of trials **(5)** into the **Trials** box. Enter the probability of success on any given trial **(0.20)** into the **Probability_s** box. To obtain the individual probability, $P(x = 2)$, enter **false** into the **Cumulative** box. To obtain the cumulative probability, $P(x \leq 2)$, enter **true** into the **Cumulative** box. Click **OK.**

II. Excel procedure for a complete set of individual or cumulative binomial probabilities

1. Enter the column labels into **A1:C1** as shown in the printout above. Enter the lowest number of successes in the set (in this case, **0**) into **A2.** Enter **1** into **A3.** Click on cell **A2** and highlight cells **A2:A3.** Click at the lower right corner of **A3** (the cursor will turn to a "+" sign) and drag downward so cells **A2:A7** will contain the values 0 through 5.
2. To get the set of individual probabilities: Click on cell **B2** (this is where the first individual probability will go). Follow the remainder of the steps in Excel procedure I, except enter **A2** into the **Number_s** (number of successes) box and enter **false** into the **Cumulative** box. The first probability now appears in **B2.** Click to select cell **B2.** Click the lower right corner of **B2** (the cursor turns to a "+" sign) and drag down until cells **B2:B7** contain the set of individual probabilities.
3. To get the set of cumulative probabilities: Click on cell **C2** (this is where the first cumulative probability will go). Follow the remainder of the steps in Excel procedure I, except enter **A2** into the **Number_s** (number of successes) box and enter **true** into the **Cumulative** box. The first probability now appears in **C2.** Click to select cell **C2.**

(continued)

Click the lower right corner of **C2** (the cursor turns to a "+" sign) and drag down until cells **C2:C7** contain the set of cumulative probabilities.

MINITAB

```
Probability Density Function      Cumulative Distribution Function

Binomial with n = 5 and p = 0.2   Binomial with n = 5 and p = 0.2

x   P( X = x )                    x   P( X <= x )
0     0.32768                     0       0.32768
1     0.40960                     1       0.73728
2     0.20480                     2       0.94208
3     0.05120                     3       0.99328
4     0.00640                     4       0.99968
5     0.00032                     5       1.00000
```

I. Minitab procedure for just one individual or cumulative binomial probability

1. Click **Calc.** Select **Probability Distributions.** Click **Binomial.**
2. In the **Binomial Distribution** menu, enter **5** into the **Number of trials** box. Enter **0.20** into the **Probability of success** box. Select **Input constant** and enter the desired *x* value (e.g., **2**) into the box. Select either **Probability** or **Cumulative probability,** depending on whether you want an individual probability or a cumulative probability. Click **OK.**

II. Minitab procedure for a complete set of individual or cumulative binomial probabilities

Enter the desired *x* values (in this case **0** through **5**) into column **C1.** Follow steps 1 and 2 above, except select **Input column** and enter **C1** into the box.

Additional Comments on the Binomial Distribution

Under certain conditions, the binomial distribution can be approximated by other probability distributions. This is especially useful since there is a practical limit to the sheer volume of printed tables that can be included in a statistical appendix. A later section of this chapter discusses how another discrete probability distribution called the *Poisson distribution* can be used to approximate the binomial. Introduced in the next chapter, a continuous distribution called the *normal distribution* is especially useful as a satisfactory substitute in many practical applications.

EXERCISES

6.16 What is necessary for a process to be considered a Bernoulli process?

6.17 When we are sampling without replacement, under what conditions can we assume that the constant π assumption has been satisfactorily met?

6.18 Seven trials are conducted in a Bernoulli process in which the probability of success in a given trial is 0.7. If x = the number of successes, determine the following:

a. $E(x)$ b. σ_x c. $P(x = 3)$
d. $P(3 \leq x \leq 5)$ e. $P(x > 4)$

6.19 Twelve trials are conducted in a Bernoulli process in which the probability of success in a given trial is 0.3. If x = the number of successes, determine the following:

a. $E(x)$ b. σ_x c. $P(x = 3)$
d. $P(2 \leq x \leq 8)$ e. $P(x > 3)$

6.20 A city law-enforcement official has stated that 20% of the items sold by pawn shops within the city have been stolen. Ralph has just purchased 4 items from one of the city's pawn shops. Assuming that the official is correct, and for x = the number of Ralph's purchases that have been stolen, determine the following:

a. $P(x = 0)$ b. $P(2 \leq x)$
c. $P(1 \leq x \leq 3)$ d. $P(x \leq 2)$

6.21 The regional manager in Exercise 6.10 plans to evaluate the speed of service at five local franchises of the restaurant. For the five servers, determine the probability that

a. none will receive a movie coupon.
b. two of the five will receive a coupon.
c. three of the five will receive a coupon.
d. all five will receive a coupon.

6.22 According to the National Marine Manufacturers Association, 50.0% of the population of Vermont were boating participants during the most recent year. For a randomly selected sample of 20 Vermont residents, with the discrete random variable x = the number in the sample who were boating participants that year, determine the following: Source: National Marine Manufacturers Association, *NMMA Boating Industry Report*, p. 3.

a. $E(x)$ b. $P(x \leq 8)$ c. $P(x = 10)$
d. $P(x = 12)$ e. $P(7 \leq x \leq 13)$

6.23 In the town of Hickoryville, an adult citizen has a 1% chance of being selected for jury duty next year. If jury members are selected at random, and the Anderson family includes three adults, determine the probability that

a. none of the Andersons will be chosen.
b. exactly one of the Andersons will be chosen.
c. exactly two of the Andersons will be chosen.
d. all three Andersons will be chosen.

6.24 It has been reported that the overall graduation rate for football players at Division I-A colleges and universities is 65%. Source: "How Football Rates," *USA Today*, December 20, 2005, p. 12C.

a. If five players are randomly selected from the entire population of Division I-A football players, what is the probability that at least 2 of the 5 will graduate?
b. In part (a), what is the probability that all 5 of the players will graduate?

6.25 An association of coffee producers has estimated that 50% of adult workers drink coffee at least occasionally while on the job. Given this estimate:

a. For a randomly selected group of 4 adult workers, what is the probability that no more than 3 of them drink coffee at least occasionally while on the job?
b. In part (a), what is the probability that at least 2 of these individuals drink coffee at least occasionally while on the job?

6.26 The U.S. Department of Labor has reported that 30% of the 2.1 million mathematical and computer scientists in the United States are women. If 3 individuals are randomly selected from this occupational group, and x = the number of females, determine $P(x = 0)$, $P(x = 1)$, $P(x = 2)$, and $P(x = 3)$. Source: Bureau of the Census, *Statistical Abstract of the United States 2002*, p. 381.

6.27 Alicia's schedule includes three Tuesday/Thursday courses in which each professor uses a single coin flip to decide whether to give a short quiz each class day. Alicia is active on the university's student advisory council, works two part-time jobs, and is helping to build a float for the upcoming homecoming parade. As a result of these activities and slightly inadequate planning on her part, she has had very little time to prepare for her Thursday classes being held tomorrow. What is the probability that Alicia will be "lucky" and have no surprise quizzes tomorrow? What is the probability that the worst will occur and she will have surprise quizzes in all three classes? What is the probability that she will escape with minimal damage and have a quiz in only one of the three classes?

6.28 OfficeQuip is a small office supply firm that is currently bidding on furniture and office equipment contracts with four different potential customers who are of comparable size. For each contract, OfficeQuip would gain a profit of \$50,000 if that contract were accepted, so the company could make as little as \$0 or as much as \$200,000. The four potential customers are making independent decisions, and in each case the probability that OfficeQuip will receive the contract is 0.40. When all the decisions have been made, what is the probability that OfficeQuip will receive none of the contracts? Exactly one of the contracts? Exactly two of the contracts? Exactly three of the contracts? All four contracts? Overall, what is OfficeQuip's expected profit in this business-procurement venture?

6.29 Four wheel bearings are to be replaced on a company vehicle, and the mechanic has selected the four replacement parts from a large supply bin. Unknown to the mechanic, 10% of the bearings are defective and will fail within the first 100 miles after installation. What is the probability that the company vehicle will have a breakdown due to defective wheel bearings before it completes tomorrow's 200-mile delivery route?

6.30 It has been estimated that one in five Americans suffers from allergies. The president of Hargrove University plans to randomly select 10 students from the undergraduate population of 1400 to attend a dinner at his home. Assuming that Hargrove students are typical in terms of susceptibility to allergies and that the college president's home happens to contain just about every common allergen to which afflicted persons react, what is the probability that at least 8 of the students will be able to stay for the duration of the dinner event? Source: *World Almanac and Book of Facts 2009*, p. 187.

6.31 Airlines book more seats than are actually available, then "bump" would-be passengers whenever more people show up than there are seats. Through the first quarter of 2009, the rate at which passengers were bumped was 1.40 per 1000 passengers. Assuming that, on average, the probability of any given passenger being bumped is 1.40/1000, or 0.00140: Source: U.S. Bureau of Transportation Statistics, June 2009.

a. Emily is flying nonstop to visit Besco, Inc., for a job interview. What is the probability that she will be bumped?
b. Ten members of the Besco board of directors will be flying nonstop to the firm's annual meeting, and their flights are independent of each other. What is the probability that at least one of them will be bumped?
c. Besco sales personnel made a total of 220 nonstop flights last year, and they experienced a total of four bumpings. Could the company's experience be considered as very unusual?

6.32 Every day, artists at Arnold's House of Fine Figurines produce 5 miniature statues that must be sanded and painted. Past experience has shown that 10% of the statues have a defect that does not show up until after the statue has been sanded and painted. Whenever a statue is found to have this defect, Arnold must pay a specialist \$20 to correct the defect and repaint the statue. Describe the probability distribution for the daily amount that Arnold must pay the specialist, and determine the average daily amount the specialist will receive.

THE HYPERGEOMETRIC DISTRIBUTION (6.3)

Description and Applications

The **hypergeometric probability distribution** is similar to the binomial in focusing on the number of successes in a given number of consecutive trials, but there are two major differences: First, *the consecutive trials are not independent;* that is, the outcome in a trial can be affected by the outcomes in earlier trials, and can in turn affect the outcomes in later trials. Second, *the probability of success does not remain constant from one trial to the next.* Unlike the binomial distribution, the hypergeometric distribution does not rely on a Bernoulli process.

The distinction between the binomial and hypergeometric distributions can also be viewed in terms of sampling with replacement versus sampling without replacement. In sampling without replacement, the object selected from the population is not returned to the population. When sampling is done without replacement and the population is not very large, the probability of a success can change quite drastically from one trial to the next.

Situations like these can be analyzed by means of tree diagrams, along with their marginal, conditional, and joint probabilities, but the hypergeometric distribution is a lot easier. The distribution can be described as follows:

The hypergeometric distribution:

The probability of exactly x successes in n trials that are not independent is

$$P(x) = \frac{\binom{s}{x} \cdot \binom{N-s}{n-x}}{\binom{N}{n}}$$

where N = size of the population
n = size of the sample
s = number of successes in the population
x = number of successes in the sample

- Mean: $$\mu = E(x) = \frac{ns}{N}$$
- Variance: $$\sigma^2 = E[(x-\mu)^2] = \frac{ns(N-s)}{N^2} \cdot \frac{N-n}{N-1}$$

EXAMPLE

Hypergeometric Distribution

Twenty businesses located in a small community have filed tax returns for the preceding year. Six of the returns were filled out incorrectly. An Internal Revenue Service (IRS) auditor randomly selects four returns for further examination, and the random variable is x = the number of examined returns that are incorrectly filled out.

What Is the Expected Value of x?

For a sample size of $n = 4$, with $s = 6$ incorrect returns in the population, and a population size of $N = 20$, the expected number of incorrect returns in the sample is

$$E(x) = \frac{ns}{N} = \frac{4(6)}{20} = 1.20 \text{ incorrect returns}$$

What Is the Probability That There Will Be Exactly Two Incorrect Returns Among the Four That Are Chosen for Further Examination?

The quantity to be determined is $P(x = 2)$, with $N = 20$ businesses in the population, $s = 6$ incorrect returns in the population, and $n = 4$ returns randomly selected from the population. Using the hypergeometric formula for $P(x)$:

$$P(x) = \frac{\binom{s}{x} \cdot \binom{N-s}{n-x}}{\binom{N}{n}} \quad \text{and} \quad P(2) = \frac{\binom{6}{2} \cdot \binom{20-6}{4-2}}{\binom{20}{4}} = \frac{\binom{6}{2} \cdot \binom{14}{2}}{\binom{20}{4}}$$

so

$$P(2) = \frac{\dfrac{6!}{2!(6-2)!} \cdot \dfrac{14!}{2!(14-2)!}}{\dfrac{20!}{4!(20-4)!}} = \frac{\dfrac{6 \cdot 5}{2 \cdot 1} \cdot \dfrac{14 \cdot 13}{2 \cdot 1}}{\dfrac{20 \cdot 19 \cdot 18 \cdot 17}{4 \cdot 3 \cdot 2 \cdot 1}} = \frac{1365}{4845} = 0.2817$$

The probability $P(x = 2)$ is 0.2817 that exactly two of the four returns will be incorrect. If we were to repeat this procedure for the other possible values of x, we would find the following probabilities for $x = 0$ through 4 incorrect returns in the sample:

x = Number of incorrect returns:	0	1	2	3	4
$P(x)$:	0.2066	0.4508	0.2817	0.0578	0.0031

As the preceding example illustrates, the calculation of hypergeometric probabilities is rather cumbersome. Unfortunately, with all the possible combinations of the determining characteristics, N, n, and s, tables of hypergeometric probabilities are not a realistic alternative. It is much easier to use the computer, and Computer Solutions 6.2 describes how we can use Excel or Minitab to determine individual or cumulative hypergeometric probabilities.

COMPUTER 6.2 SOLUTIONS

Hypergeometric Probabilities

These procedures show how to get hypergeometric probabilities associated with x = the number of successes in $n = 4$ trials from a population of $N = 20$ when there are $s = 6$ successes within the population. The trials are without replacement. For each software package:

Procedure I: Individual or cumulative probabilities for $x = 2$ successes.

Procedure II: Complete set of individual or cumulative probabilities for $x = 0$ successes through $x = 4$ successes.

EXCEL

	A	B	C
1	x	P(X = x)	P(X <= x)
2	0	0.2066	0.2066
3	1	0.4508	0.6574
4	2	0.2817	0.9391
5	3	0.0578	0.9969
6	4	0.0031	1.0000

I. Excel procedure for obtaining an individual hypergeometric probability

1. Click on the cell into which Excel is to place the individual probability. From the **Formulas** ribbon, click on the **Insert Function** symbol at the extreme left. Click **Statistical** in the **Category** box. Click **HYPGEOMDIST** in the **Function** box. Click **OK.**
2. Enter the number of successes in the sample (**2**) into the **Sample_s** box. Enter the number of trials (**4**) into the **Number_sample** box. Enter the number of successes in the population (**6**) into the **Population_s** box. Enter the size of the population (**20**) into the **Number_pop** box. Click **OK.** The probability, $P(x = 2) = 0.2817$, will appear in the cell selected in step 1.

[Alternatively, you can use Excel worksheet template **tmhyper**.]

II. Excel procedure for a complete set of individual or cumulative hypergeometric probabilities

1. Enter the column labels into **A1:C1** as shown in the printout above. Enter the lowest possible number of successes in the sample (in this case, **0**) into **A2.** Enter **1** into **A3.** Click on cell **A2** and highlight cells **A2:A3.** Click at the lower right corner of **A3** (the cursor will turn to a "+" sign) and drag downward so cells **A2:A6** will contain the values 0 through 4. (Note that the highest possible number of successes in the sample is 4.)
2. To get the set of individual probabilities: Click on cell **B2** (this is where the first individual probability will go). From the **Formulas** ribbon, click on the **Insert Function** symbol at the extreme left. Click **Statistical** in the **Category** box. Click **HYPGEOMDIST** in the **Function** box. Click **OK.** Now follow step 2 in Excel procedure I, except enter **A2** into the **Sample_s** (number of successes in the sample) box. The first individual probability now appears in **B2.**
3. Click to select cell **B2.** Click the lower right corner of **B2** (the cursor turns to a "+" sign) and drag down until cells **B2:B6** contain the set of individual probabilities from $P(x = 0)$ through $P(x = 4)$.
4. To get the set of cumulative probabilities: Copy the value in cell **B2** to cell **C2** (this will be the first cumulative probability). Click on cell **C3** and enter the equation **=C2+B3.** Click the lower right corner of cell **C3** (the cursor turns to a "+" sign) and drag down until cells **C2:C6** contain the set of cumulative probabilities.

(continued)

MINITAB

```
Probability Density Function                  Cumulative Distribution Function

Hypergeometric N = 20, M = 6, and n = 4       Hypergeometric N = 20, M = 6, and n = 4

x   P( X = x )                                x   P( X <= x )
0    0.206605                                 0   0.20660
1    0.450774                                 1   0.65738
2    0.281734                                 2   0.93911
3    0.057792                                 3   0.99690
4    0.003096                                 4   1.00000
```

I. Minitab procedure for just one individual or cumulative hypergeometric probability

1. Click **Calc**. Select **Probability Distributions.** Click **Hypergeometric.**
2. In the **Hypergeometric Distribution** menu, enter **20** into the **Population size** (**N**) box. Enter **6** into the **Event count in population** (**M**) box. Enter **4** into the **Sample size** (**n**) box. Select **Input constant** and enter the desired *x* value (e.g., **2**) into the box. Select either **Probability** or **Cumulative probability,** depending on whether you want an individual probability or a cumulative probability. Click **OK.**

II. Minitab procedure for a complete set of individual or cumulative hypergeometric probabilities

Enter the desired *x* values (in this case, **0 through 4**) into column **C1**. Follow steps 1 and 2 above, except select **Input column** and enter **C1** into the box.

EXERCISES

6.33 Under what circumstances should the hypergeometric distribution be used instead of the binomial distribution?

6.34 Using the hypergeometric distribution, with $N = 4$, $n = 2$, and $s = 3$, determine the following:
a. $P(x = 0)$ b. $P(x = 1)$ c. $P(x = 2)$

6.35 Using the hypergeometric distribution, with $N = 5$, $n = 2$, and $s = 3$, determine the following:
a. $P(x = 0)$ b. $P(x = 1)$ c. $P(x = 2)$

6.36 In a criminal trial, there are 25 persons who have been approved by both parties for possible inclusion in the eventual jury of 12. Of those who have been approved, there are 14 women and 11 men. If the judge forms the final jury of 12 by randomly selecting individuals from the approved listing, what is the probability that at least half of the eventual jurors will be males?

6.37 A computer firm must send a group of three specialists to give a technical presentation to a potential customer. Five such individuals are anxious to attend, but only two are considered to have strong interpersonal skills. Since the specialists have equal levels of seniority, they have agreed to place their names in a hat and allow management to randomly select the three to attend the presentation. What is the probability that there will be at least one person with strong interpersonal skills among the group that is sent?

6.38 Unknown to a rental car office, 3 of the 12 subcompact models now available for rental are subject to a safety recall soon to be announced by the National Highway Traffic Safety Administration. Five subcompacts will be rented today, and the cars will be randomly selected from those available in the pool. What is the probability that exactly one of the recall-affected cars will be rented today? What is the probability that all three of the recall-affected cars will be rented today?

THE POISSON DISTRIBUTION (6.4)

Description and Applications

The **Poisson distribution** is a discrete probability distribution that is applied to events for which the probability of occurrence over a given span of time, space, or distance is extremely small. The discrete random variable, x, is *the number of times* the event occurs over the given span, and x can be 0, 1, 2, 3, and so on, with (theoretically) no upper limit. Besides its ability to approximate the binomial distribution when n is large and π is small, the Poisson distribution tends to describe phenomena like these:

- Customer arrivals at a service point during a given period of time, such as the number of motorists approaching a tollbooth, the number of hungry persons entering a McDonald's restaurant, or the number of calls received by a company switchboard. In this context it is also useful in a management science technique called queuing (waiting-line) theory.
- Defects in manufactured materials, such as the number of flaws in wire or pipe products over a given number of feet, or the number of knots in wooden panels for a given area
- The number of work-related deaths, accidents, or injuries over a given number of production hours
- The number of births, deaths, marriages, divorces, suicides, and homicides over a given period of time

Although it is closely related to the binomial distribution, the Poisson distribution has a number of characteristics that make it unique. Like the binomial distribution, it is a *family* of distributions, but its shape is determined by just *one* descriptor, its mean. In the Poisson distribution, the mean is called λ (the Greek lowercase letter *lambda*) instead of μ. Also, the mean of the distribution is numerically equal to its variance. The distribution can be summarized as follows:

The Poisson distribution:

The probability that an event will occur exactly *x* times over a given span of time, space, or distance is

$$P(x) = \frac{\lambda^x \cdot e^{-\lambda}}{x!}$$

where λ = the mean, or $E(x)$; the expected number of occurrences over the given span

e = the mathematical constant, 2.71828 (e is the base of the natural logarithm system)

(*Note:* In the Poisson distribution, the mean and the variance are equal.)

NOTES

1. When using a pocket calculator to make the computation, remember that $2.71828^{-\lambda}$ is the same as 1 divided by 2.71828^{λ}.
2. Appendix A also contains tables for individual and cumulative Poisson probabilities. The application of these tables will be discussed after we use an example in demonstrating how the Poisson distribution works.

EXAMPLE

Poisson Distribution

In an urban county, health care officials anticipate that the number of births this year will be the same as last year, when 438 children were born—an average of 438/365, or 1.2 births per day. Daily births have been distributed according to the Poisson distribution.

SOLUTION

What Is the Mean of the Distribution?

$E(x) = \lambda$ can be expressed in a number of ways, because it reflects here the number of occurrences over a span of time. Accordingly, the "span of time" can be months ($\lambda = 438/12 = 36.5$ births per month), weeks ($\lambda = 438/52 = 8.42$ births per week), or days ($\lambda = 438/365 = 1.2$ births per day). For purposes of our example, we will be using $\lambda = 1.2$ births per day in describing the distribution.

For Any Given Day, What Is the Probability That No Children Will Be Born?

Using the Poisson formula, with $\lambda = 1.2$ births per day, to find $P(x = 0)$:

$$P(x) = \frac{\lambda^x \cdot e^{-\lambda}}{x!}$$

$$\text{and} \quad P(0) = \frac{1.2^0 \cdot e^{-1.2}}{0!} = \frac{1 \cdot 0.30119}{1} = \boxed{0.3012}$$

Calculate Each of the Following Probabilities: $P(x = 1)$, $P(x = 2)$, $P(x = 3)$, $P(x = 4)$, and $P(x = 5)$

Using the same approach as in the preceding calculations, with $\lambda = 1.2$ mean births per day and values of x from 1 to 5 births per day:

$$P(x = 1 \text{ birth}) = \frac{1.2^1 \cdot e^{-1.2}}{1!} = \frac{(1.2000)(0.30119)}{1} = \boxed{0.3614}$$

$$P(x = 2 \text{ births}) = \frac{1.2^2 \cdot e^{-1.2}}{2!} = \frac{(1.4400)(0.30119)}{2 \cdot 1} = \boxed{0.2169}$$

$$P(x = 3 \text{ births}) = \frac{1.2^3 \cdot e^{-1.2}}{3!} = \frac{(1.7280)(0.30119)}{3 \cdot 2 \cdot 1} = \boxed{0.0867}$$

$$P(x = 4 \text{ births}) = \frac{1.2^4 \cdot e^{-1.2}}{4!} = \frac{(2.0736)(0.30119)}{4 \cdot 3 \cdot 2 \cdot 1} = \boxed{0.0260}$$

$$P(x = 5 \text{ births}) = \frac{1.2^5 \cdot e^{-1.2}}{5!} = \frac{(2.4883)(0.30119)}{5 \cdot 4 \cdot 3 \cdot 2 \cdot 1} = \boxed{0.0062}$$

If we were to continue, we'd find that $P(x = 6) = 0.0012$ and $P(x = 7) = 0.0002$, with both rounded to four decimal places. Including these probabilities

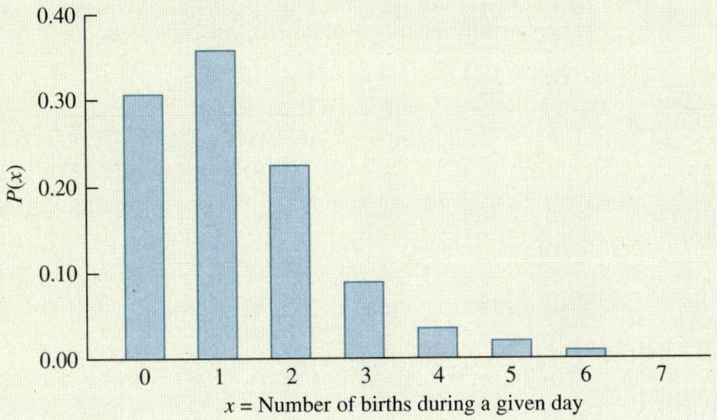

FIGURE 6.4

For the example discussed in the text, the Poisson probability distribution for the number of births per day is skewed to the right. The mean of the distribution is 1.2 births per day, and this descriptor is all that is needed to determine the Poisson probability for each value of x.

with those just calculated, the result would be the following Poisson probability distribution. As Figure 6.4 shows, the distribution is positively skewed.

x	0	1	2	3	4	5	6	7
$P(x)$	0.3012	0.3614	0.2169	0.0867	0.0260	0.0062	0.0012	0.0002

What Is the Probability That No More Than One Birth Will Occur on a Given Day?

Since the events are mutually exclusive, this can be calculated as $P(x = 0) + P(x = 1)$, or $0.3012 + 0.3614$, and $P(x \leq 1) = 0.6626$.

EXAMPLEEXAMPLE

Using the Poisson Tables

The Poisson calculations are much easier than those for the binomial. For example, you need compute $e^{-\lambda}$ only one time for a given value of λ, since the same term is used in finding the probability for each value of x. Using either the individual or the cumulative Poisson distribution table in Appendix A (Tables A.3 and A.4, respectively) makes the job even easier. Table 6.3 shows a portion of Table A.3, while Table 6.4 shows its cumulative counterpart. If the problem on which you are working happens to include one of the λ values listed, you need only refer to the $P(x = k)$ or the $P(x \leq k)$ values in the appropriate table.

In the previous example, λ was 1.2 births per day. Referring to the $\lambda = 1.2$ column of Table 6.3, we see the same probabilities just calculated for $P(x = 1)$ through $P(x = 7)$. For example, in the fifth row of the $\lambda = 1.2$ column, the entry for $P(x = 4)$ is 0.0260. To quickly find the probability that there would be no more than 2 births on a given day, we can refer to the cumulative Poisson probabilities in Table 6.4 and see that $P(x \leq 2)$ is 0.8795.

In some cases, the sum of the individual probabilities in a particular Poisson table will not add up to 1.0000. This is due to (1) rounding and (2) the fact that there is no upper limit to the possible values of x. Some values not listed in the table have $P(x) > 0$, but would appear as 0.0000 because of their extremely small chance of occurrence.

TABLE 6.3

A portion of the table of *individual* Poisson probabilities from Appendix A.

Poisson Distribution, Individual Probabilities for x = number of occurrences, prob ($x = k$)

k \ λ	1.1	1.2	1.3	1.4	1.5	1.6	1.7	1.8	1.9	2.0
0	0.3329	0.3012	0.2725	0.2466	0.2231	0.2019	0.1827	0.1653	0.1496	0.1353
1	0.3662	0.3614	0.3543	0.3452	0.3347	0.3230	0.3106	0.2975	0.2842	0.2707
2	0.2014	0.2169	0.2303	0.2417	0.2510	0.2584	0.2640	0.2678	0.2700	0.2707
3	0.0738	0.0867	0.0998	0.1128	0.1255	0.1378	0.1496	0.1607	0.1710	0.1804
4	0.0203	0.0260	0.0324	0.0395	0.0471	0.0551	0.0636	0.0723	0.0812	0.0902
5	0.0045	0.0062	0.0084	0.0111	0.0141	0.0176	0.0216	0.0260	0.0309	0.0361
6	0.0008	0.0012	0.0018	0.0026	0.0035	0.0047	0.0061	0.0078	0.0098	0.0120
7	0.0001	0.0002	0.0003	0.0005	0.0008	0.0011	0.0015	0.0020	0.0027	0.0034
8	0.0000	0.0000	0.0001	0.0001	0.0001	0.0002	0.0003	0.0005	0.0006	0.0009
9			0.0000	0.0000	0.0000	0.0000	0.0001	0.0001	0.0001	0.0002
10							0.0000	0.0000	0.0000	0.0000

For λ = 1.2 births per day, the probability that there will be exactly 3 births on a given day is $P(x = 3)$, or 0.0867.

TABLE 6.4

A portion of the table of *cumulative* Poisson probabilities from Appendix A.

Poisson Distribution, Cumulative Probabilities for x = number of occurrences, prob ($x \leq k$)

k \ λ	1.1	1.2	1.3	1.4	1.5	1.6	1.7	1.8	1.9	2.0
0	0.3329	0.3012	0.2725	0.2466	0.2231	0.2019	0.1827	0.1653	0.1496	0.1353
1	0.6990	0.6626	0.6268	0.5918	0.5578	0.5249	0.4932	0.4628	0.4337	0.4060
2	0.9004	0.8795	0.8571	0.8335	0.8088	0.7834	0.7572	0.7306	0.7037	0.6767
3	0.9743	0.9662	0.9569	0.9463	0.9344	0.9212	0.9068	0.8913	0.8747	0.8571
4	0.9946	0.9923	0.9893	0.9857	0.9814	0.9763	0.9704	0.9636	0.9559	0.9473
5	0.9990	0.9985	0.9978	0.9968	0.9955	0.9940	0.9920	0.9896	0.9868	0.9834
6	0.9999	0.9997	0.9996	0.9994	0.9991	0.9987	0.9981	0.9974	0.9966	0.9955
7	1.0000	1.0000	0.9999	0.9999	0.9998	0.9997	0.9996	0.9994	0.9992	0.9989
8			1.0000	1.0000	1.0000	1.0000	0.9999	0.9999	0.9998	0.9998
9							1.0000	1.0000	1.0000	1.0000

For λ = 1.2 births per day, the probability that there will be no more than 3 births on a given day is $P(x \leq 3)$, or 0.9662.

As with the binomial distribution, Excel and Minitab can provide either individual or cumulative Poisson probabilities. The procedures and the results are shown in Computer Solutions 6.3.

The Poisson Approximation to the Binomial Distribution

When n is relatively large and π (the probability of a success in a given trial) is small, the binomial distribution can be closely approximated by the Poisson distribution. As a rule of thumb, the binomial distribution can be satisfactorily approximated by the Poisson whenever $n \geq 20$ and $\pi \leq 0.05$. Under these conditions, we can just use $\lambda = n\pi$ and find the probability of each value of x using the Poisson distribution.

COMPUTER 6.3 SOLUTIONS

Poisson Probabilities

These procedures show how to get probabilities associated with a Poisson distribution with mean = 1.2.

For each software package:

Procedure I: Individual or cumulative probabilities for $x = 2$ occurrences.

Procedure II: Complete set of individual or cumulative probabilities for $x = 0$ occurrences through $x = 8$ occurrences.

EXCEL

	A	B	C
1	x	P(X = x)	P(X <= x)
2	0	0.3012	0.3012
3	1	0.3614	0.6626
4	2	0.2169	0.8795
5	3	0.0867	0.9662
6	4	0.0260	0.9923
7	5	0.0062	0.9985
8	6	0.0012	0.9997
9	7	0.0002	1.0000
10	8	0.0000	1.0000

I. Excel procedure for just one individual or cumulative Poisson probability

1. Click on the cell into which Excel will place the probability. From the **Formulas** ribbon, click on the **Insert Function** symbol at the extreme left. Click **Statistical** in the **Category** box. Click **POISSON** in the **Function** box. Click **OK.**
2. Enter the number of occurrences (**2**) into the **X** box. Enter the mean of the distribution (**1.2**) into the **Mean** box. To obtain the individual probability, $P(x = 2)$, enter **false** into the **Cumulative** box. To obtain the cumulative probability, $P(x \leq 2)$, enter **true** into the **Cumulative** box. Click **OK.**

(continued)

II. Excel procedure for a complete set of individual or cumulative Poisson probabilities

1. Enter the column labels into **A1:C1** as shown in the printout. Enter the lowest number of occurrences in the set (in this case, **0**) into **A2.** Enter **1** into **A3.** Click on cell **A2** and highlight cells **A2:A3.** Click at the lower right corner of **A3** (the cursor will turn to a "+" sign) and drag downward so cells **A2:A10** will contain the values 0 through 8.
2. To get the set of individual probabilities: Click on cell **B2** (this is where the first individual probability will go). Follow the remainder of the steps in Excel procedure I, except enter **A2** into the **X** (number of occurrences) box and enter **false** into the **Cumulative** box. The first probability now appears in **B2.** Click to select cell **B2.** Click the lower right corner of **B2** (the cursor turns to a "+" sign) and drag down until cells **B2:B10** contain the set of individual probabilities.
3. To get the set of cumulative probabilities: Click on cell **C2** (this is where the first cumulative probability will go). Follow the remainder of the steps in Excel procedure I, except enter **A2** into the **X** (number of occurrences) box and enter **true** into the **Cumulative** box. The first probability now appears in **C2.** Click to select cell **C2.** Click the lower right corner of **C2** (the cursor turns to a "+" sign) and drag down until cells **C2:C10** contain the set of cumulative probabilities.

MINITAB

```
Probability Density Function          Cumulative Distribution Function
Poisson with mean = 1.2               Poisson with mean = 1.2
x   P( X = x )                        x   P( X <= x )
0     0.301194                        0       0.30119
1     0.361433                        1       0.66263
2     0.216860                        2       0.87949
3     0.086744                        3       0.96623
4     0.026023                        4       0.99225
5     0.006246                        5       0.99850
6     0.001249                        6       0.99975
7     0.000214                        7       0.99996
8     0.000032                        8       1.00000
```

I. Minitab procedure for just one individual or cumulative Poisson probability

1. Click **Calc.** Select **Probability Distributions.** Click **Poisson.**
2. In the **Poisson Distribution** menu, enter **1.2** into the **Mean** box. Select **Input constant** and enter the desired *x* value (e.g., **2**) into the box. Select either **Probability** or **Cumulative probability,** depending on whether you want an individual probability or a cumulative probability. Click **OK.**

II. Minitab procedure for a complete set of individual or cumulative binomial probabilities

Enter the desired *x* values (in this case, **0** through **8**) into column **C1.** Follow steps 1 and 2 above, except select **Input column** and enter **C1** into the box.

EXAMPLEEXAMPLEE

EXAMPLE

Poisson Approximation to Binomial

Past experience has shown that 1.0% of the microchips produced by a certain firm are defective. A sample of 30 microchips is randomly selected from the firm's production. If x = the number of defective microchips in the sample, determine $P(x = 0)$, $P(x = 1)$, $P(x = 2)$, $P(x = 3)$, and $P(x = 4)$.

SOLUTION

The Binomial Approach

The descriptors for the binomial distribution are $\pi = 0.01$, the chance that any given microchip will be defective, and $n = 30$, the number of microchips in the sample. Although we will skip the actual binomial calculations, we would use $\pi = 0.01$ and $n = 30$ in arriving at the probabilities shown below.

The Poisson Approximation Approach

The only descriptor for the Poisson distribution is $\lambda = n\pi = 30(0.01)$, or $\lambda = 0.3$ microchips expected to be defective in the sample. Referring to the Poisson table for $\lambda = 0.3$, we find the following probabilities for $x = 0$ through $x = 4$. Note the very close similarity between each binomial probability and its Poisson approximation:

	P(x), Using the	
x = Number of Defective Microchips in the Sample	Binomial Distribution	Poisson Approximation
0	0.7397	0.7408
1	0.2242	0.2222
2	0.0328	0.0333
3	0.0031	0.0033
4	0.0002	0.0003

EXAMPLEEXAMPLEEXAMPLEEXAMPLEEXAMPLE

EXERCISES

6.39 For a discrete random variable that is Poisson distributed with $\lambda = 2.0$, determine the following:
a. $P(x = 0)$ b. $P(x = 1)$ c. $P(x \le 3)$
d. $P(x \ge 2)$

6.40 For a discrete random variable that is Poisson distributed with $\lambda = 9.6$, determine the following:
a. $P(x = 7)$ b. $P(x = 9)$ c. $P(x \le 12)$
d. $P(x \ge 10)$

6.41 In 2006, there were about 490 motor vehicle thefts for every 100,000 registrations. Assuming (1) a Poisson distribution, (2) a community with a comparable theft rate and 1000 registered motor vehicles, and (3) x = the number of vehicles stolen during the year in that community, determine the following: Source: *Statistical Abstract of the United States 2009*, pp. 188, 672.
a. $E(x)$ b. $P(x = 3)$ c. $P(x = 5)$
d. $P(x \le 8)$ e. $P(3 \le x \le 10)$

6.42 Arrivals at a walk-in optometry department in a shopping mall have been found to be Poisson distributed with a mean of 2.5 potential customers arriving per hour. If x = number of arrivals during a given hour, determine the following:
a. $E(x)$ b. $P(x = 1)$ c. $P(x = 3)$
d. $P(x \le 5)$ e. $P(2 \le x \le 6)$

6.43 The U.S. divorce rate has been reported as 3.6 divorces per 1000 population. Assuming that this rate applies to a small community of just 500 people and is Poisson distributed, and that x = the number of divorces in this community during the coming year, determine the following: Source: *New York Times Almanac 2009*, p. 294.
a. $E(x)$ b. $P(x = 1)$ c. $P(x = 4)$
d. $P(x \le 6)$ e. $P(2 \le x \le 5)$

6.44 During the 12 P.M.–1 P.M. noon hour, arrivals at a curbside banking machine have been found to be Poisson distributed with a mean of 1.3 persons per minute. If x = number of arrivals during a given minute, determine the following:
a. $E(x)$ b. $P(x = 0)$ c. $P(x = 1)$
d. $P(x \le 2)$ e. $P(1 \le x \le 3)$

6.45 During the winter heating season in a northern state, Howard's Heating and Cooling Service receives an average of 3.1 emergency service calls per day from heating customers in their area. Howard has increased personnel and equipment resources to better handle such calls, and his company is now able to satisfy a maximum of 8 emergency service calls per day. During a typical day during the heating season, and assuming the daily number of emergency service calls to be Poisson distributed, what is the probability that Howard will receive more emergency service calls tomorrow than his company is able to handle? Considering this probability, would it seem advisable for Howard to increase or to decrease the level of personnel and equipment he devotes to handling such calls?

6.46 Over the past year, a university's computer system has been struck by a virus at an average rate of 0.4 viruses per week. The university's information technology managers estimate that each time a virus occurs, it costs the university $1000 to remove the virus and repair the damages it has caused. Assuming a Poisson distribution, what is the probability that the university will have the good fortune of being virus-free during the upcoming week? During this same week, what is the expected amount of money that the university will have to spend for virus removal and repair?

6.47 According to the Mortgage Bankers Association of America, the foreclosure rate on home mortgages in 2007 was 2%. Assuming that this rate is applicable to a community where 500 homes have mortgages, use the Poisson approximation to the binomial distribution to determine the following for x = the number of foreclosures in this community during the coming year: Source: *Statistical Abstract of the United States 2009*, p. 727.

a. $E(x)$ b. $P(x = 7)$ c. $P(x = 10)$
d. $P(x \leq 12)$ e. $P(8 \leq x \leq 15)$

6.48 The Federal Aviation Administration reports that American Airlines got 6.06 complaints of mishandled baggage per 1000 passengers in 2008. Assuming this rate applies to the next 500 passengers who depart from Portland, Oregon, on American Airlines flights, what is the probability that at least 5 people will have their baggage mishandled? That no more than 3 people will have their baggage mishandled? Source: *The New York Times Almanac 2009*, p. 426.

6.49 Taxicab drivers are more likely to be murdered on the job than members of any other occupation, even police officers and security guards. The annual murder rate for cab drivers is 30.0 homicides per 100,000 workers, compared to just 0.6 per 100,000 for all occupations. Assume that taxicab companies in a large Northeastern city employ a total of 3000 drivers and that the family of any driver murdered on the job receives $50,000 from the city. What is the probability that at least two of the city's taxicab drivers will be murdered on the job during the next year? What is the approximate expected value for the amount of money the city will be providing to families of murdered taxicab drivers during the next year? Source: "Occupational Homicide Rate Is Highest Among Taxi Drivers," *USA Today*, May 2, 2000, p. 3A.

6.50 A manufacturing plant's main production line breaks down an average of 2.4 times per day. Whenever the line goes down, it costs the company $500 in maintenance, repairs, and lost production. What is the probability that the production line will break down at least 3 times tomorrow? What is the approximate expected value for the amount of money that production line breakdowns will cost the company each day?

6.51 The company in Exercise 6.50 has disciplined a worker who was suspected of pilfering tools and supplies from the plant. The very next day, the production line broke down 9 times. Management has confronted the union with accusations of sabotage on the line, but the union president says it's just a coincidence that the production line happened to break down so many times the day after the worker was disciplined. Using probabilities appropriate to your discussion, comment on how much of a coincidence this high number of breakdowns would appear to be.

6.5 SIMULATING OBSERVATIONS FROM A DISCRETE PROBABILITY DISTRIBUTION

Besides describing the exact probability distribution for a discrete random variable, statistical software can also randomly select observations from the distribution itself. For example, we can use Excel or Minitab to randomly select observations from a given Poisson distribution, then display them in the form of a frequency distribution. Computer Solutions 6.4 describes the necessary procedures. Because we are using two different software packages, there will be two different sets of randomly selected data. We will display and discuss the results for Minitab.

Table 6.5 shows the individual $P(x)$ probabilities as computed by Minitab for three different Poisson distributions ($\lambda = 0.1$, $\lambda = 1.2$, and $\lambda = 5.0$). Next to each

COMPUTER 6.4 SOLUTIONS

Simulating Observations from a Discrete Probability Distribution

These procedures simulate 10,000 observations from a Poisson distribution with mean = 1.2. The procedures for simulating observations from a binomial distribution will differ only slightly from those described here.

EXCEL

1. Start with a blank worksheet. From the **Data** ribbon, click **Data Analysis** in the rightmost menu section. Click **Random Number Generation.** Click **OK.**
2. Enter **1** into the **Number of Variables** box. Enter **10000** into the **Number of Random Numbers** box. Within the **Distribution** box, select **Poisson.** Enter the mean (**1.2**) into the **Lambda** box. Select **Output Range** and enter **A1** into the box. Click **OK.** The 10,000 simulated observations will be located in A1:A10000.

MINITAB

1. Click **Calc.** Select **Random Data.** Click **Poisson.** Enter **10000** into the **Number of rows of data to generate** box. Enter **C1** (the destination column) into the **Store in columns** box. Enter **1.2** into the **Mean** box. Click **OK.** The 10,000 simulated observations will be located in **C1.**
2. To obtain the frequencies for the different values of *x*, click **Stat,** select **Tables,** and click **Tally Individual Variables.** Enter **C1** into the **Variables** box and select **Counts.** Click **OK.** The relative frequencies shown in part B of the right side of Table 6.5 are obtained by dividing each of the frequencies by 10,000.

The frequency distributions in Table 6.6 can be generated using the Minitab procedure described previously in Computer Solutions 2.1. As shown in this earlier procedure, we have the option of displaying the frequencies on the chart istelf.

theoretical probability is the relative frequency with which the x value occurred when Minitab randomly selected 10,000 observations from that distribution.

Note that the relative frequencies for the randomly selected x values are very close, but not necessarily the same as the theoretical probabilities listed in the probability distribution. As we discussed in the previous chapter, $P(\text{heads}) = 0.5$ for a coin flip, but flipping a coin 10 times doesn't mean that you're certain to get heads exactly 5 times. However, due to the law of large numbers, observed relative frequencies will more closely approach the theoretical probability distribution as the number of observations grows larger.

Parts A through C of Table 6.6 show Minitab frequency distributions for 10,000 observations randomly selected from the corresponding probability distribution of Table 6.5. In comparing the three frequency distributions of Table 6.6, notice how they become more symmetrical as λ is increased when we proceed from A to C. Although the Poisson distribution is always positively skewed, it becomes more symmetrical as its mean, λ, becomes larger.

The ability to select random observations from a known or assumed probability distribution also has applications in a number of management science areas. Such applications typically involve using one or more such distributions to mathematically represent real-world phenomena of interest. In general, their purpose is to help the management scientist gain insight into how the processes

TABLE 6.5

Besides describing a probability distribution, the computer can also randomly select observations from the distribution itself.

Minitab output. Poisson probability distribution for the discrete random variable, x:

When Minitab randomly selected 10,000 observations from the distribution, the relative frequency with which each x value appeared:

A. Probability Density Function

Poisson with mean = 0.1

x	P(X = x)	Relative frequency
0	0.904837	0.9056
1	0.090484	0.0892
2	0.004524	0.0050
3	0.000151	0.0002
4	0.000004	0.0000

B. Probability Density Function

Poisson with mean = 1.2

x	P(X = x)	Relative frequency
0	0.301194	0.3060
1	0.361433	0.3588
2	0.216860	0.2085
3	0.086744	0.0918
4	0.026023	0.0273
5	0.006246	0.0063
6	0.001249	0.0011
7	0.000214	0.0002
8	0.000032	0.0000

C. Probability Density Function

Poisson with mean = 5

x	P(X = x)	Relative frequency
0	0.006738	0.0067
1	0.033690	0.0343
2	0.084224	0.0848
3	0.140374	0.1424
4	0.175467	0.1824
5	0.175467	0.1693
6	0.146223	0.1433
7	0.104445	0.1059
8	0.065278	0.0637
9	0.036266	0.0345
10	0.018133	0.0195
11	0.008242	0.0086
12	0.003434	0.0029
13	0.001321	0.0010
14	0.000472	0.0005
15	0.000157	0.0001
16	0.000049	0.0001

TABLE 6.6

Minitab frequency distributions for 10,000 observations randomly taken from each of the three Poisson probability distributions in Table 6.5.

A. 10,000 random observations from Poisson distribution with mean $\lambda = 0.1$

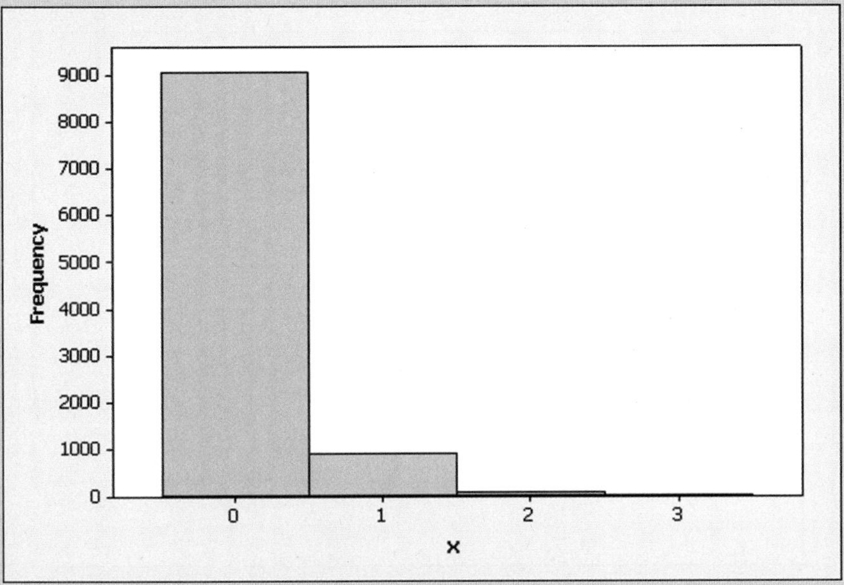

B. 10,000 random observations from Poisson distribution with mean $\lambda = 1.2$

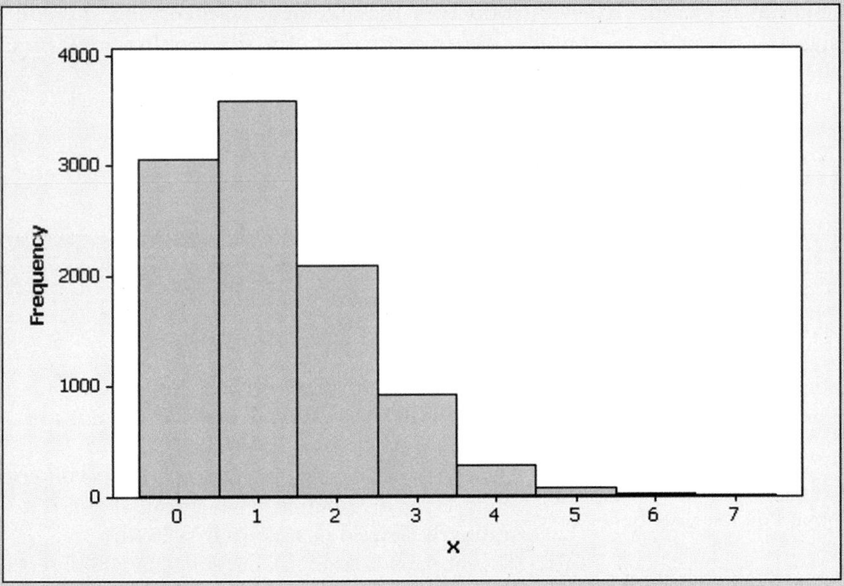

(continued)

C. 10,000 random observations from Poisson distribution with mean λ = 5.0

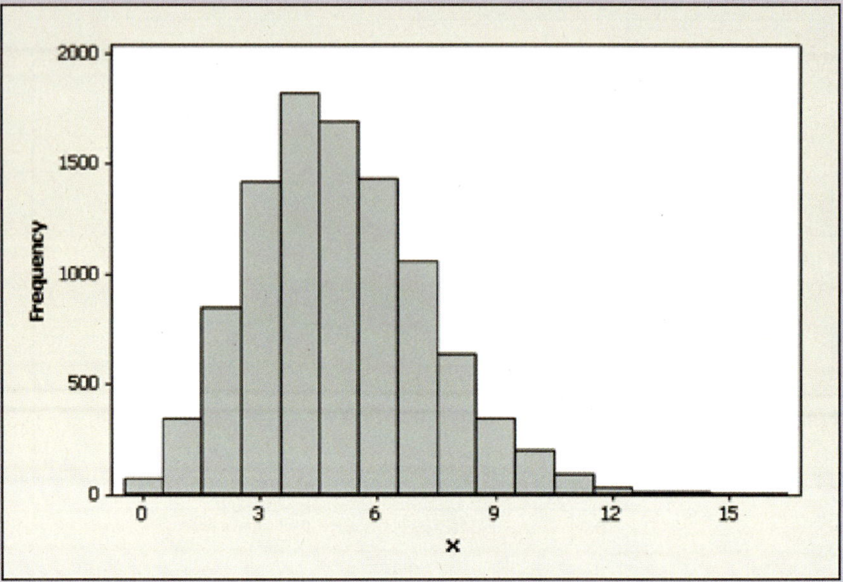

interact, as well as to determine how these interactions might be influenced by managerial decisions. Most introductory management science texts include extensive treatment of the nature, the construction, and the applications of such mathematical models of the real world.

EXERCISES

Note: These exercises require a computer and statistical software capable of selecting random observations from discrete probability distributions.

6.52 Using your statistical software package:

a. Simulate 100 observations from a Poisson distribution having $\lambda = 2.35$.
b. Generate the individual Poisson probabilities for a Poisson distribution having $\lambda = 2.35$.
c. Repeat part (a).
d. Repeat part (b).
e. When part (a) is repeated, the results are now different. However, when part (b) is repeated, the results are the same. What does this say about frequency distributions versus probability distributions?

6.53 Using your statistical software package:

a. Simulate 50 observations from a binomial distribution for which $n = 5$ and $\pi = 0.2$.
b. Generate a frequency distribution for these observations. Is the distribution symmetrical? If not, is it positively skewed or negatively skewed?
c. Repeat parts (a) and (b), but simulate 500 observations. Is the frequency distribution more symmetrical or less symmetrical than the one previously obtained?
d. Comment on how the shape of the distribution would change if you were to continue increasing the number of simulated observations.

SUMMARY

6.6

• Discrete random variable and its probability distribution

A discrete random variable can take on only certain values along a range or continuum, and there are gaps between these possible values. On the other hand, a continuous random variable can have any value along a range or continuum. This chapter has focused on discrete random variables, while the next concentrates on continuous random variables.

The probability distribution for a discrete random variable lists all possible values the variable can assume, along with their respective probabilities of occurrence. There are several types of discrete distributions, each with its own characteristics and applications.

• Binomial distribution and the Bernoulli process

The binomial distribution deals with consecutive trials, each of which has only two possible outcomes, generally referred to as *success* or *failure*. This distribution assumes a Bernoulli process, in which the trials are independent of each other, and the probability of success remains constant from one trial to the next.

• Hypergeometric distribution

Like the binomial distribution, the hypergeometric distribution deals with consecutive trials, and there are only two possible outcomes for each trial. The trials are not statistically independent, however, and the probability of a success is not constant from one trial to the next.

• Poisson distribution

The Poisson distribution applies to events for which the probability of occurrence over a given span of time, space, or distance is extremely small. In this distribution, the random variable is the number of occurrences in the specified span of time, space, or distance. For a Bernoulli process in which the number of trials is large and the probability of success on any given trial is very small, the Poisson distribution can be used as an approximation to the binomial distribution.

• Determining probabilities for a discrete random variable

Appendix A contains tables for the binomial and Poisson distributions. However, computer statistical packages enable the user to determine the probabilities associated with a given distribution without the need to refer to such tables.

• Simulation of observations from a discrete probability distribution

Some computer packages can randomly select a sample of observations from a probability distribution that has already been specified. In addition to being useful in more advanced management science applications such as simulation models intended to represent real-world phenomena, this capability enables the user to easily view the approximate shape of a given probability distribution. Based on the law of large numbers, the observed relative frequencies will more closely approach the theoretical probability distribution as the number of observations grows larger.

EQUATIONS

Characteristics of a Discrete Probability Distribution

- For any value of x, $0 \le P(x) \le 1$.
- The values of x are *exhaustive:* The probability distribution includes all possible values.
- The values of x are *mutually exclusive:* Only one value can occur for a given experiment.
- The sum of their probabilities is 1, or $\Sigma P(x_i) = 1.0$.

General Formulas: Mean and Variance of a Discrete Probability Distribution

- Mean:

$$\mu = E(x) \quad \text{or} \quad \mu = \Sigma x_i P(x_i) \quad \text{for all possible values of } x$$

- Variance:

$$\sigma^2 = E[(x - \mu)^2] \quad \text{or} \quad \sigma^2 = \Sigma(x_i - \mu)^2 P(x_i)$$

for all possible values of x, and the standard deviation is $\sigma = \sqrt{\sigma^2}$

The Binomial Probability Distribution

The probability of exactly x successes in n trials is

$$P(x) = \frac{n!}{x!(n-x)!}\pi^x(1-\pi)^{n-x}$$

where n = number of trials
x = number of successes
π = the probability of success in any given trial
$(1 - \pi)$ = the probability of failure in any given trial

- Mean:

$$\mu = E(x) = n\pi$$

- Variance:

$$\sigma^2 = E[(x - \mu)^2] = n\pi(1 - \pi)$$

The Hypergeometric Distribution

The probability of exactly x successes in n trials that are not independent is

$$P(x) = \frac{\binom{s}{x} \cdot \binom{N-s}{n-x}}{\binom{N}{n}}$$

where N = size of the population
n = size of the sample
s = number of successes in the population
x = number of successes in the sample

- Mean:

$$\mu = E(x) = \frac{ns}{N}$$

- Variance:

$$\sigma^2 = E[(x - \mu)^2] = \frac{ns(N - s)}{N^2} \cdot \frac{N - n}{N - 1}$$

The Poisson Probability Distribution

The probability that an event will occur exactly x times over a given span of time, space, or distance is

$$P(x) = \frac{\lambda^x \cdot e^{-\lambda}}{x!}$$

where λ = the mean, or $E(x)$; the expected number of occurrences over the given span

e = the mathematical constant, 2.71828 (e is the base of the natural logarithm system)

(*Note:* In the Poisson distribution, the mean and the variance are equal.)

CHAPTER EXERCISES

6.54 The owner of a charter fishing boat has found that 12% of his passengers become seasick during a half-day fishing trip. He has only two beds below deck to accommodate those who become ill. About to embark on a typical half-day trip, he has 6 passengers on board. What is the probability that there will be enough room below deck to accommodate those who become ill?

6.55 During 2007, the crash rate for commuter air carriers was 1.0 per 100,000 flying hours. Assuming this rate continues, what is the probability that there will be no more than one crash in the next 50,000 flying hours? Source: *Statistical Abstract of the United States 2009,* p. 662.

6.56 Bill has to sell three more cars this month in order to meet his quota. Tonight he has after-dinner appointments with five prospective customers, each of whom happens to be interested in a different car. If he has a 30% chance of success with each customer, what is the probability that he will meet his quota by tomorrow morning?

6.57 In Exercise 6.56, suppose that three of the customers are interested in the same car, and that they will go elsewhere if it has already been sold. Would it be appropriate to use the binomial distribution under these conditions? Why or why not?

6.58 During the 2009 tax filing season, 15.8% of all individual U.S. tax returns were prepared by H&R Block. For a random selection of 3 tax returns, describe the probability distribution for x = the number in the sample whose returns were prepared by H&R Block. Source: H&R Block, Inc., *Fast Facts 2009.*

6.59 Twenty percent of the population of Maryland describe themselves as binge drinkers. For a randomly selected group of 5 Maryland residents, use the appropriate statistical table to describe the probability distribution for x = the number in the sample who describe themselves as binge drinkers. Source: *Statistical Abstract of the United States 2009,* p. 130.

6.60 A company has installed five intrusion-detection devices in an office building. The devices are very sensitive and, on any given night, each one has a 0.1 probability of mistakenly sounding an alarm when no intruder is present. If two or more of the devices are activated on the same night, the intrusion system automatically sends a signal to the local police. If no intruder is present on a given night, what is the probability that the system will make the mistake of calling the police?

6.61 A mining company finds that daily lost-work injuries average 1.2. If the local union contract has a clause requiring that the mine be shut down as soon as three workers incur lost-work injuries, on what percentage of the days will the mine be operational throughout the day?

6.62 Unknown to a quality-control inspector, 20% of a very large shipment of electric switches are defective. The inspector has been told to reject the shipment if, in a sample of 15 switches, 2 or more are defective. What is the probability that the shipment will be rejected?

6.63 Repeat Exercise 6.29, but under the assumption that there are only 20 bearings in the supply bin, 2 of which are defective.

6.64 A town council includes 10 Democrats and 15 Republicans. In response to a developer's request for a variance that would allow her to build a convenience store in an area of town currently zoned as residential, the mayor is going to randomly select 5 council members for a committee to examine the issue. What is the probability that there will be at least 3 Republicans on the committee?

6.65 Fred and Emily are among the 50 finalists in a lottery in which 5 winners will be randomly selected to receive a free vacation in Hawaii. Using the hypergeometric distribution, determine the probability that exactly one of these two individuals will win a free vacation.

6.66 Seven of the 15 campus police officers available for assignment to the auditorium in which a local politician is to speak have received advanced training in crowd control. If 5 officers are randomly selected for service during the speech, what is the probability that exactly 2 of them will have had advanced training in crowd control? What is the probability that at least 3 of them will have had advanced training?

6.67 A producer of copper tubing has found that, on the average, five flaws occur per 2000 feet of tubing. The tubing is sold in continuous rolls of 200 feet each. What is the probability that a randomly selected 200-foot roll will have no flaws?

6.68 A large plumbing contractor buys eight 200-foot rolls from the producer described in Exercise 6.67. Determine the probability that no more than three of the eight rolls will be flawless.

6.69 In 2004, 50% of all U.S. families had a net worth less than $93,000. What is the probability that exactly three of a randomly selected sample of five families would have a net worth less than $93,000? Source: *Statistical Abstract of the United States 2009*, p. 459.

6.70 A trucking company has found that its trucks average 0.2 breakdowns during round trips from New York to Los Angeles.

a. What is the probability that a single truck will make the complete trip without experiencing a breakdown?
b. If 3 trucks are assigned to a NY/LA round trip, what is the probability that at least 2 of them will make the complete trip without experiencing a breakdown?

6.71 The Bureau of Labor Statistics says that baggage handlers, trash haulers, and coal miners are among the occupations having the highest risk of a debilitating injury or illness. For coal miners working over the course of a year, the injury rate has been reported as 3.6 injuries per 100 coal miners. For a company that employs 200 coal miners, what is the probability that at least 10 of its workers will experience a debilitating injury or illness over the next year? Source: "Workers at Risk," *USA Today*, March 15, 2006, p. 6D.

6.72 According to the U.S. Centers for Disease Control and Prevention, Americans 85 years of age and older experience flu-associated respiratory and circulatory illness at an annual rate of approximately 1200 cases per 100,000 persons. If a rural county has 500 seniors in this age group, what is the probability that no more than 5 of them will experience such an illness during the coming year? If, in such a county, 15 seniors in this age group were to come down with such an illness during the coming year, would this be considered rather unusual? Explain your conclusion. Source: Anita Manning, "Tens of Thousands More People Are Hospitalized for Flu Than Expected," *USA Today*, September 15, 2004, p. 5D.

6.73 Although they have no commercials, public television stations periodically conduct campaigns during which programming is occasionally interrupted by appeals for subscriber donations. A local public television station has found that, during these breaks, the number of calls averages 3 per minute. It has recently upgraded the switchboard so it can handle as many as 5 calls per minute. The station is now in the middle of a program break. What is the probability that the switchboard will be overloaded during the next minute?

6.74 A gorilla once made headlines in 400 newspapers and appeared on three national television shows after correctly "selecting" the winner of 13 out of 20 football games. He had made his "choice" for each game by grabbing one of two pieces of paper on which the names of the teams were written. If a great many gorillas had the same opportunity, what percentage of them would be at least this lucky? Source: "Gone Ape," *Pittsburgh Press*, June 8, 1986.

6.75 It has been estimated that over 90% of the world's large corporations are actively involved in data warehousing. Using the 90% figure, and considering a random sample of 10 large corporations, what is the probability that at least 7 of them are actively involved in data warehousing?

6.76 A country club has found that the number of complete rounds played by a member during a given outing averages 1.3 rounds. What is the probability that a golfer selected at random will complete at least 2 rounds of golf on his or her next outing?

6.77 A computer manufacturer that sells notebook computers directly to customers has been having problems with one of its models—it seems that about 10% of the units are inoperable upon arrival to the customer. They are functional when tested at the factory, but there is some problem associated with the shipping/handling process that the company's engineers have not yet been able to identify and correct. A highly valued customer places an emergency order for 8 of these notebooks, and wants them delivered overnight and ready for use in the field by tomorrow. Wanting to be honest with this customer, and wishing to help ensure that the customer receives at least 8 working computers, the manufacturer offers to send 9 units instead of the 8 that the customer needs. The customer accepts this gesture of goodwill and 9 units are sent. What is the probability that the customer will end up with at least 8 working computers?

6.78 It has been estimated that 5% of people who are hospitalized pick up infections while in the hospital, and that 2.2 million people a year get sick and 88,000 die because of these infections. Experts advise patients to be sure that health care personnel have washed their hands, to select surgeons with higher levels of experience, and to minimize the amount of time that they are treated with catheterization. Grove Tree Mountain Hospital anticipates admitting 100 patients this month for surgical and other treatments. Suppose that 14 of the patients in the sample actually end up getting a hospital-borne infection. Treating the 100 patients as a random sample of all hospital patients, and applying the infection rate cited above, what is the probability that a typical hospital would have 14 or more infections in a sample of this size? Considering this probability, comment on whether Grove Tree Mountain might not be a typical hospital in terms of its patients' likelihood of picking up an infection. Source: Anita Manning, "Hospital Infections Jump as Outpatient Care Rises," *USA Today*, March 6, 2000, p. 8D.

6.79 A tire manufacturer has advertised that only 2% of its tires have surface blemishes. Further, the manufacturer argues that tires with such blemishes are perfectly safe, and with only the physical appearance suffering very slightly. A consumer magazine randomly purchases 40 of the tires for an upcoming tire performance comparison report. Upon examining the tires, the magazine's test engineers find that 5 of them have surface blemishes. If it were true that just 2% of the tires produced have surface blemishes, what would be the probability of getting 5 or more with such blemishes in a random sample of 40 tires? Considering this probability, comment on the believability of the tire manufacturer's advertising claim.

6.80 J.D. Power and Associates' Initial Quality Study reports that the industry-average problem rate for vehicles is 108 problems per 100 vehicles. The highest-rated brand was Lexus, with a rate of just 84 problems per 100 vehicles. On the other hand, BMW's Mini was reported as having 165 problems per 100 vehicles. Larry is on his way to pick up his brand-new Lexus. At the same time, Mark is on his way to pick up his brand-new Mini. What is the probability that Larry's Lexus will have exactly 2 problems? What is the probability that Mark's Mini will have no more than 1 problem? What is the probability that neither Larry nor Mark will have a problem with his new car? Source: Sharon Silke Carty, "U.S. Autos Power Forward with Gains in Quality Survey," *USA Today*, June 23, 2009, p. 3B.

INTEGRATED CASE

Thorndike Sports Equipment

Arriving at the office a bit early, Ted finds Luke Thorndike in a fit of rage. Mr. Thorndike is sipping coffee and looking over Ted's notes on the defective racquets that were constructed with aluminum purchased from each of the company's three suppliers.

In rummaging through the basement crate, Ted has found a total of 30 defective racquets. Of these, 5 were made of aluminum supplied by the Snowmet Corporation and 5 were made of aluminum purchased from Barstow Aluminum, Inc. These were the two suppliers from whom Thorndike Sports Equipment had been buying aluminum for many years, and with good results.

The cause of Mr. Thorndike's anger is the total of 20 defective racquets made from aluminum purchased from the company's newest and lowest-priced supplier, Darwood Discount Metals, Inc. Luke recently placed a big order with Darwood, and he is counting the minutes until 10 A.M. so that he can call the West Coast firm, cancel the order, and give Mr. Darwood a big piece of his mind. The arrival of the morning mail only serves to heighten Luke's anger, as three more complaints are received from disgruntled customers demanding immediate replacements for their broken racquets.

Ten o'clock finally rolls around, and Luke places a person-to-person collect call to Mr. Darwood. After berating Mr. Darwood and his products, Luke demands that the order be canceled immediately. Mr. Darwood does not appreciate the elder Thorndike's tirade, but he is quite cognizant of the fact that 10% of his company's profits come from Thorndike Sports Equipment. Though irritated, he patiently tries to reason with Mr. Thorndike.

According to Mr. Darwood, independent metallurgists have conducted a lot of tests in which they found Darwood aluminum to be every bit as good as the product supplied by Snowmet and Barstow. He suggests that the unusually high number of defective racquets found by Luke is merely a fluke. Already operating on a short fuse, Mr. Thorndike responds that he doesn't much care for silly rhymes in the middle of the morning, and he warns Darwood not to load any more aluminum onto Thorndike-bound trucks until he and Ted have a chance to further examine the information Ted collected over the weekend. He promises to call Mr. Darwood with a final decision by 3 P.M., Pacific Time.

Ted spends a very busy morning and skips lunch, but by 2 P.M. he comes up with some data that might prove useful. Most important, he has uncovered a research study in which it was found that 0.8% of all aluminum racquets end up being returned as defective. The number of Snowmet and Barstow racquets in the "defectives" crate is about 1% of those produced. However, of the 1200 racquets made from Darwood aluminum, 20 (about 1.7%) are defective. Ted decides to consider the situation as a binomial probability distribution in which the probability of a defect on a given trial is 0.008, corresponding to the research finding that 0.8% of all the aluminum racquets produced are defective. Using the estimate for the entire industry, the *expected number of defectives* among the 1200 racquets made of Darwood aluminum would be $n\pi = 1200(0.008)$, or 9.6, *but 20 defectives were observed.*

To determine how unusual this result really is, Ted uses Excel and $n = 1200$, $\pi = 0.008$ to find the cumulative probabilities for the number of defects. The results are shown in Table 6.7.

Receiving a telephone call from home and finding out that the water pipes have just burst and his house is flooding, Ted has to flee the office. On his way out, he asks you to use the information he has found, along with Table 6.7, in giving advice to Mr. Thorndike before he makes his call to Darwood.

Table 6.7 Ted Thorndike has used Excel in generating these cumulative probabilities for a binomial distribution in which the number of trials is 1200 and the probability of a defect in any given trial is 0.8%, or 0.008.

	A	B	C	D	E
1	Binomial probabilities for				
2	trials (n), 1200, and				
3	probability of success (pi), 0.008				
4					
5	k	p(x<=k)		k	p(x<=k)
6	0	0.0001		13	0.8927
7	1	0.0007		14	0.9365
8	2	0.0037		15	0.9644
9	3	0.0136		16	0.9811
10	4	0.0373		17	0.9904
11	5	0.0830		18	0.9954
12	6	0.1564		19	0.9979
13	7	0.2574		20	0.9991
14	8	0.3788		21	0.9996
15	9	0.5086		22	0.9998
16	10	0.6331		23	0.9999
17	11	0.7419		24	1.0000
18	12	0.8287			

CHAPTER 7

Continuous Probability Distributions

ATM ARRIVAL TIMES: A CONTINUOUS DISTRIBUTION

In the opening vignette to Chapter 6, we described how the number of automatic teller machine (ATM) customers over a 5-minute period represents a discrete probability distribution, with x = the number of arrivals (0, 1, 2, etc.) during the period.

Without getting into trouble with campus security, stake out your local ATM as you did for the opener to Chapter 6—except now use the chronograph feature of your (or a friend's) watch to record the times between the arrivals of successive customers. For example, x_1 is the time between the arrival of customer A and customer B, x_2 is the time between customers B and C, and so on. What you're doing now is creating a continuous probability distribution, the topic of this chapter.

As a matter of fact, our discussion here is the flip side of the vignette that opened Chapter 6. The inverse of arrivals-per-minute is minutes-between-arrivals. Where the discrete Poisson distribution of Chapter 6 focused on the former, the continuous exponential distribution of this chapter is concerned with the latter. This unique relationship between the Poisson and exponential distributions will be examined later in this chapter. Measuring and predicting time between arrivals can be important to planners and service providers in many real-world situations, especially those dealing with emergency medical services.

Some things look a little like a normal curve.

LEARNING OBJECTIVES

After reading this chapter, you should be able to:

- Understand the nature and the applications of the normal distribution.
- Use the standard normal distribution and *z*-scores to determine probabilities associated with the normal distribution.
- Use the normal distribution to approximate the binomial distribution.
- Understand the nature and the applications of the exponential distribution, including its relationship to the Poisson distribution of Chapter 6.
- Use the computer in determining probabilities associated with the normal and exponential distributions.

7.1 INTRODUCTION

Chapter 6 dealt with probability distributions for *discrete* random variables, which can take on only certain values along an interval, with the possible values having gaps between them. This chapter presents several **continuous probability distributions;** these describe probabilities associated with random variables that are able to assume *any* of an infinite number of values along an interval.

Discrete probability distributions can be expressed as histograms, where the probabilities for the various x values are expressed by the heights of a series of vertical bars. In contrast, continuous probability distributions are smooth curves, where probabilities are expressed as areas under the curves. The curve is a function of x, and $f(x)$ is referred to as a **probability density function.** Since the continuous random variable x can be in an infinitely small interval along a range or continuum, the probability that x will take on any exact value may be regarded as zero. Therefore, we can speak of probabilities only in terms of the probability that x will be within a specified *interval* of values. For a continuous random variable, the probability distribution will have the following characteristics:

The probability distribution for a continuous random variable:

1. **The vertical coordinate is a function of x, described as $f(x)$ and referred to as the probability density function.**
2. **The range of possible x values is along the horizontal axis.**
3. **The probability that x will take on a value between a and b will be the area under the curve between points a and b, as shown in Figure 7.1.**

FIGURE 7.1

For a continuous random variable, the probability distribution is described by a curve called the probability density function, $f(x)$. The total area beneath the curve is 1.0, and the probability that x will take on some value between a and b is the area beneath the curve between points a and b.

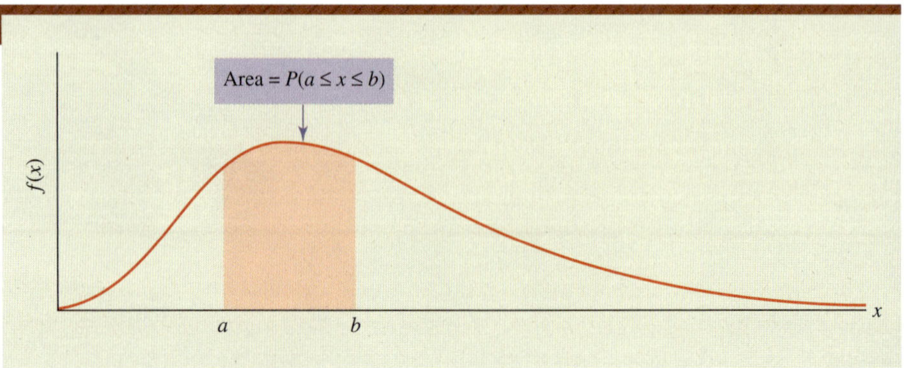

STATISTICS IN ACTION 7.1

Navigating Normal Curves with Thorndike Sports Equipment

I was flattered but kind of surprised when Dr. Weiers asked Thorndike Sports Equipment to do a Statistics in Action commentary for his textbook. Doesn't he realize we're just imaginary? After all, he made us up in the first place. Anyway, I usually don't spend a lot of time thinking about the normal curve, but preparing this commentary helped open my eyes to how often it comes up.

In many facets of our business, we find a normal curve closely approximates the data we've collected. For example, in the "bounce test" we use in monitoring the liveliness of our golf balls, we find the bounce distances tend to be normally distributed. Although we don't like to talk about it much, we try to make sure our sponsored professionals get balls that are from the most lively 0.5% of the output. My grandson Ted tells me that corresponds to $z = +2.58$ standard deviation units from the mean.

According to our human resources manager, a lot of the aptitude and performance tests we give potential and current employees also tend to result in scores that are distributed according to that classic "bell-shaped" curve. She's been with us for many years, and she finds it very interesting to compare the means and standard deviations today with the ones she got ten or twenty years ago.

Ted says we also rely on the normal distribution in constructing and interpreting the statistical process control charts that have helped Thorndike racquets gain their reputation for consistently high quality. Here at Thorndike Sports Equipment, we greatly appreciate the applications and value of the normal distribution, and we are happy to pass our experience on to you.

Luke Thorndike
President and Founder
Thorndike Sports
Equipment

The probability density function, *f*(*x*), for a given continuous distribution is expressed in algebraic terms, and the areas beneath are obtained through the mathematics of calculus. However, tables are provided in the text for readily identifying or calculating such areas.

4. The total area under the curve will be equal to 1.0.

The first continuous probability distribution we will examine is the normal distribution. As suggested by the commentary in Statistics in Action 7.1, this is a very important continuous distribution in both the study of statistics and its application to business. The normal distribution is a bell-shaped, symmetrical curve, the use of which is facilitated by a standard table listing the cumulative areas beneath.

As demonstrated in Chapter 6, binomial probabilities can involve either tedious computations or extensive tables. We will describe a method by which the normal distribution can be used as a convenient approximation for determining the probabilities associated with discrete random variables that are of the binomial distribution.

The next topic will be the exponential distribution. This is related to the (discrete) Poisson distribution of the previous chapter but is a continuous probability distribution describing probabilities associated with the (continuous) time or distance intervals between occurrences of the so-called rare events of a Poisson process.

Finally, we will use Excel and Minitab to randomly select observations from a specified continuous probability distribution. This is similar to the simulated samples obtained from selected discrete probability distributions in Chapter 6.

EXERCISES

7.1 What is the difference between a continuous probability distribution and a discrete probability distribution?

7.2 What is a probability density function and how is it relevant to the determination of probabilities associated with a continuous random variable?

7.3 Why is the total area beneath a probability density function equal to 1.0?

7.4 What is the probability that a continuous random variable will take on any specific value? Explain your answer.

7.2 THE NORMAL DISTRIBUTION

Description and Applications

The **normal distribution** is the most important continuous distribution in statistics. It occupies this position for three key reasons: (1) many natural and economic phenomena tend to be approximately normally distributed; (2) it can be used in approximating other distributions, including the binomial; and (3) as we will see in Chapter 8, sample means and proportions tend to be normally distributed whenever repeated samples are taken from a given population of any shape. This latter characteristic is related to the *central limit theorem*, a concept also discussed in the next chapter.

The normal distribution is really a family of distributions, each member of which is bell shaped and symmetrical (the left side is a mirror image of the right). As Figure 7.2 shows, (1) the mean, median, and mode are all at the same position on the horizontal axis; (2) the curve is *asymptotic*, approaching the horizontal axis at both ends, but never intersecting with it; and (3) the total area beneath the curve is equal to 1.0. The specific member of the family depends on just two descriptors:

FIGURE 7.2

The normal distribution is actually a family of bell-shaped distributions, each of which has these characteristics. The specific member of the family depends on just two descriptors, the mean (μ) and the standard deviation (σ).

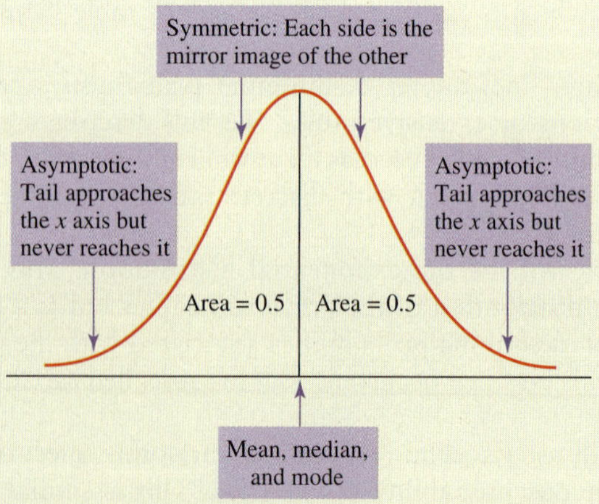

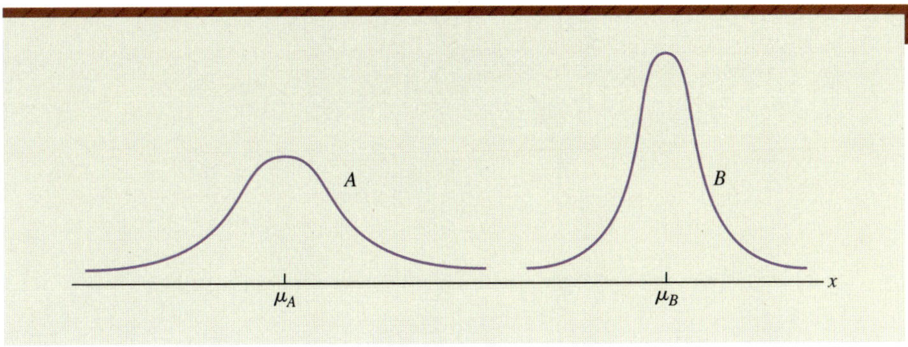

FIGURE 7.3

The shape of a normal curve and its position on the *x* axis will depend on its mean (μ) and standard deviation (σ). Although curve *B* has a greater mean, its standard deviation is smaller.

the mean (μ) and the standard deviation (σ). The normal distribution can be described as follows:

The normal distribution for the continuous random variable, *x* (with $-\infty \le x \le +\infty$):

$$f(x) = \frac{1}{\sigma\sqrt{2\pi}} e^{-(1/2)[(x-\mu)/\sigma]^2}$$

where μ = mean
σ = standard deviation
e = the mathematical constant, 2.71828
π = the mathematical constant, 3.14159

NOTE

In this equation, π is the geometric constant for the ratio of the circumference of a circle to its diameter, and is not related to the symbol used for a population proportion.

Not all normal curves will look exactly like the one in Figure 7.2. The shape of the curve and its location along the *x* axis will depend on the values of the standard deviation and mean, respectively. You can demonstrate this by using Seeing Statistics Applet 4 at the end of the chapter. Figure 7.3 shows two normal curves representing output diameters for two tubing extrusion machines:

- The average diameter of tubing from machine *B* exceeds that from machine *A*. Looking along the *x* axis, it can be seen that $\mu_B > \mu_A$.
- The variability of the output from machine *A* is greater than that from machine *B*. In symbolic terms, this can be expressed as $\sigma_A > \sigma_B$.

Areas Beneath the Normal Curve

Regardless of the shape of a particular normal curve, the areas beneath it can be described for any interval of our choice. This makes it unnecessary to get involved with calculus and the complexities of the probability density function presented earlier. For *any* normal curve, the areas beneath the curve will be as follows (see also part (a) in Figure 7.4):

- About 68.3% of the area is in the interval $\mu - \sigma$ to $\mu + \sigma$.
- About 95.5% of the area is in the interval $\mu - 2\sigma$ to $\mu + 2\sigma$.
- Nearly all of the area (about 99.7%) is in the interval $\mu - 3\sigma$ to $\mu + 3\sigma$.

FIGURE 7.4

For any normal distribution, the approximate areas shown in part (a) will lie beneath the curve for the intervals shown. In part (b), the areas show how these intervals could apply to the amount of time logged by general-aviation aircraft during the year. Assuming $\mu = 120$ hours and $\sigma = 30$ hours, about 95.5% of the planes logged between 60 and 180 hours of flying time.

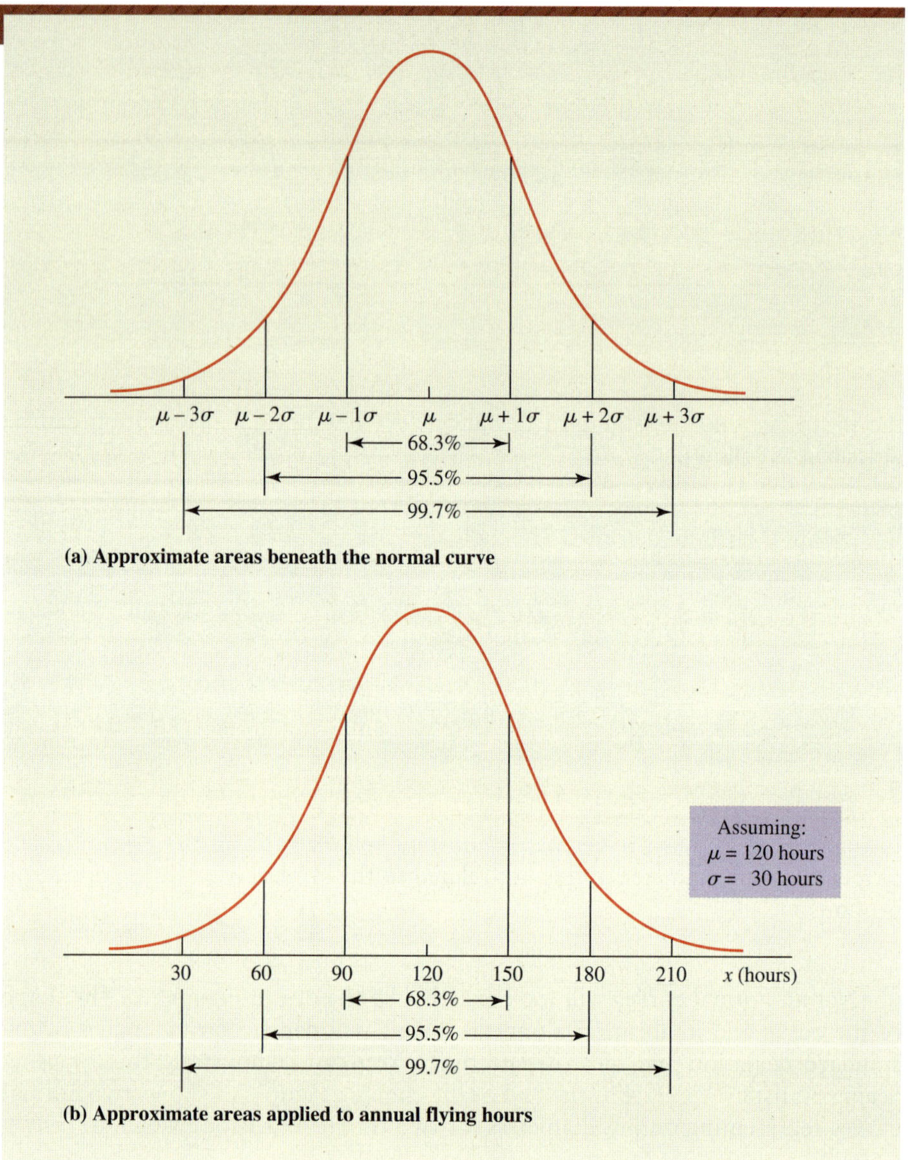

EXAMPLE

Normal Distribution

The General Aviation Manufacturers Association has reported the average annual number of hours flown by general-aviation aircraft to be approximately 120 hours.[1] For purposes of our example, we will assume a normal distribution and a standard deviation of 30 hours for the flying hours of aircraft within this population.

[1]Source: General Aviation Manufacturers Association, *General Aviation Statistical Databook, 2008 Edition*, pp. 28–29.

EXAMPLEEXAMPLEEXAMP

EXAMPLEEXAMPLEEXAMPLE

SOLUTION

Based on the reported mean, $\mu = 120$ hours, and the assumed standard deviation, $\sigma = 30$ hours, we can use the known properties of the normal distribution to determine that about 95.5% of these aircraft logged between $(120 - 2[30])$ and $(120 + 2[30])$ hours, or between 60 and 180 hours of flying time during that year. Likewise, we could determine that about 68.3% of the aircraft logged between 90 and 150 $(\mu \pm \sigma)$ hours and that about 99.7% logged between 30 and 210 $(\mu \pm 3\sigma)$ hours. These intervals and approximate areas are shown in part (b) of Figure 7.4.

EXERCISES

7.5 It has been stated that the normal distribution is really a "family" of distributions. Explain.

7.6 In the normal distribution, the probability that x will exceed $(\mu + 2\sigma)$ is the same as the probability that x will be less than $(\mu - 2\sigma)$. What characteristic of the normal distribution does this reflect?

7.7 Sketch two different normal distributions along a single x axis so that both of the following conditions are satisfied: (a) $\mu_a = \mu_b$ and (b) $\sigma_a > \sigma_b$.

7.8 If x is normally distributed with $\mu = 20.0$ and $\sigma = 4.0$, determine the following:

a. $P(x \geq 20.0)$
b. $P(16.0 \leq x \leq 24.0)$
c. $P(x \leq 12)$
d. $P(x = 22.0)$
e. $P(12.0 \leq x \leq 28.0)$
f. $P(x \geq 16)$

7.9 If x is normally distributed with $\mu = 25.0$ and $\sigma = 5.0$, determine the following:

a. $P(x \geq 25.0)$
b. $P(20.0 \leq x \leq 30.0)$
c. $P(x \leq 30)$
d. $P(x = 26.2)$
e. $P(15.0 \leq x \leq 25.0)$
f. $P(x \geq 15)$

7.10 The Canada Urban Transit Association has reported that the average revenue per passenger trip during a given year was \$1.55. If we assume a normal distribution and a standard deviation of $\sigma = \$0.20$, what proportion of passenger trips produced a revenue of Source: American Public Transit Association, *APTA 2009 Transit Fact Book*, p. 35.

a. less than \$1.55?
b. between \$1.15 and \$1.95?
c. between \$1.35 and \$1.75?
d. between \$0.95 and \$1.55?

7.11 In 2007, the average conventional first mortgage for new single-family homes was \$360,000. Assuming a normal distribution and a standard deviation of $\sigma = \$30{,}000$, what proportion of the mortgages were Source: Bureau of the Census, *Statistical Abstract of the United States 2009*, p. 726.

a. more than \$360,000?
b. between \$300,000 and \$420,000?
c. between \$330,000 and \$390,000?
d. more than \$270,000?

7.12 In 2009, the average charge for tax preparation by H&R Block, Inc. was \$187. Assuming a normal distribution and a standard deviation of $\sigma = \$20$, what proportion of H&R Block's tax preparation fees were Source: hrblock.com, July 14, 2009.

a. more than \$187?
b. between \$147 and \$227?
c. between \$167 and \$207?
d. more than \$227?

7.13 It has been reported that the average hotel check-in time, from curbside to delivery of bags into the room, is 12.0 minutes. An Li has just left the cab that brought her to her hotel. Assuming a normal distribution with a standard deviation of 2.0 minutes, what is the probability that the time required for An Li and her bags to get to the room will be:

a. greater than 14.0 minutes?
b. between 10.0 and 14.0 minutes?
c. less than 8.0 minutes?
d. between 10.0 and 16.0 minutes?

7.14 The average American family of four spends \$5000 per year on food prepared at home. Assuming a normal distribution with a standard deviation of \$1000 and a randomly selected American family of four, what is the probability that the family's annual spending for food prepared at home will be: Source: "Business Briefing," *Pittsburgh Tribune-Review*, July 5, 2009, p. E1.

a. more than \$8000?
b. between \$5000 and \$7000?
c. less than \$6000?
d. between \$3000 and \$6000?

7.15 On average, commuters in the Los Angeles, California, area require 30.0 minutes to get to work. Assume a normal distribution with a standard deviation of 5.0 minutes and a randomly selected Los Angeles-area commuter named Jamal. Source: *The World Almanac and Book of Facts 2006*, p. 474.

a. What is the probability that Jamal will require more than 45.0 minutes to get to work on any given day?

b. Jamal has just left home and must attend an important meeting with the CEO in just 25.0 minutes. If the CEO routinely fires employees who are tardy, what is the probability that Jamal will be going to work tomorrow?

7.3 THE STANDARD NORMAL DISTRIBUTION

Description and Applications

Because there is a different normal curve for every possible pair of μ and σ, the number of statistical tables would be limitless if we wished to determine the areas corresponding to possible intervals within all of them. Fortunately, we can solve this dilemma by "standardizing" the normal curve and expressing the original x values in terms of their number of standard deviations away from the mean. The result is referred to as a **standard** (or **standardized**) **normal distribution,** and it allows us to use a single table to describe areas beneath the curve. The key to the process is the ***z*-score:**

The *z*-score for a standard normal distribution:

$$z = \frac{x - \mu}{\sigma}$$

where z = the distance from the mean, measured in standard deviation units
x = the value of x in which we are interested
μ = the mean of the distribution
σ = the standard deviation of the distribution

In the aircraft flying hours example discussed earlier, the mean and (assumed) standard deviation were $\mu = 120$ hours and $\sigma = 30$ hours, respectively. Using the z-score equation, we can convert this distribution into a standard normal distribution in which we have z values instead of x values. For example:

***x* Value in the Original Normal Distribution**	**Corresponding *z* Value in the Standard Normal Distribution**
$x = 120$ hours	$z = \frac{x - \mu}{\sigma} = \frac{120 - 120}{30} = 0.00$
$x = 160$ hours	$z = \frac{x - \mu}{\sigma} = \frac{160 - 120}{30} = 1.33$
$x = 90$ hours	$z = \frac{x - \mu}{\sigma} = \frac{90 - 120}{30} = -1.00$

In Figure 7.5, there are two scales beneath the curve: an x (hours of flying time) scale and a z (standard deviation units) scale. Note that the mean of the z scale is 0.00. Regardless of the mean and standard deviation of any normal distribution, we can use the z-score concept to express the original values in terms of standard deviation multiples from the mean. It is this transformation that allows us to use the standard normal table in determining probabilities associated with any normal distribution.You can use Seeing Statistics Applet 5, at the end of the chapter, to see how the normal distribution areas respond to changes in the beginning and ending z values associated with Figure 7.5.

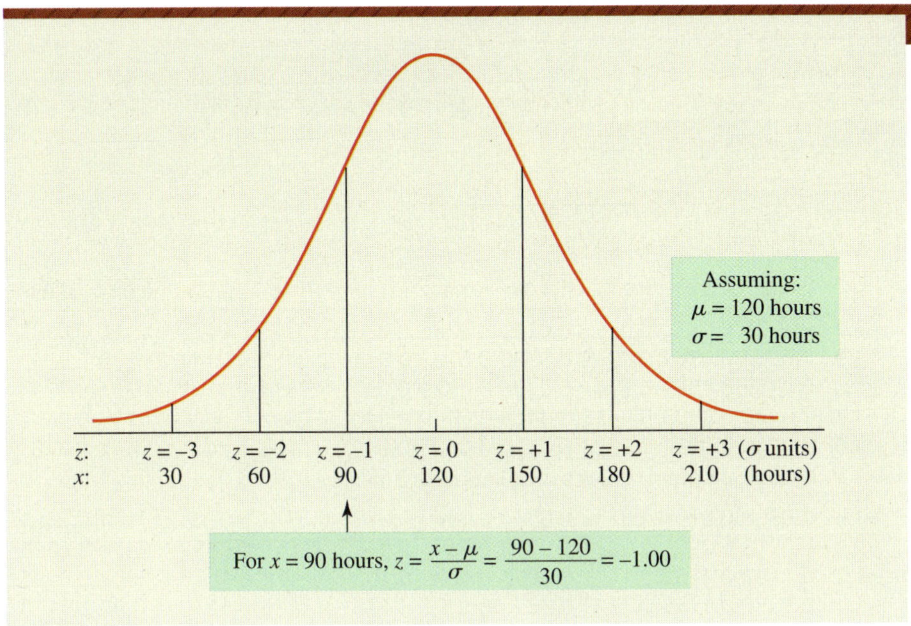

FIGURE 7.5

For the distribution of annual flying hours in part (b) of Figure 7.4, the upper of the two horizontal scales shown here represents the standardized normal distribution, in which x values (hours) have been converted to z-scores (standard deviation units from the mean). The use of z-scores makes it possible to use a single table for all references to the normal distribution, regardless of the original values of μ and σ.

Using the Standard Normal Distribution Table

Regardless of the units involved (e.g., pounds, hours, dollars, and so on) in the original normal distribution, the standard normal distribution converts them into standard deviation distances from the mean. As we proceed through the next example, our original units will be grams.

In demonstrating the use of the standard normal distribution table, we will rely on a situation involving the weight added to balance a generator shaft. For each of the related questions, we will go through these steps:

1. Convert the information provided into one or more z-scores.
2. Use the standard normal table to identify the area(s) corresponding to the z-score(s). A portion of the standard normal distribution table is shown in Table 7.1. The table provides cumulative areas to the z value of interest.
3. Interpret the result in such a way as to answer the original question.

The cumulative standard normal distribution table is on the rear endsheet of the text, and an abbreviated portion of it is shown in Table 7.1. The row labels (e.g., 1.4) refer to the integer and first decimal place of z, and the column labels refer to the second decimal place (e.g., 0.05); thus, we can look in row 1.4 and column 0.05 and see that the cumulative probability (or area) to the left of $z = 1.45$ is 0.9265. Here are a few more examples of how to find cumulative areas for given z values:

- *Cumulative area associated with $z = -1.37$:* Referring to the −1.3 row and the 0.07 column of Table 7.1, we find this probability is 0.0853.
- *Cumulative area associated with $z = -0.08$:* Referring to the −0.0 row and the 0.08 column of Table 7.1, we find this probability is 0.4681.
- *Cumulative area associated with $z = 0.09$:* Referring to the 0.0 row and the 0.09 column of Table 7.1, we see this probability is 0.5359.
- *Cumulative area associated with $z = 1.73$:* Referring to the 1.7 row and the 0.03 column of Table 7.1, we see this probability is 0.9582.

TABLE 7.1

A portion of the cumulative standard normal distribution table from the back endsheet

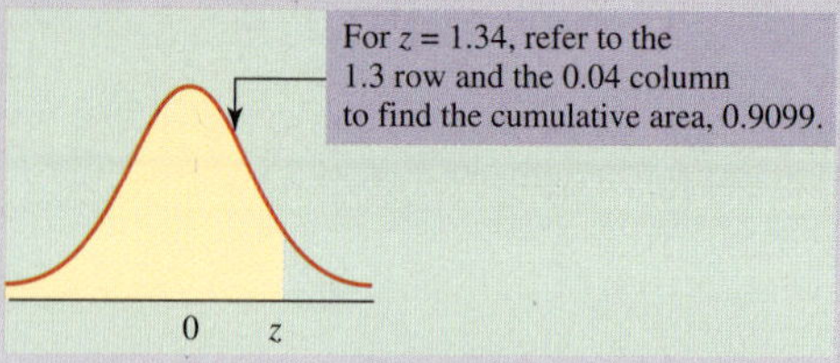

z	0.00	0.01	0.02	0.03	0.04	0.05	0.06	0.07	0.08	0.09
−2.0	0.0228	0.0222	0.0217	0.0212	0.0207	0.0202	0.0197	0.0192	0.0188	0.0183
−1.9	0.0287	0.0281	0.0274	0.0268	0.0262	0.0256	0.0250	0.0244	0.0239	0.0233
−1.8	0.0359	0.0351	0.0344	0.0336	0.0329	0.0322	0.0314	0.0307	0.0301	0.0294
−1.7	0.0446	0.0436	0.0427	0.0418	0.0409	0.0401	0.0392	0.0384	0.0375	0.0367
−1.6	0.0548	0.0537	0.0526	0.0516	0.0505	0.0495	0.0485	0.0475	0.0465	0.0455
−1.5	0.0668	0.0655	0.0643	0.0630	0.0618	0.0606	0.0594	0.0582	0.0571	0.0559
−1.4	0.0808	0.0793	0.0778	0.0764	0.0749	0.0735	0.0721	0.0708	0.0694	0.0681
−1.3	0.0968	0.0951	0.0934	0.0918	0.0901	0.0885	0.0869	0.0853	0.0838	0.0823
−1.2	0.1151	0.1131	0.1112	0.1093	0.1075	0.1056	0.1038	0.1020	0.1003	0.0985
−1.1	0.1357	0.1335	0.1314	0.1292	0.1271	0.1251	0.1230	0.1210	0.1190	0.1170
−1.0	0.1587	0.1562	0.1539	0.1515	0.1492	0.1469	0.1446	0.1423	0.1401	0.1379
−0.9	0.1841	0.1814	0.1788	0.1762	0.1736	0.1711	0.1685	0.1660	0.1635	0.1611
−0.8	0.2119	0.2090	0.2061	0.2033	0.2005	0.1977	0.1949	0.1922	0.1894	0.1867
−0.7	0.2420	0.2389	0.2358	0.2327	0.2296	0.2266	0.2236	0.2206	0.2177	0.2148
−0.6	0.2743	0.2709	0.2676	0.2643	0.2611	0.2578	0.2546	0.2514	0.2483	0.2451
−0.5	0.3085	0.3050	0.3015	0.2981	0.2946	0.2912	0.2877	0.2843	0.2810	0.2776
−0.4	0.3446	0.3409	0.3372	0.3336	0.3300	0.3264	0.3228	0.3192	0.3156	0.3121
−0.3	0.3821	0.3783	0.3745	0.3707	0.3669	0.3632	0.3594	0.3557	0.3520	0.3483
−0.2	0.4207	0.4168	0.4129	0.4090	0.4052	0.4013	0.3974	0.3936	0.3897	0.3859
−0.1	0.4602	0.4562	0.4522	0.4483	0.4443	0.4404	0.4364	0.4325	0.4286	0.4247
−0.0	0.5000	0.4960	0.4920	0.4880	0.4840	0.4801	0.4761	0.4721	0.4681	0.4641
0.0	0.5000	0.5040	0.5080	0.5120	0.5160	0.5199	0.5239	0.5279	0.5319	0.5359
0.1	0.5398	0.5438	0.5478	0.5517	0.5557	0.5596	0.5636	0.5675	0.5714	0.5753
0.2	0.5793	0.5832	0.5871	0.5910	0.5948	0.5987	0.6026	0.6064	0.6103	0.6141
0.3	0.6179	0.6217	0.6255	0.6293	0.6331	0.6368	0.6406	0.6443	0.6480	0.6517
0.4	0.6554	0.6591	0.6628	0.6664	0.6700	0.6736	0.6772	0.6808	0.6844	0.6879
0.5	0.6915	0.6950	0.6985	0.7019	0.7054	0.7088	0.7123	0.7157	0.7190	0.7224
0.6	0.7257	0.7291	0.7324	0.7357	0.7389	0.7422	0.7454	0.7486	0.7517	0.7549
0.7	0.7580	0.7611	0.7642	0.7673	0.7704	0.7734	0.7764	0.7794	0.7823	0.7852
0.8	0.7881	0.7910	0.7939	0.7967	0.7995	0.8023	0.8051	0.8078	0.8106	0.8133
0.9	0.8159	0.8186	0.8212	0.8238	0.8264	0.8289	0.8315	0.8340	0.8365	0.8389
1.0	0.8413	0.8438	0.8461	0.8485	0.8508	0.8531	0.8554	0.8577	0.8599	0.8621
1.1	0.8643	0.8665	0.8686	0.8708	0.8729	0.8749	0.8770	0.8790	0.8810	0.8830
1.2	0.8849	0.8869	0.8888	0.8907	0.8925	0.8944	0.8962	0.8980	0.8997	0.9015
1.3	0.9032	0.9049	0.9066	0.9082	0.9099	0.9115	0.9131	0.9147	0.9162	0.9177
1.4	0.9192	0.9207	0.9222	0.9236	0.9251	0.9265	0.9279	0.9292	0.9306	0.9319
1.5	0.9332	0.9345	0.9357	0.9370	0.9382	0.9394	0.9406	0.9418	0.9429	0.9441
1.6	0.9452	0.9463	0.9474	0.9484	0.9495	0.9505	0.9515	0.9525	0.9535	0.9545
1.7	0.9554	0.9564	0.9573	0.9582	0.9591	0.9599	0.9608	0.9616	0.9625	0.9633
1.8	0.9641	0.9649	0.9656	0.9664	0.9671	0.9678	0.9686	0.9693	0.9699	0.9706
1.9	0.9713	0.9719	0.9726	0.9732	0.9738	0.9744	0.9750	0.9756	0.9761	0.9767
2.0	0.9772	0.9778	0.9783	0.9788	0.9793	0.9798	0.9803	0.9808	0.9812	0.9817

In applications involving the standard normal table, actual areas of interest may lie to the left of the mean or to the right of the mean, and may even require that one cumulative probability be subtracted from another. Because of (1) the variety of possibilities, and (2) the importance of knowing how to use the standard normal table, we will examine a number of applications that relate to the following example.

EXAMPLE

Normal Probabilities

Following their production, industrial generator shafts are tested for static and dynamic balance, and the necessary weight is added to predrilled holes in order to bring each shaft within balance specifications. From past experience, the amount of weight added to a shaft has been normally distributed, with an average of 35 grams and a standard deviation of 9 grams.

SOLUTIONS

What Is the Probability That a Randomly Selected Shaft Will Require Between 35 and 40 Grams of Weight for Proper Balance? [$P(35 \leq x \leq 40)$]

1. Conversion of $x = 40$ grams to z-score:

$$z = \frac{x - \mu}{\sigma} = \frac{40 - 35}{9} = 0.56$$

Conversion of $x = 35$ grams to z-score:

$$z = \frac{x - \mu}{\sigma} = \frac{35 - 35}{9} = 0.00$$

2. Reference to the standard normal table: Referring to the 0.5 row and the 0.06 column of the table, we find the cumulative area to $z = 0.56$ is 0.7123. Referring to the 0.0 row and the 0.00 column, we can verify that the cumulative area to $z = 0.00$ (the mean) is 0.5000.

3. Solution: The probability that the randomly selected shaft will require between 35 and 40 grams is $0.7123 - 0.5000 = 0.2123$, as illustrated in Figure 7.6.

What Is the Probability That a Randomly Selected Shaft Will Require at Least 50 Grams of Weight for Proper Balance? [$P(x \geq 50)$]

1. Conversion of $x = 50$ grams to z-score:

$$z = \frac{x - \mu}{\sigma} = \frac{50 - 35}{9} = 1.67$$

2. Reference to the standard normal table: Referring to the 1.6 row and the 0.07 column, we find the cumulative area to $z = 1.67$ is 0.9525. Also, we must consider that the total cumulative area (i.e., to z approaching positive infinity) is, by definition, 1.0000.

3. Solution: The probability that the randomly selected shaft will require at least 50 grams is $1.0000 - 0.9525 = 0.0475$, as illustrated in Figure 7.7.

EXAMPLEEXAMPLEEXAMPLEEXAMPLEEXAMPLEEXAMPLEEXAMPLEEXAMP

FIGURE 7.6

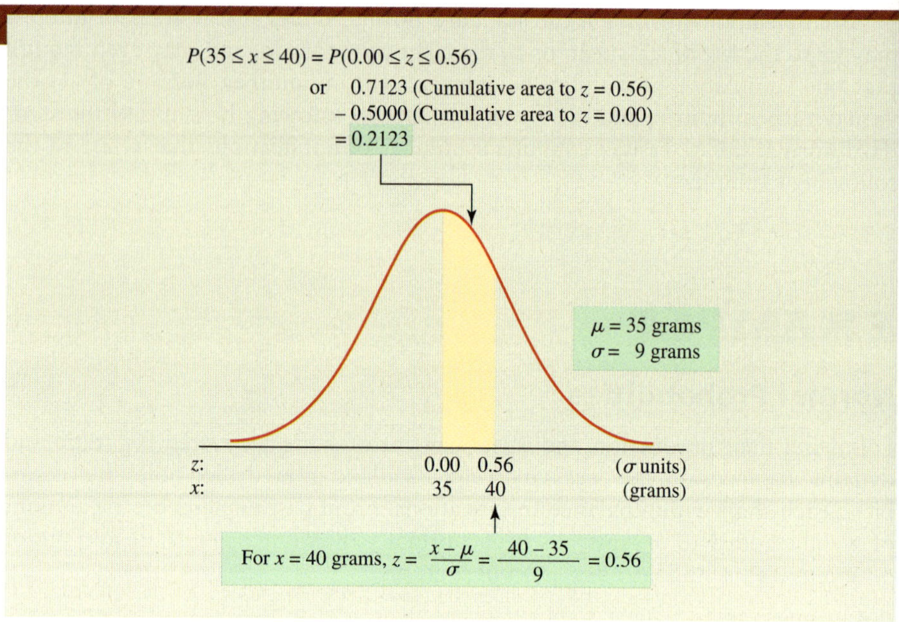

The probability is 0.2123 that a randomly selected generator shaft will require between 35 and 40 grams of weight in order to achieve proper balance.

FIGURE 7.7

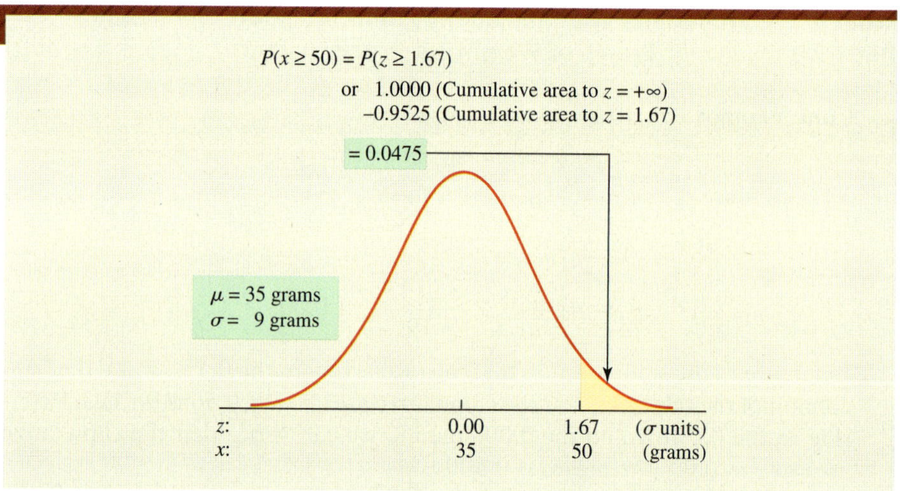

The probability is 0.0475 that a randomly selected generator shaft will require at least 50 grams for proper balance.

What Is the Probability That a Randomly Selected Shaft Will Require Between 41 and 49 Grams of Weight for Proper Balance? [$P(41 \leq x \leq 49)$]

1. Conversion of $x = 49$ grams to z-score:

$$z = \frac{x - \mu}{\sigma} = \frac{49 - 35}{9} = 1.56$$

   Conversion of $x = 41$ grams to z-score:

$$z = \frac{x - \mu}{\sigma} = \frac{41 - 35}{9} = 0.67$$

2. Reference to the standard normal table: Referring to the 1.5 row and the 0.06 column of the table, we find the cumulative area to $z = 1.56$ is 0.9406. Referring to the 0.6 row and the 0.07 column of the table, the cumulative area to $z = 0.67$ is 0.7486.

3. Solution: The probability that the randomly selected shaft will require between 41 and 49 grams is $0.9406 - 0.7486 = 0.1920$, as illustrated in Figure 7.8.

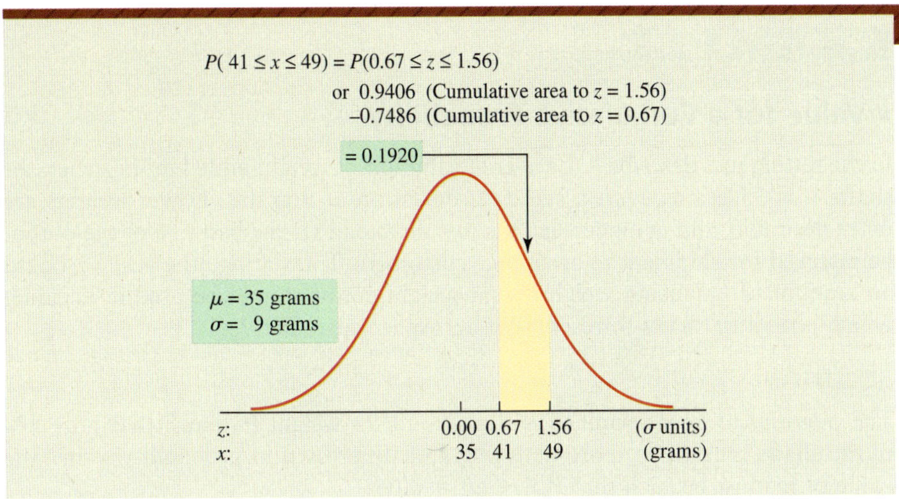

FIGURE 7.8

The probability is 0.1920 that a randomly selected generator shaft will require between 41 and 49 grams for proper balance.

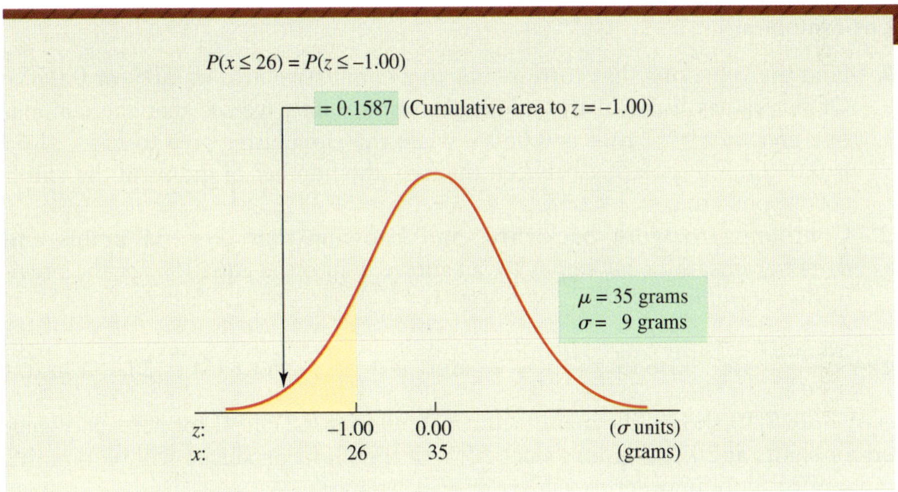

FIGURE 7.9

The probability is 0.1587 that a randomly selected generator shaft will require no more than 26 grams for proper balance.

What Is the Probability That a Randomly Selected Shaft Will Require No More Than 26 Grams of Weight for Proper Balance? [$P(x \le 26)$]

1. Conversion of x = 26 grams to z-score:

$$z = \frac{x - \mu}{\sigma} = \frac{26 - 35}{9} = -1.00$$

2. Reference to the standard normal table: Referring to the −1.0 row and the 0.00 column of the table, we find the cumulative area to $z = -1.00$ is 0.1587.
3. Solution: As Figure 7.9 illustrates, the probability that a randomly selected shaft will require no more than 26 grams is 0.1587.

In the preceding cases, we proceeded from x values to the determination of a corresponding area beneath the standard normal curve. However, it is also possible to specify an area, then identify a range of x values that corresponds to this area.

EXAMPLEEXAMPLEEXAMPLEEXAMPLEEXAMPLEEXAMPLEEXAMP

EXAMPLE

x Value for a Cumulative Probability

In the setting just described, the average amount of weight added to the generator shafts was 35 grams, with a standard deviation of 9 grams. Lower weights are more desirable, and management has just directed that the best 5% of the output be reserved for shipment to aerospace customers. Translating "the best 5%" into an amount of balancing weight, what weight cutoff should be used in deciding which generator shafts to reserve for aerospace customers?

SOLUTION

The quantity to be determined is the amount of weight (w) such that only 5% of the shafts require no more than w grams. Expressed in probability terms, the value of w must be such that $P(x \leq w) = 0.05$.

The solution is to find out what value of z corresponds to a cumulative area of 0.05, then convert this z value into w grams. Reversing the steps of the previous examples:

1. Find the value of z that corresponds to a cumulative area of 0.05, or 0.0500: Of the values listed in the standard normal table, we see that the cumulative area to $z = -1.64$ is 0.0505, while the cumulative area to $z = -1.65$ is 0.0495. Interpolating between the two, the number of grams in the cutoff corresponds to $z = -1.645$.
2. Continuing to work backward, and knowing that $\mu = 35$ grams and $\sigma = 9$ grams, the next step is to substitute w into the equation for the previously determined z score, $z = -1.645$:

$$z = \frac{w - \mu}{\sigma} \quad \text{or} \quad -1.645 = \frac{w - 35}{9} \quad \text{and} \quad w = 35 - 9(1.645) = 20.195 \text{ grams}$$

As Figure 7.10 shows, the manufacturer should set the cutoff for aerospace generator shafts at 20.195 grams, since 95% of the shafts produced will require this amount or more in order to achieve proper balance.

FIGURE 7.10

If the best 5% of the output is to be reserved for aerospace customers, shafts requiring no more than 20.195 grams for balancing would be allocated to this industry.

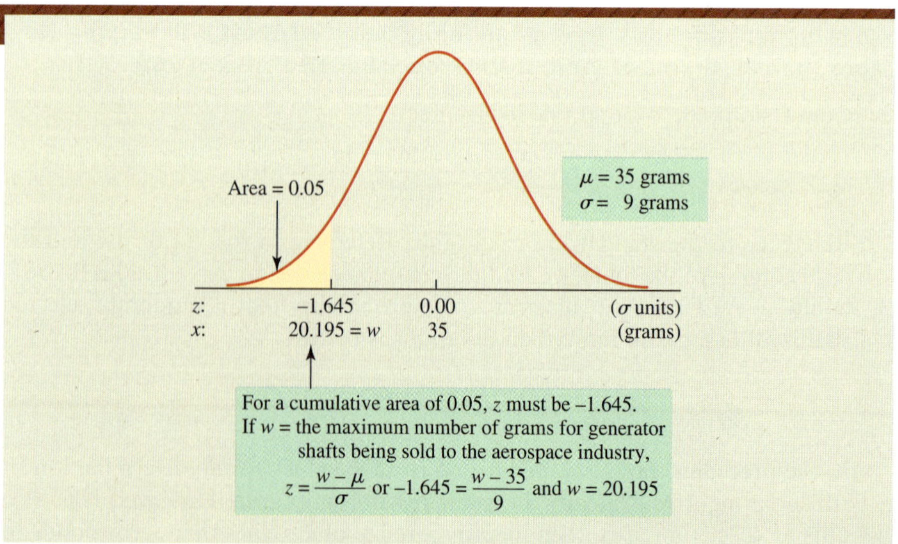

Excel and Minitab can be used to obtain cumulative normal probabilities associated with specified values for x—the procedures and results are shown in Computer Solutions 7.1. These software packages can also be used in doing the reverse—that is, specifying a cumulative probability, then finding the x value associated with it. These procedures and results are shown in Computer Solutions 7.2.

EXAMPLEEXAMPLE

COMPUTER 7.1 SOLUTIONS

Normal Probabilities

These procedures show how to get cumulative probabilities.

For each software package:

Procedure I: Getting the cumulative probability associated with one *x* value.
Procedure II: Getting cumulative probabilities associated with a set of *x* values.

EXCEL

	A	B
1	Normal Distribution	
2	with mean = 35,	
3	std dev. = 9.	
4	x	Area to L.
5	45	0.8667
6	35	0.5000
7	25	0.1333

I. Excel procedure for the cumulative probability associated with one *x* value

1. Click on the cell into which Excel will place the probability. From the **Formulas** ribbon, click on the **Insert Function** symbol at the extreme left. Click **Statistical** in the **Category** box. Click **NORMDIST** in the **Function** box. Click **OK.**
2. Enter the *x* value of interest (e.g., **45**) into the **X** box. Enter the mean of the distribution (e.g., **35**) into the **Mean** box. Enter the standard deviation (e.g., **9**) into the **Standard_dev** box. Enter **True** into the **Cumulative** box. Click **OK.**

II. Excel procedure for cumulative probabilities associated with a set of *x* values

1. Enter the labels into **A4:B4** as shown in the printout above. Enter the *x* values of interest (e.g., **45, 35, 25**) starting with cell **A5.** Click on cell **B5.**
2. From the **Formulas** ribbon, click on the **Insert Function** symbol at the extreme left. Click **Statistical** in the **Category** box. Click **NORMDIST** in the **Function** box. Click **OK.**
3. Enter **A5** into the **X** box. Enter the mean of the distribution (e.g., **35**) into the **Mean** box. Enter the standard deviation (e.g., **9**) into the **Standard_dev** box. Enter **True** into the **Cumulative** box. Click **OK.** The first cumulative probability will appear in cell B5.
4. Click at the lower right corner of **B5** (the cursor will turn to a "+" sign) and drag downward so cells **B5:B7** will contain the set of cumulative probabilities.

(continued)

MINITAB

```
Cumulative Distribution Function
Normal with mean = 35 and standard deviation = 9
 x   P( X <= x )
45      0.866740
35      0.500000
25      0.133260
```

I. Minitab procedure for cumulative probability associated with one *x* value

1. Click **Calc.** Select **Probability Distributions.** Click **Normal.**
2. In the **Normal Distribution** menu, enter the mean of the distribution (e.g., **35**) into the **Mean** box. Enter the standard deviation (e.g., **9**) into the **Standard deviation** box. Select **Input constant** and enter the *x* value of interest (e.g., **45**) into the box. Select **Cumulative.** Click **OK.**

II. Minitab procedure for cumulative probabilities associated with a set of *x* values

Enter the desired *x* values (e.g., **45, 35, 25**) into column **C1.** Follow steps 1 and 2 above, except select **Input column** and enter **C1** into the box.

COMPUTER 7.2 SOLUTIONS

Inverse Normal Probabilities

These procedures show how to get *x* values associated with cumulative normal probabilities.

For each software package:

Procedure I: Getting the *x* value associated with one cumulative probability.
Procedure II: Getting the *x* values associated with a set of cumulative probabilities.

EXCEL

	A	B
1	Normal Distribution	
2	with mean = 35,	
3	std. dev. = 9.	
4	Area to L.	x
5	0.9750	52.64
6	0.5000	35.00
7	0.2500	28.93

I. Excel procedure for the *x* value associated with one cumulative probability

1. Click on the cell into which Excel will place the *x* value. From the **Formulas** ribbon, click on the **Insert Function** symbol at the extreme left. Click **Statistical** in the **Category** box. Click **NORMINV** in the **Function** box. Click **OK.**
2. Enter the cumulative probability of interest (e.g., **0.9750**) into the **Probability** box. Enter the mean of the distribution (e.g., **35**) into the **Mean** box. Enter the standard deviation (e.g., **9**) into the **Standard_dev** box. Click **OK.**

II. Excel procedure for *x* values associated with a set of cumulative probabilities

1. Enter the labels into **A4:B4** as shown in the printout above. Enter the cumulative probabilities of interest (e.g., **0.9750, 0.5000, 0.2500**) starting with cell **A5.** Click on cell **B5.**
2. From the **Formulas** ribbon, click on the **Insert Function** symbol at the extreme left. Click **Statistical** in the **Category** box. Click **NORMINV** in the **Function** box. Click **OK.**
3. Enter **A5** into the **Probability** box. Enter the mean of the distribution (e.g., **35**) into the **Mean** box. Enter the standard deviation (e.g., **9**) into the **Standard_dev** box. Click **OK.** The first cumulative probability will appear in cell B5.
4. Click at the lower right corner of **B5** (the cursor will turn to a "+" sign) and drag downward so cells **B5:B7** will contain the set of cumulative probabilities.

MINITAB

```
Inverse Cumulative Distribution Function
Normal with mean = 35 and standard deviation = 9
P( X <= x )        x
      0.975  52.6397
      0.500  35.0000
      0.250  28.9296
```

I. Minitab procedure for the *x* value associated with one cumulative probability

1. Click **Calc.** Select **Probability Distributions.** Click **Normal.**
2. In the **Normal Distribution** menu, enter the mean of the distribution (e.g., **35**) into the **Mean** box. Enter the standard deviation (e.g., **9**) into the **Standard deviation** box. Select **Input constant** and enter the cumulative probability of interest (e.g., **0.9750**) into the box. Select **Inverse cumulative probability.** Click **OK.**

II. Minitab procedure for the *x* values associated with a set of cumulative probabilities

Enter the cumulative probabilities of interest (e.g., **0.9750, 0.5000, 0.2500**) into column **C1.** Follow steps 1 and 2 above, except select **Input column** and enter **C1** into the box.

EXERCISES

7.16 The normal distribution is really a family of distributions. Is the standard normal distribution also a family of distributions? Explain.

7.17 In the standard normal distribution, approximately what values of z correspond to the

a. first and third quartiles?
b. first and ninth deciles?
c. 23rd and 77th percentiles?

7.18 A continuous random variable, x, is normally distributed with a mean of \$1000 and a standard deviation of \$100. Convert each of the following x values into its corresponding z-score:

a. $x = \$1000$ b. $x = \$750$
c. $x = \$1100$ d. $x = \$950$
e. $x = \$1225$

7.19 A continuous random variable, x, is normally distributed with a mean of 200 grams and a standard deviation of 25 grams. Convert each of the following x values into its corresponding z-score:

a. $x = 150$ b. $x = 180$
c. $x = 200$ d. $x = 285$
e. $x = 315$

7.20 Using the standard normal table, find the following probabilities associated with z:
a. $P(0.00 \leq z \leq 1.25)$
b. $P(-1.25 \leq z \leq 0.00)$
c. $P(-1.25 \leq z \leq 1.25)$

7.21 Using the standard normal table, find the following probabilities associated with z:
a. $P(0.00 \leq z \leq 1.10)$
b. $P(z \geq 1.10)$
c. $P(z \leq 1.35)$

7.22 Using the standard normal table, find the following probabilities associated with z:
a. $P(-0.36 \leq z \leq 0.00)$
b. $P(z \leq -0.36)$ c. $P(z \geq -0.43)$

7.23 Using the standard normal table, find the following probabilities associated with z:
a. $P(-1.96 \leq z \leq 1.27)$
b. $P(0.29 \leq z \leq 1.00)$
c. $P(-2.87 \leq z \leq -1.22)$

7.24 Using the standard normal table, determine a z value (to two decimal places) such that the
a. cumulative area to z is 0.7486.
b. cumulative area to z is 0.0735.
c. area between z and positive infinity is 0.3508.
d. area between z and positive infinity is 0.0212.

7.25 Using the standard normal table, determine a z value (to two decimal places) such that the
a. cumulative area to z is 0.70.
b. cumulative area to z is 0.10.
c. area between z and negative infinity is 0.54.
d. area between z and positive infinity is 0.30.

7.26 For the normal distribution described in Exercise 7.10, what is the probability that a randomly selected passenger trip would have produced a revenue
a. between \$1.55 and \$1.80?
b. over \$1.90? c. under \$1.25?

7.27 For the normal distribution described in Exercise 7.11, what is the probability that a randomly selected first mortgage would have been for an amount
a. between \$305,000 and \$325,000?
b. over \$325,000? c. under \$390,000?

7.28 For the normal distribution described in Exercise 7.12, what is the probability that a randomly selected tax preparation customer would have paid a fee
a. between \$150 and \$180?
b. under \$140? c. over \$220?

7.29 For the normal distribution described in Exercise 7.15, what is the probability that Jamal's commuting time to work today will be
a. less than 29 minutes?
b. between 29 and 33 minutes?
c. more than 32 minutes?

7.30 For the normal distribution described in Exercise 7.15, what commuting time will be exceeded on only 10% of Jamal's commuting days?

7.31 For the normal distribution described in Exercise 7.11, what first-mortgage amount would have been exceeded by only 5% of the mortgage customers?

7.32 For the normal distribution described in Exercise 7.12, what tax preparation fee would have been exceeded by 90% of the tax preparation customers?

7.33 It has been reported that households in the West spend an annual average of \$6050 for groceries. Assume a normal distribution with a standard deviation of \$1500. Source: David Stuckey and Sam Ward, "Home Cooked," *USA Today*, July 25, 2006, p. 1A.
a. What is the probability that a randomly selected Western household spends more than \$6350 for groceries?
b. How much money would a Western household have to spend on groceries per year in order to be at the 99th percentile (i.e., only 1% of Western households would spend more on groceries)?

7.34 Andre is a fearless circus performer who gets shot from a special cannon during the grand finale of the show and is supposed to land on a safety net at the other side of the arena. The distance he travels varies, but is normally distributed with a mean of 150 feet and a standard deviation of 10 feet. The landing net is 30 feet long.
a. To maximize Andre's probability of landing on the net, how far away from the cannon should he position the nearest edge of the net?
b. Given the net position in part (a), what is the probability that Andre will be able to return for tomorrow night's show?

7.35 Drying times for newly painted microwave oven cabinets are normally distributed with a mean of 2.5 minutes and a standard deviation of 0.25 minutes. After painting, each cabinet is mated with its electronic modules and mechanical components. The production manager must decide how much time to allow after painting before these other components are installed. If the time is too short, the paint will smudge and the unit will have to be refinished. If the time is too long, production efficiency will suffer. A consultant has concluded that the time delay should be just enough to allow 99.8% of the cabinets to dry completely, with just 0.2% ending up being smudged and sent back for refinishing. Given this information, for what time setting should the production manager set the automatic timer that pauses the production line while each cabinet dries?

7.36 KleerCo supplies an under-hood, emissions-control air pump to the automotive industry. The pump is vacuum powered and works while the engine is operating, cleaning the exhaust by pumping extra oxygen into the exhaust system. If a pump fails before the vehicle in which it is installed has covered 50,000 miles, federal emissions regulations require that it be replaced at no cost to the vehicle owner. The company's current air pump lasts an average of 63,000 miles, with a standard deviation of 10,000 miles. The number of miles a pump operates before becoming ineffective has been found to be normally distributed.

a. For the current pump design, what percentage of the company's pumps will have to be replaced at no charge to the consumer?
b. What percentage of the company's pumps will fail at exactly 50,000 miles?
c. What percentage of the company's pumps will fail between 40,000 and 55,000 miles?
d. For what number of miles does the probability become 80% that a randomly selected pump will no longer be effective?

7.37 The KleerCo company in Exercise 7.36 would like to design a more durable pump so that no more than 2% of original-equipment pumps are returned under free warranty, and the standard deviation continues to be 10,000 miles. What will the new average service life have to be in order for the new pump design to comply with this requirement?

7.38 A battery manufacturer has just begun distribution of its SuperVolt, a 6-volt lantern battery for outdoor enthusiasts. Prototype tests of the new battery found the average lifetime in continuous use to be 8 hours, with a standard deviation of 2 hours. Battery lifetimes in such applications have been approximately normally distributed. Edgar Evans, the consumer reporter for a large metropolitan newspaper, has purchased one of the batteries to take along on a camping trip. On the first day of his trip, Edgar goes fishing in his rowboat, becomes disoriented, and gets lost among the many small islands in the large lake near his camp. Fortunately, Edgar has brought along his trusty flashlight and its new SuperVolt battery. At 9 P.M., Edgar turns on his flashlight and shines it into the air, hoping that someone will see his signal and rescue him. Unfortunately, it is not until 3 A.M. that searchers begin their flight over the lake. If Edgar's light is still shining as they become airborne, they will easily spot it from the air. When Edgar finally gets back to the city, what is the probability that he will have an exciting story to tell about how his flashlight and its SuperVolt battery were two of the heroes in his rescue?

THE NORMAL APPROXIMATION TO THE BINOMIAL DISTRIBUTION (7.4)

The binomial distribution, discussed in Chapter 6, is symmetrical whenever the population proportion, π, is 0.5 and approaches symmetry for values that are close to 0.5. In addition, whenever the number of trials, n, becomes larger, the binomial distribution takes on a bell-shaped appearance that resembles very closely the normal distribution discussed in this chapter.

Although the binomial distribution is discrete and the normal distribution is continuous, the normal distribution is a very good approximation to the binomial whenever both $n\pi$ and $n(1 - \pi)$ are ≥ 5. This can be quite useful when n and π are not among the values listed in the binomial probability tables. As we have seen, such tables are already quite extensive for even relatively small upper limits for n. In addition, calculating binomial probabilities in the absence of such tables is a tedious proposition.

The Binomial Distribution: A Brief Review

The Bernoulli Process

The binomial distribution assumes a *Bernoulli process,* which has the following characteristics:

1. There are two or more consecutive trials.
2. On each trial, there are just two possible outcomes—denoted as "success" or "failure."
3. The trials are statistically independent; that is, the outcome in any trial is not affected by the outcomes of earlier trials, and it does not affect the outcomes of later trials.

4. The probability of a success remains the same from one trial to the next. This is satisfied if either we sample with replacement or the sample is small compared to the size of the population. As a rule of thumb, if we are sampling without replacement and the population is at least 20 times as large as the number of trials, the constant π assumption can be considered to have been met.

The Mean and Standard Deviation

The mean and standard deviation of the binomial distribution are the basis for the normal distribution that will be used in making the approximation. As described in Chapter 6, these are

- Mean:

$$\mu = n\pi$$

- Standard deviation:

$$\sigma = \sqrt{n\pi(1 - \pi)}$$

where n = the number of trials
π = the probability of success on any given trial
x = the number of successes in n trials

Correction for Continuity

Because the binomial distribution has gaps between possible values of x, while the normal distribution is continuous, the normal approximation to the binomial involves a **correction for continuity.** The correction consists of expanding each possible value of the discrete variable, x, by 0.5 in each direction. In continuous distributions like the normal, the probability of x taking on any exact value (e.g., $x = 12.000$) is zero. This is why we must consider each of the discrete values of x as being an *interval.*

The continuity correction process where the discrete random variable, *x*, can take on integer values from 0 to 15:

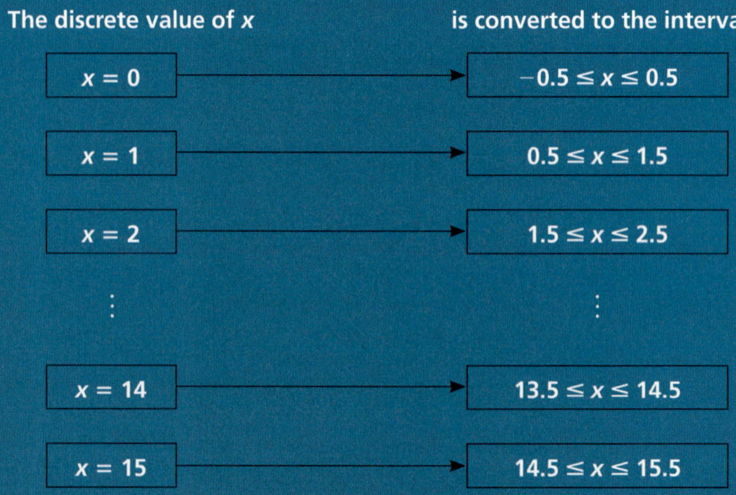

The Approximation Procedure

In using the normal distribution to approximate the binomial, we first determine the mean and standard deviation for the binomial distribution, then use these values as μ and σ in finding the area(s) of interest beneath the standard normal curve. In particular, we can find

1. the probability for an individual value for x, now expressed in terms of the interval from $[x - 0.5]$ to $[x + 0.5]$, or
2. the probability that x will lie within or beyond a given range of values. (*Note:* In this application, the correction for continuity becomes less important whenever x can take on a large number of possible values.)

EXAMPLE

Normal Approximation

This example demonstrates the use of the normal distribution to approximate the binomial. It represents a situation in which both $n\pi$ and $n(1 - \pi)$ are ≥ 5, which is requisite for using the approximation. For purposes of our example, we will assume that 60% of the Americans who exercise each week by walking are women. A randomly selected group comprises 15 persons who exercise each week by walking, with x = the number of women in the group.

SOLUTION

Determine the Mean and Standard Deviation of *x*.
The expected number of females in the group is 9.0.

$$\mu = n\pi \quad \text{or} \quad 15(0.60) = 9$$

$$\sigma = \sqrt{n\pi(1 - \pi)} = \sqrt{15(0.60)(1 - 0.60)} = 1.897$$

What Is the Probability That There Will Be Exactly 11 Females in the Group?
Using the normal distribution to approximate the binomial, with $\mu = 9.0$ and $\sigma = 1.897$, this probability is expressed as $P(10.5 \leq x \leq 11.5)$. The z-score associated with $x = 11.5$ is

$$z = \frac{x - \mu}{\sigma} = \frac{11.5 - 9.0}{1.897} = 1.32$$

The z-score associated with $x = 10.5$ is

$$z = \frac{x - \mu}{\sigma} = \frac{10.5 - 9.0}{1.897} = 0.79$$

Referring to the standard normal table, we find the cumulative area to $z = 1.32$ is 0.9066 and the cumulative area to $z = 0.79$ is 0.7852. As Figure 7.11 shows, $P(10.5 \leq x \leq 11.5)$ is the difference between these areas, or $0.9066 - 0.7852 = 0.1214$.

Note that this approximation is quite close to the actual binomial probability of 0.1268, obtained from the binomial probability tables in Appendix A. Keep in mind, however, that this example was one in which both π and n happened to be available in these tables. This will not always be the case, hence the usefulness of being able to use the normal distribution as an approximation to the binomial.

FIGURE 7.11

Normal approximation to a binomial distribution with $n = 15$ and $\pi = 0.60$. The continuity correction represents $x = 11$ as the interval from 10.5 to 11.5, then the standard normal table is used in obtaining the approximate value of 0.1214 for $P(x = 11)$.

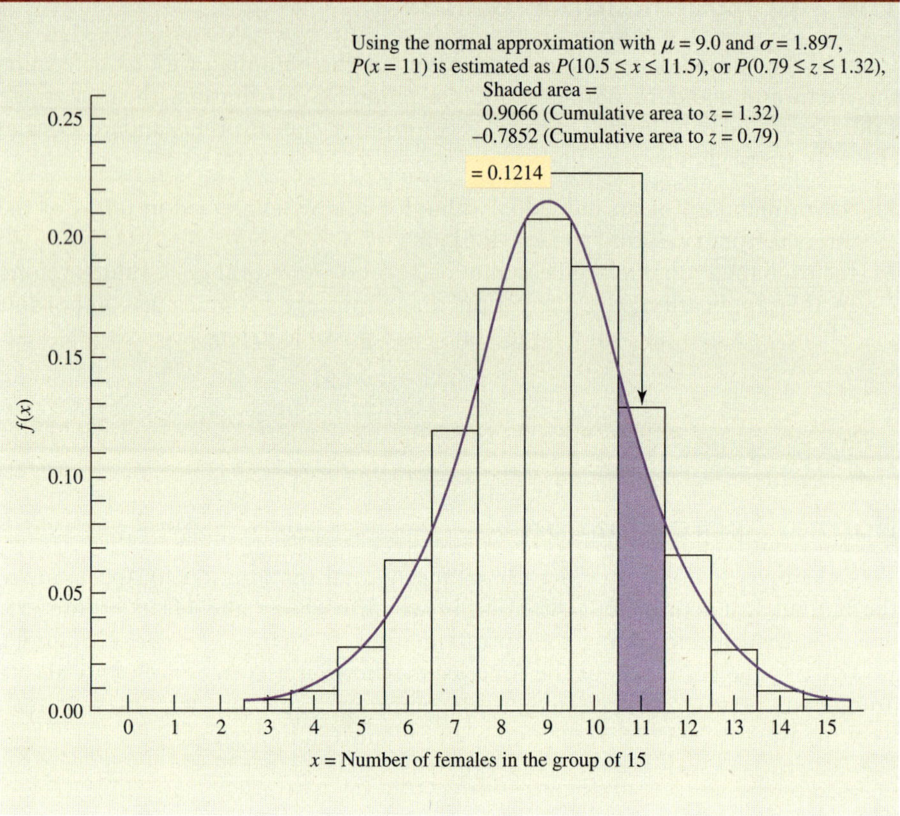

EXAMPLEEXAMPLEEXAMPLE

What Is the Probability That There Will Be at Least 10 Females in the Group of 15?

Again using the normal distribution to approximate the binomial, the probability of interest is now expressed as $P(x \geq 9.5)$. (The continuity correction has replaced the discrete value of 10 with the interval from 9.5 to 10.5.) The z-score associated with $x = 9.5$ is

$$z = \frac{x - \mu}{\sigma} = \frac{9.5 - 9.0}{1.897} = 0.26$$

Referring to the standard normal table, we see that the cumulative area to $z = 0.26$ is 0.6026. Because the total area beneath the normal curve is 1.0000, the area to the right of $z = 0.26$ is $(1.0000 - 0.6026) = P(x \geq 9.5) = 0.3974$. The normal approximation is again very close to the actual probability (0.4032) obtained from the binomial tables.

The normal approximation to the binomial distribution is quite close whenever both $n\pi$ and $n(1 - \pi)$ are ≥ 5. However, the approximation becomes even better as the value of n increases and whenever π is closer to 0.5. Also, when we are determining probabilities for x being within or beyond a given interval, the correction for continuity becomes less important when x can take on a great many different values. For example, if $n = 1000$ and $\pi = 0.3$, adding and subtracting 0.5 from each x value will make virtually no difference in the results. Refer to Seeing Statistics Applet 6, at the end of the chapter, for a visual demonstration of the normal approximation to the binomial distribution.

EXERCISES

7.39 What is the correction for continuity and why is it employed when the normal distribution is used in approximating the binomial distribution?

7.40 Under what circumstances is it permissible to use the normal distribution in approximating the binomial distribution?

7.41 In a certain binomial distribution, $\pi = 0.25$ and $n = 40$. In using the normal approximation,

a. What are the mean and standard deviation of the corresponding normal distribution?
b. If x = the number of "successes" among the 40 observations, determine the following: $P(x = 8)$, $P(12 \leq x \leq 16)$, $P(10 \leq x \leq 12)$, $P(x \geq 14)$.

7.42 In a certain binomial distribution, $\pi = 0.30$ and $n = 20$. In using the normal approximation,

a. What are the mean and standard deviation of the corresponding normal distribution?
b. If x = the number of "successes" among the 20 observations, determine the following: $P(x = 5)$, $P(4 \leq x \leq 7)$, $P(1 \leq x \leq 5)$, $P(x \geq 7)$.

7.43 Approximately 80% of American families have some form of life insurance coverage. For a randomly selected sample of $n = 15$ families, and the discrete random variable x = the number of families in the sample who are covered by life insurance: Source: American Council of Life Insurers, *2008 Life Insurers Fact Book,* p. 78.

a. Calculate the mean and the standard deviation of the binomial distribution.
b. Using the binomial distribution, calculate the probability that x will be exactly 12 families.
c. Repeat part (b), using the normal approximation to the binomial. (*Hint:* The area of interest under the normal curve will be between $x = 11.5$ and $x = 12.5$.)
d. Using the normal approximation to the binomial, find the probability that at least 10 families in the sample will be covered by some form of life insurance.

7.44 About 40% of new single-family homes completed in the United States during 2007 were heated with electricity. For a randomly selected sample of 20 new single-family homes completed during that year and the discrete random variable, x = the number of homes in this group heated with electricity: Source: *Statistical Abstract of the United States 2009,* p. 593.

a. Calculate the mean and standard deviation of the binomial distribution.
b. Using the binomial distribution tables, determine the probability that x will be exactly 8 homes heated with electricity.
c. Using the binomial distribution tables, determine the probability that x will be at least 6 but no more than 9 homes.
d. Repeat part (b), using the normal approximation to the binomial distribution.
e. Repeat part (c), using the normal approximation to the binomial distribution.

7.45 Of all individual tax returns filed in the United States during the 2009 tax filing season, 15.8% were prepared by H&R Block. For a randomly selected sample of 900 tax returns filed during this period, use the normal approximation to the binomial distribution in determining the probability that between 110 and 140 of these returns were prepared by H&R Block. Source: hrblock.com, July 14, 2009.

7.46 The Electronic Industries Association reports that about 50% of U.S. households have a camcorder. For a randomly selected sample of 800 U.S. households, use the normal approximation to the binomial distribution in determining the probability that at least 410 of these households have a camcorder. Source: *New York Times Almanac 2009*, p. 409.

7.47 Repeat Exercise 7.45, but without using the correction for continuity (i.e., the discrete random variable, x, is not replaced by the interval $x - 0.5$ to $x + 0.5$). Does your answer differ appreciably from that obtained in Exercise 7.45? If not, comment on the need for using the correction for continuity when x can take on a large number of possible values.

7.48 Repeat Exercise 7.46, but without using the correction for continuity (i.e., the discrete random variable, x, is not replaced by the interval $x - 0.5$ to $x + 0.5$). Does your answer differ appreciably from that obtained in Exercise 7.46? If not, comment on the need for using the correction for continuity when x can take on a large number of possible values.

7.5 THE EXPONENTIAL DISTRIBUTION

Description and Applications

Chapter 6 discussed the Poisson distribution, in which the discrete random variable was the number of *rare events* occurring during a given interval of time, space, or distance. For a Poisson process, a distribution called the **exponential distribution** describes the continuous random variable, x = the amount of time, space, or distance between occurrences of these rare events. For example, in the context of arrivals at a service counter, x will be a continuous random variable describing the time between successive arrivals.

As with the Poisson distribution, the exponential distribution has application to the management science topic of queuing, or waiting-line theory. Like the Poisson distribution, it also assumes that the arrivals are independent from one another. The distribution can be summarized as follows:

The exponential distribution:

- **For x = the length of the interval between occurrences:**

  $f(x) = \lambda e^{-\lambda x}$ for $x > 0$ and $\lambda > 0$ where λ = the mean and variance of a Poisson distribution

  $1/\lambda$ = the mean and standard deviation of the corresponding exponential distribution

  e = the mathematical constant, 2.71828, the base of the natural logarithm system

- **Areas beneath the curve:**

  $P(x \geq k) = e^{-\lambda k}$ where k = the time, space, or distance until the next occurrence

Like the Poisson distribution, the exponential distribution is really a family of distributions, and the particular member of the family is determined by just one descriptor: its mean. Actually, the mean of the exponential distribution is the *inverse* of that of the Poisson. For example, if the average number of arrivals at a tollbooth is $\lambda = 5$ persons per minute, the mean of the corresponding exponential distribution will be $1/\lambda = 1/5$, or 0.20 minutes between persons. Note that the Poisson random variable is discrete (number of persons), whereas the random variable for the exponential distribution is continuous, in this case, *time*.

NOTE

It is important that we be consistent in our use of physical-measurement units when we invert λ to arrive at the mean of the corresponding exponential distribution. For example, if arrivals tend to occur at the rate of 15 vehicles per hour, we can express this as 15 arrivals divided by 60 minutes, or 0.25 arrivals per minute. If we then consider $\lambda = 0.25$ arrivals per minute as the Poisson mean, the mean of the corresponding exponential distribution must also be based on minutes, and $1/\lambda$ will be $1/0.25$, or 4 minutes between arrivals.

EXAMPLE

Exponential Distribution

Calls to the "911" emergency number of a large community have been found to be Poisson distributed with an average of 10 calls per hour.

SOLUTIONS

Determine the Mean and Standard Deviation of the Corresponding Exponential Distribution

The mean of the Poisson distribution is $\lambda = 10$ calls per hour, and the mean of the corresponding exponential distribution is the inverse of this, $1/\lambda = 1/10$, or 0.10 hours between calls. The standard deviation is numerically equal to the mean.

For greater clarity in answering the questions that follow, we will express the means of the Poisson and exponential distributions using minutes instead of hours. Accordingly, the mean of the Poisson distribution is 10 calls per 60 minutes, or $\lambda = 1/6$ calls per minute. The mean of the corresponding exponential distribution is $1/\lambda = 6$ minutes between calls.

What Is the Probability That the Next Call Will Occur at Least 5 Minutes from Now?

Referring to the continuous probability distribution of Figure 7.12, this probability corresponds to the area to the right of $k = 5$ minutes on the x axis. Applying the formula for determining areas beneath the curve:

$$P(x \geq k) = e^{-\lambda k} \quad \text{where } k = 5 \text{ minutes}$$

or

$$P(x \geq 5) = e^{-(1/6)(5)} = 2.71828^{-0.833} \quad \text{or} \quad 0.4347$$

There is a 0.4347 chance that at least 5 minutes will elapse before the next call is received. Alternatively, the probability that the next call will occur *within* the next 5 minutes is (1.000 − 0.4347), or 0.5653. This is because (1) the total area beneath the curve is 1.0, and (2) the exponential distribution is continuous

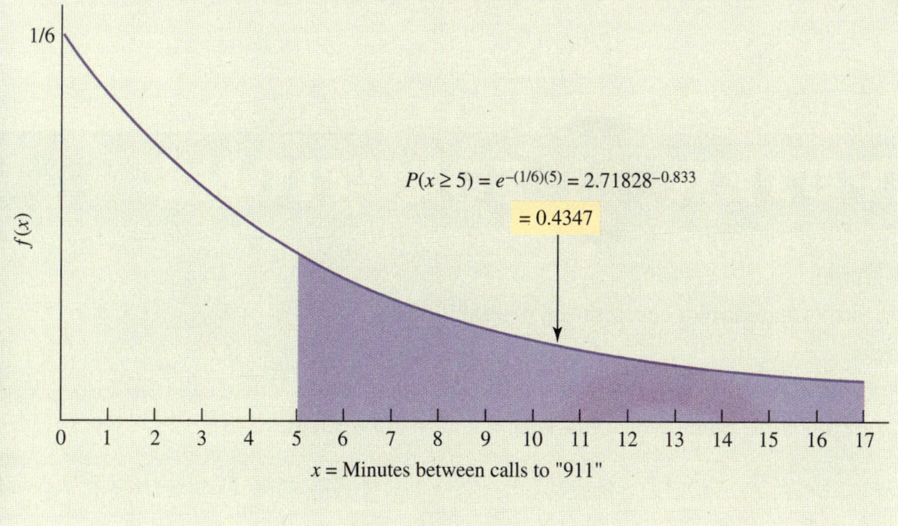

FIGURE 7.12

An exponential distribution in which the mean time between occurrences is 6 minutes. The probability is 0.4347 that at least 5 minutes will elapse before the next call to the "911" switchboard.

and the probability is zero that a continuous random variable will take on any exact value [i.e., $P(x = 5)$ is 0.0].

What Is the Probability That the Next Call Will Occur Between 3 and 8 Minutes from Now?

The area beneath the curve in the interval $3 \le x \le 8$ can be determined by subtracting the area representing $x \ge 8$ from the area representing $x \ge 3$. This is shown in Figure 7.13 and is calculated as

$$P(3 \le x \le 8) = P(x \ge 3) - P(x \ge 8) = e^{-(1/6)(3)} - e^{-(1/6)(8)}$$
$$= 2.71828^{-0.500} - 2.71828^{-1.333} \quad \text{or} \quad 0.6065 - 0.2637 = 0.3428$$

Excel and Minitab can be used to obtain cumulative exponential probabilities associated with specified values for x—the procedures and results are shown in Computer Solutions 7.3. These software packages can also be used in doing the reverse—that is, specifying a cumulative probability, then finding the x value associated with it. These procedures and results are shown in Computer Solutions 7.4.

FIGURE 7.13

For the exponential distribution in which the mean time between calls is 6 minutes, the probability is 0.3428 that between 3 and 8 minutes will elapse before the next call to the "911" switchboard.

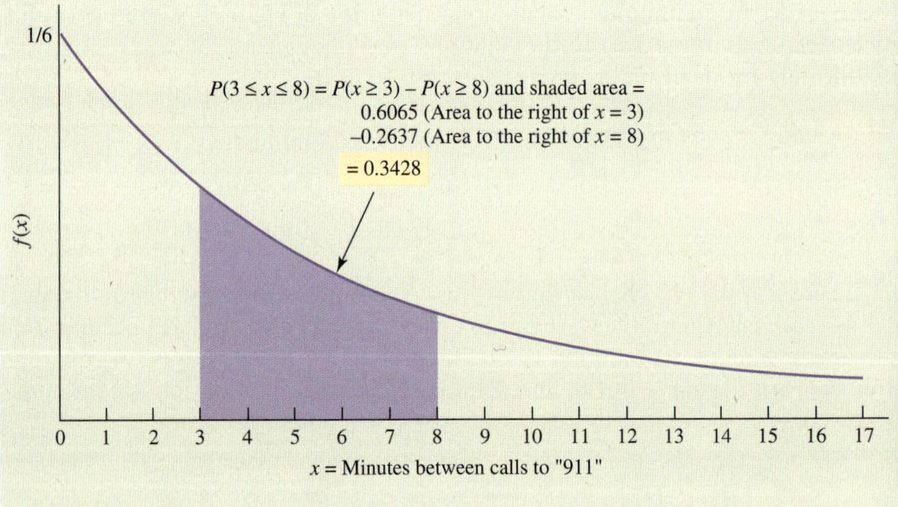

COMPUTER 7.3 SOLUTIONS

Exponential Probabilities

These procedures show how to get cumulative exponential probabilities.

For each software package:

Procedure I: Getting the cumulative probability associated with one x value.

Procedure II: Getting cumulative probabilities associated with a set of x values.

EXCEL

	A	B	C
1	Exponential Distribution		
2	with mean = 6.0.		
3	x	Area to L.	
4	7.0	0.6886	
5	6.0	0.6321	
6	5.0	0.5654	

I. Excel procedure for the cumulative probability associated with one *x* value

1. Click on the cell into which Excel will place the probability. From the **Formulas** ribbon, click on the **Insert Function** symbol at the extreme left. Click **Statistical** in the **Category** box. Click **EXPONDIST** in the **Function** box. Click **OK.**
2. Enter the *x* value of interest (e.g., **5**) into the **X** box. Enter the *inverse* of the exponential mean (e.g., if the mean is 6 minutes between calls, enter 1/6, or **0.16667**) into the **Lambda** box. Enter **True** into the **Cumulative** box. Click **OK.**

II. Excel procedure for cumulative probabilities associated with a set of *x* values

1. Enter the labels into **A3:B3** as shown in the printout above. Enter the *x* values of interest (e.g., **7, 6, 5**) starting with cell **A4.** Click on cell **B4.**
2. From the **Formulas** ribbon, click on the **Insert Function** symbol at the extreme left. Click **Statistical** in the **Category** box. Click **EXPONDIST** in the **Function** box. Click **OK.**
3. Enter **A4** into the **X** box. Enter the *inverse* of the exponential mean (e.g., 1/6, or **0.16667**) into the **Lambda** box. Enter **True** into the **Cumulative** box. Click **OK.** The first cumulative probability will appear in cell B4.
4. Click at the lower right corner of **B4** (the cursor will turn to a "+" sign) and drag downward so cells **B4:B6** will contain the set of cumulative probabilities.

MINITAB

```
Cumulative Distribution Function
Exponential with mean = 6
x  P( X <= x )
7      0.688597
6      0.632121
5      0.565402
```

I. Minitab procedure for cumulative probability associated with one *x* value

1. Click **Calc.** Select **Probability Distributions.** Click **Exponential.**
2. In the **Exponential Distribution** menu, enter the mean of the distribution (e.g., **6**) into the **Mean** or **Scale** box. Select **Input constant** and enter the *x* value of interest (e.g., **5**) into the box. Select **Cumulative probability.** Click **OK.**

II. Minitab procedure for cumulative probabilities associated with a set of *x* values

Enter the desired *x* values (e.g., **7, 6, 5**) into column **C1.** Follow steps 1 and 2 above, except select **Input column** and enter **C1** into the box.

COMPUTER 7.4 SOLUTIONS

Inverse Exponential Probabilities

For Minitab only. Excel does not currently offer inverse exponential probabilities.

MINITAB

```
Inverse Cumulative Distribution Function
Exponential with mean = 6
P( X <= x )          x
      0.975    22.1333
      0.500     4.1589
      0.250     1.7261
```

I. Minitab procedure for the *x* value associated with one cumulative probability

1. Click **Calc.** Select **Probability Distributions.** Click **Exponential.**
2. In the **Exponential Distribution** menu, enter the mean of the distribution (e.g., **6**) into the **Mean** or **Scale** box. Select **Input constant** and enter the cumulative probability of interest (e.g., **0.8000**) into the box. Select **Inverse cumulative probability.** Click **OK.**

II. Minitab procedure for the *x* values associated with a set of cumulative probabilities

Enter the cumulative probabilities of interest (e.g., **0.9750, 0.5000, 0.2500**) into column **C1.** Follow steps 1 and 2 above, except select **Input column** and enter **C1** into the box.

EXERCISES

7.49 What is the relationship between the Poisson distribution and the exponential distribution?

7.50 Every day, drivers arrive at a tollbooth. If the Poisson distribution were applied to this process, what would be an appropriate random variable? What would be the exponential-distribution counterpart to this random variable?

7.51 The main switchboard at the Home Shopping Network receives calls from customers. If the Poisson distribution were applied to this process, what would be an appropriate random variable? What would be the exponential-distribution counterpart to this random variable?

7.52 A random variable is Poisson distributed with $\lambda = 1.5$ occurrences per hour. For the corresponding exponential distribution, and x = hours until the next occurrence, identify the mean of x and determine the following:

a. $P(x \geq 0.5)$ b. $P(x \geq 1.0)$
c. $P(x \geq 1.5)$ d. $P(x \geq 2.0)$

7.53 A random variable is Poisson distributed with $\lambda = 0.02$ occurrences per minute. For the corresponding exponential distribution, and x = minutes until the next occurrence, identify the mean of x and determine the following:

a. $P(x \geq 30.0)$ b. $P(x \geq 40.0)$
c. $P(x \geq 50.0)$ d. $P(x \geq 60.0)$

7.54 A random variable is Poisson distributed with $\lambda = 0.50$ arrivals per minute. For the corresponding exponential distribution, and x = minutes until the next arrival, identify the mean of x and determine the following:

a. $P(x \leq 0.5)$ b. $P(x \leq 1.5)$
c. $P(x \geq 2.5)$ d. $P(x \geq 3.0)$

7.55 The owner of a self-service carwash has found that customers take an average of 8 minutes to wash and dry their cars. Assuming that the self-service times tend to be exponentially distributed, what is the probability that a customer will require more than 10 minutes to complete the job?

7.56 A taxi dispatcher has found that successive calls for taxi service are exponentially distributed, with a mean time between calls of 5.30 minutes. The dispatcher must disconnect the telephone system for 3 minutes in order to have the push-button mechanism repaired. What is the probability that a call will be received while the system is out of service?

7.57 During 2008, U.S. general aviation pilots had 1.20 fatal crashes per 100,000 flying hours. Harriet Arnold is president of Arnold's Flying Service, a company that operates a total of 50 sightseeing planes based in 20 regions of the United States. Altogether, the planes in this fleet are in the air about 40,000 hours per year. Ever since a fatal crash last year involving one of the company's planes that took off under questionable weather conditions, the company has been getting close scrutiny from both government safety officials and the national media. Harriet believes her planes are just as safe as anyone else's, but is concerned that even one more fatal accident within the coming year might be too much for the now-struggling company to overcome. Assuming an exponential distribution for x = thousands of flying hours between fatal crashes, what is the probability that Arnold's Flying Service will not experience a fatal crash until at least a year from today? Until at least 2 years from today? What number of flying hours would be associated with a 90% probability of experiencing no crashes within that number of flying hours from today? Source: General Aviation Manufacturers Association, *General Aviation Statistical Databook, 2008 Edition,* p. 58.

7.58 The Bureau of Labor Statistics reports the fatality rate for workers in the rail transportation industry is 0.006 fatalities per 200,000 worker-hours. During 2007, rail transportation workers put in a total of 545 million worker-hours. Assuming that (1) an exponential distribution applies for the number of worker-hours between fatalities in this industry, and (2) the annual 545 million worker-hours of activity continues in the industry, what is the expected number of worker-hours until the next fatality occurs in this industry? If a given rail transportation company requires 1,000,000 worker-hours each year, what is the probability that the next fatal injury in this company will not occur until at least 30 years from now? Source: Bureau of Labor Statistics, *Census of Fatal Occupational Injuries, 2007.*

SIMULATING OBSERVATIONS FROM A CONTINUOUS PROBABILITY DISTRIBUTION 7.6

Although practically all statistical packages can provide probabilities like those in the Computer Solutions earlier in this chapter, some can also randomly select observations from the continuous distribution itself. For example, Computer Solutions 7.5 lists the procedures and shows the results when Excel and Minitab are used in randomly selecting 10,000 observations from a normal distribution. Both sets of simulated observations are from a normal distribution examined earlier in the chapter—in this situation, the continuous random variable is x = the amount of weight added so that a generator shaft would have proper static and dynamic balance. In the probability distribution, the average amount of weight added is $\mu = 35$ grams and the standard deviation is $\sigma = 9$ grams.

In each of the printouts in Computer Solutions 7.5, the mean and standard deviation of the simulated observations are very close to the actual mean and standard deviation of the theoretical distribution from which the 10,000 observations were drawn. In addition to differing slightly from the theoretical values, the Excel and Minitab sample means and standard deviations also differ slightly

COMPUTER 7.5 SOLUTIONS

Simulating Observations from a Continuous Probability Distribution

These procedures simulate 10,000 observations from a normal distribution with mean = 35 and standard deviation = 9. The procedures for simulating observations from an exponential distribution will differ only slightly from those described here.

EXCEL

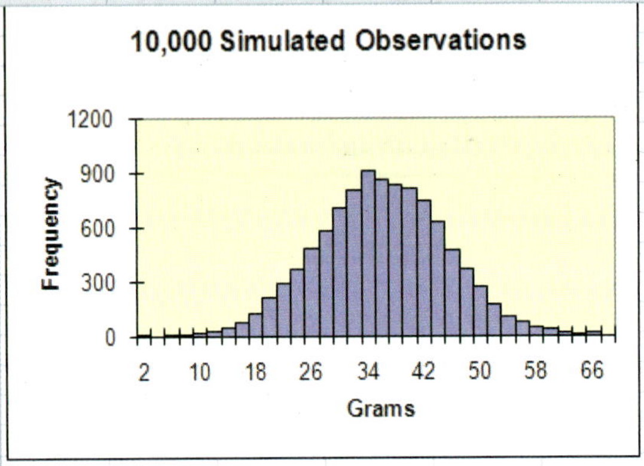

Grams	
Mean	34.965
Standard Error	0.090
Median	34.987
Mode	29.216
Standard Deviation	8.994
Sample Variance	80.897
Kurtosis	-0.002
Skewness	0.002
Range	63.870
Minimum	1.326
Maximum	65.196
Sum	349645.570
Count	10000

1. Start with a blank worksheet. From the **Data** ribbon, click **Data Analysis** in the rightmost menu section. Click **Random Number Generation.** Click **OK.**
2. Enter **1** into the **Number of Variables** box. Enter **10000** into the **Number of Random Numbers** box. Select **Normal** within the **Distribution** box. Enter **35** into the **Mean** box. Enter **9** into the **Standard Deviation** box. Select **Output Range** and enter **A1** into the box. Click **OK.** The 10,000 simulated observations will be located in A1:A10000.
3. The Excel descriptive statistics and histogram above can be generated using the procedures in Computer Solutions 2.1 and 3.1.

MINITAB

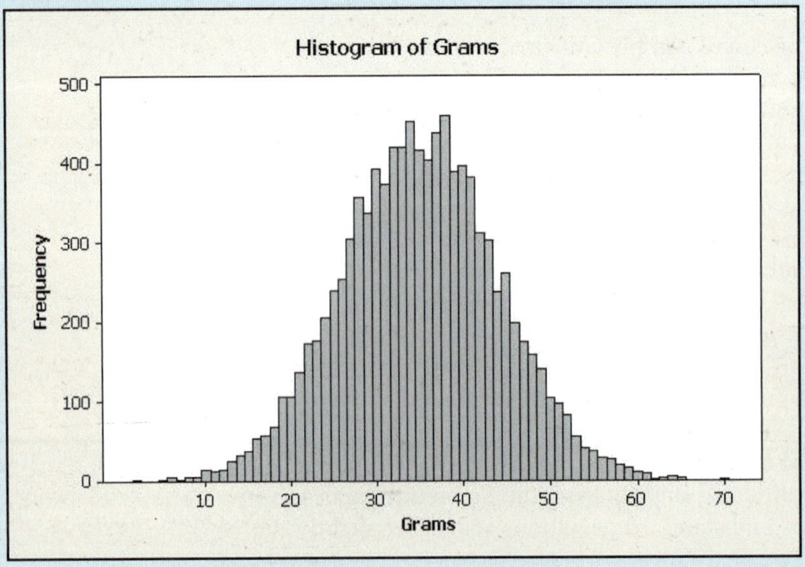

Descriptive Statistics: Grams

Variable	N	N*	Mean	SE Mean	StDev	Minimum	Q1	Median	Q3
Grams	10000	0	34.995	0.091	9.078	2.055	28.835	35.020	41.079

Variable	Maximum
Grams	70.164

1. Click **Calc.** Select **Random Data.** Click **Normal.** Enter **10000** into the **Number of rows of data to generate** box. Enter **C1** (the destination column) into the **Store in columns** box. Enter **35** into the **Mean** box. Enter **9** into the **Standard deviation** box. Click **OK.** The 10,000 simulated observations will be located in column C1.
2. The Minitab descriptive statistics and histogram shown above can be generated using the procedures in Computer Solutions 2.1 and 3.1.

from each other. This is because each package generated its own sample of 10,000 observations, and each sample had a different set of observations from the other.

According to the law of large numbers, histograms like those in Computer Solutions 7.5 will more closely approach the theoretical distribution from which they were drawn whenever the number of observations becomes larger. The topic of how samples can differ from each other, and from the population from which they were drawn, will be an important topic in the next two chapters.

EXERCISES

7.59 A computer statistical package has simulated 1000 random observations from a normal distribution with $\mu = 50$ and $\sigma = 10$. Sketch the approximate box-and-whisker display for the resulting data.

7.60 A computer statistical package has simulated 2000 random observations from a normal distribution with $\mu = 80$ and $\sigma = 20$. Sketch the approximate box-and-whisker display for the resulting data.

7.61 If a computer statistical package were to simulate 2000 random observations from a normal distribution with $\mu = 150$ and $\sigma = 25$, what percentage of these observations would you expect to have a value of 140 or less? Do you think the actual number in the "≤140" range would equal the expected number in this range? If so, why? If not, why not?

7.62 If a computer statistical package were to simulate 1000 random observations from a normal distribution with $\mu = 200$ and $\sigma = 50$, what percentage of these observations would you expect to have a value of 300 or more? Do you think the actual number in the "≥300" range would equal the expected number in this range? If so, why? If not, why not?

SUMMARY (7.7)

• Nature and applications of the normal distribution

Continuous probability distributions describe probabilities associated with random variables that can take on any value along a given range or continuum and for which there are no gaps between these possible values. The curve $f(x)$ is a function of the continuous random variable, and $f(x)$ is called the probability density function. For any specified interval of values, the probability that x will assume a value within the interval is the area beneath the curve between the two points describing the interval. The total area beneath the curve is 1.0, and the probability that x will take on any specific value is 0.0.

The most important continuous distribution is the normal distribution. This is because many natural and economic phenomena tend to be approximately normally distributed; it can be used to approximate other distributions, such as the binomial; and the distribution of sample means and proportions tends to be normally distributed whenever repeated samples are taken from a given population of any shape.

The normal distribution is a symmetrical, bell-shaped curve, and the mean, median, and mode are located at the same position. The curve is asymptotic in that each end approaches the horizontal (x) axis, but never reaches it. The normal distribution is actually a family of such distributions, and the specific member of the family is determined by the mean (μ) and standard deviation (σ). Regardless of the shape of a particular normal curve, about 68.3% of the area is within the interval $\mu \pm \sigma$, 95.5% within $\mu \pm 2\sigma$, and 99.7% within $\mu \pm 3\sigma$.

• Standard normal distribution and *z*-scores

When the original x values are expressed in terms of the number of standard deviations from the mean, the result is the standard (or standardized) normal distribution in which the z value (or z-score) replaces x. The standard normal distribution has a mean of 0 and a standard deviation of 1, and cumulative areas beneath its curve are readily available from the standard normal table.

• Normal distribution to approximate the binomial distribution

Although the binomial distribution is discrete and the normal distribution is continuous, the normal distribution is a good approximation to the binomial whenever both $n\pi$ and $n(1 - \pi)$ are ≥ 5. The normal approximation is useful whenever suitable binomial tables are not available, and makes it unnecessary to engage in the tedious computations that the binomial distribution would otherwise require.

Because the binomial distribution has gaps between possible values of x, the normal approximation to the binomial involves a correction for continuity, which consists of "expanding" each possible value of x by 0.5 in each direction. This correction becomes less important whenever the number of trials is large and we are finding the probability that x lies within or beyond a given range of values.

• Exponential distribution

For a Poisson process, the exponential distribution describes the continuous random variable, x = the amount of time, space, or distance between occurrences of the rare events of interest. The exponential distribution is a family of distributions, all of which are positively skewed, and the specific member is determined by the value of the mean.

• Computer-generated probabilities and simulations

As with discrete probability distributions, computer statistical packages can determine the probabilities associated with possible values of x. In the case of continuous distributions, this includes finding the exact area beneath a specified portion of the curve, or probability density function, and eliminates the need for calculations or table references. Some packages, such as Excel and Minitab, can also simulate the random selection of observations from the distribution itself.

EQUATIONS

The Normal Distribution

$$f(x) = \frac{1}{\sigma\sqrt{2\pi}} e^{-(1/2)[(x-\mu)/\sigma]^2}$$

where μ = mean
σ = standard deviation
e = the mathematical constant, 2.71828
π = the mathematical constant, 3.14159

The *z*-Score for a Standard Normal Distribution

$$z = \frac{x - \mu}{\sigma}$$

where z = the distance from the mean, measured in standard deviation units
x = the value of x in which we are interested
μ = the mean of the distribution
σ = the standard deviation of the distribution

The Normal Approximation to the Binomial Distribution

- Mean:

$$\mu = n\pi$$

- Standard deviation:

$$\sigma = \sqrt{n\pi(1 - \pi)}$$

where n = the number of trials
π = the probability of success on any given trial
x = the number of successes in n trials

(*Note*: This approximation is used when $n\pi$ and $n(1 - \pi)$ are both ≥ 5.)

The Exponential Distribution

- For x = the length of the interval between occurrences:

$$f(x) = \lambda e^{-\lambda x} \quad \text{for } x > 0 \text{ and } \lambda > 0$$

where λ = the mean and variance of a Poisson distribution
$1/\lambda$ = the mean and standard deviation of the corresponding exponential distribution
e = the mathematical constant, 2.71828, the base of the natural logarithm system

- Areas beneath the curve:

$$P(x \geq k) = e^{-\lambda k}$$

where k = the time, space, or distance until the next occurrence

CHAPTER EXERCISES

7.63 According to a recent survey, 35% of the adult citizens of Mayville have a charge account at Marcy's Department Store. Mr. Marcy's nephew is going to trial for embezzlement, and nine jurors are to be randomly selected from the adult citizenry. What is the probability that at least four of the jurors will be charge-account holders?

7.64 A food expert claims that there is no difference between the taste of two leading soft drinks. In a taste test involving 200 persons, however, 55% of the subjects say they like soft drink A better than B. If the food expert is correct, we would expect that 50% of the persons would have preferred soft drink A. Under the assumption that the food expert is correct, what would be the probability of a "taste-off" in which 110 or more of the subjects preferred soft drink A?

7.65 In Exercise 7.64, the maker of soft drink A claims its product to be "superior" to soft drink B. Given the results obtained in Exercise 7.64, do they appear to have a solid basis for their claim?

7.66 During an annual heating season, the average gas bill for customers in a New England community heating their homes with gas was $457. Assuming a normal distribution and a standard deviation of $80:

a. What proportion of homes heating with gas had a gas bill over $382?
b. What proportion of homes heating with gas had a gas bill between $497 and $537?
c. What amount was exceeded by only 2.5% of the homes heating with gas?
d. What amount was exceeded by 95% of the homes heating with gas?

7.67 It has been reported that the average monthly cell phone bill is $50. Assuming a normal distribution and a standard deviation of $10, what is the probability that a randomly selected cell phone subscriber's bill last month was less than $35? More than $70? Source: *Statistical Abstract of the United States 2009*, p. 706.

7.68 During 2004, the average number of flying hours for aircraft operated by regional airlines in the United States was 2389. Assume a normal distribution and a standard deviation of 300 hours. Source: *Statistical Abstract of the United States 2006*, p. 701.

a. What proportion of aircraft flew more than 2200 hours during the year?
b. What proportion of aircraft flew between 2000 and 2400 hours during the year?
c. What number of annual flying hours was exceeded by only 15% of the aircraft?
d. What number of annual flying hours was exceeded by 75% of the aircraft?

7.69 During fiscal 2008, the average daily volume for FedEx Corporation was 7,000,000 packages per day. Assuming a normal distribution and a standard deviation of 800,000 packages per day, on what proportion of the days was the volume between 6,000,000 and 6,500,000 packages? Source: *FedEx Corporation, 2008 Annual Report*, p. 28.

7.70 For itemized tax returns in the $60,000–$75,000 income group for the most recent year reported, the average charitable contribution was $1935. Assume a normal distribution and a standard deviation of $400. Source: *Statistical Abstract of the United States 2002*, p. 360.

a. For a randomly selected household from this income group, what is the probability that the household made charitable contributions of at least $1600?
b. For a randomly selected household from this income group, what is the probability that the household made charitable contributions of between $2200 and $2400?
c. What level of contributions was exceeded by only 20% of the itemized returns in this income group?
d. What level of contributions was exceeded by 40% of the itemized returns in this income group?

7.71 Of the 1.0 million persons of voting age in Maine, 72% voted in the 2008 presidential election. For a randomly selected group of 30 Maine residents who were of voting age at that time, what is the probability that at least 20 of these persons voted in the presidential election? Source: *nonprofitvote.org*, July 15, 2009.

7.72 A public relations agency tells its client that 80% of the residents in a 50-mile radius believe the company is "an industry leader." Skeptical, the company commissions a survey in which just 320 of the 500 persons interviewed felt the company was an industry leader. *If the public relations agency's claim is correct,* what is the probability that interviews with 500 people would yield 320 or fewer who felt the company was an industry leader?

7.73 Given the results of Exercise 7.72, evaluate the claim made by the public relations agency.

7.74 A researcher is studying a particular process and has identified the following times, in minutes, between consecutive occurrences: 1.8, 4.3, 2.3, 2.1, 5.6, 1.5, 4.5, 1.8, 2.4, 2.6, 3.2, 3.5, 1.6, 2.1, and 1.9. Given this information, what would be the researcher's best estimate for the mean of the exponential distribution? What would be her best estimate for the mean of the corresponding Poisson distribution?

7.75 A safety researcher has found that fourth to sixth graders engaging in unorganized activities on school grounds experienced injuries at the rate of 2.4 injuries per 100 thousand student-days. On the average, how many student-days elapse between injuries to fourth to sixth graders in unorganized activities at school, and what is the probability that the next such injury will occur before 45 thousand student-days have passed?

7.76 The number of defects in rolls of aluminum sheet average 3.5 defects per 100 feet. The company president is accompanying overseas visitors on a tour of the plant, and they have just arrived at the aluminum-sheet machine. If they remain at the machine long enough to observe 32 feet of aluminum being produced, what is the probability that they will have left the area before the next defect occurs?

7.77 The mileage death rate for motorcycle riders has been estimated to be about 39 deaths per 100 million miles of motorcycle travel. A national motorcycling association has 1200 members who travel a total of 2 million miles each year on their motorcycles. What is the probability that the next highway fatality experienced by the association will not occur until more than a year from now? Until more than 2 years from now?
Source: USAToday.com, July 30, 2006.

7.78 The "20-ounce" package of mozzarella cheese distributed by a food company actually has a weight that is normally distributed, with $\mu = 21$ ounces and $\sigma = 0.2$ ounces.

a. What is the probability that a randomly selected package of this cheese will actually contain at least 20.5 ounces?
b. What is the probability that a randomly selected package of this cheese will actually contain between 20.5 and 21.3 ounces?
c. A shopper randomly selects 8 of these packages. What is the probability that 3 or more of them will contain at least 21.2 ounces of cheese?

7.79 State Police Academy graduates from a large graduating class have scored an average of 81.0 on a proficiency test, and the scores are normally distributed with $\sigma = 8.5$. Graduates are randomly paired and assigned to barracks throughout the state. Madeleine has scored 83.0 on the test. What is the probability that she will be paired with someone whose score was within 5.0 points of her own?

7.80 The trainer for a professional football team has found that, on the average, his team experiences 2.3 sprained ankles per 1000 game minutes. If a game lasts 60 minutes, and the number of game minutes between ankle sprains is exponentially distributed, determine the probability that the next ankle sprain will occur

a. during the team's next game.
b. sometime during the team's next ten games.
c. at the moment the 2-minute warning is sounded in the next game.

7.81 The precooked weight of hamburgers at a gourmet hamburger restaurant has been normally distributed with a mean of 5.5 ounces and a standard deviation of 0.15 ounces. A dining-column journalist from the local newspaper enters the restaurant and orders a hamburger. What is the probability that he will receive a hamburger that had a precooked weight less than 5.3 ounces? If, in a separate visit to the restaurant, the journalist is accompanied by three colleagues who also order a hamburger, what is the probability that at least 2 of these 4 customers will receive a hamburger that had a precooked weight greater than 5.7 ounces?

7.82 Boxes are filled with sugar by a machine that is set to deliver an average of 20.3 ounces. The weights are normally distributed with a standard deviation of 0.3 ounces. If a consumer advocate buys a package of the product and then proceeds to weigh the contents, what is the probability that the package she has purchased will have a content weight between 20 and 21 ounces? If the consumer advocate plans to purchase 100 boxes of the product, then file suit against the company if more than 5 of the boxes have content weights below the 20-ounce stated weight on the package, what is the probability that she will end up filing suit against the company?

7.83 In Exercise 7.82, the company would like to put in a greater weight "cushion" to help protect itself from consumer advocates. If the company wants to have just 2% of the packages contain less than 20 ounces, to what average weight must the filling machine be set?

7.84 Discount Micros, a computer mail-order company, has received 2000 desktop computers from a major manufacturer. Unknown to the mail-order company, 250 of the computers have a hard disk that was improperly installed at the factory. A consulting firm purchases 40 computers from Discount Micros. If the order is filled by random selection from the 2000 computers just received, what is the probability that the shipment to the consulting firm will include no more than 2 defective computers? At least 1, but no more than 3 defective computers?

7.85 The U-Drive car rental corporation has found that the cars in the company's rental fleet experience punctures at the rate of 1.25 punctures per 10,000 miles, and the number of punctures per distance traveled is Poisson distributed. Ed and Harriet are taking advantage of a U-Drive promotion in which there is no mileage charge. During their vacation, they plan to drive from Boston to Los Angeles and back, a total distance of 6164 miles.

a. What is the probability that Ed and Harriet will not have to change any tires during their vacation?
b. While traveling west, what is the probability that Ed and Harriet will experience a punctured tire before they make it to Denver, 2016 miles away?
c. For what mileage is the probability 0.80 that Ed and Harriet will experience a puncture before they reach this distance?

7.86 Bonnie Rogers, the customer relations manager for a large computer manufacturer, has recently studied the duration of telephone calls received by the company's technical support department. The call lengths are exponentially distributed, with a mean duration of 8 minutes. Experience has taught Bonnie that lengthy calls usually can be more efficiently handled by a technical support manager, rather than by the less experienced personnel who receive the initial technical inquiry. To avoid having the higher-paid managers virtually take over the task of handling technical support, Bonnie would like to have only the longest 10% of the calls redirected to a technical support manager. How long must a technical support call last before it qualifies for redirection to a technical support manager?

7.87 A storage warehouse in a relatively remote part of a county is protected by a homemade burglary detection system. Once actuated by an intruder, the system sounds a horn and flashes a light for 15 minutes, then saves battery power by shutting down and resetting itself until the next intrusion is detected. Police in three patrol cars have this part of the county as one of their responsibilities, and they routinely drive by the warehouse facility during the hours from dark to dawn. On average, a patrol car passes the warehouse every 20 minutes, and the time between patrol car arrivals is exponentially distributed. It's the middle of the night and a burglar has just broken into the storage warehouse and set off the alarm. What is the probability that the alarm will shut off before the next police patrol car drives by?

INTEGRATED CASES

Thorndike Sports Equipment (Corresponds to Thorndike Video Unit Three)

Since Thorndike "Graph-Pro" racquetball racquets were introduced several years ago, no attempt has ever been made to offer a lightweight version. Ted has done some research and found that some of the other manufacturers offer a lighter version of their standard racquet. In addition, some of the racquetball players he has talked to have said a lighter racquet would probably improve their game.

Talking with production engineers, Ted learns that the racquets coming off the line don't all weigh the same anyway. According to the chief engineer, the racquets weigh an average of 240 grams, but the weights "vary all over the lot," and he takes advantage of the situation to press for a high-tech $700,000 machine that would produce racquets of equal quality while holding this variation within narrower limits.

Old Luke isn't too keen on buying a $700,000 machine, especially since Ted might have a better solution. Ted has dusted off the literature that accompanied the racquet-production machine and found that the machine generates output that is approximately normally distributed, with a standard deviation of 10 grams. This is a little sloppy these days, but maybe Thorndike Sports Equipment can turn this weakness into a strength.

Conversing further with Luke, talking to three sporting-goods retailers, and interviewing players at five different racquetball facilities, Ted comes back to the office with a terrific idea. Asking Luke to sit down, and bringing him some coffee, Ted begins his pitch: "Say, grandfather, why don't we take the racquets at the lighter end of the range and label them 'Graph-Pro Light,' with the ones more in the middle being 'Graph-Pro Regular,' and the ones at the heavier end can be 'Graph-Pro Stout'?"

Luke responds that just because this works for the beer companies, this doesn't mean it will work for Thorndike Sports Equipment. However, he's willing to give it a try. After accompanying Ted on another round of conversations with retailers and players, the elder Thorndike tells Ted that for now, he'd like to see 15% of the production consist of "Graph-Pro Light," with 80% "Graph-Pro Regular," and 5% "Graph-Pro Stout."

Ted's next task is to tell the production people how to go about selecting which racquets are to be put into these three different categories. He's also trying to follow his grandfather's directive to come up with a better name for the "Stout" model, but he figures he can do that after lunch.

Thorndike Golf Products Division

During an 8-hour shift of golf-ball production, one golf ball is randomly selected from each 2 minutes' worth of output. The ball is then tested for "liveliness" by rolling it down a grooved, stainless-steel surface. At the bottom, it strikes a smooth iron block and bounces backward, up a gradually sloped, grooved surface inscribed with distance markings. The higher the ball gets on the rebound surface, the more lively it is.

The production machine is set so that the average rebound distance will be $\mu = 30.00$ inches, and the standard deviation is $\sigma = 2.00$ inches. The scores are normally distributed.

SEEING STATISTICS: APPLET 4

Size and Shape of Normal Distribution

This applet has two normal curves, a blue one and a red one. The blue one stays fixed with a mean of 0 and a standard deviation of 1. Using the sliders at the bottom, we can change the mean and standard deviation of the red curve and observe how these changes affect its position and its shape. Moving the top slider to the left or right will decrease or increase the mean of the red normal curve, and moving the bottom slider to the left or right will decrease or increase its standard deviation.

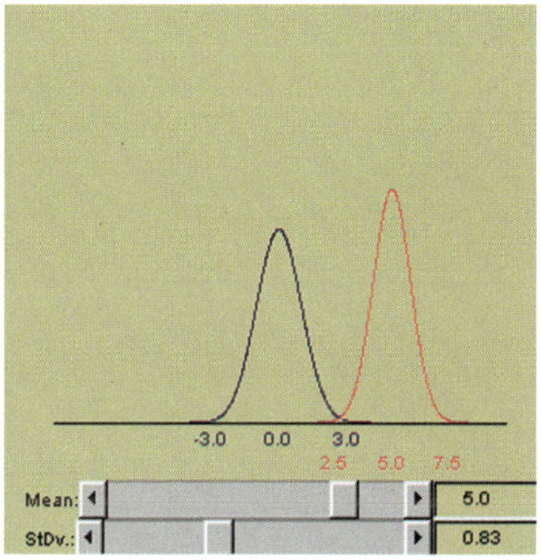

Applet Exercises

4.1 Position the top slider at the far left, then gradually move it to the far right. What effect does this have on the mean of the distribution?

4.2 Using the bottom slider, change the standard deviation so that it is greater than 1. How does the shape of the red curve compare to that of the blue curve?

4.3 Using the two sliders, position the red curve so that its mean is as small as possible and its shape is as narrow as possible. Compare the numerical values of the mean and standard deviation to those of the blue curve (mean = 0, standard deviation = 1).

Try out the Applets: http://www.cengage.com/bstatistics/weiers

1. A driving range has just purchased 100 golf balls. Use the computer and the binomial distribution in determining the individual and cumulative probabilities for x = the number of balls that scored at least 31.00 inches on the "bounce" test.
2. Repeat part (1), but use the normal approximation to the binomial distribution. Do the respective probabilities appear to be very similar?
3. Given the distribution of bounce test scores described above, what score value should be exceeded only 5% of the time? For what score value should only 5% of the balls do more poorly?
4. What is the probability that, for three consecutive balls, all three will happen to score below the lower of the two values determined in part (3)?

Two hundred forty golf balls have been subjected to the bounce test during the most recent 8-hour shift. From the 1st through the 240th, their scores are provided in computer data file **CDB07** as representing the continuous random variable, BOUNCE. The variable BALL is a sequence from 1 to 240.

5. Using the computer, generate a line graph in which BOUNCE is on the vertical axis and BALL is on the horizontal axis.
6. On the graph obtained in part (5), draw two horizontal lines—one at each of the BOUNCE scores determined in part (3).
7. Examining the graph and the horizontal lines drawn in part (6), does it appear that about 90% of the balls have BOUNCE scores between the two horizontal lines you've drawn, as the normal distribution would suggest when $\mu = 30.00$ and $\sigma = 2.00$?
8. Comparing the BOUNCE scores when BALL = 1 through 200 to those when BALL = 201 through 240, does it appear that the process may have changed in some way toward the end of the work shift? In what way? Does it appear that the machine might be in need of repair or adjustment? If so, in what way should the adjustment alter the process as it appeared to exist at the end of the work shift?

Note: Items (5) through (8) contain elements of the topic known as *statistical process control*. It is an important technique that will be examined more completely in Chapter 20.

SEEING STATISTICS: APPLET 5

Normal Distribution Areas

This applet allows us to conveniently find the area corresponding to a selected interval beneath a normal curve that has been specified. We can specify the mean by typing it into the box at the lower left (be sure to press the Return or Enter key while the cursor is still in the text box). Likewise, we can specify the standard deviation of the normal curve by using the text box at the lower right.

With the distribution described by its mean and standard deviation, we can click on the shaded portion of the curve and change the left and right endpoints for the interval. The resulting *z* values and the probability corresponding to the shaded area are shown at the top of the applet.

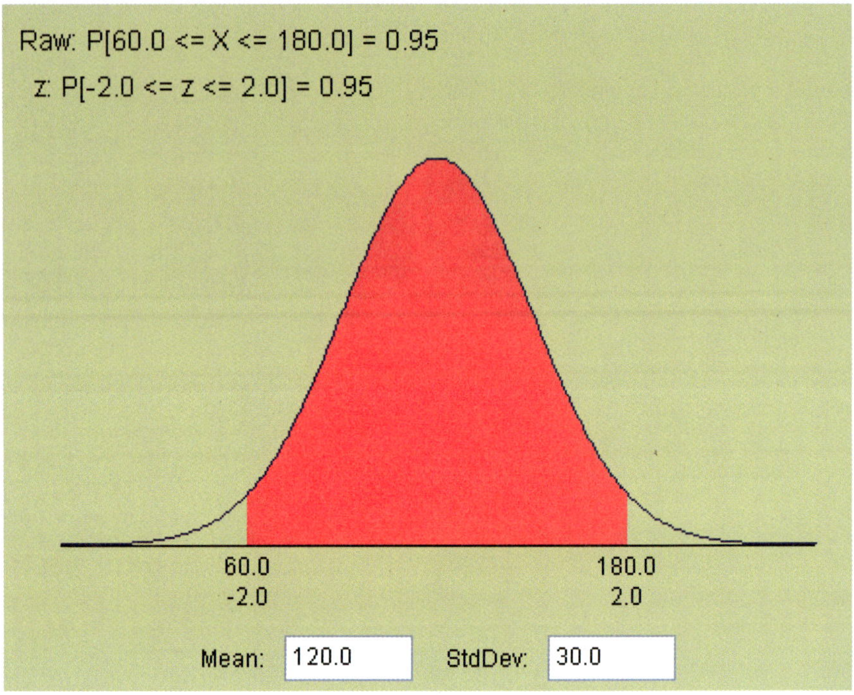

Applet Exercises

5.1 Using the text boxes, ensure that the mean is 120 and the standard deviation is 30, as in the distribution of flying hours in Figure 7.5. (Don't forget to hit the **Enter** or **Return** key while the cursor is still in the text box.) Next, ensure that the left and right boundaries of the shaded area are 60 and 180, respectively; if necessary, click at the left and right edges of the shaded area and drag them. Observing the display at the top, what is the probability that a randomly selected plane will have flown between 60 and 180 hours during the year? What values of *z* correspond to 60 hours and 180 hours, respectively?

5.2 With the mean and standard deviation set at 120 and 30, respectively, drag the edges of the shaded area so that its left boundary is at 120 and its right boundary is at 180. What is the probability that a randomly selected plane will have flown between 120 and 180 hours during the year? What values of *z* correspond to 120 hours and 180 hours, respectively?

5.3 By dragging the left and right edges of the shaded area, determine the probability that a randomly selected plane will have flown between 130 and 160 hours during the year. What values of *z* correspond to 130 hours and 160 hours, respectively?

5.4 Drag the left and right edges of the shaded area so that the left boundary corresponds to $z = 0$ and the right boundary corresponds to $z = +1.0$. What probability is now associated with the shaded area?

Try out the Applets: http://www.cengage.com/bstatistics/weiers

SEEING STATISTICS: APPLET 6

Normal Approximation to Binomial Distribution

This applet demonstrates the normal approximation to the binomial distribution. We specify the desired binomial distribution by entering the sample size (n) and the population proportion (π, shown in the applet as **pi**). We enter the k value of interest into the text box at the lower left. Once again, when using the text boxes, remember to press **Enter** or **Return** while the cursor is still within the box. Once we have specified n, π, and k, we see two results: (1) the binomial probability $P(x \leq k)$, shown in the **Prob** text box, and (2) the corresponding approximation using the normal distribution. Please note that probabilities shown in the applet may sometimes differ very slightly from those we might calculate using the pocket calculator and printed tables. This is because the applet is using more exact values than we are able to show within printed tables like the binomial and normal tables in the text.

Applet Exercises

6.1 With $n = 15$ and $\pi = 0.6$, as in the example in Section 7.4, what is the probability that there are no more than $k = 9$ females in the sample of 15 walkers? How does this compare with the corresponding probability using the normal approximation to the binomial distribution?

6.2 With $n = 5$ and $\pi = 0.6$, what is the actual binomial probability that there are no more than $k = 3$ females in the sample of 5? How does this compare with the corresponding probability using the normal approximation? What effect has the smaller sample size had on the closeness of the approximation?

6.3 With $n = 100$ and $\pi = 0.6$, what is the actual binomial probability that there are no more than $k = 60$ females in the sample of 100? How does this compare with the corresponding probability using the normal approximation? What effect has the larger sample size had on the closeness of the approximation?

6.4 Repeat Applet Exercise 6.1 for k values of 6 through 12, in each case identifying the actual binomial probability that there will be no more than k females in the sample. For each value of k, also note the normal approximation result and its closeness to the actual binomial probability.

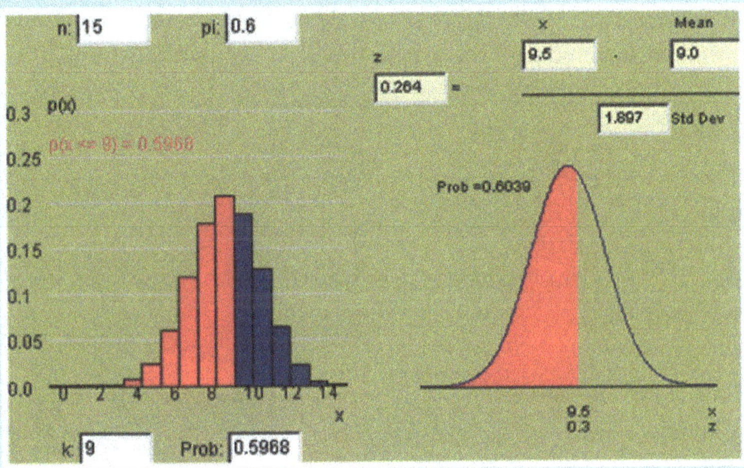

Try out the Applets: http://www.cengage.com/bstatistics/weiers

CHAPTER 15

Simple Linear Regression and Correlation

WINNING FOR DOLLARS

In sports, some teams are referred to as "the best that money can buy." If team payrolls and wins during the 2006 Major League Baseball season are any indication, there may be some truth to that adage. Although the highest-paid team didn't win the World Series, there was a very strong relationship between the size of a team's payroll and the number of wins it achieved during the regular season.

This relationship can be described by a linear equation fitted to the data using a technique called *simple linear regression*. This popular method for data analysis involves both the *scatterplot* discussed in Chapter 2 and the *coefficient of correlation* that was briefly introduced in Chapter 3. Using linear regression and correlation, we can make interesting observations like these: (1) variation in team payrolls explains approximately 29% of the variation in team wins; (2) on average, each additional win cost an extra $5.99 million in team payroll; and (3) the Minnesota Twins, one of the lowest-paid teams in baseball, pleased their fans and owners by winning about 17 more games than their payroll level would have predicted. Later in the chapter, in Statistics in Action 15.2, we'll delve a little further into how regression and correlation analysis applies to the payrolls and win totals of the 2006 Major League Baseball season.

Some slopes are a little steeper than others.

LEARNING OBJECTIVES

After reading this chapter, you should be able to:

- Explain the individual terms in the simple linear regression model, and describe the assumptions that the model requires.
- Determine the least-squares regression equation, and make point and interval estimates for the dependent variable.
- Determine and interpret the value of the coefficient of correlation.
- Describe the meaning of the coefficient of determination.
- Construct confidence intervals and carry out hypothesis tests involving the slope of the regression line.
- Test the significance of the correlation coefficient.
- Use residual analysis in examining the appropriateness of the linear regression model and the extent to which underlying assumptions are met.

15.1 INTRODUCTION

In Section 3.6 of Chapter 3, we briefly introduced the coefficient of correlation as a way of examining whether two interval- or ratio-scale variables could be related, and in Section 13.4 of Chapter 13, we used chi-square analysis to examine whether two nominal-scale variables could be related. Although these approaches can suggest that a relationship exists, they do not reveal *in exactly what way* the variables are related. The techniques of this chapter are applicable to two interval- or ratio-scale variables and describe both the nature and the strength of the relationship between the variables.

Regression analysis provides a "best-fit" mathematical equation for the values of the two variables. The equation may be linear (a straight line) or curvilinear, but we will be concentrating on the linear type. **Correlation analysis** measures the strength of the relationship between the variables.

The topic of this chapter is known as *simple* linear regression and correlation because there are just two variables, y and x. These are called the **dependent (y)** and **independent (x) variables,** since a typical purpose for this type of analysis is to estimate or predict what y will be for a given value of x. In the next chapter we'll cover *multiple* linear regression and correlation, in which two or more independent variables are used in estimating a dependent variable. The following are examples of dependent and independent variables that might be the subject of simple regression and correlation:

Dependent Variable (y)	Independent Variable (x)
Sales	Advertising
Starting salary	Grade point average
Product recalls	Quality assurance expenditures
Cost of a magazine ad	Circulation of the magazine

One way to examine whether two variables might be linearly related is to construct a scatter diagram, as we did in Section 2.5 of Chapter 2, then use the "eyeball method" to fit a straight line roughly approximating the points in the diagram. In the next section, we will introduce the simple linear regression model and employ a more quantitative approach that relies on what is known as the least-squares criterion.

THE SIMPLE LINEAR REGRESSION MODEL 15.2

Model and Assumptions

The *simple linear regression model* is a linear equation having a y-intercept and a slope, with estimates of these population parameters based on sample data and determined by standard formulas. The model is described in terms of the population parameters as follows:

The simple linear regression model:

$y_i = \beta_0 + \beta_1 x_i + \epsilon_i$ where y_i = a value of the dependent variable, y
x_i = a value of the independent variable, x
β_0 = the y-intercept of the regression line
β_1 = the slope of the regression line
ϵ_i = random error, or residual

NOTES

1. For a given value of x, the expected value of y is given by the linear equation, $\mu_{y.x} = \beta_0 + \beta_1 x_i$. The term $\mu_{y.x}$ can be stated as "the mean of y, given a specific value of x."
2. The difference between the actual value of y and the expected value of y is the error, or residual, $\epsilon_i = y_i - (\beta_0 + \beta_1 x_i)$.

According to this model, the y-intercept for the population of (x_i, y_i) pairs is β_0 and the slope is β_1. The ϵ_i term is the random error, or **residual,** for the ith observation or measurement—this residual is the difference between the actual value (y_i) and the expected value, $\mu_{y.x} = \beta_0 + \beta_1 x_i$, from the regression line. The y values may be scattered above and below the regression line, but the *expected* value of y for a given value of x will be given by the linear equation, $\mu_{y.x} = \beta_0 + \beta_1 x_i$.

Three assumptions underlie the simple linear regression model:

1. For any given value of x, the y values are normally distributed with a mean that is on the regression line, $\mu_{y.x} = \beta_0 + \beta_1 x_i$.
2. Regardless of the value of x, the standard deviation of the distribution of y values about the regression line is the same. The assumption of equal standard deviations about the regression line is called **homoscedasticity.**
3. The y values are statistically independent of each other. For example, if a given y value happens to exceed $\mu_{y.x} = \beta_0 + \beta_1 x_i$, this does not affect the probability that the next y value observed will also exceed $\mu_{y.x} = \beta_0 + \beta_1 x_i$.

Figure 15.1 shows the variation of y values above and below a population regression line. There is a "family" of such distributions (one for each possible x value). Each distribution has $\mu_{y.x}$ as its mean, and the standard deviations are the same $(\sigma_{y.x} = \sigma)$.

The three assumptions can also be expressed in terms of the error, or residual, component (ϵ_i) in the simple linear regression model: (1) For any given value of x, the population of ϵ_i values will be normally distributed with a mean of zero and a standard deviation of σ; (2) this standard deviation will be the same regardless of the value of x; and (3) the ϵ_i values are statistically independent of each other.

FIGURE 15.1

For any given value of x, the y values are assumed to be normally distributed about the population regression line and to have the same standard deviation, σ. The regression line based on sample data is an estimate of this "true" line. Likewise, $s_{y.x}$ is our sample estimate of σ.

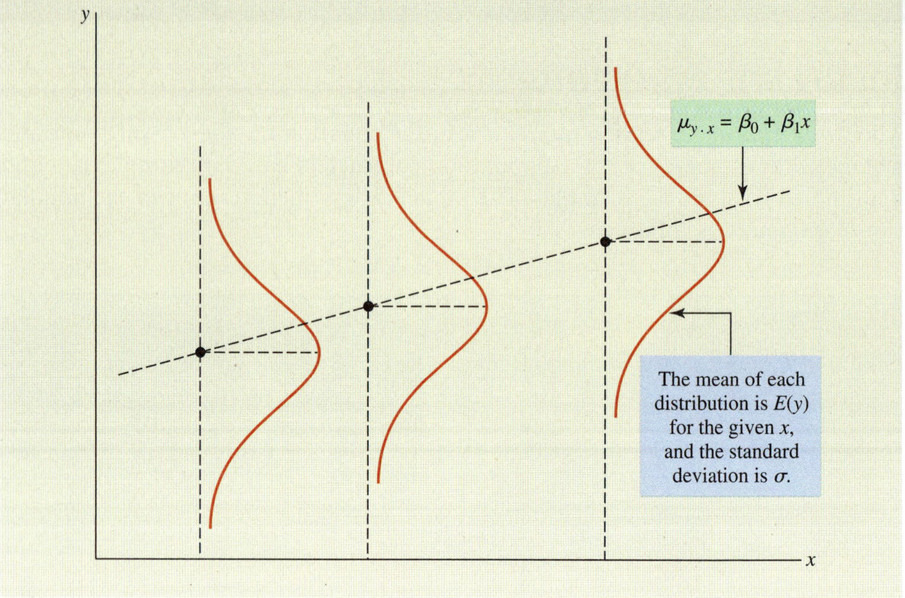

Based on the sample data, the y-intercept and slope of the population regression line can be estimated. The result is the sample regression line:

Sample regression line:

$\hat{y} = b_0 + b_1 x$ where $\hat{y}$ = the estimated value of the dependent variable (y) for a given value of x

b_0 = the y-intercept; this is the value of y where the line intersects the y-axis whenever $x = 0$.

b_1 = the slope of the regression line

x = a value for the independent variable

The cap (^) over the y indicates that it is an estimate of the (unknown) "true" value of y. The equation is completely described by the y-intercept (b_0) and slope (b_1), which are sample estimates of their population counterparts, β_0 and β_1, respectively. An infinite number of possible equations can be fitted to a given scatter diagram, and each equation will have a unique combination of values for b_0 and b_1. However, only one equation will be the "best fit" as defined by the least-squares criterion we are going to use.

The Least-Squares Criterion

The **least-squares criterion** requires that the sum of the squared deviations between y values in the scatter diagram and y values predicted by the equation be minimized. In symbolic terms:

Least-squares criterion for determining the best-fit equation:

The equation must be such that $\sum(y_i - \hat{y}_i)^2$ is minimized

where y_i = the observed value of y for the given value of x
$\hat{y}_i$ = the predicted value of y for that x value, as determined from the regression equation

To show how the least-squares criterion works, consider parts (a) and (b) of Figure 15.2. In part (a) the sum of the squared deviations between observed and predicted y values is 100.0, while in part (b) the sum is only 67.0. According to the least-squares criterion, the line in part (b) is a better fit to the data than the line in part (a).

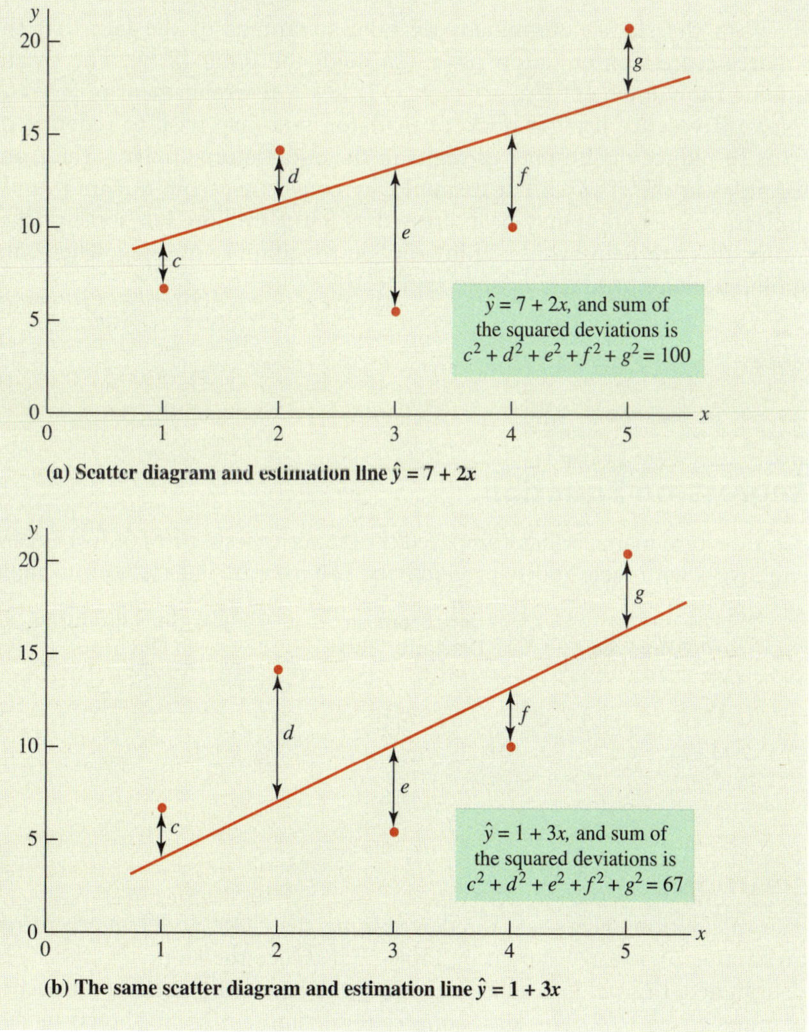

FIGURE 15.2

Using the least-squares criterion, the line fitted in part (b) is a better fit to the data than the line in part (a).

Determining the Least-Squares Regression Line

Equations have been developed for proceeding from a set of data to the least-squares regression line. They are based on the methods of calculus and provide values for b_0 and b_1 such that the least-squares criterion is met. The least-squares regression line may also be referred to as the *least-squares regression equation* or as simply the **regression line:**

Least-squares regression line, $\hat{y} = b_0 + b_1x$:

- **Slope**

$$b_1 = \frac{(\sum x_i y_i) - n\bar{x}\bar{y}}{(\sum x_i^2) - n\bar{x}^2} \quad \text{where } n = \text{number of data points}$$

- **y-intercept**

$$b_0 = \bar{y} - b_1\bar{x}$$

With the slope determined, we take advantage of the fact that the least-squares regression equation passes through the point $(\bar{x}, \bar{y})$. The equation for finding the y-intercept ($b_0 = \bar{y} - b_1\bar{x}$) is just a rearrangement of $\bar{y} = b_0 + b_1\bar{x}$. (*Note:* If you are using a pocket calculator, you may wish to redefine the units so as to reduce the number of digits in the data before applying these and other formulas in the chapter. For example, by converting from dollars to millions of dollars, \$30,500,000 may be expressed as \$30.5 million. For some pocket calculators, this can help avoid a "blow-up" in the number of digits when calculating summations of products or squares.)

EXAMPLEEXAMPLEEXAMPLEEXAMPLE

EXAMPLE

Regression Equation

A production manager has compared the dexterity test scores of five assembly-line employees with their hourly productivity. The data are in CX15DEX and shown here

Employee	x = Score on Dexterity Test	y = Units Produced in One Hour
A	12	55
B	14	63
C	17	67
D	16	70
E	11	51

SOLUTION

The calculations necessary for determining the slope and y-intercept of the regression equation are shown in Table 15.1. Once the slope ($b_1 = 3.0$) has been determined, this value is substituted into the equation for the y-intercept, and b_0 is found to be 19.2. The least-squares regression equation, shown in the scatter diagram of Figure 15.3, is $\hat{y} = 19.2 + 3.0x$.

TABLE 15.1

Data and calculations for determining the least-squares regression line for the example involving dexterity test score (*x*) and units produced per hour (*y*).

Data and Preliminary Calculations

Employee	x_i = Score on Dexterity Test	y_i = Units Produced in One Hour	$x_i y_i$	x_i^2	y_i^2
A	12	55	660	144	3025
B	14	63	882	196	3969
C	17	67	1139	289	4489
D	16	70	1120	256	4900
E	11	51	561	121	2601
	70	306	4362	1006	18,984
	$\sum x_i$	$\sum y_i$	$\sum x_i y_i$	$\sum x_i^2$	$\sum y_i^2$

$\bar{x} = 70/5 = 14.0$ $\bar{y} = 306/5 = 61.2$

Calculations for Slope and *y*-Intercept of Least-Squares Regression Line

$$\text{slope, } b_1 = \frac{(\sum x_i y_i) - n\bar{x}\bar{y}}{(\sum x_i^2) - n\bar{x}^2} = \frac{4362 - 5(14.0)(61.2)}{1006 - 5(14.0)^2} = \frac{78.0}{26.0} = 3.0$$

$$y\text{-intercept, } b_0 = \bar{y} - b_1\bar{x} = 61.2 - 3.0(14.0) = 61.2 - 42.0 = 19.2$$

The least-squares regression line is $\hat{y} = 19.2 + 3.0x$

where $\hat{y}$ = estimated units produced per hour
x = score on manual dexterity test

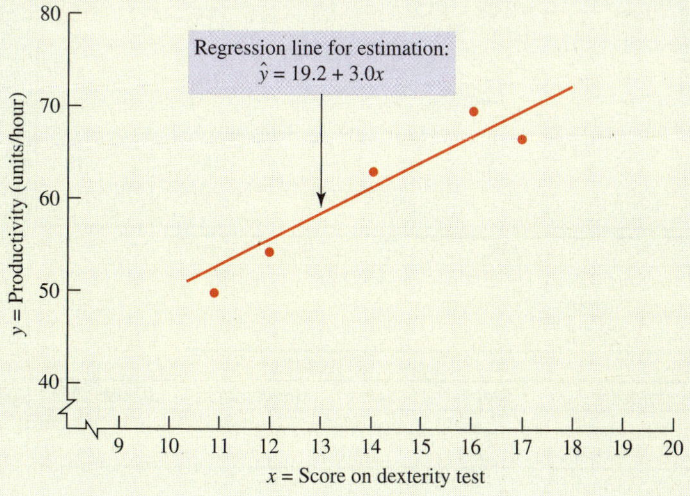

FIGURE 15.3

Scatter diagram and least-squares regression line for the data of Table 15.1.

The slope of the regression line is positive, suggesting a direct relationship between dexterity test score and productivity. The value of the slope ($b_1 = 3.0$) indicates that each one-point increase in the dexterity test score will increase the estimated productivity by 3.0 units per hour.

Point Estimates Using the Regression Line

Making point estimates based on the regression line is simply a matter of substituting a known or assumed value of x into the equation, then calculating the estimated value of y. For example, if a job applicant were to score $x = 15$ on the manual dexterity test, we would predict this person would be capable of producing 64.2 units per hour on the assembly line. This is calculated as shown below. You can also use Seeing Statistics Applet 18, at the end of the chapter, to demonstrate how different inputs for the dexterity test score will produce different point estimates for productivity.

Estimated productivity, units per hour

The applicant's score on the dexterity test

$$\hat{y} = 19.2 + 3.0(15) = 64.2 \text{ units/hour, estimated productivity}$$

Computer Solutions 15.1 describes the procedures and shows the results when Excel and Minitab are used in performing a simple linear regression analysis on

COMPUTER 15.1 SOLUTIONS

Simple Linear Regression

EXCEL

	E	F	G	H	I	J	K
1	SUMMARY OUTPUT						
2							
3	*Regression Statistics*						
4	Multiple R	0.955					
5	R Square	0.911					
6	Adjusted R Square	0.882					
7	Standard Error	2.757					
8	Observations	5					
9							
10	ANOVA						
11		*df*	*SS*	*MS*	*F*	*Significance F*	
12	Regression	1	234	234.0	30.789	0.012	
13	Residual	3	22.8	7.6			
14	Total	4	256.8				
15							
16		*Coefficients*	*Standard Error*	*t Stat*	*P-value*	*Lower 95%*	*Upper 95%*
17	Intercept	19.2	7.669	2.504	0.087	-5.206	43.606
18	Dexscore	3.0	0.541	5.549	0.012	1.279	4.721

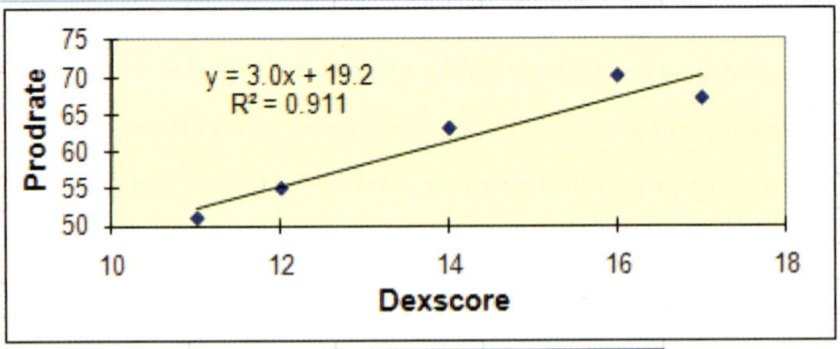

Excel Simple Linear Regression

1. The data for the primary example in this chapter are in Excel file **CX15DEX**. The dexterity test scores (Dexscore) are in A2:A6, the production rates (Prodrate) are in B2:B6, and the labels are in A1 and B1. From the **Data** ribbon, click **Data Analysis.** Select **Regression.** Click **OK.**
2. Enter **B1:B6** into the **Input Y Range** box. Enter **A1:A6** into the **Input X Range** box. Select **Labels.** Enter **E1** into the **Output Range** box. Check the **Line Fit Plots** box. Click **OK.**
3. Right-click within the legend descriptions to the right of the chart and click **Delete.** Right-click on any one of the linearly-arranged squares that describe predicted values for the data points and click **Delete.**
4. Right-click on any one of the data points in the scatter diagram and click **Add Trendline.** Select the **Linear** option. Click to select the **Display Equation on chart** and **Display R-squared value on chart** options. Click **Close.** Make formatting adjustments to the chart.

MINITAB

Minitab Simple Linear Regression

```
Regression Analysis: Prodrate versus Dexscore

The regression equation is Prodrate = 19.2 + 3.00 Dexscore

Predictor     Coef  SE Coef     T      P
Constant    19.200    7.669  2.50  0.087
Dexscore    3.0000   0.5407  5.55  0.012

S = 2.75681   R-Sq = 91.1%   R-Sq(adj) = 88.2%

Analysis of Variance

Source          DF      SS      MS      F      P
Regression       1  234.00  234.00  30.79  0.012
Residual Error   3   22.80    7.60
Total            4  256.80
```

1. Using Minitab file **CX15DEX**, with dexterity test scores in C1 and production rates in C2: Click **Stat.** Select **Regression.** Click **Regression.** Enter **C2** (Prodrate) into the **Response** box. Enter **C1** (Dexscore) into the **Predictors** box. Click **OK.**
2. Alternatively, we can generate a scatter diagram and line plot (not shown): Click **Stat.** Select **Regression.** Click **Fitted Line Plot.** Enter **C2** (Prodrate) into the **Response** box. Enter **C1** (Dexscore) into the **Predictor** box. Select **Linear.** Click **OK.**
3. Another alternative is to use the equation to estimate *y* for a given value of *x*: In step 1, click **Options.** Enter the desired *x* value (e.g., **15**) into the **Prediction intervals for new observations** box. Click **OK.** The estimated value of *y* (not shown here) will be 64.2.

the productivity data in our example. We will be referring back to these printouts in later sections of the chapter. In the Excel printout, the intercept (19.2) and slope (3.0) are shown in cells F17:F18. Minitab shows the intercept and slope in the context of the least-squares equation.

EXERCISES

15.1 Explain each term in the linear regression model, $y_i = \beta_0 + \beta_1 x_i + \epsilon_i$.

15.2 What assumptions are required in using the linear regression model?

15.3 In the linear regression equation, $\hat{y} = b_0 + b_1 x$, why is the term at the left given as $\hat{y}$ instead of simply y?

15.4 What is the least-squares criterion, and what does it have to do with obtaining a regression line for a given set of data?

15.5 A scatter diagram includes the data points ($x = 2$, $y = 10$), ($x = 3$, $y = 12$), ($x = 4$, $y = 20$), and ($x = 5$, $y = 16$). Two regression lines are proposed: (1) $\hat{y} = 10 + x$, and (2) $\hat{y} = 8 + 2x$. Using the least-squares criterion, which of these regression lines is the better fit to the data? Why?

15.6 Determine the least-squares regression line for the data in Exercise 15.5, then calculate the sum of the squared deviations between y and $\hat{y}$.

15.7 A scatter diagram includes the data points ($x = 3$, $y = 8$), ($x = 5$, $y = 18$), ($x = 7$, $y = 30$), and ($x = 9$, $y = 32$). Two regression lines are proposed: (1) $\hat{y} = 5 + 3x$, and (2) $\hat{y} = -2 + 4x$. Using the least-squares criterion, which of these regression lines is the better fit to the data? Why?

15.8 Determine the least-squares regression line for the data in Exercise 15.7, then calculate the sum of the squared deviations between y and $\hat{y}$.

15.9 For a sample of 8 employees, a personnel director has collected the following data on ownership of company stock versus years with the firm.

x = Years	y = Shares
6	300
12	408
14	560
6	252
9	288
13	650
15	630
9	522

a. Determine the least-squares regression line and interpret its slope.
b. For an employee who has been with the firm 10 years, what is the predicted number of shares of stock owned?

15.10 The following data represent x = boat sales and y = boat trailer sales from 1995 through 2000. Source: Bureau of the Census, *Statistical Abstract of the United States 1999*, p. 269 and *Statistical Abstract 2002*, p. 755.

Year	Boat Sales (Thousands)	Boat Trailer Sales (Thousands)
1995	649	207
1996	619	194
1997	596	181
1998	576	174
1999	585	168
2000	574	159

a. Determine the least-squares regression line and interpret its slope.
b. Estimate, for a year during which 500,000 boats are sold, the number of boat trailers that would be sold.
c. What reasons might explain why the number of boat trailers sold per year is less than the number of boats sold per year?

15.11 A media analyst studying popular movies has generated the accompanying table, which shows gross ticket sales during the initial 2 weeks following release compared to total gross sales during the movie's run. Given these data, determine the least-squares equation for predicting total gross ticket sales and interpret its slope. What would be the estimated total gross sales for a film that had $100 million in ticket sales during the first 2 weeks of its run?

After 2 Weeks ($ millions)	Total Gross Ticket Sales During Run ($ millions)
104.5	207.9
125.0	250.3
64.7	191.8
88.2	205.0
75.4	175.1

(**DATA SET**) *Note:* Exercises 15.12–15.14 require a computer and statistical software.

15.12 For the top law firms in the world in terms of profit per equity partner, data file **XR15012** lists the number of equity partners and the gross revenue ($ millions) for the most recent fiscal year. Given these data, determine the least-squares equation for predicting gross revenue on the basis of the number of equity partners, then interpret its slope. What would be the estimated gross revenue for a firm with 200 equity partners? Source: law.com, August 8, 2009.

15.13 The Florida ecosystem is heavily reliant on wet and dry seasons, and is particularly dependent on the winter and spring rainfall that helps reduce damage due to wildfires. Data file **XR15013** lists the number of acres burned and the number of inches of rainfall during January–May for 1988–2000. Given these data, determine the least-squares equation for predicting acreage burned during the January–May period as a function of rainfall during that period, then interpret its slope. What would be the estimated number of acres burned during January–May if there were 18 inches of rainfall during the same period? Source: "Rainfall, Fires Linked In Florida," *USA Today*, June 9, 2000, p. 4A.

15.14 For the leading health maintenance organization (HMO) firms, data file **XR15014** lists the number of persons enrolled in the HMO along with the number of plans the firm administers. Given these data, determine the least-squares equation for predicting the total enrollment as a function of the number of plans administered. What would be the estimated total enrollment for an HMO firm that administered 20 plans? Source: *The Wall Street Journal Almanac 1999*, p. 565.

INTERVAL ESTIMATION USING THE SAMPLE REGRESSION LINE

15.3

In the preceding example, we found that an applicant scoring $x = 15$ on the dexterity test would be predicted to produce 64.2 units per hour. This is just a point estimate, however, and there will be some uncertainty as to how high or low the individual's productivity might actually be. Using the techniques of this section, we can draw some conclusions regarding the upper and lower limits within which this individual's actual productivity is likely to fall.

The Standard Error of Estimate

To develop interval estimates for the dependent variable, we must first determine the **standard error of estimate,** $s_{y.x}$. This is a standard deviation describing the dispersion of data points above and below the regression line. The formula for the standard error of estimate is shown below and is very similar to that for determining a sample standard deviation, s.

Standard error of estimate

$$s_{y.x} = \sqrt{\frac{\Sigma(y_i - \hat{y}_i)^2}{n - 2}}$$

where y_i = each observed value of y in the data

$\hat{y}_i$ = the value of y that would have been estimated from the regression line

n = the number of data points

The number of degrees of freedom associated with $s_{y.x}$ is $n - 2$, since both the y-intercept b_0 and the slope b_1 are estimated from sample data. For the dexterity test example of Table 15.1 and Figure 15.3, the standard error of estimate can be calculated as shown below. In each row, the predicted value ($\hat{y}$) is estimated by the regression line, $\hat{y} = 19.2 + 3.0x$, and the deviation between the observed and predicted values is the *residual*. Each residual is the vertical distance from the regression line to an observed data point.

Employee	x_i = Test Score	Number of Units/Hour Actual (y_i)	Number of Units/Hour Predicted $(\hat{y}_i)$	Error, or Residual, $(y_i - \hat{y}_i)$	Error, or Residual, Squared, $(y_i - \hat{y}_i)^2$
A	12	55	55.2	−0.2	0.04
B	14	63	61.2	1.8	3.24
C	17	67	70.2	−3.2	10.24
D	16	70	67.2	2.8	7.84
E	11	51	52.2	−1.2	1.44
					$\Sigma(y_i - \hat{y}_i)^2$ = 22.80

Having calculated the sum of the squared deviations (i.e., squared residuals) about the regression line as $\Sigma(y_i - \hat{y}_i)^2 = 22.80$, we may now calculate the standard error of estimate as

$$s_{y.x} = \sqrt{\frac{\Sigma(y_i - \hat{y}_i)^2}{n-2}} = \sqrt{\frac{22.80}{5-2}} = \sqrt{7.6} = 2.757 \text{ units/hour}$$

This is the same value provided in the Excel and Minitab portions of Computer Solutions 15.1, shown previously.

As with any standard deviation, larger values of the standard error of estimate reflect a greater amount of *scatter* in the data. If every data point in the scatter diagram were to fall exactly on the regression line, the standard error of estimate would be zero. In this event, there would be no variability at all above and below the regression line.

Confidence Interval for the Mean of *y* Given a Specific *x* Value

Given a specific value of x, we can make two kinds of interval estimates regarding y: (1) a confidence interval for the (unknown) true mean of y, and (2) a prediction interval for an individual y observation. These applications are examined in this section and the next.

For a given value of x, the best point estimate we can make about the mean of y is obtained from the regression line, or $\mu_{y.x} = \hat{y} = b_0 + b_1x$. However, the regression line itself is based on sample data, and b_0 is an estimate of the true y-intercept (β_0) and b_1 is an estimate of the true slope (β_1). As such, both b_0 and b_1 are subject to error. Estimates improve when they are based on an x value that is closer to $\bar{x}$. [Remember that the regression line for a set of data always includes the point $(\bar{x}, \bar{y})$.] The following formula also considers the number of data points, since regression lines based on fewer data points (smaller n) tend to result in greater errors of estimation:

Confidence interval for the mean of y, given a specific value of x:

$$\hat{y} \pm t(s_{y.x})\sqrt{\frac{1}{n} + \frac{(x \text{ value} - \bar{x})^2}{\left(\sum x_i^2\right) - \left(\sum x_i\right)^2/n}}$$

where $\hat{y}$ = the estimated value of y for the given value of x
n = the number of data points
$s_{y.x}$ = the standard error of estimate
t = t value for confidence level desired and $df = n - 2$
x value = the given value of x

EXAMPLE

Confidence Interval

For persons scoring $x = 15$ on the dexterity test, what is the 95% confidence interval for their mean productivity?

SOLUTION

The x value given is $x = 15$, and the standard error of estimate has been calculated as $s_{y.x} = 2.757$ units/hour. The correction factor (the square root symbol and the terms within) will end up being slightly larger than if we were making an interval estimate based on an x value closer to the mean of x.

For $x = 15$, $\hat{y} = 19.2 + 3.0(15)$, or 64.2 units per hour. This will be the midpoint of the interval. The mean of x is $\bar{x} = 14.0$, there are $n = 5$ data points, and the Σx_i^2 and Σx_i terms have been calculated in Table 15.1 as 1006 and 70, respectively. For the 95% level of confidence and $df = n - 2 = 3$, $t = 3.182$ and the 95% confidence interval can now be calculated as

$$\hat{y} \pm t(s_{y.x})\sqrt{\frac{1}{n} + \frac{(x \text{ value} - \bar{x})^2}{\left(\sum x_i^2\right) - \left(\sum x_i\right)^2/n}}$$

$$= 64.2 \pm 3.182(2.757)\sqrt{\frac{1}{5} + \frac{(15.0 - 14.0)^2}{1006 - (70)^2/5}} = 64.2 \pm 8.773\sqrt{0.2385}$$

$$= 64.2 \pm 8.773(0.488) = 64.2 \pm 4.281, \quad \text{or between } 59.919 \text{ and } 68.481$$

Based on these calculations, we have 95% confidence that the mean productivity for persons scoring $x = 15$ on the dexterity test will be between 59.919 and 68.481 units per hour.

EXAMPLEEXAMPLEEXAMPLEEXAMPLEEXAMPLE

Prediction Interval for an Individual y Observation

For a given value of x, the estimation interval for an individual y observation is called the **prediction interval.** For the dexterity-test regression line, we'll demonstrate this concept by constructing a prediction interval for the productivity of an individual job applicant who has achieved a given score on the dexterity test. The formula is similar to that just presented, except that a 1 is included beneath the square root symbol:

Prediction interval for an individual y, given a specific value of x:

$$\hat{y} \pm t(s_{y.x})\sqrt{1 + \frac{1}{n} + \frac{(x \text{ value} - \bar{x})^2}{(\Sigma x_i^2) - (\Sigma x_i)^2/n}}$$

where $\hat{y}$ = estimated value of y for the given value of x
n = number of data points
$s_{y.x}$ = the standard error of estimate
t = t value for the confidence level desired and $df = n - 2$
x value = the given value of x

EXAMPLE

Prediction Interval

A prospective employee has scored $x = 15$ on the dexterity test. What is the 95% prediction interval for his productivity?

SOLUTION

For $x = 15$, y is estimated as $\hat{y} = 19.2 + 3.0(15)$, or $\hat{y} = 64.2$. As in the preceding example, this will be the midpoint of the interval. Applying the prediction interval formula, the interval is described by

$$\hat{y} \pm t(s_{y.x})\sqrt{1 + \frac{1}{n} + \frac{(x \text{ value} - \bar{x})^2}{(\Sigma x_i^2) - (\Sigma x_i)^2/n}}$$

$$= 64.2 \pm 3.182(2.757)\sqrt{1 + \frac{1}{5} + \frac{(15.0 - 14.0)^2}{1006 - (70)^2/5}}$$

$$= 64.2 \pm 8.773\sqrt{1.2385}$$

$$= 64.2 \pm 8.773(1.113)$$

$$= 64.2 \pm 9.764, \quad \text{or between } 54.436 \text{ and } 73.964$$

For this applicant, we have 95% confidence that his productivity as an employee would be between 54.436 and 73.964 units per hour. The width of this interval is 19.528 [i.e., 2(9.764)].

For purposes of illustration, we have used the same dexterity test score ($x = 15$) in determining both the 95% confidence interval for the mean and the 95% prediction interval for the productivity of an individual. Given that $x = 15$, notice that the 95% *confidence* interval (64.2 ± 4.281 units per hour) is much narrower than the 95% *prediction* interval (64.2 ± 9.764 units per hour). This reflects that we will have less uncertainty about the *mean productivity of all persons* who score $x = 15$ on the dexterity test than about the *productivity of one individual* who has scored $x = 15$ on the dexterity test.

Had the applicant scored $x = 14$ on the dexterity test, the prediction interval would be slightly narrower. This is because $\bar{x} = 14.0$, and there is less error in making interval estimates based on x values that are closer to the mean. For additional practice, you may wish to repeat the preceding calculations for $x = 14$. If so, you will find the interval to be 61.2 ± 9.610, for a total interval

EXAMPLEEXAMPLEEXAMPLEEXAMPLEEXAMPLEEXAMPLEEXAMP

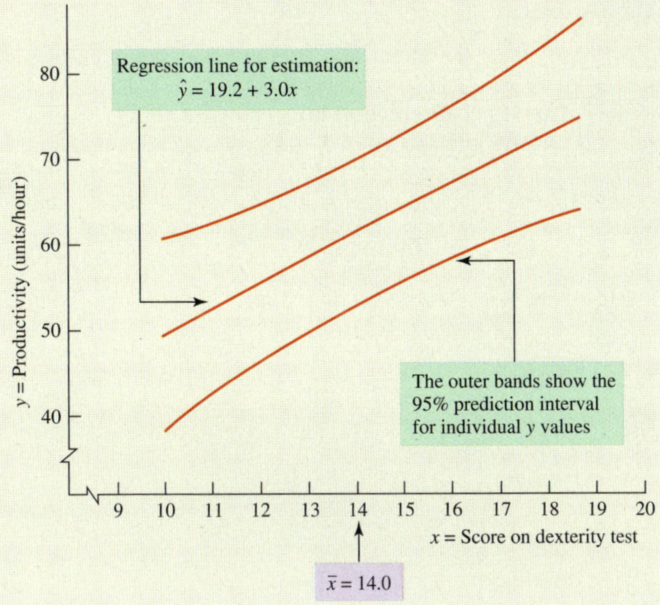

FIGURE 15.4

The 95% prediction interval for individual y values becomes slightly wider whenever the interval is based on x values that are farther away from the mean of x.

width of 2(9.610), or 19.220 units per hour. Figure 15.4 shows how the 95% prediction interval becomes wider whenever the specified value of x gets farther away from $\bar{x}$ in either direction.

EXAMPLE

Computer Solutions 15.2 shows the procedures and results when we use Excel and Minitab to obtain confidence and prediction intervals for productivity when the score on the dexterity test is $x = 15$. In each case, we also get a point estimate for productivity.

COMPUTER 15.2 SOLUTIONS

Interval Estimation in Simple Linear Regression

EXCEL

	A	B	C	D
1	**Prediction Interval**			
2				
3			Prodrate	
4	Predicted value		64.2	
5				
6	Prediction Interval			
7	Lower limit		54.44	
8	Upper limit		73.96	
9				
10	Interval Estimate of Expected Value			
11	Lower limit		59.92	
12	Upper limit		68.48	

(continued)

Excel Interval Estimation in Simple Linear Regression

1. Using Excel file **CX15DEX**, with dexterity test scores in A2:A6, production rates in B2:B6, and the labels in A1 and B1, we will obtain interval estimates associated with $x = 15$: Select a convenient cell (e.g., C3) and enter **15** into **C3.**
2. From the **Add-Ins** ribbon, click **Data Analysis Plus.** Select **Prediction Interval.** Click **OK.**
3. Enter **B1:B6** into the **Input Y Range** box. Enter **A1:A6** into the **Input X Range** box. Click **Labels.** Enter **C3** into the **Given X Range** box. Enter the desired confidence level as a decimal fraction (e.g., **0.95**) into the **Confidence Level (1 – Alpha)** box. Click **OK.** When $x = 15$, the point estimate for y is 64.2, and the 95% prediction and confidence intervals will be as shown above.

MINITAB

Minitab Interval Estimation in Simple Linear Regression

```
Predicted Values for New Observations
New
Obs    Fit   SE Fit       95% CI             95% PI
  1  64.20     1.35   (59.92, 68.48)   (54.44, 73.96)

Values of Predictors for New Observations
New
Obs   Dexscore
  1       15.0
```

Using Minitab file **CX15DEX**, follow the procedure in step 1 of Computer Solutions 15.1, but select **Options.** Enter **15** into the **Prediction intervals for new observations** box. Enter the desired confidence level as a percentage (e.g., **95**) into the **Confidence level** box. Click **OK.** Click **OK.** In addition to the printout in Computer Solutions 15.1, Minitab will generate the items shown above. When $x = 15$, the point estimate for y is 64.20, and the confidence and prediction intervals will be as shown here.

EXERCISES

15.15 What is the standard error of estimate, and what role does it play in simple linear regression and correlation analysis?

15.16 Is it possible for the standard error of estimate to be equal to zero? If so, under what circumstances?

15.17 Differentiate between a confidence interval and a prediction interval.

15.18 For a given value of x, which will be wider, a 95% confidence interval or a 95% prediction interval? Explain.

15.19 What happens to the width of a prediction interval for y as the x value on which the interval estimate is based gets farther away from the mean of x? Why?

15.20 For a set of 8 data points, the sum of the squared differences between observed and estimated values of y is 34.72. Given this information, what is the standard error of estimate?

15.21 For $n = 6$ data points, the following quantities have been calculated:

$$\Sigma x = 40 \qquad \Sigma y = 76 \qquad \Sigma xy = 400$$
$$\Sigma x^2 = 346 \qquad \Sigma y^2 = 1160 \qquad \Sigma(y - \hat{y})^2 - 52.334$$

a. Determine the least-squares regression line.
b. Determine the standard error of estimate.
c. Construct the 95% confidence interval for the mean of y when $x = 7.0$.
d. Construct the 95% confidence interval for the mean of y when $x = 9.0$.
e. Compare the width of the confidence interval obtained in part (c) with that obtained in part (d). Which is wider and why?

15.22 For the summary data provided in Exercise 15.21, construct a 95% prediction interval for an individual y value whenever

a. $x = 2$. b. $x = 3$. c. $x = 4$.

15.23 For the National Football League, ratings for the all-time leading passers were as shown below. Also shown for each quarterback is the percentage of passes that were interceptions, along with the percentage of passes that were touchdowns. Source: *Sports Illustrated 2009 Almanac*, p. 158.

Player	Rating	TD%	Inter%
Steve Young	96.8	5.6	2.6
Peyton Manning	94.7	5.7	2.8
Kurt Warner	93.2	5.1	3.4
Tom Brady	92.9	5.4	2.4
Joe Montana	92.3	5.2	2.6
Carson Palmer	90.1	5.1	3.1
Daunte Culpepper	89.9	4.9	3.2

a. Determine the least-squares regression line for estimating the passer rating based on the percentage of passes that were touchdowns.
b. A quarterback says 5.0% of his passes will be touchdowns next year. Assuming this to be true, estimate his rating for the coming season.
c. Determine the standard error of estimate.
d. For the quarterback in part (b), what is the 95% prediction interval for his passer rating next year?
e. For quarterbacks similar to the one in part (b), what is the 95% confidence interval for their mean rating next year?

15.24 Repeat parts (a)–(e) of Exercise 15.23, but using the interception percentage as the independent variable, and for a quarterback who claims his interception percentage will be just 3.0% next year.

15.25 For the data in Exercise 15.9, find the standard error of estimate, then construct the 95% prediction interval for the amount of stock owned by an individual employee who has been with the firm for 5 years.

15.26 For the data in Exercise 15.10, find the standard error of estimate, then construct the 95% prediction interval for boat trailer sales during a year in which the number of boats sold is 500,000.

(DATA SET) *Note:* Exercises 15.27–15.30 require a computer and statistical software.

15.27 Information describing 98 of the top automobile dealership groups in the United States is provided in computer database file **DEALERS**. For calendar year 1999, the data include the number of dealers, the average number of retail units sold per dealer, and the total revenue for the dealership group. Determine the regression equation estimating total revenue as a function of the number of dealers. For a dealership group consisting of 100 dealers, determine and interpret the 95% confidence and prediction intervals associated with total revenue. Source: *Automotive News*, May 1, 2000, pp. 54–60.

15.28 Repeat Exercise 15.27, but with average retail units sold per dealer as the independent variable and with the 95% interval estimates assuming a dealer group for which the average number of retail units sold per dealer is 1000.

15.29 Given the information in Exercise 15.13, determine and interpret the 90% confidence and prediction intervals associated with burned acreage in Florida when the January–May rainfall in the state is 20 inches.

15.30 Given the information in Exercise 15.12, determine and interpret the 90% confidence and prediction intervals associated with gross revenue when there are 200 equity partners.

CORRELATION ANALYSIS (15.4)

Regression analysis determines the nature of the linear relationship between two interval- or ratio-scale variables, while correlation analysis measures the strength of the linear relationship between them. In particular, correlation analysis provides us with two important measures of this strength: (1) the coefficient of correlation and (2) the coefficient of determination.

The Coefficient of Correlation

Briefly introduced in Chapter 3, the **coefficient of correlation** (r, with $-1 \leq r \leq +1$) is a number that indicates both the direction and the strength of the linear relationship between the dependent variable (y) and the independent variable (x):

- **Direction of the relationship** If r is positive, y and x are directly related—i.e., when x increases, y will tend to increase. If r is negative, y and x are inversely related—i.e., when x increases, y will tend to decrease.

- **Strength of the relationship** The larger the absolute value of r, the stronger the linear relationship between y and x. If $r = -1$ or $r = +1$, the regression line will actually include all of the data points and the line will be a perfect fit.

Calculating the coefficient of correlation for a set of data involves combining the same terms that appear in Table 15.1. The formula for r can be expressed as follows:

The coefficient of correlation:

$$r = \frac{n(\sum x_i y_i) - (\sum x_i)(\sum y_i)}{\sqrt{n(\sum x_i^2) - (\sum x_i)^2} \cdot \sqrt{n(\sum y_i^2) - (\sum y_i)^2}}$$

where r = coefficient of correlation, $-1 \leq r \leq +1$

n = number of data points

Referring to the dexterity test example, the coefficient of correlation between productivity (y) and dexterity test score (x) can be computed as

$$r = \frac{n(\sum x_i y_i) - (\sum x_i)(\sum y_i)}{\sqrt{n(\sum x_i^2) - (\sum x_i)^2} \cdot \sqrt{n(\sum y_i^2) - (\sum y_i)^2}}$$

$$= \frac{5(4362) - (70)(306)}{\sqrt{5(1006) - (70)^2} \cdot \sqrt{5(18{,}984) - (306)^2}} = \frac{390}{11.402 \cdot 35.833} = 0.9546$$

The coefficient of correlation ($r = 0.9546$) is positive, reflecting that productivity (y) is directly related to dexterity test score (x). In other words, persons scoring higher on the dexterity test tend to record higher levels of productivity. This is also reflected in the positive slope of the regression line, $\hat{y} = 19.2 + 3.0x$. Using Seeing Statistics Applet 19, at the end of the chapter, you can insert data points onto a blank grid and instantly see how the correlation coefficient and best-fit regression line change in response.

The Coefficient of Determination

Another measure of the strength of the relationship is the **coefficient of determination, r^2**. Its numerical value is the proportion of the variation in y that is explained by the regression line, $\hat{y} = b_0 + b_1x$. For the dexterity test example, $r = 0.9546$, $r^2 = 0.911$, and 91.1% of the variation in productivity is explained by the dexterity test scores.

The coefficient of determination can be described in terms of the *total* variation in y versus the *unexplained* variation in y. In turn, each type of variation can be described as a sum of squared deviations. In the denominator of the fraction in the following expression, each y value is compared to the mean of y in arriving at the total sum of squares (SST). In the numerator, each y value is compared to the value of y predicted by the regression line in obtaining the unexplained, or error, sum of squares (SSE):

The coefficient of determination, r^2:

$$r^2 = 1 - \frac{\textit{Error}\text{ variation, which is }\textit{not}\text{ explained by the regression line}}{\textit{Total}\text{ variation in the }y\text{ values}}$$

$$= 1 - \frac{\sum(y_i - \hat{y}_i)^2}{\sum(y_i - \bar{y})^2} = 1 - \frac{SSE}{SST}$$

In considering the variation of y, a natural question is, "Variation from what?" The logic involved in the calculations of *SST* and *SSE* warrants further explanation:

1. Without any knowledge of x, the best estimate of y is the *mean* of y. This is why the variation of y from $\bar{y}$ is used in calculating SST = total sum of squares, or total variation.
2. When knowledge about x is available, the best estimate of y is the *conditional mean* of y (i.e., $\hat{y}$ from the regression line). When an actual value of y differs from the value predicted by the regression line, this constitutes unexplained variation and it contributes to SSE = error sum of squares.

The error variation (*SSE*) is the sum of the squared residuals calculated earlier during our determination of the standard error of estimate. These calculations are repeated below, along with those for determining the total sum of squares (*SST*) for the y values. In the following calculations, recall that the mean of y is $\bar{y} = 61.2$ units per hour:

	Number of Units/Hour					
x_i = Test Score	Actual, y_i	Predicted, $\hat{y}_i$	$y_i - \hat{y}_i$	$(y_i - \hat{y}_i)^2$	$y_i - \bar{y}$	$(y_i - \bar{y})^2$
12	55	55.2	−0.2	0.04	−6.2	38.44
14	63	61.2	1.8	3.24	1.8	3.24
17	67	70.2	−3.2	10.24	5.8	33.64
16	70	67.2	2.8	7.84	8.8	77.44
11	51	52.2	−1.2	1.44	−10.2	104.04
				22.80		256.80

$\sum(y_i - \hat{y}_i)^2$, error, or residual, sum of squares (*SSE*)

$\sum(y_i - \bar{y})^2$, total sum of squares (*SST*)

and the coefficient of determination can be expressed as

$$r^2 = 1 - \frac{SSE}{SST} = 1 - \frac{22.80}{256.80} = 0.911$$

Total variation in y (*SST*) is the sum of the variation explained by the regression line (*SSR*) plus the variation not explained by the regression line (*SSE*), or

Total variation in y values (*SST*)	=	Variation explained by regression line (*SSR*)	+	Variation *not* explained by regression line (*SSE*)

We have calculated the total sum of squares as $SST = 256.80$ and the error sum of squares as $SSE = 22.80$. The variation explained by the regression line (*SSR*) is the difference between these two quantities, and $SSR = SST - SSE$, or $256.80 - 22.80$, or $SSR = 234.0$. All three of these quantities are typically included in the output whenever regression and correlation analyses are carried out with the assistance of a computer statistical package. You can use Seeing Statistics Applet 20, at the end of the chapter, to instantly see how explained and unexplained variation will change as you "try" different lines to fit the data in this chapter's primary example.

Some statistical software packages—Excel and Minitab included—provide an r^2 that has been "adjusted" for degrees of freedom. The "adjusted" r^2 approaches the size of the "unadjusted" value as the number of data points becomes greater. For our productivity example, the printouts in Computer Solutions 15.1 show the adjusted r^2 as 0.882. The formula for obtaining the adjusted r^2 is given in the footnote.[1]

Computer Solutions 15.3 shows the procedures and results when we use Excel and Minitab to obtain the coefficient of correlation between dexterity test score (x) and productivity (y). The procedures are comparable to their counterparts presented earlier in the text, in Computer Solutions 3.5. Minitab provides a p-value for the coefficient of correlation; if the true coefficient of correlation were 0, there would be a 0.012 probability of getting a sample r this far away from 0 just by chance. Testing whether the true coefficient of correlation could be 0 is discussed in Section 15.5.

COMPUTER 15.3 SOLUTIONS

Coefficient of Correlation

EXCEL

	A	B	C	D	E
1	Dexscore	Prodrate		*Dexscore*	*Prodrate*
2	12	55	Dexscore	1	
3	14	63	Prodrate	0.955	1
4	17	67			
5	16	70			
6	11	51			

1. Open the Excel data file **CX15DEX**. The labels and data are already entered, as shown in columns A and B in the printout above. From the **Data** ribbon, click **Data Analysis.**
2. In the **Analysis Tools** box, click **Correlation.** Click **OK.** Enter **A1:B6** into the **Input Range** box. In the **Grouped By** section, click next to **Columns.** Select **Labels in First Row.** Enter **C1** into the **Output Range** box. Click **OK.**

MINITAB

```
Correlations: Dexscore, Prodrate

Pearson correlation of Dexscore and Prodrate = 0.955
P-Value = 0.012
```

1. Open the Minitab data file **CX15DEX**. The dexterity test scores are in C1 and the production rates are in C2.
2. Click **Stat.** Select **Basic Statistics.** Click **Correlation.** Enter **C1** and **C2** into the **Variables** box. Select **Display P-values.** Click **OK.**

[1]Expressed as a proportion, the adjusted r^2 typically included in computer printouts of regression analyses is calculated as adjusted $r^2 = 1 - \dfrac{SSE/(n-2)}{SST/(n-1)}$.

EXERCISES

15.31 Differentiate between the coefficients of correlation and determination. What information does each one offer that the other does not?

15.32 The values of y and x are inversely related, and 64% of the variation in y is explained by the regression equation. What is the coefficient of correlation?

15.33 The coefficient of correlation between variables y and x is -0.90. Calculate and interpret the value of the coefficient of determination.

15.34 For the data in Exercise 15.23, and with y = touchdown percentage and x = interception percentage, determine and interpret the coefficients of correlation and determination.

15.35 In regression and correlation analysis, what are the two components of the total amount of variation in the y values? Define the coefficient of determination in terms of these components.

15.36 For a set of data, the total variation or sum of squares for y is $SST = 143.0$, and the error sum of squares is $SSE = 24.0$. What proportion of the variation in y is explained by the regression equation?

15.37 The Insurance Institute for Highway Safety has listed the following ratings based on collision and comprehensive claims for 12 makes of midsize four-door cars from the 2005–2007 model years. Higher numbers reflect higher claims in the collision and comprehensive categories of coverage. Source: iihs.org, August 8, 2009.

Collision	Comp	Collision	Comp
113	89	93	79
108	91	105	97
90	74	106	86
124	92	99	71
131	108	116	98
126	108	120	93

a. Determine the least-squares regression line for predicting the rate of collision claims based on comprehensive claim rating.
b. Calculate and interpret the values of r and r^2.
c. If a new model were to have a comprehensive claim rating of 90, what would be the predicted rating for collision claims?

15.38 The following table compares U.S. population (millions) with U.S. per-capita consumption of bottled water (gallons). Source: Bureau of the Census, *Statistical Abstract of the United States 2009*, pp. 10, 134.

Year	x = U.S. Population	y = Per-Capita Consumption of Bottled Water
2002	287.9 million	20.1 gallons
2003	290.4	21.6
2004	293.2	23.2
2005	295.9	25.5
2006	298.8	27.6

a. Determine the least-squares regression line and the value of r.
b. What proportion of the variability in per-capita bottled water consumption is explained by the regression equation?
c. During a year in which the U.S. population is 320 million, what would be the prediction for per-capita consumption of bottled water?

15.39 It has been reported that the average American male consumes 3774 calories per day and that 72.2% of American males are overweight. This information, along with data for seven other countries, is listed below. Source: "A Measure of How Much Countries Eat," *USA Today*, December 16, 2005, p. 4D.

Country	Calories per Day	% Overweight
Chad	2114	10.4
China	2951	27.5
Ecuador	2754	40.2
Egypt	3338	64.5
Great Britain	3412	62.5
India	2459	15.0
Mexico	3145	64.6
United States	3774	72.2

a. Determine the least-squares regression line for predicting the percentage of males who are overweight based on the number of calories consumed per day.
b. Calculate and interpret the values of r and r^2.
c. If a nation's males were to consume an average of 3000 calories per day, what would be the predicted percentage of the nation's males who are overweight?

(**DATA SET**) *Note:* Exercises 15.40–15.43 require a computer and statistical software.

15.40 Given the information in Exercise 15.13, determine and interpret the coefficients of correlation and determination for January–May burned acreage in Florida versus January–May rainfall in the state.

15.41 Given the information in Exercise 15.12, determine and interpret the coefficients of correlation and determination for gross revenue versus the number of equity partners.

15.42 For each of 10 popular prescription drugs, file **XR15042** lists the retail price (in U.S. dollars) for the drug in several different countries, including the United States, Canada, Great Britain, and Australia. Determine and interpret the coefficients of correlation and determination for U.S. prices versus Canadian prices. Source: "Prescription Drugs Cheaper in Other Nations," *USA Today*, November 10, 1999, p. 2A.

15.43 For each of the 20 top-grossing U.S. films made during the 1990s, file **XR15043** lists the domestic gross ticket sales versus foreign gross ticket sales (both in millions of U.S. dollars). Determine the linear regression equation relating the foreign gross as a function of the domestic gross. Identify and interpret the coefficients of correlation and determination. Source: "Overseas Grosses Outdoing North America," *USA Today*, December 13, 2000.

15.5 ESTIMATION AND TESTS REGARDING THE SAMPLE REGRESSION LINE

The coefficient of correlation (r) is based on sample data and estimates the population coefficient of correlation, ρ. A high absolute value for r may be due merely to chance, especially if the number of data points is small. Likewise, the slope (b_1) of the sample regression line is an estimate of the population slope (β_1). If $\beta_1 = 0$, the population regression line is horizontal and there is no linear relationship between the variables.

Testing H_0: $\rho = 0$ is equivalent to testing H_0: $\beta_1 = 0$. If one null hypothesis is true, there is no linear relationship between the variables and the other null hypothesis will also be true. The tests have the same p-value.

Testing the Coefficient of Correlation

In this test for the significance of the linear relationship, the null and alternative hypotheses pertain to the population coefficient of correlation, ρ:

***t*-test for the population coefficient of correlation, ρ:**

- **Null hypothesis**

  H_0: $\rho = 0$ There is no linear relationship.

- **Alternative hypothesis**

  H_1: $\rho \neq 0$ A linear relationship exists.

- **Test statistic**

$$t = \frac{r}{\sqrt{\frac{1 - r^2}{n - 2}}}$$

where r = sample coefficient of correlation
n = number of data points
t = t value for the level of significance of the test and $df = n - 2$

EXAMPLE

Testing the Coefficient of Correlation

For the employee dexterity test data, the coefficient of correlation was $r = 0.9546$. At the 0.05 level of significance, test the null hypothesis that the population coefficient of correlation (ρ) is really zero.

SOLUTION

For $n = 5$ persons in the dexterity test example, with $r = 0.9546$, the test statistic is

$$t = \frac{r}{\sqrt{\frac{1 - r^2}{n - 2}}} = \frac{0.9546}{\sqrt{\frac{1 - (0.9546)^2}{5 - 2}}} = 5.55$$

For a two-tail test at the 0.05 level and $df = 5 - 2$, or 3, the critical values are $t = -3.182$ and $t = +3.182$. The calculated test statistic ($t = 5.55$) falls outside these critical values, and the null hypothesis is rejected. At the 0.05 level, the sample coefficient of correlation differs significantly from zero, and we conclude that a linear relationship could exist between the dexterity score and the production rate. If the population coefficient of correlation were really zero, there is less than a 5% chance that a sample coefficient of correlation would be this large.

As shown in Computer Solutions 15.3, Minitab provides the coefficient of correlation along with the p-value for the test described here. The exact p-value is 0.012, considerably less than the 0.05 level of significance used in the test we've just carried out.

EXAMPLEEXAMPLEEXAMPLEEXAMPLEEXAMPLEEXAMPLE

Testing and Estimation for the Slope

An equivalent method of testing the significance of the linear relationship is to examine whether the slope (β_1) of the population regression line could be zero.

***t*-test for the slope (β_1) of the population regression line:**

- **Null hypothesis**

  H_0: $\beta_1 = \beta_{10}$ The population slope is β_{10}.

- **Alternative hypothesis**

  H_1: $\beta_1 \neq \beta_{10}$ The population slope is not β_{10}.

- **Test statistic**

  $$t = \frac{b_1 - \beta_{10}}{s_{b_1}}$$

  where b_1 = slope of the sample regression line
  β_{10} = a value that has been hypothesized for the slope of the population regression line

(*continued on page 574*)

STATISTICS IN ACTION 15.1

The Beta Coefficient and Investment Decisions

Beta, the second letter in the Greek alphabet, ascends to the number one position in the alphabet of financial decisions. In finance, million-dollar investments are made with the assistance of the Capital Asset Pricing Model (CAPM), a mathematical model in which beta plays a prominent role.

The CAPM is a lot easier to understand than its name might imply. This is because it is basically a simple linear regression model. For a given stock, and for $i = 1$ through n time periods, the model is

$$y_i = \beta_0 + \beta_1 x_i + \epsilon_i$$

where y_i = rate of return for the stock during the ith time period

x_i = rate of return for the overall stock market during the ith time period

For the stock being evaluated, the slope of the regression line (i.e., β_1, the beta coefficient) is a measure of how risky the stock is in terms of having volatile swings in return versus the market as a whole. A stock with $\beta_1 = 1$ is considered to be an average risk, while $\beta_1 < 1$ and $\beta_1 > 1$ reflect more conservative and more risky stocks, respectively.

Finance managers first set guidelines on how much risk they are willing to take in making their investments. The CAPM is then used in selecting individual investments that represent a mixture of risk levels, as represented by their beta coefficients, but for which the total amount of risk falls within the guidelines originally specified.

For a given stock, the slope of the sample regression equation (b_1) is used as an estimate of its population counterpart (β_1), and estimation and testing for β_1 can be done on the basis of the $i = 1$ through n time periods for which sample data have been collected. This is similar to the approach discussed within the chapter.

Source: For more information on the beta coefficient and the Capital Asset Pricing Model, interested readers may wish to refer to finance textbooks such as J. Fred Weston and Thomas E. Copeland, *Managerial Finance*, 9th ed. (Hinsdale, Ill.: The Dryden Press, 1992).

with

$$s_{b_1} = \frac{s_{y.x}}{\sqrt{\Sigma x_i^2 - n(\bar{x})^2}}$$

s_{b_1} = the estimated standard deviation of the slope

$s_{y.x}$ = the standard error of estimate

n = number of data points

t = t value for the level of significance of the test, and $df = n - 2$

Most typically, we are interested in testing whether β_1 could be equal to zero, and the numerator of the test statistic will simply be $(b_1 - 0)$. In this case, the hypothesized value will be $\beta_{10} = 0$. However, the method described above can also be used to test whether the slope of the population regression line might be any specific, nonzero value. (For example, in the Capital Asset Pricing Model discussed in Statistics in Action 15.1, an investor may wish to test whether the slope differs significantly from 1.0.)

An important component of the preceding test is the estimated standard deviation of the sample slope b_1. This is because s_{b_1} can also be used in the construction of a confidence interval for β_1.

Confidence interval for the slope (β_1) of the population regression line:

The interval is

$b_1 \pm ts_{b_1}$ where b_1 = slope of the sample regression line
t = t value for the confidence level desired, and $df = n - 2$
s_{b_1} = the estimated standard deviation of the slope, $s_{b_1} = \dfrac{s_{y.x}}{\sqrt{\Sigma x_i^2 - n(\bar{x})^2}}$

EXAMPLE

Testing and Estimation for the Slope

For the dexterity test data, the slope of the sample regression line was $b_1 = 3.0$.

1. Using the 0.05 level of significance, examine whether the slope of the population regression line could be zero.
2. Construct the 95% confidence interval for the slope of the population regression line.

SOLUTION

Testing H_0: $\beta_1 = 0$ versus H_1: $\beta_1 \neq 0$

For the $n = 5$ persons in the dexterity test example, we have already calculated $s_{y.x} = 2.757$, $\Sigma x_i^2 = 1006$, and $\bar{x} = 14.0$. The standard deviation of the slope will be

$$s_{b_1} = \frac{s_{y.x}}{\sqrt{\Sigma x_i^2 - n(\bar{x})^2}} = \frac{2.757}{\sqrt{1006 - 5(14.0)^2}} = 0.5407$$

and the observed value of the test statistic is

$$\text{Observed } t = \frac{b_1 - 0}{s_{b_1}} = \frac{3.0 - 0}{0.5407} = 5.55$$

For the 0.05 level of significance and $df = 5 - 2 = 3$, the critical values of t are -3.182 and $+3.182$, just as they were for the test of the coefficient of correlation. Notice that the calculated test statistic ($t = 5.55$) is also equal to the t statistic calculated during that test. As pointed out earlier, the tests are equivalent and yield the same conclusions and p-values. At the 0.05 level, the test statistic is outside the critical values, and we are able to reject the null hypothesis that the slope of the population regression line could be zero. In Computer Solutions 15.1, the exact p-value is 0.012—see the "Dexscore" row of either printout.

95% Confidence Interval for the Slope of the Population Regression Line

The 95% confidence interval is centered on the slope of the sample regression line, $b_1 = 3.0$, and s_{b_1} has already been calculated as 0.5407. For $n = 5$ data points, the number of degrees of freedom for the t distribution will be $df = n - 2$, or $df = 3$. Referring to the t distribution table, $t = 3.182$ will correspond to the 95% confidence interval:

$$b_1 \pm ts_{b_1} = 3.0 \pm 3.182(0.5407), \quad \text{or from } 1.279 \text{ to } 4.721$$

EXAMPLEEXAMPLE

We have 95% confidence that the slope (β_1) of the population regression line is in the interval bounded by 1.279 and 4.721. This interval is also provided by Excel in Computer Solutions 15.1—see the "Dexscore" row of the printout. Since $\beta_1 = 0$ is not within the 95% confidence interval, we can conclude, at the 0.05 level, that a linear relationship might exist between productivity and the dexterity test score.

The Analysis of Variance Perspective

Finally, we can carry out an analysis of variance test based on the total amount of variation in y (SST) and the amount explained by the regression line (SSR). This test is equivalent to those for the coefficient of correlation and the slope conducted earlier, and its results are usually given when a computer statistical package is used in carrying out the analysis.

The test is based on the sum-of-squares values that were discussed in the context of the coefficient of determination, or

Total Variation in y values (SST)	=	Variation explained by regression line (SSR)	+	Variation *not* explained by regression line (SSE)
$\Sigma(y_i - \bar{y})^2$		$SSR = SST - SSE$		$\Sigma(y_i - \hat{y}_i)^2$

ANOVA test for the significance of a simple linear regression:

- **Null and alternative hypotheses**

  H_0: There is no linear relationship between x and y.
  H_1: A linear relationship exists between x and y.

- **Test Statistic**

  $$\text{Calculated } F = \frac{SSR/1}{SSE/(n-2)}$$

  where SSR = sum of squares explained by the regression line
  SSE = error, or residual, sum of squares
  n = number of data points

- **Decision Rule** Reject H_0 if calculated F exceeds the critical F for the level of significance of the test, with df (numerator) = 1 and df (denominator) = $n - 2$.

In the numerator of the calculated F-ratio, SSR is divided by 1, the number of degrees of freedom associated with the regression equation. In the denominator, SSE is divided by $(n - 2)$, the number of degrees of freedom for the error sum of squares. For the dexterity test example, our calculations in Section 15.4 showed that $SST = 256.80$, $SSR = 234.0$, and $SSE = 22.80$. We now substitute the SSR and SSE values into the expression for the calculated F-ratio:

$$\text{Calculated } F = \frac{SSR/1}{SSE/(n-2)} = \frac{234.0/1}{22.80/(5-2)} = 30.79$$

For df (numerator) = 1 and df (denominator) = $(n - 2) = 3$, the critical value of F for the 0.025 level is 17.44, while the critical F for the 0.01 level is 34.12. The calculated value ($F = 30.79$) exceeds the critical value for the 0.025 level but does not exceed that for the 0.01 level. Using the F-distribution tables, the most accurate statement we can make about the p-value for the test is that it is between 0.025 and 0.01. From Computer Solutions 15.1, the exact p-value is 0.012—see the ANOVA portion of either printout.

EXERCISES

15.44 Testing the null hypothesis that the slope of the true regression line equals zero is equivalent to testing whether the true coefficient of correlation could be zero. Why?

15.45 For $n = 15$ data points, $r^2 = 0.81$. At the 0.05 level of significance, can we conclude that the true coefficient of correlation could be zero?

15.46 For $n = 37$ data points, $r^2 = 0.29$. At the 0.02 level of significance, can we conclude that the true coefficient of correlation could be zero?

15.47 Based on sample data, the 90% confidence interval for the slope of the population regression line is found to be from −2.5 to 1.4. Based on this information, what is the most accurate statement that can be made about the p-value in testing H_0: $\beta_1 = 0$ versus H_1: $\beta_1 \neq 0$? Explain.

15.48 Based on sample data, the 95% confidence interval for the slope of the population regression line is found to be from 4.2 to 6.7. Based on this information, what is the most accurate statement that can be made about the p-value in testing H_0: $\beta_1 = 0$ versus H_1: $\beta_1 \neq 0$? Explain.

15.49 For the regression line developed in Exercise 15.37,

a. Use the 0.05 level in testing whether the population coefficient of correlation could be zero.
b. Use the 0.05 level in testing whether the population regression equation could have a slope of zero.
c. Construct the 95% confidence interval for the slope of the population regression equation.

15.50 For the regression line developed in Exercise 15.38,

a. Use the 0.05 level in testing whether the population coefficient of correlation could be zero.
b. Use the 0.05 level in testing whether the population regression equation could have a slope of zero.
c. Construct the 95% confidence interval for the slope of the population regression equation.

15.51 A computer analysis of 30 pairs of observations results in the least-squares regression equation $\hat{y} = 14.0 + 5.0x$, and the standard deviation of the slope is listed as $s_{b_1} = 2.25$.

a. At the 0.05 level of significance, can we conclude that no linear relationship exists within the population of x and y values?
b. Given the results in part (a), what conclusion would be reached if the same level of significance were used in testing whether the population coefficient of correlation could be zero?
c. Construct the 95% confidence interval for the population slope, β_1.

15.52 How is analysis of variance related to regression analysis?

15.53 In a regression analysis, the sum of the squared deviations between y and $\bar{y}$ is $SST = 120.0$. If the coefficient of correlation is $r = 0.7$, what are the values of SSE and SSR?

15.54 In a regression analysis, the sum of the squared deviations between y and $\bar{y}$ is $SST = 200.0$. If the sum of the squared deviations about the regression line is $SSE = 40.0$, what is the coefficient of determination?

15.55 Given the information below, calculate SST, SSR, and SSE, determine the coefficient of determination, then use ANOVA and the 0.05 level in testing whether r^2 is significantly different from zero.

x:	5	4	10	9	10	Least-squares equation:
y:	34	44	65	47	66	$\hat{y} = 20.66 + 4.02x$

15.56 Given the information below, calculate SST, SSR, and SSE, determine the coefficient of determination, then use ANOVA and the 0.05 level in testing whether r^2 is significantly different from zero.

x:	4	1	4	6	Least-squares equation:
y:	381	403	394	385	$\hat{y} = 404.94 - 3.78x$

(**DATA SET**) *Note:* Exercises 15.57–15.59 require a computer and statistical software.

15.57 The General Aviation Manufacturers Association has reported annual flying hours and fuel consumption for airplanes with a single, piston-driven engine as listed in file **XR15057**. Data are in millions of flying hours and millions of gallons of fuel, respectively. Determine the linear regression equation describing fuel consumption as a function of flying hours, then identify and interpret the slope, the coefficient of correlation, and the coefficient of determination. At the 0.05 level of significance, could the population slope and the population coefficient of correlation be zero? Determine the 95% confidence interval for the population slope. Source: General Aviation Manufacturers Association, *1999 Statistical Databook*, p. 25.

15.58 Given the information in Exercise 15.13, determine the linear regression equation describing January–May burned acreage in Florida as a function of January–May rainfall in the state, then identify and interpret the slope, the coefficient of correlation, and the coefficient of determination. At the 0.10 level of significance, could the population slope and the population coefficient of correlation be zero? Determine the 90% confidence interval for the population slope.

15.59 Computer database **GROWCO** describes the characteristics of 100 companies identified by *Fortune* as the fastest growing. Two of the variables listed are revenue and net income (millions of dollars) for the most recent four quarters. Determine the linear regression equation describing net income as a function of revenue, then identify and interpret the slope, the coefficient of correlation, and the coefficient of determination. At the 0.05 level of significance, could the population slope and the population coefficient of correlation be zero? Determine the 95% confidence interval for the population slope.

Source: "*Fortune's* 100 Fastest-Growing Companies," *Fortune*, September 4, 2000, pp. 142–158.

15.6 ADDITIONAL TOPICS IN REGRESSION AND CORRELATION ANALYSIS

Residual Analysis

If you wish, you can have your computer statistical package or spreadsheet list the residuals (observed minus estimated values of y) along with the regression printout. For example, Table 15.2 is Minitab's listing of the residuals for the dexterity data we've been examining throughout the chapter. It was generated using the Minitab procedure in Computer Solutions 15.4, which we will present shortly.

For each of the 5 data points in the dexterity example, Table 15.2 provides observed and estimated ("fitted") values of y, along with the difference between them (the residual). In Table 15.2, the residuals have also been "standardized," indicating how many standard deviation multiples each represents. The standard deviations for the estimated y values are listed in the "SE Fit" column. Note that the value differs from one row to the next—this is because estimates become more uncertain when they are based on x values that are farther away from the mean of x.

Besides simply listing the residuals, computer packages can readily provide an analysis of them. For example, some data points may be flagged as "unusual observations," or "outliers." These points are relatively distant from the regression line, and their distance may be a reflection of unique characteristics possessed by the sample members involved. For example, a person whose productivity is

TABLE 15.2

Minitab summary of the residuals and standardized residuals for the dexterity regression analysis using the procedure in Computer Solutions 15.4.

- Prodrate: Observed Productivity
- Fit: Estimated Productivity
- SE Fit: The standard error of fitted values for this particular value of Dexscore. This gets larger whenever the value of Dexscore is farther away from its mean.
- Residual: Observed Minus Estimated Productivity
- St Resid: Standardized Residuals

Obs	Dexscore	Prodrate	Fit	SE Fit	Residual	St Resid
1	12.0	55.00	55.20	1.64	-0.20	-0.09
2	14.0	63.00	61.20	1.23	1.80	0.73
3	17.0	67.00	70.20	2.04	-3.20	-1.72
4	16.0	70.00	67.20	1.64	2.80	1.26
5	11.0	51.00	52.20	2.04	-1.20	-0.65

extremely high may have done poorly on the dexterity test because he had a migraine headache on the test day. If the discrepancy between his observed and estimated productivity scores were great enough, he would be identified as an outlier.

Perhaps most important, residual analysis can also help us determine whether the assumptions of the regression model have been met. These assumptions (in terms of the errors or residuals) and the methods by which they can be examined are as follows:

Assumption 1: The population of ϵ values is normally distributed with a mean of zero.

a. **Construct a histogram of the $y - \hat{y}$ values, or residuals. The histogram should be nearly symmetric, centered near or at zero, and have a shape that is approximately normal.**

b. **Use the computer to test for normality of the residuals. Whether the number of data points is small or large, we can use the Kolmogorov-Smirnov test for normality, introduced in Chapter 14. If the number of data points is sufficiently large ($n \geq 32$), Excel and the chi-square goodness-of-fit test for normality of Chapter 13 can be applied.**

Excel and Minitab can also generate a normal probability plot for the residuals, another method introduced in Chapter 14. To the extent that the residuals are from a normal distribution, the points in the normal probability plot should fall approximately along a straight line.

c. **Plot the residuals versus the x values. The points should be distributed about the horizontal line at zero, as shown in part (a) of Figure 15.5.**

(*continued on page 582*)

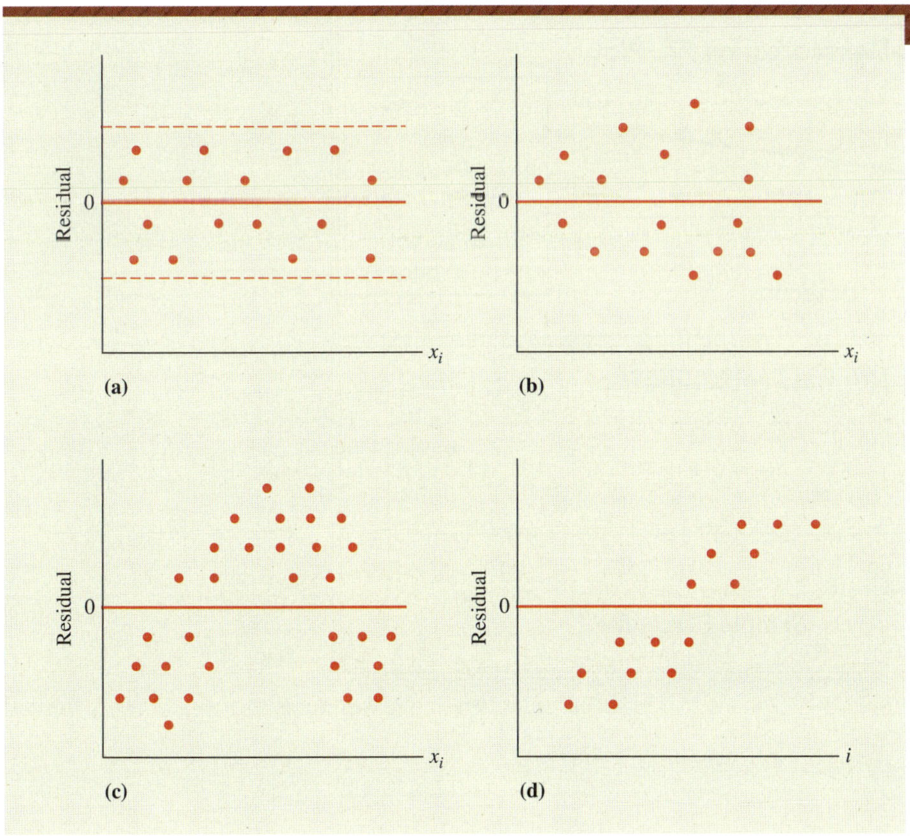

FIGURE 15.5

Selected residual plots; the interpretations are discussed in the text.

COMPUTER 15.4 SOLUTIONS

Residual Analysis

EXCEL

	O	P	Q	R
1	RESIDUAL OUTPUT			
2				
3	*Observation*	*Predicted Prodrate*	*Residuals*	*Standard Residuals*
4	1	55.2	-0.2	-0.084
5	2	61.2	1.8	0.754
6	3	70.2	-3.2	-1.340
7	4	67.2	2.8	1.173
8	5	52.2	-1.2	-0.503

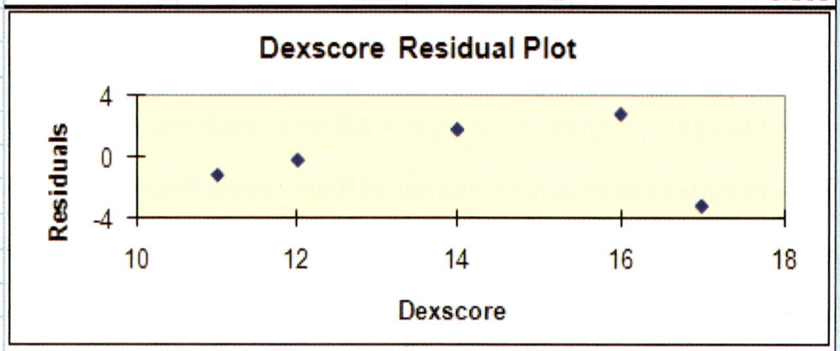

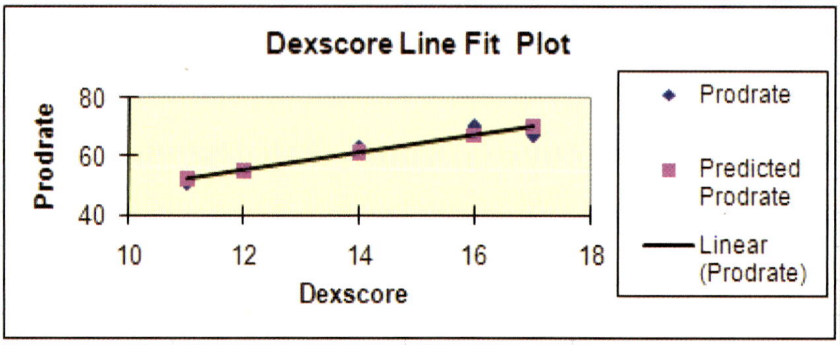

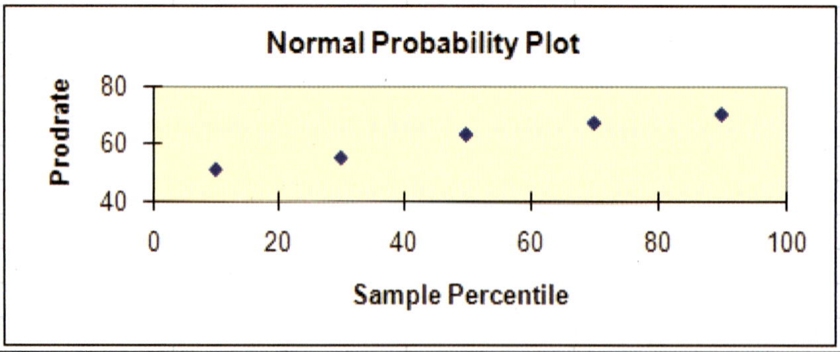

In step 2 of the procedure in Computer Solutions 15.1, select all four items in the **Residuals** section and select **Normal Probability Plot.**

MINITAB

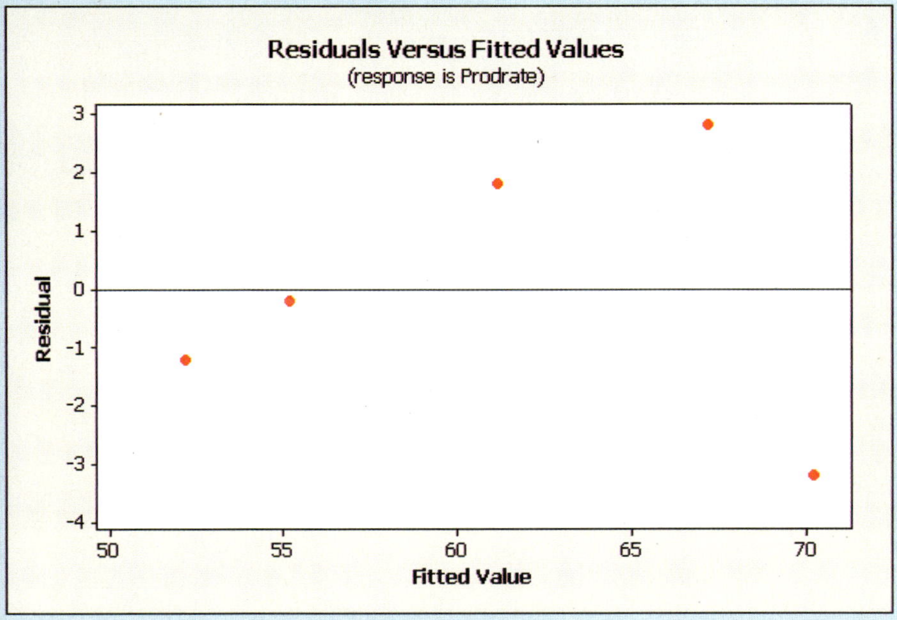

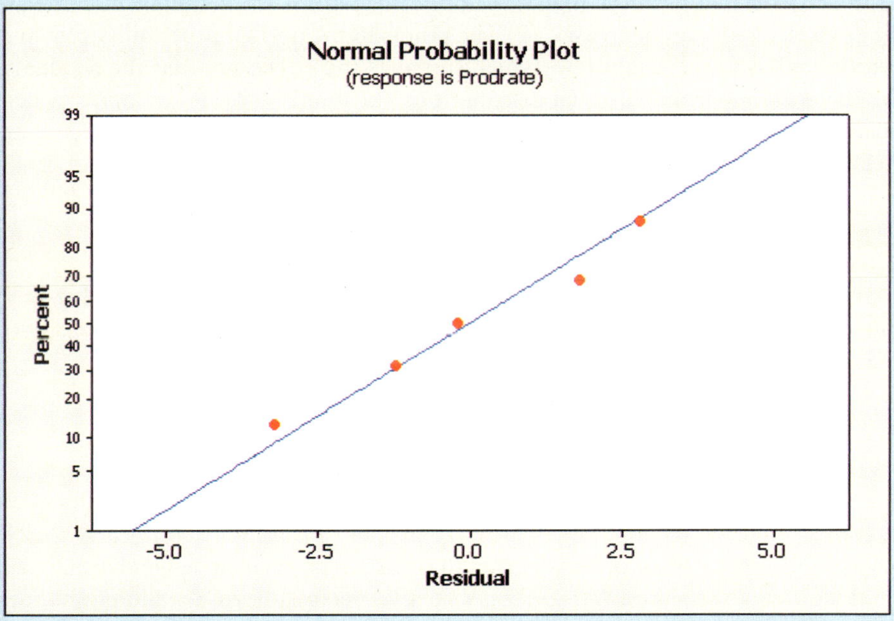

In step 2 of the procedure in Computer Solutions 15.1, click **Graphs.** In the **Residuals for plots** menu, select **Regular.** In the **Residual Plots** menu, select **Normal plot of residuals** and **Residuals versus fits.** A histogram of residuals and a plot of residuals versus order are also available, but not shown here. To obtain the printout of residuals shown in Table 15.2, in step 1 of Computer Solutions 15.1 click **Results,** then select **In addition, the full table of fits and residuals.**

Assumption 2: The standard deviation of the ϵ values is the same regardless of the given value of x.

Plot the residuals versus the x values. The amount of scatter should be approximately the same for all values of x, as shown in part (a) of Figure 15.5. In part (b) of Figure 15.5, the assumption of equal standard deviations has been violated, as the residuals tend to have a greater standard deviation for larger values of x. In part (c) of Figure 15.5, the scatter diagram of residuals is in a band that is not horizontal and centered on zero for all values of x, reflecting that the relationship between y and x may not be linear.

Assumption 3: The residuals are independent of each other.

Plot $(y_i - \hat{y}_i)$ versus the value of i. If successive observations have errors, or residuals, that are correlated, a phenomenon known as autocorrelation exists. For example, if the scatter diagram in this plot tends to have a negative slope or a positive slope, as shown in part (d) of Figure 15.5, this would suggest the presence of autocorrelation. The topic of autocorrelation is especially relevant to regression models fitted to a time series, and is discussed in Chapter 18.

Computer Solutions 15.4 describes the procedures and shows the results when Excel and Minitab are used in analyzing the residuals associated with the dexterity regression analysis. Referring to the first and third plots in the Excel portion of Computer Solutions 15.4, we make the following observations: (1) Except for the residual associated with the highest dexterity test score, the residuals increase as the dexterity test score increases; (2) The points in the normal probability plot are approximately in a straight line, which supports the possibility that the residuals could have come from a normal distribution. The number of data points is rather small and, despite the tendency noted in item (1), we are unable to conclude that the assumptions of the simple linear regression model have not been met.

Cautionary Notes

In applying regression and correlation analysis, it is not enough just to plug in y and x values, then carry out the calculations or assign them to the computer. Two major pitfalls can undermine our good intentions.

Causation Confusion

When analysis results in a high value for the coefficient of determination, the indication is that changes in x go a long way toward explaining changes in y. However, this does not mean that x causes y. In fact, y may be causing x, or both y and x may be caused by one or more other variables that have not been included in the analysis.

If we were to analyze 1990–1999 net sales (y) versus advertising and promotion (x) for Bristol-Myers Squibb Company, we would find the coefficient of correlation to be $r = 0.982$ and the coefficient of determination to be $r^2 = 0.965$ (see Figure 15.6). Statistically, changes in advertising and promotion "explain" 96.5% of the changes in net sales levels. This supports the widespread notion that more advertising tends to result in greater sales. However, higher sales allow for a larger advertising budget, and many companies actually budget their advertising as a fraction of their actual or anticipated sales. In such a case, y and x are actually causing each other!

Sometimes variables that are highly correlated have almost nothing at all to do with each other. For example, an analysis of BMW sales versus the average

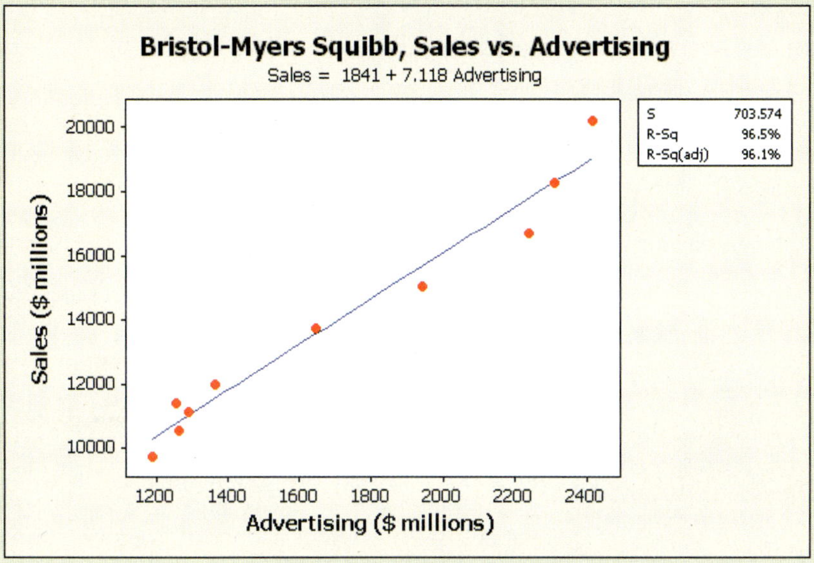

FIGURE 15.6

Minitab scatter diagram and fitted line for annual sales (y) versus annual expenditures on advertising and promotion (x) for Bristol-Myers Squibb Company for the period 1990–1999. While the coefficient of correlation is very high ($r = +0.982$), the question remains as to whether (1) higher levels of advertising cause greater sales, or (2) higher levels of sales lead to a larger advertising budget. *Source of underlying data:* Bristol-Myers Squibb Company, *1999 Annual Report,* pp. 50, 51.

salary of secondary school teachers results in a very high correlation:[2]

For the years 2002–2006:

y = Sales of BMW automobiles in the United States
x = Average salary, secondary school teachers

The regression equation is $\hat{y} = -375{,}997 + 14.126x$

$$r = +0.959 \quad \text{and} \quad r^2 = 0.920$$

Without further consideration, it might seem as though all these BMWs were being purchased by secondary school teachers, since the teachers' salary levels "explain" 92.0% of the variation in BMW sales. However, it is more likely that both y and x have simply increased largely *as a function of time.* If two variables are linearly related to time, they will also be linearly related to each other. Whenever two variables are highly correlated, but there is little evidence to suggest that either causes the other (e.g., annual flounder catch in the North Atlantic and annual bachelor's degrees granted in physics), this is known as a *spurious*, or nonsense, correlation.

Extrapolating Beyond the Range of the Data

When regression and correlation are based on a given set of data, it can be dangerous to make estimates for y based on x values outside the range of these data. Figure 15.7 shows the scatter diagram and regression line for EPA city miles per gallon (y) versus curb weight (x) for a sampling of automobiles from the 2009 model year. As expected, heavier cars tend to get fewer miles per gallon. However, if we go beyond the range of x values from which the regression line was derived, we find that *a car weighing 6591 pounds would get zero miles per gallon on the EPA city test,* a physical impossibility. Statistics in Action 15.2 discusses both causation and extrapolation beyond the data for salaries versus team performance in professional baseball.

[2]Source: autointell.com, August 8, 2009; and Bureau of the Census, *Statistical Abstract of the United States 2009,* p. 158.

STATISTICS IN ACTION 15.2

Major League Baseball: Does Money Really Win Games?

In the chapter's opening vignette, we discussed the possible relationship between how well a team is paid and how well they play. Using Excel, we've performed a simple regression analysis of the performance of Major League Baseball teams versus the size of their team payrolls during the 2006 season. Using Excel File **CX15BASE** and the methods in Computer Solutions 15.1 and 15.4, we obtained the regression equation

Wins = 67.980 + 0.167 Payroll
(payroll in millions of dollars)
$r = +0.536$ with $r^2 = 0.287$

Not unexpectedly, teams with higher payrolls tended to win more games. The slope (+0.167) indicates that an extra one million dollars in payroll tends to increase the number of wins by 0.167. Thus, on average, each additional win costs 1/0.167, or $5.99 million. Table 15.3 shows the payroll and number of wins for each team, the number of wins predicted by the regression equation, and the residual (actual wins minus predicted wins). Especially interesting in Table 15.3 is a comparison of two teams that were not among the highest-paid in the league: the Minnesota Twins and the Chicago Cubs. The Twins got a lot of wins for the money, 17.4 *more* than the equation predicted, while the Cubs were 17.79 wins *below* the performance predicted for a team with their payroll. (Which manager would *you* prefer to be if you're invited to meet with the team president the morning after the season ends?)

The upper portion of Figure 15.8 shows the scatter diagram and fitted regression line, and the lower portion is the plot of residuals (vertical axis) versus payrolls (horizontal axis). At approximately $63 million and $94 million payroll levels, respectively, the overachieving Twins and the underachieving Cubs are exceptionally distant from the regression line, but in opposite directions.

With their league-high $194.6 million payroll, one would expect the New York Yankees to have had a successful season. They did, in a way. Nobody had more regular-season wins than the Yankees, but they were eliminated in the post-season playoffs and the St. Louis Cardinals went on to defeat the Detroit Tigers in the World Series.

A final observation: If there were a team whose members received *zero* salary, the equation would nevertheless predict them to win nearly 68 games! This is misleading, however, because such an estimate would be based on a payroll that lies unrealistically far below those in the underlying data.

Data sources: usatoday.com and mlb.com, June 21, 2009.

FIGURE 15.7

Making estimates of y based on x values beyond the data is a dangerous proposition. This Excel scatter diagram represents EPA city gas mileage (y) versus vehicle weight (x) for a sample of 2009 auto models. According to the regression line, a car weighing 6591 pounds would get *zero* miles per gallon in the test. *Source of underlying data: Consumer Reports New Car Ratings & Reviews 2009.*

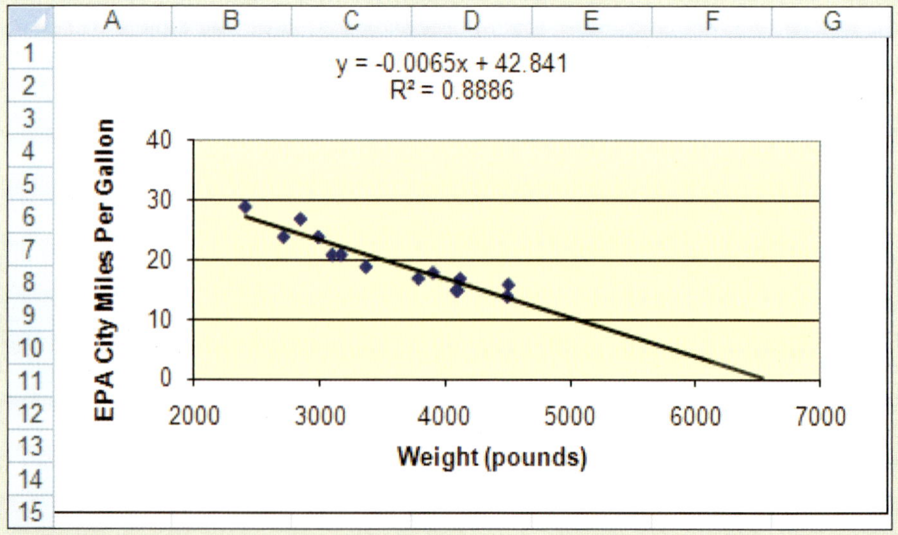

TABLE 15.3

For the 2006 Major League Baseball season discussed in Statistics in Action 15.2, team payrolls, wins, wins predicted by the regression equation, and residuals.

	A	B	C	D	E
1	Team	Payroll	Wins	Predicted Wins	Residual
2	Arizona Diamondbacks	59.684	76	77.97	-1.97
3	Atlanta Braves	90.157	79	83.08	-4.08
4	Baltimore Orioles	72.586	70	80.13	-10.13
5	Boston Red Sox	120.100	86	88.09	-2.09
6	Chicago Cubs	94.424	66	83.79	-17.79
7	Chicago White Sox	102.751	90	85.19	4.81
8	Cincinnati Reds	60.910	80	78.18	1.82
9	Cleveland Indians	56.032	78	77.36	0.64
10	Colorado Rockies	41.233	76	74.88	1.12
11	Detroit Tigers	82.613	95	81.81	13.19
12	Florida Marlins	14.999	78	70.49	7.51
13	Houston Astros	92.552	82	83.48	-1.48
14	Kansas City Royals	47.294	62	75.90	-13.90
15	Los Angeles Angels	103.472	89	85.31	3.69
16	Los Angeles Dodgers	98.447	88	84.46	3.54
17	Milwaukee Brewers	57.568	75	77.62	-2.62
18	Minnesota Twins	63.396	96	78.60	17.40
19	New York Mets	101.085	97	84.91	12.09
20	New York Yankees	194.633	97	100.57	-3.57
21	Oakland Athletics	62.243	93	78.40	14.60
22	Philadelphia Phillies	88.273	85	82.76	2.24
23	Pittsburgh Pirates	46.718	67	75.80	-8.80
24	St. Louis Cardinals	88.891	83	82.86	0.14
25	San Diego Padres	69.896	88	79.68	8.32
26	San Francisco Giants	90.056	76	83.06	-7.06
27	Seattle Mariners	87.960	78	82.71	-4.71
28	Texas Rangers	68.229	80	79.40	0.60
29	Tampa Bay Devil Rays	35.418	61	73.91	-12.91
30	Toronto Blue Jays	71.915	87	80.02	6.98
31	Washington Nationals	63.143	71	78.55	-7.55

EXERCISES

15.60 In regression analysis, what is a "residual"?

15.61 What is residual analysis, and what information can it provide?

15.62 What is spurious correlation? Provide a real or hypothetical example where two variables might exhibit such a relationship.

15.63 A firm finds the coefficient of correlation between y = annual sales and x = annual expenditure on research and development to be $r = +0.90$. Comment on the likely direction of causation, if any, between these variables.

15.64 "The Ford Edsel was sold only in the 1957–1959 model years, but the company made a big mistake in dropping it. If they had just fitted a regression line to the sales for these 3 years and extended it to 2006, they could have seen that it would have been one of the ten sales leaders of that year." Comment on the appropriateness of this statement.

(DATA SET) *Note:* Exercises 15.65–15.68 require a computer and statistical software.

15.65 For the y and x values listed in file **XR15065**, obtain the simple linear regression equation, then analyze

FIGURE 15.8

Excel scatter diagram, fitted regression equation, and residuals plot for the regression analysis of the payroll and wins data in Table 15.3.

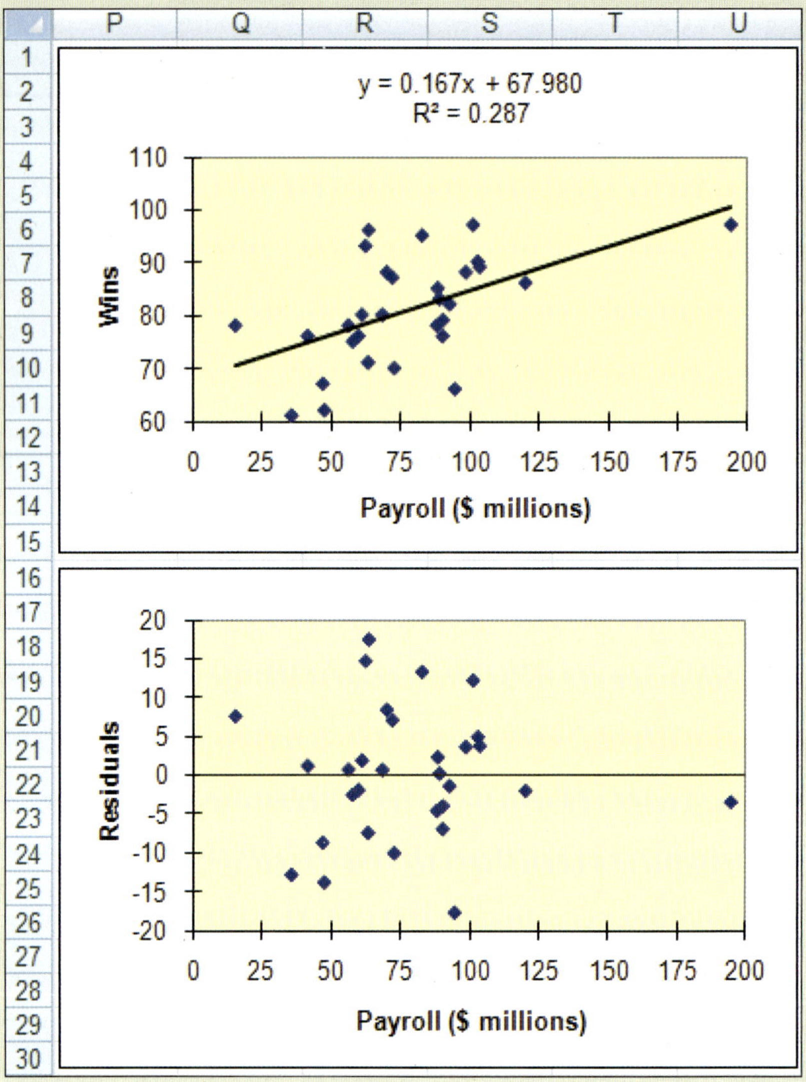

the residuals by (a) constructing a histogram, (b) using a normal probability plot, (c) plotting the residuals versus the x values, and (d) plotting the residuals versus the order in which they were observed. Do any of the assumptions of the simple linear regression model appear to have been violated?

15.66 For the y and x values listed in file **XR15066**, obtain the simple linear regression equation, then analyze the residuals by (a) constructing a histogram, (b) using a normal probability plot, (c) plotting the residuals versus the x values, and (d) plotting the residuals versus the order in which they were observed. Do any of the assumptions of the simple linear regression model appear to have been violated?

15.67 For the regression equation obtained in Exercise 15.57, analyze the residuals by (a) constructing a histogram, (b) using a normal probability plot, (c) plotting the residuals versus the x values, and (d) plotting the residuals versus the order in which they were observed. Do any of the assumptions of the simple linear regression model appear to have been violated? Comment on whether any of the years were associated with especially high or low fuel consumption for the number of flying hours during the year.

15.68 For the regression equation obtained in Exercise 15.59, analyze the residuals by (a) constructing a histogram, (b) using a normal probability plot, (c) plotting the residuals versus the x values, and (d) plotting the residuals versus the order in which they were observed. Do any of the assumptions of the simple linear regression model appear to have been violated? Comment on whether any of these high-growth companies had an especially high or low income for a firm with their level of revenue.

SUMMARY

15.7

• Linear regression model and its assumptions

In simple linear regression and correlation, regression analysis provides a "best-fit" linear equation for the data, and correlation analysis measures the strength of the linear relationship. In the linear model, $y_i = \beta_0 + \beta_1 x_i + \epsilon_i$, y and x are the dependent and independent variables, respectively. The simple linear regression model assumes the following: (1) for any given value of x, the y values are normally distributed with a mean that is on the regression line, or $\mu_{y.x} = \beta_0 + \beta_1 x_i$; (2) regardless of the value of x, the standard deviation of the y values about the regression line is the same; and (3) the y values are statistically independent of each other. In multiple linear regression and correlation (Chapter 16), there is one dependent variable, but two or more independent variables.

Variables are directly related when they increase or decrease together and inversely related when an increase in one is accompanied by a decrease in the other. The relationship may also be linear or curvilinear, depending on whether a straight line or a curve is best for describing their relationship. The emphasis of the chapter is on linear relationships.

• Determining the least-squares linear regression equation

The scatter diagram visually summarizes the data points and consists of vertical (y) and horizontal (x) axes. A ruler and the "eyeball" method can be used in fitting an approximate straight line to the data. A superior approach is to calculate the regression line, $\hat{y} = b_0 + b_1 x$, that satisfies the least-squares criterion. This criterion requires that the y-intercept (b_0) and slope (b_1) be such that the sum of the squared vertical distances between data points and the line is minimized.

Formulas are provided for the calculation of b_0 and b_1 for the least-squares regression line, $\hat{y} = b_0 + b_1 x$. Because the regression line is based on sample data, the y-intercept (b_0) and slope (b_1) are really estimates of their respective (and unknown) population values, β_0 and β_1. For a given value of x, the point estimate of y will be $\hat{y}$, calculated from the regression line. Interval estimates for y can also be developed, and the first step is to calculate the standard error of estimate, $s_{y.x}$. The standard error of estimate is the extent to which y values are scattered or dispersed above and below the regression line. For a given x value, the difference between the observed and estimated values of y is the error, or residual.

• Point and interval estimates

Given a specific value of x, two kinds of interval estimates can be made regarding y: (1) a confidence interval for the (unknown) true mean of y, and (2) a prediction interval for an individual observation of y. Each approach relies on a correction formula that takes into account the standard error of estimate, the number of data points in the sample, and the distance between $\bar{x}$ and the specific value of x on which the interval is based.

• Coefficients of correlation and determination

Two important measures of the strength of the linear relationship between y and x are the coefficient of correlation (r) and the coefficient of determination (r^2). The coefficient of correlation will be no less than $r = -1$ and no greater than $r = +1$, and the larger the absolute value of r, the stronger the linear relationship between the variables. Whenever $r = 0$, there is no linear relationship at all, and the slope of the regression line will be $b_1 = 0$. The coefficient of determination (r^2) is the proportion of the variation in y that is explained by the regression line, $\hat{y} = b_0 + b_1 x$, and r^2 will always be ≤ 1.

• Testing the linear relationship

The significance of the linear relationship can be tested by examining whether the coefficient of correlation differs significantly from zero, or by testing whether the slope differs significantly from zero. The tests are equivalent to each other as well as to an analysis of variance test that examines the amount of variation in y that is explained by the equation.

• Computer-assisted regression analysis

Computer statistical packages provide detailed regression and correlation analysis, and their output typically includes the descriptors, tests, and interval estimates discussed in the chapter. Both Excel and Minitab Computer Solutions procedures are provided that correspond to the principal example used throughout the chapter.

With statistical software, it is easy to analyze the residuals and examine the assumptions of the model. Computer statistical packages also identify data points that are relatively distant from the regression line. Such "unusual" observations may reflect the possession of unusual characteristics by the sample members that these points represent. Residual analysis is especially useful in examining whether various assumptions of the regression model may have been violated.

• Potential problems with simple linear regression

In applying regression and correlation analysis, it is important to avoid confusing a strong linear relationship between y and x with the conclusion that y is actually caused by x. In reality, y could be causing x, or both y and x might be heavily influenced by another variable that has not been included in the analysis. Another precaution is to avoid extrapolating beyond the range of the data. It can be extremely misleading to make point or interval estimates of y based on an x value that lies beyond the range of data points from which the regression line has been derived.

EQUATIONS

The Simple Linear Regression Model

$$y_i = \beta_0 + \beta_1 x_i + \epsilon_i$$

where y_i = a value of the dependent variable, y
x_i = a value of the independent variable, x
β_0 = the y-intercept of the regression line
β_1 = the slope of the regression line
ϵ_i = random error, or residual

NOTES

1. For a given value of x, the expected value of y is given by the linear equation, $\mu_{y.x} = \beta_0 + \beta_1 x_i$. The term $\mu_{y.x}$ can be stated as "the mean of y, given a specific value of x."
2. The difference between the actual value of y and the expected value of y is the error, or residual, $\epsilon_i = y_i - (\beta_0 + \beta_1 x_i)$.

The Sample Regression Line

$$\hat{y} = b_0 + b_1 x$$

where $\hat{y}$ = the estimated value of the dependent variable (y) for a given value of x
b_0 = the y-intercept; this is the value of y where the line intersects the y-axis whenever $x = 0$
b_1 = the slope of the regression line
x = a value for the independent variable

Least-Squares Criterion for the Best-Fit Regression Line

The equation must be such that $\sum(y_i - \hat{y}_i)^2$ is minimized

where y_i = the observed value of y for the given value of x
$\hat{y}_i$ = the predicted value of y for the x value, as determined from the regression equation

Least-Squares Regression Line, $\hat{y} = b_0 + b_1x$

- Slope:

$$b_1 = \frac{\left(\sum x_i y_i\right) - n\bar{x}\bar{y}}{\left(\sum x_i^2\right) - n\bar{x}^2}$$

where n = number of data points

- y-intercept:

$$b_0 = \bar{y} - b_1\bar{x}$$

Standard Error of Estimate

$$s_{y.x} = \sqrt{\frac{\sum(y_i - \hat{y}_i)^2}{n - 2}}$$

where y_i = each observed value of y in the data
$\hat{y}_i$ = the value of y that would have been estimated from the regression line
n = the number of data points

Confidence Interval for the Mean of *y*, Given a Specific Value of *x*

$$\hat{y} \pm t(s_{y.x})\sqrt{\frac{1}{n} + \frac{(x \text{ value} - \bar{x})^2}{\left(\sum x_i^2\right) - \left(\sum x_i\right)^2/n}}$$

where $\hat{y}$ = the estimated value of y for the given value of x
n = the number of data points
$s_{y.x}$ = the standard error of estimate
t = t value for confidence level desired and $df = n - 2$
x value = the given value of x

Prediction Interval for an Individual *y*, Given a Specific Value of *x*

$$\hat{y} \pm t(s_{y.x})\sqrt{1 + \frac{1}{n} + \frac{(x \text{ value} - \bar{x})^2}{\left(\sum x_i^2\right) - \left(\sum x_i\right)^2/n}}$$

where $\hat{y}$ = the estimated value of y for the given value of x
n = number of data points
$s_{y.x}$ = the standard error of estimate
t = t value for the confidence level desired and $df = n - 2$
x value = the given value of x

The Coefficient of Correlation

$$r = \frac{n\left(\sum x_i y_i\right) - \left(\sum x_i\right)\left(\sum y_i\right)}{\sqrt{n\left(\sum x_i^2\right) - \left(\sum x_i\right)^2} \cdot \sqrt{n\left(\sum y_i^2\right) - \left(\sum y_i\right)^2}}$$

where r = coefficient of correlation, $-1 \le r \le +1$
n = number of data points

The Coefficient of Determination, r^2

$$r^2 = 1 - \frac{\textit{Error} \text{ variation, which is } \textit{not} \text{ explained by the regression line}}{\textit{Total} \text{ variation in the } y \text{ values}}$$

$$= 1 - \frac{\sum(y_i - \hat{y}_i)^2}{\sum(y_i - \bar{y})^2} = 1 - \frac{SSE}{SST}$$

t-Test for the Population Coefficient of Correlation, ρ

- Null hypothesis

$$H_0: \quad \rho = 0 \qquad \text{There is no linear relationship.}$$

- Alternative hypothesis

$$H_1: \quad \rho \neq 0 \qquad \text{A linear relationship exists.}$$

- Test statistic

$$t = \frac{r}{\sqrt{\dfrac{1 - r^2}{n - 2}}}$$

where r = sample coefficient of correlation
n = number of data points
t = t value for the level of significance of the test and $df = n - 2$

(Testing H_0: $\rho = 0$ is equivalent to testing H_0: $\beta_1 = 0$, described below.)

t-Test for the Slope (β_1) of the Population Regression Line

- Null hypothesis

$$H_0: \quad \beta_1 = \beta_{10} \qquad \text{The population slope is } \beta_{10}.$$

- Alternative hypothesis

$$H_1: \quad \beta_1 \neq \beta_{10} \qquad \text{The population slope is not } \beta_{10}.$$

- Test statistic

$$t = \frac{b_1 - \beta_{10}}{s_{b_1}}$$

where b_1 = slope of the sample regression line
β_{10} = a value that has been hypothesized for the slope of the population regression line

with

$$s_{b_1} = \frac{s_{y.x}}{\sqrt{\sum x_i^2 - n(\overline{x})^2}}$$

where s_{b_1} = the estimated standard deviation of the slope
$s_{y.x}$ = the standard error of estimate
n = the number of data points
t = t value for the level of significance of the test, and $df = n - 2$

Confidence Interval for the Slope (β_1) of the Population Regression Line

The interval is

$$b_1 \pm t s_{b_1}$$

where b_1 = slope of the sample regression line
t = t value for the confidence level desired, and $df = n - 2$
s_{b_1} = the estimated standard deviation of the slope, $s_{b_1} = \dfrac{s_{y.x}}{\sqrt{\sum x_i^2 - n(\overline{x})^2}}$

Components of the Total Variation in the Dependent Variable, y

Total variation in y values (SST)	=	Variation explained by regression line (SSR)	+	Variation *not* explained by regression line (SSE)
$\sum(y_i - \overline{y})^2$		$SSR = SST - SSE$		$\sum(y_i - \hat{y}_i)^2$

ANOVA Test for the Significance of a Simple Linear Regression

- Null and alternative hypotheses

H_0: There is no linear relationship between x and y.
H_1: A linear relationship exists between x and y.

- Test statistic

$$\text{Calculated } F = \frac{SSR/1}{SSE/(n-2)}$$

where SSR = sum of squares explained by the regression line
SSE = error, or residual, sum of squares
n = number of data points

- **Decision rule** Reject H_0 if calculated F exceeds the critical F for the level of significance of the test, with df (numerator) = 1 and df (denominator) = $n - 2$.

CHAPTER EXERCISES

15.69 For households in a community, many different variables can be observed or measured. If y = monthly mortgage payment, x = annual income would probably be directly related to y. For y = monthly mortgage payment, provide an example of an x variable likely to be (a) directly related to y, (b) inversely related to y, and (c) unrelated to y.

15.70 For high school seniors entering college, and for y = grade point average for the freshman year, x = Scholastic Aptitude Test score would tend to be directly related to y. Provide an example of another personal variable that might be directly related to y along with an example of a personal variable that you think would tend to be inversely related to y.

15.71 McDonald's Corporation has reported the following values for total revenues and net income during the 1998 to 2005 period. All data are in billions of dollars:
Source: McDonald's Corporation, *2005 Annual Report*.

	1998	1999	2000	2001
Net Income	1.55	1.95	1.98	1.64
Total Revenues	12.42	13.26	14.24	14.87

	2002	2003	2004	2005
Net Income	0.89	1.47	2.28	2.60
Total Revenues	15.41	17.14	19.07	20.46

Determine the least-squares regression equation line for estimating net income and interpret its slope. For a year in which total revenues are $18.0 billion, estimate the net income for that year.

15.72 Prior to being hired, the five salespersons for a computer store were given a standard sales aptitude test. For each individual, the score achieved on the aptitude test and the number of computer systems sold during the first 3 months of their employment are shown here.

	John	Cora	Sam	Cindy	Henry
x = Score on aptitude test	80	70	45	90	20
y = Units sold in 3 months	25	15	10	40	5

Determine the least-squares regression line and interpret its slope. Estimate, for a new employee who scores 60 on the sales aptitude test, the number of units the new employee will sell in her first 3 months with the company.

15.73 As of mid-June 2009, five U.S. Government Bond mutual funds were reported as having generated the 1-year and 3-year (annualized) rates of return shown below. Source: *Money*, August 2009, p. 113.

U.S. Government Bond Fund	x = 3-Year Annualized Rate of Return	y = 1-Year Rate of Return
American Funds, AMUSX	6.3%	7.1%
Fidelity, FGOVX	7.4	8.4
Wells Fargo, STVSX	6.5	7.3
Morgan Stanley, USGBX	2.4	1.5
MFS Govt. Securities, MFGSX	7.2	9.0

a. Determine the least-squares regression line and interpret its slope.
b. For a U.S. Government Bond mutual fund with a 3-year annualized rate of return of 5.0%, estimate the 1-year rate of return.
c. For a U.S. Government Bond mutual fund with a 3-year annualized rate of return of 7.0%, estimate the 1-year rate of return.

15.74 Peoples Energy, Inc. reports the following data for operating revenue and net income for 2001–2005. Source: Peoples Energy, *2005 Annual Report*, pp. 83–84.

Year	x = Operating Revenue (Millions)	y = Net Income (Millions)
2001	2270.2	96.9
2002	1482.5	89.1
2003	2138.4	103.9
2004	2260.2	81.6
2005	2600.0	78.1

Determine the least-squares regression line and interpret its slope. Estimate the net income if the operating revenue figure is $2500 million.

15.75 For 2004 through 2008, Coca-Cola annual operating revenues and net incomes were as shown below. Source: *Coca-Cola 2008 Annual Review*, p. 7.

Year	x = Operating Revenue	y = Net Income
2004	$21.74 billion	$4.85 billion
2005	23.10	4.87
2006	24.09	5.08
2007	28.86	5.98
2008	31.94	5.81

Determine the least-squares regression line and interpret its slope. Estimate net income if the operating revenue is $30 billion.

15.76 The data in the table describe retail furniture store sales ($ billions) versus housing starts over a 5-year time period. Source: *Statistical Abstract of the United States 2009*, pp. 594, 644.

	2001	2002	2003	2004	2005
Retail Furniture Store Sales ($ billions)	50.6	51.3	52.1	56.5	59.8
Housing Starts (millions)	1.27	1.36	1.50	1.61	1.72

a. Determine the least-squares regression equation for predicting retail furniture store sales on the basis of housing starts.
b. Determine and interpret the coefficients of correlation and determination.
c. If there were 1.80 million housing starts in a year, what would be the prediction for retail furniture store sales?

15.77 In the early 1900s, population, prices, wages, and individual life insurance policy face amounts were all incredibly low compared to their counterparts of today. The following data show the number of life insurance policies (millions) and their total face amounts ($ millions) for five years selected from this long-ago era that was about to experience the Great Depression. Source: American Council of Life Insurers, *Life Insurers Fact Book 2008*, p. 71.

	1900	1905	1910	1915	1920
Individual Life Insurance Policies (millions)	14	22	29	41	64
Total Individual Life Insurance Face Amounts ($ millions)	7573	11,863	14,908	20,929	38,966

a. Determine the least-squares regression equation for predicting total value of individual life insurance on the basis of the number of individual life insurance policies in force. Interpret the slope of the equation.
b. Determine and interpret the coefficients of correlation and determination.
c. If there were 30 million individual life insurance policies in force during a given year, what would be the prediction for the total face value of these policies?

15.78 For the regression analysis of the data in Exercise 15.77:
a. Use the 0.10 level in testing whether the population coefficient of correlation could be zero.
b. Use the 0.10 level in testing whether the population regression equation could have a slope of zero.
c. Construct the 90% confidence interval for the slope of the population regression equation. Discuss the interval in terms of the results of part (b).

15.79 The following data describe U.S. motor vehicle travel and fuel consumption from 2002 through 2006. The data represent billions of gallons consumed and billions of miles traveled, respectively. Source: *Statistical Abstract of the United States 2009*, p. 676.

	2002	2003	2004	2005	2006
Fuel Consumed	168.7	170.0	173.5	174.8	174.9
Miles Traveled	2856	2890	2965	2989	3014

a. Determine the least-squares equation line for predicting fuel consumption on the basis of driving miles.
b. Determine and interpret the coefficients of correlation and determination.
c. If motor vehicle travel during a year were 3100 billion miles, what level of fuel consumption would be predicted?

15.80 For the regression analysis of the data in Exercise 15.79:

a. Use the 0.05 level in testing whether the population coefficient of correlation could be zero.
b. Use the 0.05 level in testing whether the population regression equation could have a slope of zero.
c. Construct the 95% confidence interval for the slope of the population regression equation. Discuss the interval in terms of the results of part (b).

15.81 As part of its ongoing testing and reporting program, the Insurance Institute for Highway Safety has crashed vehicles into a barrier intended to represent another vehicle's bumper, then recorded the cost of repairs necessary to fix whatever damage has occurred. In these tests, the simulated bumper was set at 16 inches from the ground in the corner tests, and at 18 inches from the ground in the full-width tests. In crash-testing a sample of mini and micro-cars at 6 mph, the repair costs were as shown in the table. Source: Insurance Institute for Highway Safety, *Status Report*, June 11, 2009, p. 2.

Vehicle Tested	Rear Full-Width	Rear Corner
Smart Fortwo	$631	$507
Chevrolet Aveo	1370	612
Mini Cooper	929	743
Toyota Yaris	3345	474
Honda Fit	3648	999
Hyundai Accent	2057	831
Kia Rio	3148	773

a. Based on these data, determine the least-squares regression equation for predicting rear full-width crash repair cost on the basis of rear corner crash repair cost.
b. Determine and interpret the coefficients of correlation and determination.
c. Based on these data, if a mini or micro-car like the ones tested were to incur $800 in damages in the rear corner test, what level of repairs would be predicted for the rear full-width test?

15.82 For the regression analysis of the data in Exercise 15.81:

a. Use the 0.10 level in testing whether the population coefficient of correlation could be zero.
b. Use the 0.10 level in testing whether the population regression equation could have a slope of zero.
c. Construct the 90% confidence interval for the slope of the population regression equation. Discuss the interval in terms of the results of part (b).

15.83 The following Minitab output describes the results of a production experiment in which $n = 10$ different steel bars were subjected to tempering times ranging from 6 to 10 seconds, then measured for tensile strength (thousands of pounds):

```
Regression Analysis

The regression equation is
STRENGTH = 60.0 + 10.5 TEMPTIME

Predictor      Coef      StDev        T       P
Constant      60.02      29.36     2.04   0.075
TEMPTIME     10.507      3.496     3.01   0.017

S = 14.03      R-Sq = 53.0%     R-Sq(adj) = 47.2%

Analysis of Variance

Source       DF       SS        MS       F        P
Regression    1   1777.4    1777.4    9.03    0.017
Error         8   1574.6     196.8
Total         9   3352.0
```

a. To the greatest number of decimal places in the printout, what is the least-squares regression line?
b. What proportion of the variation in strength is explained by the regression line?
c. At what level of significance does the slope of the line differ from zero? What type of test did Minitab use in reaching this conclusion?
d. At what level of significance does the coefficient of correlation differ from zero? Compare this with the level found in part (c) and explain either why they are different or why they are the same.
e. Construct the 95% confidence interval for the slope of the population regression line.

15.84 For the ANOVA portion of the printout shown in Exercise 15.83, explain the exact meaning of each number in the "SS" and "F" columns.

15.85 A tire company has carried out tests in which rolling resistance (pounds) and inflation pressure (pounds per square inch, or psi) have been measured for psi values ranging from 20 to 45. The regression analysis is summarized in the following Minitab printout:

```
Regression Analysis

The regression equation is
ROLRESIS = 9.45 - 0.0811 PSI

Predictor      Coef      StDev        T       P
Constant      9.450      1.228     7.69   0.000
PSI        -0.08113    0.03416    -2.38   0.029

S = 0.8808     R-Sq = 23.9%     R-Sq(adj) = 19.6%

Analysis of Variance

Source       DF       SS        MS       F        P
Regression    1   4.3766    4.3766    5.64    0.029
Error        18  13.9657    0.7759
Total        19  18.3422
```

a. To the greatest number of decimal places in the printout, what is the least-squares regression line?
b. What proportion of the variation in rolling resistance is explained by the regression line?
c. At what level of significance does the slope of the line differ from zero? What type of test did Minitab use in reaching this conclusion?
d. At what level of significance does the coefficient of correlation differ from zero? Compare this with the level

found in part (c) and explain either why they are different or why they are the same.

e. Construct the 95% confidence interval for the slope of the population regression line.

15.86 For the ANOVA portion of the printout shown in Exercise 15.85, explain the exact meaning of each number in the "SS" and "F" columns.

15.87 Given the data and regression analysis for Exercise 15.72, construct and interpret the 95% confidence and prediction intervals associated with a score of 60 on the sales aptitude test.

15.88 Given the data and regression analysis for Exercise 15.73, construct and interpret the 90% confidence and prediction intervals associated with a 3-year annualized rate of return of 7.0%.

15.89 Given the data and regression analysis for Exercise 15.76, construct and interpret the 95% confidence and prediction intervals associated with 1.4 million housing starts.

15.90 A physical-fitness researcher tests 25 professional football players, finding that the time required to run 40 yards (y, in seconds) is related to weight (x, in pounds) by the least-squares equation $\hat{y} = -2.0 + 0.038x$. According to this equation, a person weighing 200 pounds would run the 40-yard dash in an estimated time of 5.6 seconds. However, a child weighing 55 pounds would have an estimated time of just 0.09 seconds to run 40 yards. Briefly explain how this unlikely result could have been obtained.

(DATA SET) *Note:* Exercises 15.91–15.96 require a computer and statistical software.

15.91 A university administrator has collected data on total Scholastic Aptitude Test (SAT) score versus freshman grade point average (GPA) for a sample of students who recently finished their first year at the university. The data are listed in file **XR15091**. Given these data:

a. Determine the least-squares regression equation for estimating the freshman grade point average based on total SAT score, then estimate the freshman grade point average for an applicant who has scored a total of 1100 on his SAT examination.

b. What percentage of the variation in freshman GPA is explained by SAT total?

c. Given an SAT total of 1100, construct and interpret the 99% confidence and prediction intervals associated with freshman GPA.

15.92 For the regression equation obtained in Exercise 15.91, analyze the residuals by (a) constructing a histogram, (b) using a normal probability plot, (c) plotting the residuals versus the x values, and (d) plotting the residuals versus the order in which they were observed. Do any of the assumptions of the simple linear regression model appear to have been violated?

15.93 The National Association of Realtors has reported data on housing affordability for 1990–1999, and values for two of the variables are listed in file **XR15093**. Representing affordability is y = average monthly mortgage payment as a percentage of median household income. Representing the cost of a mortgage is x = the average mortgage rate, as a percentage. Source: *World Almanac and Book of Facts 2001*, p. 743.

a. Determine the least-squares regression equation for y as a function of x, then estimate y if the average mortgage rate is $x = 8.0\%$.

b. What percentage of the variation in affordability (y) is explained by the average mortgage rate (x)? At the 0.05 level, does the slope of the equation differ significantly from 0?

c. Given $x = 8.0\%$, construct and interpret the 95% confidence and prediction intervals associated with y.

15.94 For the regression equation obtained in Exercise 15.93, analyze the residuals by (a) constructing a histogram, (b) using a normal probability plot, (c) plotting the residuals versus the x values, and (d) plotting the residuals versus the order in which they were observed. Do any of the assumptions of the simple linear regression model appear to have been violated?

15.95 Computer database **GROWCO** describes the characteristics of 100 companies identified by *Fortune* as among the fastest growing. Two of the variables listed are (1) estimated stock price/earnings ratio next year and (2) annual revenue growth (as a percentage). Source: *"Fortune's* 100 Fastest-Growing Companies," *Fortune,* September 4, 2000, pp. 142–158.

a. Determine the least-squares regression equation for y = next year's price/earning ratio estimate as a function of x = annual revenue growth. What would be the predicted value of y for a company whose annual revenue growth has been 150%?

b. What percentage of the variation in the magazine's price/earnings estimates (y) is explained by the annual revenue growth percentage (x)? At the 0.05 level, does the slope of the equation differ significantly from 0?

c. Given $x = 150\%$, construct and interpret the 95% confidence and prediction intervals associated with y.

15.96 For the regression equation obtained in Exercise 15.95, analyze the residuals by (a) constructing a histogram, (b) using a normal probability plot, (c) plotting the residuals versus the x values, and (d) plotting the residuals versus the order in which they were observed. Do any of the assumptions of the simple linear regression model appear to have been violated?

INTEGRATED CASES

Thorndike Sports Equipment

Ted Thorndike's friend Mary Stuart teaches a statistics course at the local university, and she has asked Ted to stop in and talk to the class about statistics as it relates to sports. Among the topics that Ted thinks might be interesting is a discussion of how Olympic swimming performances for both men and women have improved over the years.

According to Ted, swimmers and other athletes have been getting bigger, faster, and stronger. In addition, training technologies and equipment have improved by leaps and bounds. Among the data that Ted has collected are the winning times for men and women in the 400-meter freestyle swimming event for Olympics held from 1924 through 2008. The times are shown here and are also listed in data file **THORN15**. Source: *Time Almanac 2009*, pp. 841, 842.

1. Using the women's winning time as the dependent variable and year as the independent variable, determine and interpret both the regression equation and the coefficient of correlation. For the year 2016, determine the point estimate and the 95% prediction interval for the winning time in the women's 400-meter freestyle.
2. Repeat question 1, using the men's winning time as the dependent variable.

	400-Meter Freestyle Swimming, Winning Time for	
Year	**Women**	**Men**
1924	6.037 minutes	5.070 minutes
1928	5.713	5.027
1932	5.475	4.807
1936	5.440	4.742
1948	5.297	4.683
1952	5.202	4.512
1956	4.910	4.455
1960	4.843	4.305
1964	4.722	4.203
1968	4.530	4.150
1972	4.324	4.005
1976	4.165	3.866
1980	4.146	3.855
1984	4.118	3.854
1988	4.064	3.783
1992	4.120	3.750
1996	4.121	3.800
2000	4.097	3.676
2004	4.089	3.718
2008	4.054	3.698

3. Using the women's winning time as the dependent variable and the men's winning time as the independent variable, determine and interpret both the regression equation and the coefficient of correlation. If the men's winning time were 3.600 minutes in a given Olympics, determine the point estimate and the 95% prediction interval for the winning time in the women's event.
4. For the regression equation obtained in question 3, use a test of your choice in examining the significance of the linear relationship between the variables.

Springdale Shopping Survey

This exercise involves the following variables from data file **SHOPPING**, the Springdale Shopping Survey:

- **Variables 7–9** Overall attitude toward the shopping area. The highest possible rating is 5. In each analysis, one of these will be the dependent variable.
- **Variables 18–25** The importance (highest rating = 7) placed on each of 8 attributes that a shopping area might possess. In each analysis, one of these will be the independent variable.

Information like that gained from this database exercise could provide management with useful insights into how respondents form their overall attitude toward a shopping area. For example, management can find out whether people who place more importance on "a lot of bargain sales" tend to have a higher perception or a lower perception of a shopping area.

1. With variable 7 (attitude toward Springdale Mall) as the dependent variable, perform two separate regression analyses—one for each of the independent variables listed below. In each case, determine and interpret the regression equation and the coefficients of correlation and determination, then use the 0.05 level of significance in reaching a conclusion on the significance of the linear relationship. If possible on your computer statistical package, retain the residuals for analysis in question 2.
   a. Variable 21 (good variety of sizes/styles).
   b. Variable 25 (a lot of bargain sales).
2. For each of the regression analyses in question 1, examine the residuals by using a histogram, a normal probability plot, and a plot of the residuals against the values of the independent variable. In each case, comment on whether an assumption of the linear regression model may have been violated.
3. Repeat questions 1 and 2, using variable 9 (attitude toward West Mall) as the dependent variable.

BUSINESS CASE

Pronto Pizza (B)

In the Pronto Pizza case, visited previously in Chapter 10, the company was facing stiff competition in its pizza delivery service, and the owners were particularly concerned about being able to provide call-in customers with a guarantee as to how quickly their pizza would arrive. Manager Tony Scapelli had collected a month's worth of data that included such variables as preparation time, wait time, travel time, and delivery distance. In this chapter, we will use simple linear regression and correlation to help Tony examine the number of travel minutes required for a delivery versus the number of miles involved in making the delivery. Using the **PRONTO** data file and the variables TRAVEL_TIME and DISTANCE:

1. Determine the coefficient of correlation between TRAVEL_TIME and DISTANCE. Is the sign of the coefficient of correlation positive or negative? Is this the sign you would expect to be associated with this coefficient? What percentage of the variation in the driving time to deliver a pizza is explained by the number of miles for the trip?
2. Determine the best-fit linear regression equation for estimating TRAVEL_TIME on the basis of DISTANCE. Identify and interpret the slope of the equation in the context of this situation.
3. What would be the estimated time of travel for a pizza delivery that involved a 5-mile trip?
4. Determine and interpret the 95% confidence and prediction intervals associated with a pizza delivery that involved a 5-mile trip.
5. Were there any deliveries for which the time for travel was "flagged" by your computer statistical package as being unusually long or short compared to the time that would have been estimated by the equation? What managerial implications could this have, especially considering the possibility of litigation if a delivery person were to be involved in an accident while trying to meet a delivery guarantee set by the company?

SEEING STATISTICS: APPLET 18

Regression: Point Estimate for *y*

When this applet loads, we see the best-fit linear regression equation for the dexterity test example featured throughout this chapter. Moving the slider at the bottom allows us to select a dexterity test score and immediately see the estimated productivity for someone with that score.

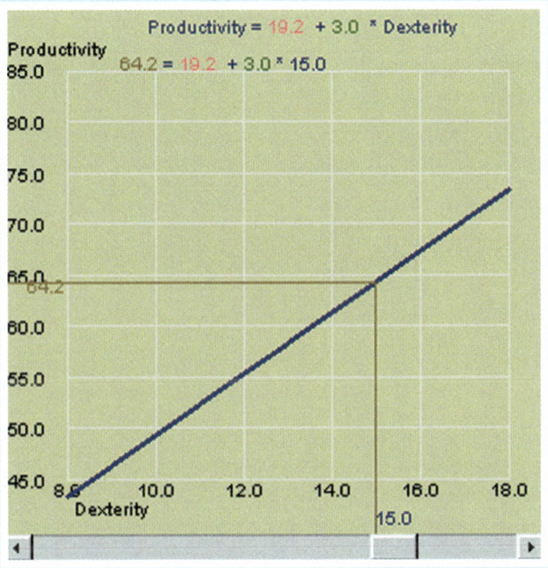

Applet Exercises

18.1 Move the slider so that the dexterity test score is 10. What is the point estimate of the productivity for someone with this score on the dexterity test?

18.2 Move the slider so that the dexterity test score is 11. What is the point estimate of the productivity for someone with this score on the dexterity test?

18.3 In Applet Exercise 18.2, we increased the dexterity test score by 1. By how much did the point estimate for productivity increase? Compare this increase with the slope of the regression equation.

18.4 By how much should the point estimate for productivity increase if we were to increase the dexterity test score by 5 points? Verify your answer by comparing the point estimates for dexterity test scores of 10 and 15.

Try out the Applets: http://www.cengage.com/bstatistics/weiers

SEEING STATISTICS: APPLET 19

Point Insertion Diagram and Correlation

This applet loads with a blank grid. By clicking the mouse, we can insert points anywhere within the grid. For the points we have chosen, the applet automatically updates the coefficient of correlation and the best-fit regression line. To start over again with a blank grid, just click on the "Reset" button.

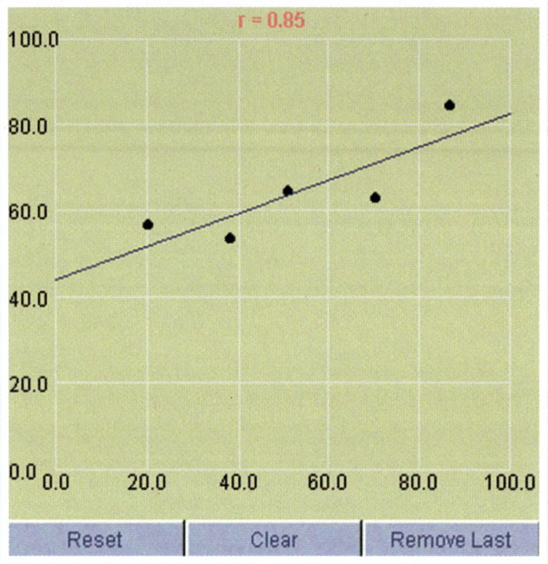

Applet Exercises

19.1 Insert any three points so that the best-fit line will slope upward. What is the sign of r?

19.2 Reset to a clear screen and insert a pattern so that r will be between $+0.7$ and $+0.9$. Does the regression line seem to be a very good fit to the points? Next, insert some additional points that will make $r < 0.7$.

19.3 Reset to a clear screen and insert two points that form an upward-sloping diagonal line. What is the value of r?

19.4 Reset to a clear screen and insert a pattern of at least 10 points so that r will be very close to -1.

19.5 Reset to a clear screen and insert a pattern of at least 10 points so that r will be very close to 0.

Try out the Applets: http://www.cengage.com/bstatistics/weiers

SEEING STATISTICS: APPLET 20

Regression Error Components

This applet loads with a scatter diagram containing the points in the dexterity test example used in this chapter. By clicking on the line, we can make it better or worse as an estimator of productivity. For each data point, the error is the vertical distance between that point and the line we have selected. In the display on the left, the sum of the squared errors is shown at the bottom and the value of r^2 is shown at the top. They are represented by the green and red rectangles, respectively. By clicking the "Find Best Model" button, we obtain the best-fit regression equation for the data.

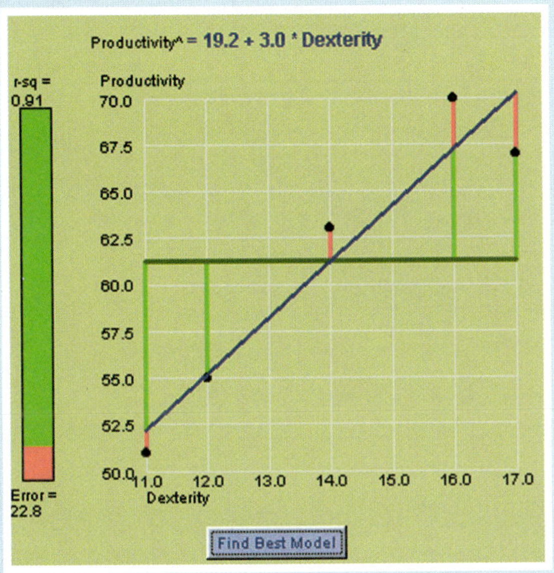

Applet Exercises

20.1 Click on the regression line and move it so that it is as close as possible to horizontal. Identify the value of r^2 and interpret its meaning.

20.2 Click on the regression line and move it so that the slope of the equation is approximately 1.0. What effect has this had on r^2?

20.3 Click on the regression line and gradually move it so that the slope of the equation increases from approximately 1.0 to approximately 5.0. What happens to r^2 as you are shifting the line?

20.4 Click on the "Find the Best Model" button, then identify and interpret both the slope of the equation and the resulting r^2 value in the context of the dexterity test example in the text.

Try out the Applets: http://www.cengage.com/bstatistics/weiers

CHAPTER 16

Multiple Regression and Correlation

ACCIDENTS DON'T TAKE HOLIDAYS OFF

Statistics often represent positive events and pleasant experiences. For example, 44.0 million people visited Walt Disney World in 2007, and Americans bought 7.6 million new cars the same year.

On the other hand, it's often necessary to use statistics to keep track of things that are not so positive. A few examples from 2006: the 119 million visits to hospital emergency rooms (leading reason: stomach and abdominal pain), the 21,200 who died as the result of falls, and the 44,700 who perished in automobile accidents.

A disproportionate share of the auto accident fatalities tend to occur during major holiday periods, such as Thanksgiving, Christmas, and New Year's. Using multiple regression, an important technique covered in this chapter, we find a strong relationship exists between (1) the number of fatalities that occur during New Year's and (2) the numbers that occurred during the preceding Thanksgiving and Christmas periods.

As we'll examine more closely in Statistics in Action 16.1, a multiple regression analysis of holiday traffic fatalities for a 23-year period revealed that nearly 80% of the variation in New Year's fatalities was explained by a linear equation linking New Year's fatalities (y) with the numbers occurring during the preceding Thanksgiving and Christmas (x_1 and x_2, respectively). Although driving during all holiday periods tends to be a little more dangerous than everyday driving, holidays during some years would appear to be worse than holidays in others. If you're reading this either before or during a major holiday, be careful out there.

Some outcomes depend on many variables.

Sources: *TEA/ERA Attraction Attendance 2007*, p. vii, via parkworld-online.com; and *World Almanac and Book of Facts 2009*, pp. 122, 183, 216.

- Explain how the scatter diagram and least-squares concepts apply to multiple regression.
- Obtain and interpret the multiple regression equation; then make point and interval estimates regarding the dependent variable.
- Interpret the value of the coefficient of multiple determination and carry out a hypothesis test for its significance.
- Construct confidence intervals and carry out hypothesis tests involving the partial regression coefficients.
- Explain both the meaning and the applicability of a dummy variable.
- Use residual analysis in examining the appropriateness of the multiple regression model and the extent to which underlying assumptions are met.

LEARNING OBJECTIVES

After reading this chapter, you should be able to:

INTRODUCTION (16.1)

In Chapter 15, we examined the linear relationship between a dependent variable (y) and just one independent variable (x). More realistically, the value of y will be related to two or more independent variables (x_1, x_2, and so on). The inclusion of more than one independent variable is the reason for the *multiple* descriptor in the chapter title. **Multiple regression** and **correlation** analysis could prove useful with variables like these:

Dependent Variable (y)	Independent Variables (x's)
Annual unit sales	Advertising expenditure Size of sales force Price of the company's product
Selling price of a home	Number of rooms Square feet of living space Size of the surrounding lot

The purpose of this chapter is to determine, interpret, and examine the strength of linear relationships between a single y and two or more x's. As in Chapter 15, regression analysis is used in determining and interpreting the linear relationship, while correlation analysis measures its strength.

In previous chapters, the basics of a technique were presented through small-scale problems for which relevant calculations could easily be performed with the pocket calculator. The calculations in multiple regression and correlation analysis are especially intensive. Computer statistical packages are readily available for reducing the number-crunching drudgery, and the computer will be our partner through much of the chapter.

We'll begin by discussing the multiple regression model and examining the sample regression equation for an example that will continue throughout the chapter. The complete Excel and Minitab printouts will be presented and interpreted; then selected segments of these printouts will be featured as they become relevant to the topics under discussion. Later in the chapter, we'll revisit the full printouts for an overview of their format and interpretation. Once again, for multiple regression and correlation, the term "computer assisted" is a distinct understatement. It's like saying a jet pilot is "aircraft assisted."

As in the preceding chapter, point and interval estimates for y are discussed, as are tests of significance involving the regression and correlation analysis itself. The basic principles are similar to their counterparts from Chapter 15. However, specific topics and methods of analysis and interpretation will differ slightly in order to accommodate the greater number of independent variables.

EXERCISES

16.1 What is the difference between simple linear regression and multiple regression? Under what circumstances would a multiple regression analysis be preferable?

16.2 In multiple regression and correlation analysis, what is the purpose of the regression component? Of the correlation component?

16.3 For y = annual household expenditure for auto maintenance and repair, what independent variables could help explain the amount of money a household spends per year for this purpose?

16.4 A personnel director is using multiple regression analysis to examine employee absence records. He wants to find an estimation equation that will improve his understanding of variables that might explain the extent of such absences. For y = the number of days an employee called in sick last year, what x variables might be both useful to this analysis and likely to be available in each employee's personnel file?

16.2 THE MULTIPLE REGRESSION MODEL

Model and Assumptions

The multiple regression model is an extension of the simple linear regression model of Chapter 15. However, there are *two or more independent variables* instead of just one. As in Chapter 15, estimates of the population parameters in the model are made on the basis of sample data. Described in terms of its population parameters, the multiple regression model is as follows:

The multiple regression model:

$$y_i = \beta_0 + \beta_1 x_{1i} + \beta_2 x_{2i} + \cdots + \beta_k x_{ki} + \epsilon_i$$

where y_i = a value of the dependent variable, y

β_0 = a constant

$x_{1i}, x_{2i}, \ldots, x_{ki}$ = values of the independent variables, $x_1, x_2, \ldots, x_k$

$\beta_1, \beta_2, \ldots, \beta_k$ = partial regression coefficients for the independent variables, $x_1, x_2, \ldots, x_k$

ϵ_i = random error, or residual

For a given set of x values, the expected value of y is given by the regression equation, $E(y) = \beta_0 + \beta_1 x_{1i} + \beta_2 x_{2i} + \cdots + \beta_k x_{ki}$. For each x term in the equation, the corresponding β term is referred to as the **partial regression coefficient.** Each β_i ($i = 1, 2, \ldots, k$) is a slope relating changes in $E(y)$ to changes in one x variable whenever all of the other x's are held constant.

As in Chapter 15, the difference between an observed value of y and the value that was expected is the residual, ϵ_i. In terms of the residual component of the model, the following assumptions underlie multiple regression:

1. For any given set of values for the independent variables, the population of ϵ_i values will be normally distributed with a mean of zero and a standard deviation of σ.
2. The standard deviation of the ϵ_i values is the same regardless of the combination of values taken on by the independent variables.
3. The ϵ_i values are statistically independent of each other.

Determining the Sample Regression Equation

The sample regression equation is based on observed values for the dependent and independent variables. An extension of its counterpart from Chapter 15, it has the form $\hat{y} = b_0 + b_1x_1 + b_2x_2 + \cdots + b_kx_k$. The constants in the sample regression equation ($b_0, b_1, b_2, \ldots, b_k$) are estimates of their population counterparts ($\beta_0, \beta_1, \beta_2, \ldots, \beta_k$).

Determination of the "best-fit" multiple regression equation is according to the least-squares criterion, in which the sum of the squared deviations between observed and estimated values of y is minimized. The following information, and analyses related to it, will be used as our medium for discussing multiple regression and correlation in the sections to follow.

EXAMPLE

Multiple Regression Equation

The president of a large chain of fast-food restaurants has randomly selected 10 franchises and recorded for each franchise the following information on last year's net profit and sales activity. The data are also in file **CX16REST**.

Franchise Number	Net Profit, y	Counter Sales, x_1	Drive-Through Sales, x_2
1	$1.5 million	$8.4 million	$7.7 million
2	0.8	3.3	4.5
3	1.2	5.8	8.4
4	1.4	10.0	7.8
5	0.2	4.7	2.4
6	0.8	7.7	4.8
7	0.6	4.5	2.5
8	1.3	8.6	3.4
9	0.4	5.9	2.0
10	0.6	6.3	4.1

For these data, there will be one dependent variable (y = net profit) and two independent variables (x_1 = counter sales; x_2 = drive-through sales). The form of the sample regression equation will be $\hat{y} = b_0 + b_1x_1 + b_2x_2$.

EXAMPLEEXAMPLEEXAMPLEEXAMPLEEXAMPLEEXAMPLEEX

TABLE 16.1

These portions of the Excel and Minitab printouts for the fast-food franchise analysis show the *y*-intercept and the partial regression coefficients for the least-squares multiple regression equation. The full printouts and the procedures for obtaining them are shown at the end of this section, in Computer Solutions 16.1.

	E	F
16		Coefficients
17	Intercept	-0.2159
18	$Counter	0.0855
19	$Driveup	0.1132

```
The regression equation is

$Profit = - 0.216 + 0.0855 $Counter + 0.113 $Driveup
```

SOLUTION

Table 16.1 shows Excel and Minitab printout segments describing the multiple regression equation for the restaurant data. The full printouts and the steps to obtain them are given in Computer Solutions 16.1, at the end of this section. In later sections of the chapter, we will examine other portions of the printouts. At this stage, our primary concern is the sample regression equation itself. To four decimal places, it is

Estimated net profit → $\hat{y}$; *y*-intercept → −0.2159; Partial regression coefficients → 0.0855 and 0.1132

$$\hat{y} = -0.2159 + 0.0855x_1 + 0.1132x_2$$

Counter sales → x_1; Drive-through sales → x_2

This least-squares regression equation is actually a two-dimensional plane, and for each possible value of x_1 and x_2, the estimated value of y lies on the surface of the plane. Imagine the data points in Figure 16.1 as helium-filled balloons in a room, with each balloon fastened to the floor at its particular (x_1, x_2) location. The length of the string holding each balloon represents the observed y value for that individual data point. The height where each string would intersect the regression plane is the estimated y value for that combination of x_1 and x_2 values. Some of the balloons (data points) will be above the plane, others below it. As in Chapter 15, the distances between observed and predicted positions are residuals, or errors of estimation. Because the least-squares criterion is also used in multiple regression, the sum of the squared residuals will be minimized.

When there are only two variables (y and x), the least-squares equation is a straight line. With (y, x_1, and x_2), it is a plane fitted to a three-dimensional scatter diagram like Figure 16.1. However, when we have more than three variables, there is no counterpart to the scatter diagram. With four or more variables, the regression equation becomes a mathematical entity called a **hyperplane.** Scatter diagrams in three dimensions are difficult to draw; with four or more dimensions, such visual summarization becomes impossible.

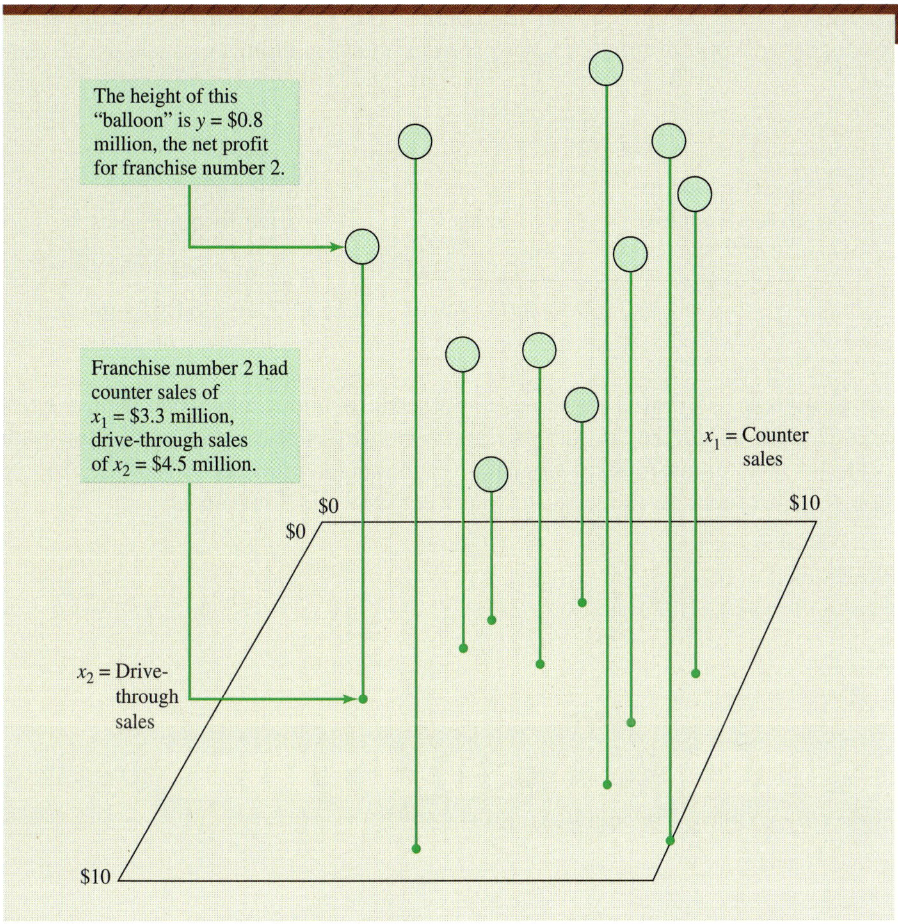

FIGURE 16.1

The scatter diagram for the fast-food data can be visualized as a room where each of 10 helium-filled balloons is held to the floor by a string. The length of each string is the observed value of *y* for that data point. The least-squares regression equation (not shown) passes through the data and takes the form of a two-dimensional surface, or plane.

Interpreting the Regression Equation

The *y*-intercept ($b_0 = -0.2159$) in the fast-food regression equation is the estimated value of y (net profit) whenever x_1 (counter sales) and x_2 (drive-through sales) are both zero. In other words, a franchise with no customers would be estimated as losing $215,900 for the year. (*Caution:* None of the 10 franchises in the study even approaches having zero sales in both departments, and we must be skeptical of estimates based on x values that lie beyond the range of the underlying data.)

The partial regression coefficient for x_1 ($b_1 = 0.0855$) indicates that, for a given level of drive-through sales, the estimated net profit will increase by $0.0855 for each additional $1.00 of counter sales.

The partial regression coefficient for x_2 ($b_2 = 0.1132$) shows that, for a given level of counter sales, the estimated net profit will increase by $0.1132 for each additional $1.00 of drive-through sales. Both partial regression coefficients are positive, reflecting that estimated net profit will increase regardless of whether an additional $1.00 of sales happens to occur over the counter or at the drive-through window.

Point Estimates Using the Regression Equation

Point estimates are made by substituting a set of x values into the regression equation and calculating the estimated value of y. For example, if a franchise had sold x_1 = \$5.0 million over the counter and x_2 = \$7.4 million at the drive-through, we would estimate its net profit as \$1.05 million:

The estimated net profit for the franchise | \$5.0 million in counter sales | \$7.4 million in drive-through sales

$$\hat{y} = -0.2159 + 0.0855(5.0) + 0.1132(7.4) = \$1.05 \text{ million}$$

Computer Solutions 16.1 shows the procedures and results when we use Excel and Minitab to perform a multiple regression analysis of the restaurant data. Both printouts contain information we will be examining in other sections of the chapter. For now, just focus on the segments that were shown in Table 16.1.

COMPUTER 16.1 SOLUTIONS

Multiple Regression

EXCEL

	E	F	G	H	I	J	K
1	SUMMARY OUTPUT						
2							
3	*Regression Statistics*						
4	Multiple R	0.8786					
5	R Square	0.7719					
6	Adjusted R Square	0.7067					
7	Standard Error	0.2419					
8	Observations	10					
9							
10	ANOVA						
11		*df*	*SS*	*MS*	*F*	*Significance F*	
12	Regression	2	1.3864	0.6932	11.8448	0.0057	
13	Residual	7	0.4096	0.0585			
14	Total	9	1.7960				
15							
16		*Coefficients*	*Standard Error*	*t Stat*	*P-value*	*Lower 95%*	*Upper 95%*
17	Intercept	-0.2159	0.2643	-0.8169	0.4409	-0.8409	0.4091
18	\$Counter	0.0855	0.0438	1.9516	0.0920	-0.0181	0.1890
19	\$Driveup	0.1132	0.0385	2.9368	0.0218	0.0220	0.2043

Excel Multiple Regression

1. The data for the primary example in this chapter are in Excel file **CX16REST**. The profits ($Profit) are in A2:A11, the counter sales ($Counter) are in B2:B11, the drive-through sales ($Driveup) are in C2:C11, and the labels are in the first rows. Note that the independent variables must be in adjacent columns. From the **Data** ribbon, click **Data Analysis.** Select **Regression.** Click **OK.**
2. Enter **A1:A11** into the **Input Y Range** box. Enter **B1:C11** into the **Input X Range** box. Select **Labels.** Enter **E1** into the **Output Range** box. Click **OK.** After slight adjustments to the column widths and the number of decimal places, the results are as shown in the printout.

MINITAB

```
Regression Analysis: $Profit versus $Counter, $Driveup

The regression equation is
$Profit = - 0.216 + 0.0855 $Counter + 0.113 $Driveup

Predictor      Coef   SE Coef      T      P
Constant    -0.2159    0.2643  -0.82  0.441
$Counter    0.08547   0.04380   1.95  0.092
$Driveup    0.11315   0.03853   2.94  0.022

S = 0.241912   R-Sq = 77.2%   R-Sq(adj) = 70.7%

Analysis of Variance
Source           DF        SS        MS      F      P
Regression        2   1.38635   0.69318  11.84  0.006
Residual Error    7   0.40965   0.05852
Total             9   1.79600

Source     DF    Seq SS
$Counter    1   0.88160
$Driveup    1   0.50475

Unusual Observations
Obs  $Counter  $Profit     Fit   SE Fit  Residual  St Resid
  8       8.6   1.3000  0.9039   0.1468    0.3961     2.06R

R denotes an observation with a large standardized residual.
```

Using Minitab file **CX16REST**, with $Profit in C1, $Counter in C2, and $Driveup in C3: Click **Stat.** Select **Regression.** Click **Regression.** Enter **C1** ($Profit) into the **Response** box. Enter **C2-C3** ($Counter and $Driveup) into the **Predictors** box. Click **OK.**

EXERCISES

16.5 Explain each of the terms in the multiple regression model.

16.6 What assumptions are required in using the multiple regression model?

16.7 In simple linear regression, the regression equation is a straight line. In multiple regression, what geometric form is taken by the regression equation when there are two independent variables? When there are three or more independent variables?

16.8 For the multiple regression equation $\hat{y} = 100 + 20x_1 - 3x_2 + 120x_3$:

a. Identify the *y*-intercept and partial regression coefficients.

b. If $x_1 = 12$, $x_2 = 5$, and $x_3 = 10$, what is the estimated value of y?
c. If x_3 were to increase by 4, what change would be necessary in x_2 in order for the estimated value of y to remain unchanged?

16.9 Last year's utility bill (in dollars) for a sample of homes is estimated by the multiple regression equation $\hat{y} = 300 + 7x_1 + 13x_2$, where x_1 is the number of occupants and x_2 is the number of rooms in the home.
a. Identify the y-intercept and partial regression coefficients.
b. If 3 people live in a 6-room home, what is the estimated bill?

16.10 A study describes operating costs for comparable manufacturing plants operating in 15 different U.S. metropolitan areas. From the costs obtained in the study, the following multiple regression equation was developed: $\hat{y} = 1.0 + 1.1x_1 + 2.8x_2$, with $\hat{y}$ = total operating cost, x_1 = labor cost, and x_2 = electric power cost. All costs are in millions of dollars and on an annual basis.
a. Interpret the y-intercept and the partial regression coefficients.
b. If labor costs \$6 million and electric power costs \$0.3 million, what is the estimated annual cost to operate the plant?

(DATA SET) *Note:* Exercises 16.11–16.14 require a computer and statistical software.

16.11 The owner of a large chain of health spas has selected eight of her smaller clubs for a test in which she varies the size of the newspaper ad and the amount of the initiation fee discount to see how this might affect the number of prospective members who visit each club during the following week. The results are shown in the table below.

Club	New Visitors, y	Ad Column-Inches, x_1	Discount Amount, x_2
1	23	4	\$100
2	30	7	20
3	20	3	40
4	26	6	25
5	20	2	50
6	18	5	30
7	17	4	25
8	31	8	80

a. Determine the least-squares multiple regression equation.
b. Interpret the y-intercept and partial regression coefficients.
c. What is the estimated number of new visitors to a club if the size of the ad is 5 column-inches and a \$75 discount is offered?

16.12 In testing 9 sedans, an automotive publication rated each on 13 different characteristics, including ride, handling, and driver comfort. Each vehicle also received an overall rating. Scores for each vehicle were as follows.

Car	y = Overall Rating	x_1 = Ride	x_2 = Handling	x_3 = Driver Comfort
1	85	8	7	7
2	88	8	8	8
3	85	6	8	7
4	85	8	7	9
5	97	9	9	9
6	86	8	8	9
7	90	9	6	9
8	84	7	8	7
9	94	8	9	8

a. Determine the least-squares multiple regression equation.
b. Interpret the y-intercept and partial regression coefficients.
c. What is the estimated overall rating for a vehicle that scores 6 on ride, 9 on handling, and 7 on driver comfort?

16.13 The makers of a pie crust mix have baked test pies and collected the following data on y = amount of crispness (a rating of 1 to 100), x_1 = minutes in the oven, and x_2 = oven temperature (degrees Fahrenheit).

y	x_1	x_2
68	6.0	460
76	8.9	430
49	8.8	360
99	7.8	460
90	7.3	390
32	5.3	360
96	8.8	420
77	9.0	350
94	8.0	450
82	8.2	400
97	6.4	450

a. Determine the least-squares multiple regression equation.
b. Interpret the y-intercept and partial regression coefficients.
c. What is the estimated crispness rating for a pie that is baked for 5 minutes at 200 degrees?

16.14 The following data have been reported for a sample of 10 major U.S. zoological parks:

City	y = Budget	x_1 = Attendance	x_2 = Acres	x_3 = Number of Species
1	\$19.5 million	0.6 million	210	271
2	40.0	2.0	216	400
3	11.9	0.4	70	377
4	14.0	1.0	125	277
5	11.6	1.5	55	721
6	22.2	1.3	80	400
7	20.5	1.3	42	437
8	26.0	2.5	91	759
9	17.0	0.9	125	270
10	14.6	1.1	92	260

a. Determine the least-squares multiple regression equation.
b. Interpret the y-intercept and partial regression coefficients.
c. What is the estimated budget for a zoological park that draws an annual attendance of 2.0 million, occupies 150 acres, and has 600 species?

16.15 During its manufacture, a product is subjected to four different tests, each giving a score of 200 to 800. An efficiency expert claims the fourth test to be unnecessary since its results can be predicted based on the first three. The following Minitab printout is an analysis of scores received by a sample of 12 units subjected to all four tests.

```
Regression Analysis

The regression equation is
TEST04 = 12.0 + 0.274 TEST01 + 0.376 TEST02 + 0.326 TEST03

Predictor        Coef       StDev          T        P
Constant        11.98       80.50       0.15    0.885
TEST01         0.2745      0.1111       2.47    0.039
TEST02        0.37619     0.09858       3.82    0.005
TEST03        0.32648     0.08084       4.04    0.004

S = 52.72        R-Sq = 87.2%      R-Sq(adj) = 82.4%

Analysis of Variance

Source       DF          SS          MS         F         P
Regression    3      151417       50472     18.16     0.001
Error         8       22231        2779
Total        11      173648
```

a. Interpret the multiple regression equation.
b. An individual unit from the assembly area receives ratings of 350, 400, and 600 on the first three tests. Using the estimation equation in the printout, what is the unit's estimated score on test 4?

INTERVAL ESTIMATION IN MULTIPLE REGRESSION 16.3

Given a specific combination of x values, the multiple regression equation provides our single best estimate for both the mean of y and an individual y value for that set of x values. However, as in simple linear regression, the estimate is subject to uncertainty. The first step in quantifying this uncertainty is to determine the multiple standard error of estimate.

The Multiple Standard Error of Estimate

The **multiple standard error of estimate (s_e)** is analogous to the standard error of estimate in Chapter 15, and it will be part of the output when the regression analysis is computer assisted. The numerical value of s_e reflects the amount of scatter, or dispersion, of data points about the plane (or hyperplane) represented by the multiple regression equation. For two independent variables, it describes the amount of scatter above and below the two-dimensional surface passing through data points like those shown in Figure 16.1. Regardless of the number of independent variables, the multiple standard error of estimate can be calculated as

The multiple standard error of estimate:

$$s_e = \sqrt{\frac{\Sigma(y_i - \hat{y}_i)^2}{n - k - 1}}$$

where y_i = each observed value of y in the data
$\hat{y}_i$ = the value of y that would have been estimated from the regression equation
n = the number of data points
k = the number of independent (x) variables

TABLE 16.2

For the sample of 10 fast-food franchises, the summary results from these calculations are important in determining both the multiple standard error of estimate and the coefficients of multiple correlation and determination. (Data are in millions of dollars.)

Counter Sales, x_1	Drive-Through Sales, x_2	Net Profit: Actual, y	Net Profit: Predicted, $\hat{y}$	Error, or Residual, $(y - \hat{y})$	Error, or Residual, Squared, $(y - \hat{y})^2$	Actual y versus the Mean of y: $(y - \bar{y})$	$(y - \bar{y})^2$
\$8.4	\$7.7	\$1.5	\$1.3739	\$0.1261	0.0159	\$0.62	0.3844
3.3	4.5	0.8	0.5757	0.2243	0.0503	−0.08	0.0064
5.8	8.4	1.2	1.2309	−0.0309	0.0010	0.32	0.1024
10.0	7.8	1.4	1.5221	−0.1221	0.0149	0.52	0.2704
4.7	2.4	0.2	0.4576	−0.2576	0.0664	−0.68	0.4624
7.7	4.8	0.8	0.9858	−0.1858	0.0345	−0.08	0.0064
4.5	2.5	0.6	0.4519	0.1481	0.0219	−0.28	0.0784
8.6	3.4	1.3	0.9043	0.3957	0.1566	0.42	0.1764
5.9	2.0	0.4	0.5150	−0.1150	0.0132	−0.48	0.2304
6.3	4.1	0.6	0.7869	−0.1869	0.0349	−0.28	0.0784
		$\bar{y}$ = \$0.88			0.4096		1.7960

$\Sigma(y - \hat{y})^2$; this is the residual (error) sum of squares, or *SSE*. It is the amount of variation in y that is *not* explained by the multiple regression equation.

$\Sigma(y - \bar{y})^2$; this is the total sum of squares, or *SST*. It is the *total* amount of variation in y, both explained and unexplained.

The numerator of the term beneath the square root symbol is the sum of the squared residuals, also known as the error sum of squares, or *SSE*. For the fast-food example and its multiple regression equation, the calculation of the error sum of squares (*SSE*) is demonstrated in Table 16.2. Each row of the table represents one of the franchises in the sample and compares the actual net profit (y) with the predicted value ($\hat{y}$). When the resulting residuals (errors) are squared and totaled, the error sum of squares is $SSE = 0.4096$.

For our example, there are $k = 2$ independent variables and $n = 10$ data points. Having calculated $\Sigma(y_i - \hat{y}_i)^2$, the error sum of squares, as $SSE = 0.4096$, the multiple standard error of estimate can be calculated as

$$s_e = \sqrt{\frac{\Sigma(y_i - \hat{y}_i)^2}{n - k - 1}} = \sqrt{\frac{0.4096}{10 - 2 - 1}} = \sqrt{0.0585} = \$0.2419 \text{ million}$$

Just as b_0, b_1, and b_2 are sample estimates of their counterparts in the "true" population regression equation, $E(y) = \beta_0 + \beta_1 x_1 + \beta_2 x_2$, s_e is our sample estimate of the actual multiple standard error of estimate, σ_e.

Confidence Interval for the Conditional Mean of *y*

In Chapter 15, the least-squares regression line passed through the point $(\bar{x}, \bar{y})$, and estimation intervals were shown to become wider when based on x values more distant from $\bar{x}$. The estimation formulas in Chapter 15 took into account the difference between x and $\bar{x}$, and increased the interval widths accordingly as x values got farther away from the mean of x.

As in simple regression, the least-squares multiple regression equation passes through the point containing the means of all the variables. (To verify this, substitute the means [$\bar{y}$ = \$0.88, $\bar{x}_1$ = \$6.52, and $\bar{x}_2$ = \$4.76] into the fast-food multiple regression equation.) Likewise, interval estimates become wider as the specified x values get farther away from their respective means.

Most computer statistical packages increase interval widths according to how far away the various x values are from their means. However, if your package does not have this capability, such calculations are nearly impossible to carry out by hand. Also, you may have only summary information (e.g., n, s_e, and the regression equation itself) from an analysis performed by someone else. In either case, all is not lost. You can still construct an *approximate* estimation interval using the methods in this section and the next.

Approximate confidence interval for the conditional mean of y:

$$\hat{y} \pm t \frac{s_e}{\sqrt{n}}$$

where $\hat{y}$ = the estimated value of y based on the set of x values provided

t = t value from the t distribution table for the desired confidence level and $df = n - k - 1$ (n = number of data points, k = number of x variables)

s_e = the multiple standard error of estimate

EXAMPLE

Confidence Interval

For franchises having counter sales of \$6.0 million and drive-through sales of \$4.0 million, what is the approximate 95% confidence interval for their average net profit?

SOLUTION

For x_1 = \$6.0 and x_2 = \$4.0, $\hat{y} = -0.2159 + 0.0855(6.0) + 0.1132(4.0)$, or \$0.7499 million. This is the midpoint of the approximate confidence interval for the mean of y for these x values. As in the previous example, the number of degrees of freedom will be $n - k - 1$, or $df = 7$. The multiple standard error of estimate is s_e = \$0.2419 million, and the t value for 95% confidence and $df = 7$ will be 2.365. The approximate confidence interval for the mean of y is

$$\hat{y} \pm t\frac{s_e}{\sqrt{n}} = 0.7499 \pm 2.365\,\frac{0.2419}{\sqrt{10}} = 0.7499 \pm 0.1809,$$

or from \$0.5690 to \$0.9308 million

We have 95% confidence that the average net profit for stores having \$6.0 million in counter sales and \$4.0 million in drive-through sales will be within the approximate confidence interval bounded by \$0.5690 and \$0.9308 million. The width of this interval is 0.9308 − 0.5690, or \$0.3618 million. Because the interval is based on x values that differ from their respective means, the exact 95% confidence interval will be wider than the \$0.3618 million calculated here.

EXAMPLEEXAMPLEEXAMPLEEXAMPLEEXAM

Prediction Interval for an Individual y Observation

As with the confidence interval, we can construct an approximate prediction interval for an individual y observation. Such intervals can be approximated by using the t distribution, the multiple standard error of estimate, and $\hat{y}$ as the point estimate for y.

Approximate prediction interval for an individual y observation:

$\hat{y} \pm ts_e$ where $\hat{y}$ = the estimated value of y based on the set of x values provided

t = t value from the t distribution table for the desired confidence level and $df = n - k - 1$

(n = number of data points, k = number of x variables)

s_e = the multiple standard error of estimate

EXAMPLEEXAMPLEEXAMPLEEXAMPLEEXAMPLEEX

EXAMPLE

Prediction Interval

The president of the fast-food chain believes that a new franchise being planned for construction in a suburban neighborhood would have experienced \$6.0 million in counter sales and \$4.0 million in drive-through sales had it been in operation during the preceding year. For these sales figures, what would be the approximate 95% prediction interval for the store's net profit?

SOLUTION

The midpoint of the approximate prediction interval will be the point estimate of net profit for the counter sales (x_1 = \$6.0 million) and drive-through sales (x_2 = \$4.0 million) specified. The point estimate is

$$\hat{y} = -0.2159 + 0.0855(6.0) + 0.1132(4.0) = \$0.7499 \text{ million}$$

For $n = 10$ data points and $k = 2$ independent variables, the number of degrees of freedom associated with t will be $df = n - k - 1$, or $df = 7$. Referring to the t distribution table, $t = \pm 2.365$ includes 95% of the area beneath the curve. The approximate 95% prediction interval is

$$\hat{y} \pm ts_e = 0.7499 \pm 2.365(0.2149) = 0.7499 \pm 0.5721,$$
$$\text{or from \$0.1778 to \$1.3220 million}$$

We have 95% confidence that this store's net profit would have been within the approximate prediction interval bounded by \$0.1778 and \$1.3220 million, an interval width of 1.3220 − 0.1778, or \$1.1442 million. Since $\bar{x}_1$ = \$6.52 million and $\bar{x}_2$ = \$4.76 million (as noted earlier), this approximate prediction interval is based on x values that differ from their respective means. As such, the exact prediction interval will be slightly wider than the \$1.1442 million calculated here.

Computer Solutions 16.2 shows the procedures and results when we use Excel and Minitab to generate confidence and prediction intervals like those obtained here. Because the values for x_1 and x_2 on which the intervals are based are not the sample means for the two independent variables, the exact intervals produced by Excel and Minitab are slightly wider than their more approximate counterparts that were determined using the methods described in this section.

COMPUTER 16.2 SOLUTIONS

Interval Estimation in Multiple Regression

EXCEL

	A	B	C	D
1	**Prediction Interval**			
2				
3			$Profit	
4	Predicted value		0.750	
5				
6	Prediction Interval			
7	Lower limit		0.146	
8	Upper limit		1.353	
9				
10	Interval Estimate of Expected Value			
11	Lower limit		0.558	
12	Upper limit		0.941	

Excel Interval Estimation in Multiple Regression

1. Using Excel file **CX16REST**, with data in the fields specified in Computer Solutions 16.1, select two convenient cells (e.g., A15:B15) and enter the specified values for the two independent variables—i.e., enter **6** into **A15** and **4** into **B15.**
2. From the **Add-Ins** ribbon, click **Data Analysis Plus.** Select **Prediction Interval.** Click **OK.**
3. Enter **A1:A11** into the **Input Y Range** box. Enter **B1:C11** into the **Input X Range** box. Select **Labels.** Enter **A15:B15** into the **Given X Range** box. Enter the desired confidence level as a decimal fraction (e.g., **0.95**) into the **Confidence Level (1 – Alpha)** box. Click **OK.**

MINITAB

Minitab Interval Estimation in Multiple Regression

```
Predicted Values for New Observations

New
Obs     Fit  SE Fit        95% CI               95% PI
  1  0.7496  0.0811  (0.5578, 0.9413)  (0.1462, 1.3529)

Values of Predictors for New Observations

New
Obs  $Counter  $Driveup
  1      6.00      4.00
```

(continued)

Using Minitab file **CX16REST**, follow the procedure in Computer Solutions 16.1, but select **Options.** Enter **6 4** (include the space between them) into the **Prediction intervals for new observations** box. Enter the desired confidence level as a percentage (e.g., **95**) into the **Confidence level** box. Click **OK.** Click **OK.** In addition to the printout in Computer Solutions 16.1, Minitab will generate the items shown here.

EXERCISES

16.16 Given the printout in Exercise 16.15, determine the approximate

a. 90% confidence interval for the mean rating on test 4 for units that have been rated at 300, 500, and 400 on the first three tests.
b. 90% prediction interval for the rating a unit will get on test 4 if its ratings are 300, 500, and 400 on the first three tests.

16.17 The regression equation $\hat{y} = 5.0 + 1.0x_1 + 2.5x_2$ has been fitted to 20 data points. The means of x_1 and x_2 are 25 and 40, respectively. The sum of the squared differences between observed and predicted values of y has been calculated as $SSE = 173.5$, and the sum of squared differences between y values and the mean of y is $SST = 520.8$. Determine the following:

a. The mean of the y values in the data.
b. The multiple standard error of estimate.
c. The approximate 95% confidence interval for the mean of y whenever $x_1 = 20$ and $x_2 = 30$.
d. The approximate 95% prediction interval for an individual y value whenever $x_1 = 20$ and $x_2 = 30$.

(DATA SET) *Note:* Exercises 16.18–16.22 require a computer and statistical software. If possible with your statistical software, determine the exact, rather than the approximate, intervals.

16.18 In testing eight value notebook computers, *PC World* reported various physical and performance measures of each. Data included overall rating, street price, PC WorldBench 4 performance score, and battery life, with the results shown here. Source: "Top 15 Notebook PCs," *PC World*, July 2003, p. 135.

Computer:	1	2	3	4
Overall Rating:	80	80	76	75
Street Price:	1024	949	1499	1478
Performance Score:	102	96	111	104
Battery Life (hours):	3.03	2.97	2.00	3.05

Computer:	5	6	7	8
Overall Rating:	75	74	72	72
Street Price:	1857	1865	1699	1298
Performance Score:	100	98	104	91
Battery Life (hours):	1.88	3.00	2.38	2.33

a. From these data, obtain the multiple regression equation estimating overall rating on the basis of the other three variables; then determine and interpret the approximate 95% confidence interval for the mean overall rating of notebook computers having a \$1000 street price, a performance score of 100, and a 2.50-hour battery life.
b. Determine the 95% prediction interval for the overall rating of an individual notebook computer with a \$1000 street price, a performance score of 100, and a 2.50-hour battery life.

16.19 In considering whether to implement a program where selected entering freshmen would be required to take a precalculus refresher course before moving on to calculus, a university's business college has given mathematical proficiency tests to a sample of entering freshmen. The college hopes to use this score and the Scholastic Aptitude Test (SAT) quantitative score in predicting how well a student would end up scoring on the standardized final exam in calculus later in his or her studies. Data are available for a sample of nine entering freshmen, and their scores are shown here.

Student:	1	2	3	4	5
Math Proficiency:	72	96	68	86	70
SAT, Quantitative:	462	545	585	580	592
Calculus Final:	71	92	72	82	74

Student:	6	7	8	9
Math Proficiency:	73	91	75	76
SAT, Quantitative:	516	638	615	596
Calculus Final:	71	100	87	81

a. From these data, obtain the multiple regression equation estimating calculus final exam score on the basis of the other two variables; then determine and interpret the approximate 90% confidence interval for the mean calculus final exam score for entering freshmen who score 70 on the math proficiency test and 500 on the quantitative portion of the SAT.
b. Determine the 90% prediction interval for the calculus final exam score for an entering freshman who scored 70 on the math proficiency test and 500 on the quantitative portion of the SAT.

16.20 For the multiple regression equation obtained in Exercise 16.11, determine the approximate
a. 95% confidence interval for the mean number of new visitors for clubs using a 5-column-inch ad and offering an $80 discount.
b. 95% prediction interval for the number of new visitors to a given club when the ad is 5 column-inches and the discount is $80.

16.21 For the multiple regression equation obtained in Exercise 16.12, determine the approximate
a. 95% confidence interval for the mean overall rating of cars that receive ratings of 8 on ride, 7 on handling, and 9 on driver comfort.
b. 95% prediction interval for the overall rating of a car that has received ratings of 8 on ride, 7 on handling, and 9 on driver comfort.

16.22 For the multiple regression equation obtained in Exercise 16.13, determine the approximate
a. 95% confidence interval for the mean crispness rating of pies that are baked 5.0 minutes at 300 degrees.
b. 95% prediction interval for the crispness rating of a pie that is baked 5.0 minutes at 300 degrees.

MULTIPLE CORRELATION ANALYSIS 16.4

The purpose of the multiple correlation analysis is to measure the strength of the relationship between the dependent (y) and the set of independent (x) variables. This strength can be measured on an overall basis by the coefficient of multiple determination.

The Coefficient of Multiple Determination

The **coefficient of multiple determination** (R^2, with $0 \le R^2 \le 1$) is the proportion of the variation in y that is explained by the multiple regression equation. Its positive square root is the *coefficient of multiple correlation* (R), a measure that is less important than its counterpart from Chapter 15.

Computation of the coefficient of multiple determination is similar to that for the coefficient of determination for simple linear regression, and R^2 can be both defined and calculated in terms of the total variation in y versus the unexplained variation in y:

The coefficient of multiple determination:

$$R^2 = 1 - \frac{\text{Error variation, which is } \textit{not} \text{ explained by the regression equation}}{\textit{Total} \text{ variation in the } y \text{ values}}$$

$$= 1 - \frac{\sum(y_i - \hat{y}_i)^2}{\sum(y_i - \bar{y})^2} \quad \text{or} \quad 1 - \frac{SSE}{SST}$$

As in Chapter 15, the total variation in y is the sum of the explained variation plus the unexplained (error) variation, or

$$\underset{(SST)}{\text{Total variation in } y \text{ values}} = \underset{(SSR)}{\text{Variation explained by the regression equation}} + \underset{(SSE)}{\text{Variation not explained by the regression equation}}$$

For the fast-food multiple regression equation, both *SST* (the sum of the squared deviations of y from its mean) and *SSE* (the sum of the squared deviations between observed and predicted values of y) were calculated in Table 16.2. Substituting $SST = 1.7960$ and $SSE = 0.4096$ into the expression above, the coefficient of multiple determination is

$$R^2 = 1 - \frac{SSE}{SST} = 1 - \frac{0.4096}{1.7960} = 1 - 0.2281 = 0.7719$$

For the 10 franchises, 77.19% of the variation in net profit is explained by the multiple regression equation. This value of R^2 is listed in the computer output segments in Table 16.3. Note the inclusion of an "adjusted" R^2 that takes into account the number of data points and the number of independent variables. The formula used in obtaining the adjusted value is shown in the footnote.[1]

The printout segments in Table 16.3 list the value of the multiple standard error of estimate (s_e) calculated earlier. They also provide the regression, error (residual), and total sums of squares, along with the number of degrees of freedom associated with each. The remainder of the analysis of variance table will be discussed in the significance tests of Section 16.5.

TABLE 16.3

These portions of the Excel and Minitab printouts for the fast-food franchise analysis include the multiple standard error of estimate and the coefficient of multiple determination. The full printouts and the procedures for obtaining them are shown in Computer Solutions 16.1.

	E	F	G
3	*Regression Statistics*		
4	Multiple R	0.8786	
5	R Square	0.7719	
6	Adjusted R Square	0.7067	
7	Standard Error	0.2419	
8	Observations	10	
9			
10	ANOVA		
11		*df*	*SS*
12	Regression	2	1.3864
13	Residual	7	0.4096
14	Total	9	1.7960

```
S = 0.241912   R-Sq = 77.2%   R-Sq(adj) = 70.7%

Analysis of Variance

Source          DF       SS        MS      F      P
Regression       2  1.38635   0.69318  11.84  0.006
Residual Error   7  0.40965   0.05852
Total            9  1.79600
```

[1]Expressed as a proportion, the adjusted R^2 typically included in computer printouts of multiple regression analyses is calculated as:

$$\text{adjusted } R^2 = 1 - \frac{SSE/(n - k - 1)}{SST/(n - 1)}$$

EXERCISES

16.23 What is the coefficient of multiple determination, and what does it tell us about the relationship between y and the independent variables?

16.24 Among the results of a multiple regression analysis are the following sum-of-squares terms: *SST*, *SSR*, and *SSE*. What does each term represent, and how do the terms contribute to our understanding of the relationship between y and the set of independent variables?

16.25 Given the printout in Exercise 16.15, identify the coefficient of multiple determination and interpret its value in terms of that exercise.

(**DATA SET**) *Note:* Exercises 16.26 and 16.27 require a computer and statistical software.

16.26 For the multiple regression equation obtained in Exercise 16.12, what is the coefficient of multiple determination and exactly what does it mean?

16.27 For the multiple regression equation obtained in Exercise 16.11, what is the coefficient of multiple determination and exactly what does it mean?

SIGNIFICANCE TESTS IN MULTIPLE REGRESSION AND CORRELATION (16.5)

In the simple linear regression ($\hat{y} = b_0 + b_1x$) of Chapter 15, several tests were available regarding the relationship between the variables, but their results were the same. With just one independent variable, concluding that the population slope could be zero was equivalent to concluding that the population coefficient of correlation could be zero. However, in multiple regression there are two or more independent variables, and it is necessary to separately test (1) the overall significance of the multiple regression equation and (2) each of the partial regression coefficients in the equation.

Testing the Significance of the Regression Equation

The multiple regression equation, $\hat{y} = b_0 + b_1x_1 + b_2x_2 + \cdots + b_kx_k$, is based on sample data and is an estimate of the (unknown) population multiple regression equation, $E(y) = \beta_0 + \beta_1x_1 + \beta_2x_2 + \cdots + \beta_kx_k$. If there really is no relationship between y and any of the x variables, all of the partial regression coefficients ($\beta_1, \beta_2, \ldots, \beta_k$) in the population regression equation will be zero, and this is the basis on which we test the overall significance of our regression equation.

The test uses analysis of variance in examining the "explained" or regression sum of squares (*SSR*) versus the "unexplained" or error sum of squares (*SSE*). Some computer statistical packages carry out the entire test and report a p-value for its significance. Others may provide summary information (e.g., *SSR*, *SSE*, *SST*, and their degrees of freedom) that can be used in making the final calculations and interpreting the results. In any event, the procedure is as follows.

ANOVA test for the overall significance of the multiple regression equation:

- **Null hypothesis**

  H_0: $\beta_1 = \beta_2 = \cdots = \beta_k = 0$ **The regression equation is not significant.**

(*continued on page 618*)

- **Alternative hypothesis**

  H_1: One or more of the β_i ($i = 1, 2, \ldots, k$) $\neq 0$ The regression equation is significant.

- **Test statistic**

  $$F = \frac{SSR/k}{SSE/(n - k - 1)}$$

  where SSR = regression sum of squares
  SSE = error sum of squares
  n = number of data points
  k = number of independent (x) variables

- **Critical value of the F statistic**

  F for significance level specified and df (numerator) = k, and df (denominator) = $n - k - 1$

In the ANOVA test, the number of degrees of freedom associated with the numerator of the F statistic will be k, the number of independent variables used in estimating y. For the denominator, $df = n - k - 1$, or $n - (k + 1)$, since $(k + 1)$ quantities (b_0, along with $b_1, b_2, \ldots, b_k$) have been determined on the basis of sample data.

For the fast-food franchise data, we have calculated $SST = 1.7960$ and $SSE = 0.4096$. Since $SST = SSR + SSE$, SSR will be $1.7960 - 0.4096$, or $SSR = 1.3864$. There are $n = 10$ data points and $k = 2$ independent variables in our analysis, so the test statistic can be calculated as

$$F = \frac{SSR/k}{SSE/(n - k - 1)} = \frac{1.3864/2}{0.4096/(10 - 2 - 1)} = \frac{0.6932}{0.0585} = 11.85$$

Using the F distribution tables to identify the critical value for F, the number of degrees of freedom for the numerator will be $k = 2$; df for the denominator will be $(n - k - 1)$, or $(10 - 2 - 1) = 7$. Conducting the test at the 0.01 level of significance, the critical value is $F = 9.55$. The calculated test statistic ($F = 11.85$) exceeds the critical value, and H_0 is rejected. At the 0.01 level, the multiple regression equation is significant.

Table 16.4 shows the Excel and Minitab printout segments applicable to this test. The two printouts show the p-value for the overall significance of the equation as 0.0057 and 0.006, respectively.

Testing the Partial Regression Coefficients

In testing the significance of each partial regression coefficient (b_1, b_2, and so on), the null hypothesis is that the corresponding population value is really zero—i.e., that the observed value differs from zero merely by chance. The test is a two-tail t-test with the following format.

***t*-test for the significance of a partial regression coefficient, b_i:**

- **Null hypothesis**

  H_0: $\beta_i = 0$ The population coefficient is 0.

- **Alternative hypothesis**

  H_1: $\beta_i \neq 0$ The population coefficient is not 0.

- **Test statistic**

  $$t = \frac{b_i - 0}{s_{b_i}}$$

  where b_i = the observed value of the regression coefficient
  s_{b_i} = the estimated standard deviation of b_i

TABLE 16.4

EXCEL

	E	F	G	H	I	J	K
10	ANOVA						
11		*df*	*SS*	*MS*	*F*	*Significance F*	
12	Regression	2	1.3864	0.6932	11.8448	0.0057	
13	Residual	7	0.4096	0.0585			
14	Total	9	1.7960				
15							
16		*Coefficients*	*Standard Error*	*t Stat*	*P-value*	*Lower 95%*	*Upper 95%*
17	Intercept	-0.2159	0.2643	-0.8169	0.4409	-0.8409	0.4091
18	$Counter	0.0855	0.0438	1.9516	0.0920	-0.0181	0.1890
19	$Driveup	0.1132	0.0385	2.9368	0.0218	0.0220	0.2043

MINITAB

```
Predictor      Coef   SE Coef      T      P
Constant    -0.2159    0.2643  -0.82  0.441
$Counter    0.08547   0.04380   1.95  0.092
$Driveup    0.11315   0.03853   2.94  0.022

S = 0.241912    R-Sq = 77.2%    R-Sq(adj) = 70.7%

Analysis of Variance

Source          DF       SS       MS      F      P
Regression       2  1.38635  0.69318  11.84  0.006
Residual Error   7  0.40965  0.05852
Total            9  1.79600
```

These portions of the Excel and Minitab printouts for the fast-food franchise data show the results of tests for (1) the overall significance of the regression equation and (2) the significance of each of the partial regression coefficients. The full printouts and the procedures for obtaining them are shown in Computer Solutions 16.1.

NOTES

1. We can also test H_0: $\beta_i = \beta_{i0}$ versus H_1: $\beta_i \neq \beta_{i0}$ where β_{i0} is any value of interest. With the exception of β_{i0} replacing zero, the test is the same as described here.
2. Critical values of t are $\pm t$ for the level of significance desired and $df = n - k - 1$ (n = number of data points, k = number of independent variables).

The printout segments in Table 16.4 list the value and the standard deviation for each partial regression coefficient in the fast-food regression equation. In demonstrating these tests, we will use the number of decimal places shown in the Minitab portion of Table 16.4, and the 0.05 level of significance.

1. Test of the null hypothesis, H_0: $\beta_1 = 0$ versus H_1: $\beta_1 \neq 0$.

$$\text{Calculated test statistic, } t = \frac{b_1 - 0}{s_{b_1}} = \frac{0.08547 - 0}{0.04380} = 1.95$$

For $df = n - k - 1$, or $(10 - 2 - 1) = 7$, critical values for the 0.05 level of significance are $t = -2.365$ and $t = +2.365$. The calculated t (1.95) falls within these limits, and we are unable to reject the null hypothesis that $\beta_1 = 0$. At this level, $b_1 = 0.08547$ does not differ significantly from zero.

Had the test been done at the 0.10 level (with critical values of $t = -1.895$ and $t = +1.895$), the conclusion would be reversed, and the null hypothesis would have been rejected.

From our t distribution table, the most accurate statement we can make about the exact p-value for the test is that it is between 0.05 and 0.10. As shown in Table 16.4, the actual p-value is 0.092. If there were no linear relationship between profit and counter sales (i.e., $\beta_1 = 0$), there would be just a 0.092 probability of obtaining a sample b_1 this different from zero.

2. Test of the null hypothesis, H_0: $\beta_2 = 0$, versus H_1: $\beta_2 \neq 0$.

$$\text{Calculated test statistic, } t = \frac{b_2 - 0}{s_{b_2}} = \frac{0.11315 - 0}{0.03853} = 2.94$$

As with the preceding test, $df = n - k - 1$, or 7, and critical values for the 0.05 level of significance are $t = -2.365$ and $t = +2.365$. The calculated t exceeds these limits, and we are able to reject the null hypothesis that $\beta_2 = 0$. At this level, $b_2 = 0.11315$ differs significantly from zero. Had the test been done at the more demanding 0.02 level (having critical t values of $t = -2.998$ and $t = +2.998$), the conclusion would have been reversed, and the null hypothesis would not have been rejected.

From the t distribution table, the most accurate statement we can make about the exact p-value for this test is that it is between 0.05 and 0.02. As shown in Table 16.4, the actual p-value is 0.022. If there were no linear relationship between profit and drive-through sales (i.e., $\beta_2 = 0$), there would be just a 0.022 chance of obtaining a sample b_2 this different from zero.

We have just duplicated the procedure used by Excel and Minitab to test each of the partial regression coefficients. As expected, our results are consistent with those produced by the computer. The printouts in Table 16.4 go beyond our text discussion by including a t-test and p-value for the constant in the regression equation. The y-intercept, $b_0 = -0.2159$ does not differ very significantly from zero (p-value $= 0.441$). As noted in Chapter 15, testing the y-intercept is generally not of practical importance.

Interval Estimation for the Partial Regression Coefficients

In addition to playing a role in significance testing, the estimated standard deviation of b_i is used in constructing a confidence interval for its population counterpart β_i. The method, which is similar to the procedure used in Chapter 15, is described below.

Confidence interval for a partial regression coefficient, β_i:

The interval is

$$b_i \pm t(s_{b_i})$$

where b_i = partial regression coefficient in the sample regression equation

t = t value for the confidence level desired, and $df = n - k - 1$

s_{b_i} = estimated standard deviation of b_i

Both b_i and the estimated standard deviation of b_i will typically be included in the printout whenever the regression analysis is done with the assistance of a computer statistical package. Applying these to our fast-food example, we will construct the 95% confidence intervals for β_1 and β_2.

1. **95% confidence interval for β_1** Referring to the Minitab portion of Table 16.4, b_1 (the partial regression coefficient for x_1, or counter sales) is 0.08547, and the standard deviation of b_1 is 0.04380. There are $n = 10$ observations in the data and $k = 2$ independent variables. For $df = n - k - 1$, or 7 degrees of freedom, the value of t for a 95% confidence interval is $t = 2.365$. The confidence interval will be

$$b_1 \pm t(s_{b_1}) = 0.08547 \pm 2.365(0.04380)$$
$$= 0.08547 \pm 0.10359, \quad \text{or from } -0.01812 \text{ to } 0.18906$$

We have 95% confidence that β_1 is somewhere from -0.01812 to 0.18906. Note that $\beta_1 = 0$ falls within the interval. This is consistent with our earlier finding that, at the 0.05 level, $b_1 = 0.08547$ is not significantly different from zero. As we've seen before, an x% confidence interval and a nondirectional significance test at the $(100 - x)/100$ level will give the same conclusion regarding a hypothesized value for a population parameter.

2. **95% confidence interval for β_2** Referring to the Minitab portion of Table 16.4, b_2 (associated with drive-through sales) is 0.11315, and the standard deviation of b_2 is 0.03853. With $n = 10$ observations and $k = 2$ independent variables, $df = n - k - 1$, or 7 degrees of freedom, and the value of t for the 95% confidence interval will be $t = 2.365$. The confidence interval is

$$b_2 \pm t(s_{b_2}) = 0.11315 \pm 2.365(0.03853)$$
$$= 0.11315 \pm 0.09112, \quad \text{or from } 0.02203 \text{ to } 0.20427$$

We have 95% confidence that β_2 is somewhere from 0.02203 to 0.20427. Note that $\beta_2 = 0$ does not fall within the interval. This is consistent with our earlier finding that, at the 0.05 level, $b_2 = 0.11315$ differs significantly from zero.

If you're using Excel, confidence intervals for the partial regression coefficients are included in the output, as shown in Table 16.4. Unless you specify otherwise, 95% confidence intervals are provided. In the section exercises that follow, Excel users can readily obtain many of the requested confidence intervals without the need for further calculations. If you wish, you can use a pocket calculator and the techniques in this section to verify the confidence intervals your Excel spreadsheet package has so thoughtfully carried out for you.

EXERCISES

16.28 What is the relationship between (a) the results of the hypothesis test examining whether b_3 differs significantly from zero at the 0.02 level and (b) the 98% confidence interval for β_3?

16.29 For the multiple regression equation obtained in Exercise 16.11:

a. At the 0.05 level, is the overall regression equation significant?
b. Use the 0.05 level in concluding whether each partial regression coefficient differs significantly from zero.
c. Interpret the results of the preceding tests in the context of the variables described in that exercise.
d. Construct a 95% confidence interval for each partial regression coefficient in the population regression equation.

16.30 For the multiple regression equation obtained in Exercise 16.12:

a. At the 0.05 level, is the overall regression equation significant?
b. Use the 0.05 level in concluding whether each partial regression coefficient differs significantly from zero.
c. Interpret the results of the preceding tests in the context of the variables described in that exercise.
d. Construct a 95% confidence interval for each partial regression coefficient in the population regression equation.

16.31 Given the printout in Exercise 16.15, (a) determine the 90% confidence interval for each partial regression coefficient, and (b) interpret each significance test in the context of that exercise.

(DATA SET) *Note:* Exercises 16.32–16.35 require a computer and statistical software.

16.32 Referring to the least-squares regression equation and printout obtained in Exercise 16.18:

a. At the 0.10 level, is the overall regression equation significant?
b. Use the 0.10 level in concluding whether each partial regression coefficient differs significantly from zero.

16.33 Referring to the least-squares regression equation and printout obtained in Exercise 16.19:

a. At the 0.05 level, is the overall regression equation significant?
b. Use the 0.05 level in concluding whether each partial regression coefficient differs significantly from zero.

16.34 The computer database **GROWCO** describes the characteristics of 100 companies identified by *Fortune* as among the fastest growing. Three of the variables listed are (1) estimated stock price/earnings ratio next year, (2) annual revenue growth (as a percentage), and (3) annual earnings per share growth (as a percentage). Source: "*Fortune's* 100 Fastest-Growing Companies," *Fortune*, September 4, 2000, pp. 142–158.

a. Determine the least-squares regression equation for y = next year's price/earnings ratio estimate as a function of x_1 = annual revenue growth and x_2 = annual earnings per share growth. Interpret each of the partial regression coefficients.
b. At the 0.05 level, is the overall regression equation significant?
c. Use the 0.05 level in concluding whether each partial regression coefficient differs significantly from zero.
d. Construct a 95% confidence interval for each partial regression coefficient in the population regression equation.

16.35 Information describing 98 of the top automobile dealership groups in the United States is provided in computer database file **DEALERS**. For each group, the data include (1) total group revenue during 1999, (2) the number of retail vehicles sold that year, and (3) the number of dealerships in the group. Source: *Automotive News*, May 1, 2000, pp. 54–60.

a. Determine the least-squares regression equation for y = total group revenue as a function of x_1 = number of retail vehicles sold and x_2 = number of dealers in the group. Interpret each of the partial regression coefficients.
b. At the 0.02 level, is the overall regression equation significant?
c. Use the 0.02 level in concluding whether each partial regression coefficient differs significantly from zero.
d. Construct a 98% confidence interval for each partial regression coefficient in the population regression equation.

16.6 OVERVIEW OF THE COMPUTER ANALYSIS AND INTERPRETATION

A Summary of the Results

In the preceding sections, it was useful to segment the Excel and Minitab printouts and discuss each portion as it became relevant. We'll now reverse the process and put the pieces back together.

Table 16.5 is the Excel multiple regression printout for the fast-food example. In addition to the standard Excel printout shown in Computer Solutions 16.1, it includes the confidence and prediction interval estimates that were generated in Computer Solutions 16.2.

The Minitab printout in Table 16.6 includes the segments in Computer Solutions 16.1 and 16.2, plus two standard Minitab items we have not discussed. The first one is the "Unusual Observations" comment at the bottom, where franchise number 8 is identified as a data point that is especially distant from the regression equation. We'll expand on this shortly.

TABLE 16.5

Excel printout for the multiple regression analysis of the fast-food franchise data. In addition to the standard Excel printout from Computer Solutions 16.1, it includes the confidence and prediction intervals generated in Computer Solutions 16.2.

	E	F	G	H	I	J	K
1	SUMMARY OUTPUT			**Prediction Interval**			
2						$Profit	
3	*Regression Statistics*			**Predicted value**		0.750	
4	Multiple R	0.8786		**Prediction Interval**			
5	R Square	0.7719		Lower limit		0.146	
6	Adjusted R Square	0.7067		Upper limit		1.353	
7	Standard Error	0.2419		**Interval Estimate of Expected Value**			
8	Observations	10		Lower limit		0.558	
9				Upper limit		0.941	
10	ANOVA						
11		*df*	*SS*	*MS*	*F*	*Significance F*	
12	Regression	2	1.3864	0.6932	11.8448	0.0057	
13	Residual	7	0.4096	0.0585			
14	Total	9	1.7960				
15							
16		*Coefficients*	*Standard Error*	*t Stat*	*P-value*	*Lower 95%*	*Upper 95%*
17	Intercept	-0.2159	0.2643	-0.8169	0.4409	-0.8409	0.4091
18	$Counter	0.0855	0.0438	1.9516	0.0920	-0.0181	0.1890
19	$Driveup	0.1132	0.0385	2.9368	0.0218	0.0220	0.2043

TABLE 16.6

Minitab printout for the multiple regression analysis of the fast-food franchise data. It includes the standard Minitab printout in Computer Solutions 16.1 plus the confidence and prediction intervals in Computer Solutions 16.2.

```
Regression Analysis: $Profit versus $Counter, $Driveup

The regression equation is
$Profit = - 0.216 + 0.0855 $Counter + 0.113 $Driveup

Predictor      Coef    SE Coef       T      P
Constant    -0.2159    0.2643    -0.82  0.441
$Counter    0.08547    0.04380    1.95  0.092
$Driveup    0.11315    0.03853    2.94  0.022

S = 0.241912    R-Sq = 77.2%    R-Sq(adj) = 70.7%

Analysis of Variance
Source          DF       SS        MS       F      P
Regression       2  1.38635   0.69318  11.84  0.006
Residual Error   7  0.40965   0.05852
Total            9  1.79600

Source     DF   Seq SS
$Counter    1  0.88160
$Driveup    1  0.50475

Unusual Observations
Obs  $Counter  $Profit     Fit  SE Fit  Residual  St Resid
  8       8.6   1.3000  0.9039  0.1468    0.3961     2.06R
R denotes an observation with a large standardized residual.

Predicted Values for New Observations
New
Obs     Fit  SE Fit          95% CI              95% PI
  1  0.7496  0.0811  (0.5578, 0.9413)  (0.1462, 1.3529)

Values of Predictors for New Observations
New
Obs  $Counter  $Driveup
  1      6.00      4.00
```

The other portion of Table 16.6 that has not been previously discussed is the information in the "SEQ SS" column near the bottom. In this section, Minitab shows how much the error sum of squares (*SSE*) is reduced by each x variable when they are sequentially introduced into the regression equation. The first row in the "SEQ SS" column indicates that, compared to an "equation" in which only a constant is used as a predictor for y, the inclusion of x_1 reduces *SSE* by 0.88160. Once x_1 has been included, bringing x_2 into the regression equation reduces *SSE* by another 0.50475. Note that $0.88160 + 0.50475 = 1.38635$, the regression (or explained) sum of squares.

In describing our overall conclusions for the fast-food franchise regression analysis, we will use the number of decimal places shown in Table 16.5. This is to avoid confusion, since the two printouts do not always present findings to the same number of decimal places.

- **Dependent and independent variables**

  y = net profit x_1 = counter sales x_2 = drive-through sales

- **Multiple regression equation**

$$\hat{y} = -0.2159 + 0.0855x_1 + 0.1132x_2$$

- **Partial regression coefficients**

$$b_1 = 0.0855 \quad \text{and} \quad b_2 = 0.1132$$

  - For a given level of drive-through sales, another \$1.00 of counter sales increases the estimated net profit by \$0.0855.
  - For a given level of counter sales, another \$1.00 of drive-through sales increases the estimated net profit by \$0.1132.

- **Coefficient of multiple determination**

$$R^2 = 0.7719$$

  Changes in x_1 and x_2 explain 77.19% of the variation in y.

- **Significance test for the overall regression equation** The regression equation is quite significant. For the analysis of variance test of H_0: $\beta_1 = \beta_2 = 0$, the p-value is just 0.0057.

- **Significance of the individual partial regression coefficients**
  - b_1 (0.0855) differs from 0 at the p-value = 0.0920 level of significance
  - b_2 (0.1132) differs from 0 at the p-value = 0.0218 level of significance

- **Confidence intervals for the partial regression coefficients** These are provided by Excel but not by Minitab. We have 95% confidence that β_1 is between −0.0181 and 0.1890, and that β_2 is between 0.0220 and 0.2043.

- **Confidence and prediction intervals for the dependent variable** These intervals are associated with x_1 = \$6 million in counter sales and x_2 = \$4 million in drive-through sales. When x_1 = \$6 million in counter sales and x_2 = \$4 million in drive-through sales, the point estimate for y = profit is \$0.750 million. The corresponding interval estimates are:
  - Confidence interval: For all franchises having x_1 = \$6 million in counter sales and x_2 = \$4 million in drive-through sales, we are 95% confident that *the mean profit for all such franchises* is between \$0.558 and \$0.941 million.
  - Prediction interval: For any individual franchise having x_1 = \$6 million in counter sales and x_2 = \$4 million in drive-through sales, we are 95% confident that *the profit for this single franchise* is between \$0.146 and \$1.353 million.

STATISTICS IN ACTION 16.1

Predicting Holiday Traffic Fatalities

Among other data, the National Highway Traffic Safety Administration (NHTSA) reports the number of U.S. traffic fatalities during holiday periods. To see if we might be able to predict New Year's fatalities based on the number during each of the two major holidays preceding it, we used NHTSA figures and multiple regression to analyze Thanksgiving, Christmas, and New Year's fatalities during the period 1983–2005. A portion of the Minitab printout is shown.

```
Regression Analysis: New_Yrs versus Thanksgiving, Christmas

The regression equation is
New_Yrs = - 47 + 0.291 Thanksgiving + 0.690 Christmas

Predictor         Coef   SE Coef       T      P
Constant         -46.6     140.2   -0.33  0.743
Thanksgiving    0.2912    0.2451    1.19  0.249
Christmas      0.69022   0.08054    8.57  0.000

S = 55.0154   R-Sq = 79.4%   R-Sq(adj) = 77.4%

Analysis of Variance
Source          DF      SS      MS      F      P
Regression       2  233898  116949  38.64  0.000
Residual Error  20   60534    3027
Total           22  294432
```

The overall regression equation is highly significant (the p-value, rounded, is 0.000), and x_2 = Christmas fatalities (p-value = 0.000) is easily the better of the two independent variables in predicting the number that will occur during the New Year's holiday. Over this time period, 79.4% of the variation in New Year's fatalities is explained by the regression equation.

Source: U.S. National Highway Traffic Safety Administration, *Traffic Safety Facts 2006*, p. 33.

Statistics in Action 16.1, mentioned in this chapter's opening vignette, describes a multiple regression analysis in which traffic fatalities during the New Year's holiday period are estimated based on the number of fatalities during the Thanksgiving and Christmas periods. The Minitab printout is also included, and it is similar in appearance to the one we've discussed for the fast-food franchise data.

Residual Analysis

Like most computer statistical software, Excel and Minitab give you an optional summary of the residuals, as shown in Table 16.7. As discussed in Chapter 15, this information can be used in finding data points that are especially distant from the regression equation as well as in identifying residual "patterns" that suggest our model may not be valid.

Refer to the Minitab portion of Table 16.7. For each of the 10 data points, observed and estimated ("fitted") values of y are printed, along with their difference (the residual). Each residual has been "standardized," or expressed in terms

TABLE 16.7

Using the procedures in Computer Solutions 16.3, we can obtain a listing of the residuals like the Excel and Minitab printouts shown here. Minitab and Excel differ slightly in their method of determining the values of the standardized residuals.

EXCEL

	E	F	G	H
23	RESIDUAL OUTPUT			
24				
25	*Observation*	*Predicted $Profit*	*Residuals*	*Standard Residuals*
26	1	1.373	0.127	0.594
27	2	0.575	0.225	1.053
28	3	1.230	-0.030	-0.142
29	4	1.521	-0.121	-0.569
30	5	0.457	-0.257	-1.206
31	6	0.985	-0.185	-0.869
32	7	0.452	0.148	0.696
33	8	0.904	0.396	1.857
34	9	0.515	-0.115	-0.538
35	10	0.787	-0.187	-0.874

MINITAB

Obs	$Counter	$Profit	Fit	SE Fit	Residual	St Resid
1	8.4	1.5000	1.3734	0.1279	0.1266	0.62
2	3.3	0.8000	0.5754	0.1564	0.2246	1.22
3	5.8	1.2000	1.2303	0.1756	-0.0303	-0.18
4	10.0	1.4000	1.5214	0.1592	-0.1214	-0.67
5	4.7	0.2000	0.4574	0.1157	-0.2574	-1.21
6	7.7	0.8000	0.9854	0.0919	-0.1854	-0.83
7	4.5	0.6000	0.4516	0.1172	0.1484	0.70
8	8.6	1.3000	0.9039	0.1468	0.3961	2.06R
9	5.9	0.4000	0.5147	0.1228	-0.1147	-0.55
10	6.3	0.6000	0.7865	0.0797	-0.1865	-0.82

R denotes an observation with a large standardized residual.

of the number of standard deviation multiples it represents. The standard deviations for the estimated y values, listed in the "SE Fit" column, differ for the various rows. This is because estimates become more uncertain when they are based on x_1 and x_2 values that are farther away from their respective means.

Franchise number 8 has been "flagged" as an outlier because its standardized residual is more than 2 standard deviation multiples away from zero. The net profit for this franchise was considerably higher than estimated, and the franchise may have some unique characteristic that explains its strong performance versus the regression equation. Perhaps the manager is an excellent cost cutter or employee motivator, or the franchise may have a very attractive location, such as near an office building or high school.

Among many other possibilities for residual analysis, we can (1) construct a histogram of the residuals as a rough check to see whether they are approximately normally distributed, (2) examine a normal probability plot to check for normality, (3) plot the residuals versus the predicted values of y, (4) plot the residuals versus

COMPUTER 16.3 SOLUTIONS

Residual Analysis in Multiple Regression

EXCEL

Excel Residual Analysis in Multiple Regression

Follow the procedure in Computer Solutions 16.1, but in step 2 select all four items in the **Residuals** section and select **Normal Probability Plots.** The tabular results will be as shown in the Excel portion of Table 16.7. Three of the resulting Excel residual analysis plots are shown in Figure 16.2.

MINITAB

Minitab Residual Analysis in Multiple Regression

With two exceptions, the procedure is as described in Computer Solutions 16.1. (1) To get the printout of residuals shown in the Minitab portion of Table 16.7, click **Results,** then select **In addition, the full table of fits and residuals.** (2) To get the plots shown in Figure 16.3 and discussed in the text: Click **Graphs.** In the **Residuals for plots** menu, select **Regular.** In the **Residual Plots** menu, select all of the options. Enter **C2 C3** (include the space between them) into the **Residuals versus the variables** box.

the order in which the observations are arranged, (5) plot the residuals versus x_1 to see whether they exhibit some kind of cycle or pattern with respect to this variable, and (6) repeat plot 5 for each of the other independent variables.

In testing for normality, we can also apply the chi-square test that was introduced in Chapter 13 (n must be ≥ 32) or the Kolmogorov-Smirnov test that was introduced in Chapter 14. In possibility number 4 of the preceding paragraph, we are concerned about the possible presence of *autocorrelation,* in which the value of a residual is related to the values of those that precede or follow it. The topic of autocorrelation is especially relevant to regression models fitted to a time series, and is discussed further in Chapter 18.

Computer Solutions 16.3 shows the Excel and Minitab procedures for getting residuals listings like those in Table 16.7, as well as the procedures involved in generating Excel and Minitab residual analysis plots like those in Figures 16.2 and 16.3 (pages 628–630). The Minitab residual analysis plots in Figure 16.3 include the following features:

- *Histogram of residuals.* There are only 10 franchises. Given this histogram, it looks like the 10 residuals could have come from a normal distribution with a mean of 0.
- *Normal probability plot.* The normal probability plot is not terribly nonlinear, again supporting the assumption that the residuals could have come from a normal distribution with a mean of 0.
- *Residuals versus predicted y values.* The residuals seem to be somewhat randomly scattered, so it does not look like there is any relationship between the residuals and the predicted profits for the 10 franchises.
- *Residuals versus order of the observations.* There seems to be a bit of a U shape here. It would appear that the order in which the franchises are listed

FIGURE 16.2

Several of the plots that Excel provides for analysis of the residuals, generated using the procedures in Computer Solutions 16.3.

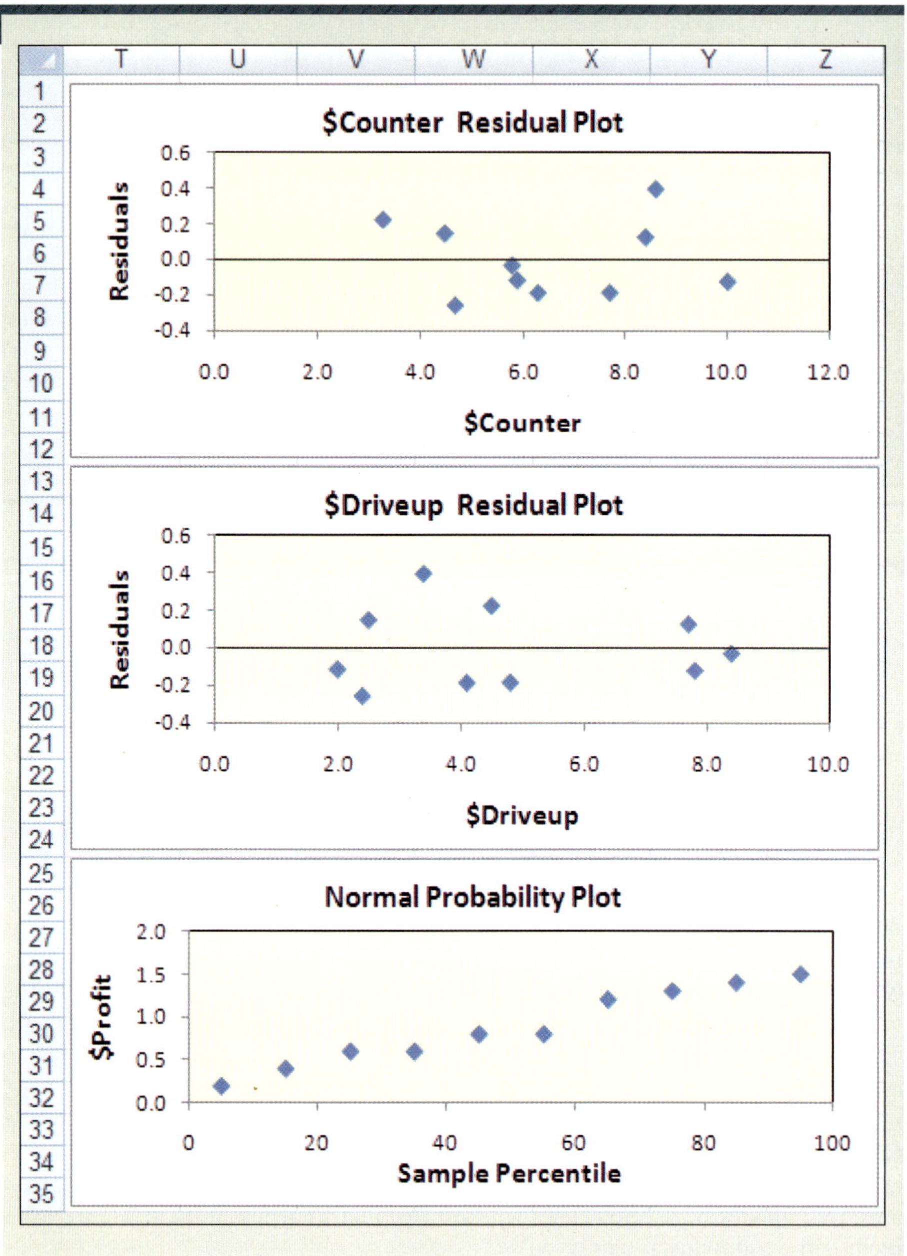

in the original data may have something to do with whether they have made (a) a higher profit than the equation predicts or (b) a lower profit than the equation predicts. For example, franchise #5 has "underperformed," earning a much lower profit ($0.2 million) than the $0.4574 million that the equation predicts, a residual of −$0.2574 million.

- *Residuals versus* x_1. This plot looks somewhat like a U-shaped curve, too. The residuals might not be independent of the level of counter sales.
- *Residuals versus* x_2. Although the residuals seem to be a bit more scattered for franchises with lower levels of drive-through sales, there are not very many data points, and this slight pattern could have happened by chance.

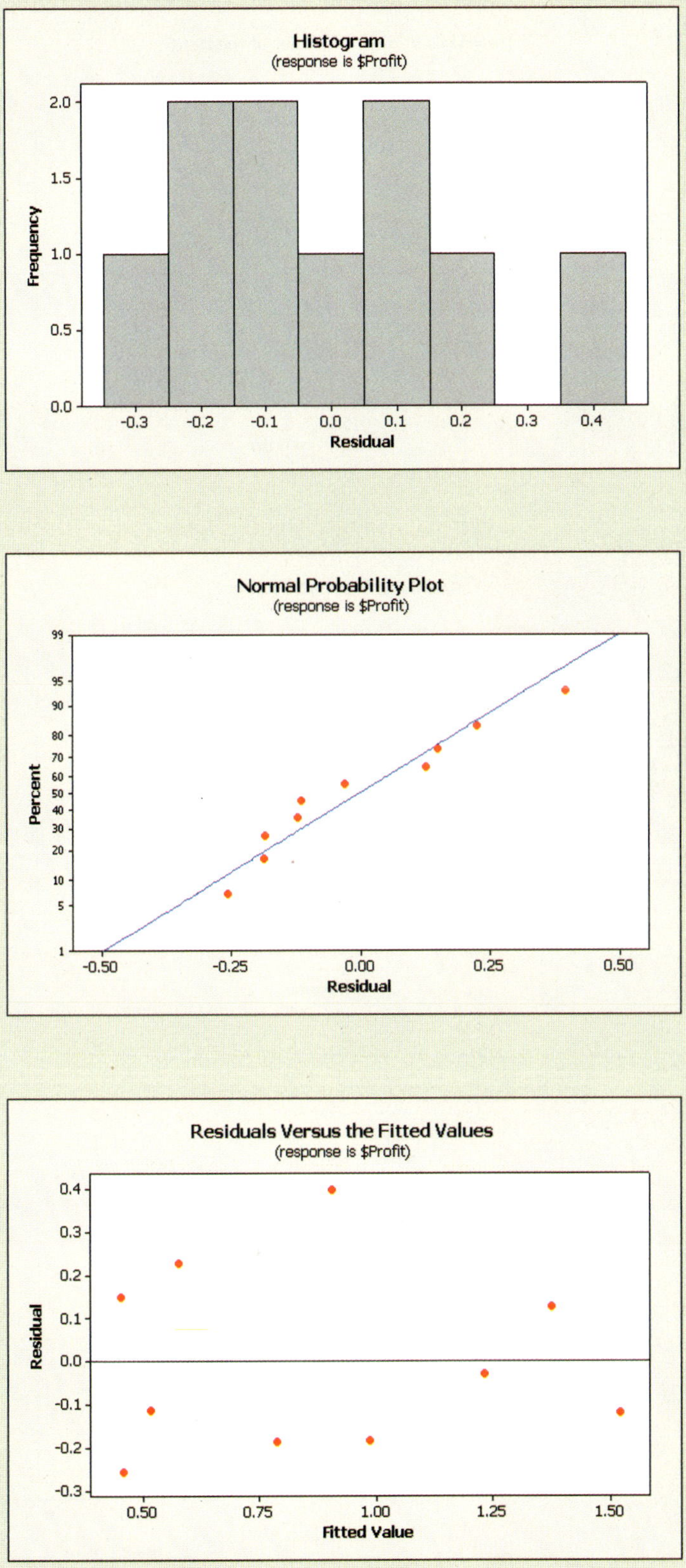

FIGURE 16.3

Minitab plots for analysis of the residuals, discussed in the text and generated using the procedures in Computer Solutions 16.3.

FIGURE 16.3
(continued)

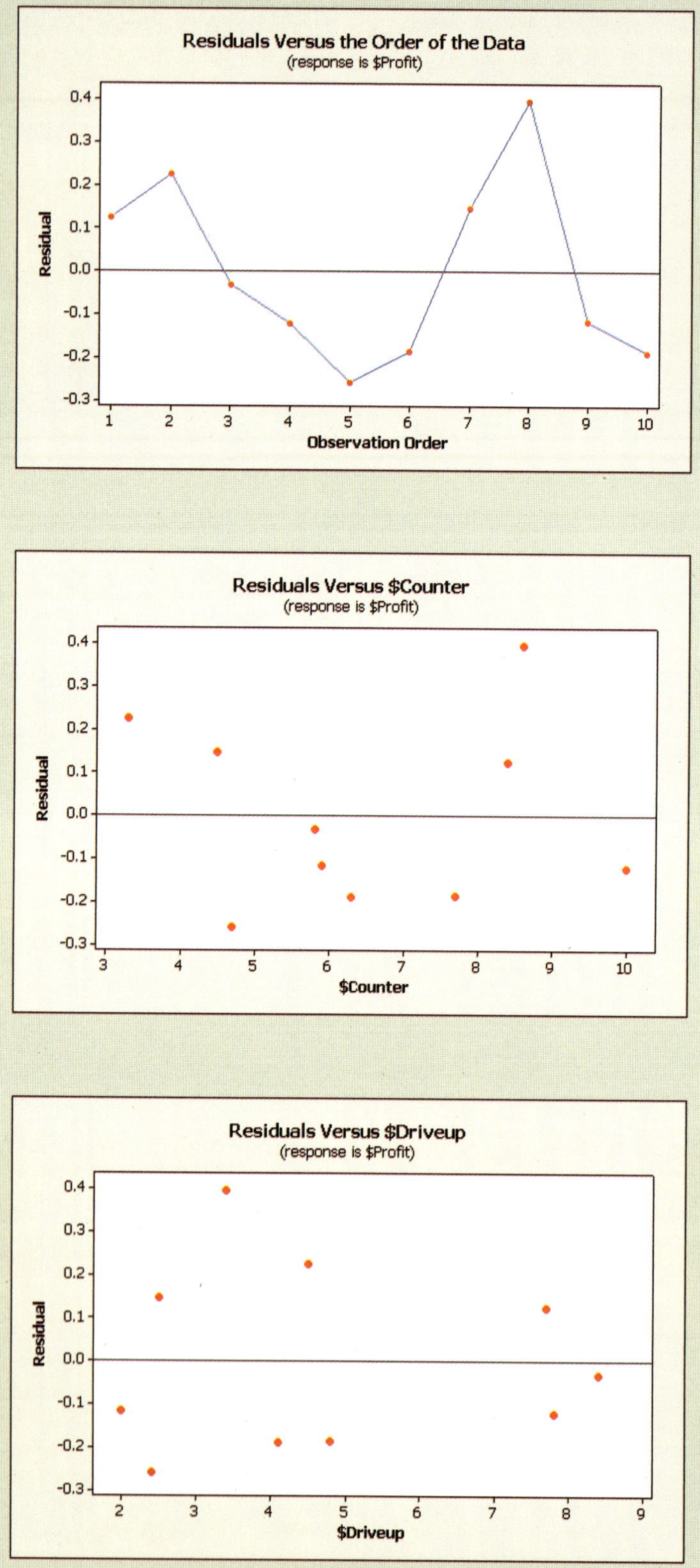

EXERCISES

16.36 What role does the normal probability plot play in examining whether the multiple regression model is appropriate for a given set of data?

16.37 Discuss residual analysis in terms of the assumptions required for using the multiple regression model.

16.38 Given the printout in Exercise 16.15, briefly summarize the overall results in the context of that exercise:

a. Identify and interpret each of the partial regression coefficients.
b. What proportion of the variation in y is explained by the equation?
c. At what p-value is the overall regression equation significant?
d. For each of the partial regression coefficients, at what p-value does the coefficient differ from zero? Which independent variables appear to be most and least useful in explaining changes in y?

(DATA SET) *Note:* Exercises 16.39–16.45 require a computer and statistical software.

16.39 For the regression equation obtained in Exercise 16.11, analyze the residuals by (a) constructing a histogram, (b) utilizing the normal probability plot, and (c) plotting the residuals versus each of the independent variables. Do any of the assumptions of the multiple regression model appear to have been violated?

16.40 For the regression equation obtained in Exercise 16.12, analyze the residuals by (a) constructing a histogram, (b) utilizing the normal probability plot, and (c) plotting the residuals versus each of the independent variables. Do any of the assumptions of the multiple regression model appear to have been violated?

16.41 For the regression equation obtained in Exercise 16.13, analyze the residuals by (a) constructing a histogram, (b) utilizing the normal probability plot, and (c) plotting the residuals versus each of the independent variables. Do any of the assumptions of the multiple regression model appear to have been violated?

16.42 For the regression equation obtained in Exercise 16.18, analyze the residuals by (a) constructing a histogram, (b) utilizing the normal probability plot, and (c) plotting the residuals versus each of the independent variables. Do any of the assumptions of the multiple regression model appear to have been violated?

16.43 For the regression equation obtained in Exercise 16.19, analyze the residuals by (a) constructing a histogram, (b) utilizing the normal probability plot, and (c) plotting the residuals versus each of the independent variables. Do any of the assumptions of the multiple regression model appear to have been violated?

16.44 An efficiency expert wants to determine whether the time an assembly worker requires to perform a standard task might be related to (1) the number of years on the job and (2) the score on a pre-employment aptitude test. The following data have been collected.

y = Time, seconds:	52	48	56	52	48	45	42
x_1 = Years on job:	7	14	8	6	10	6	10
x_2 = Test score:	73	78	71	75	78	78	80

y = Time, seconds:	51	51	53	47	45	49
x_1 = Years on job:	11	10	6	8	12	12
x_2 = Test score:	72	79	77	79	84	72

a. Determine the multiple regression equation and interpret the partial regression coefficients.
b. Determine the 95% confidence interval for the population partial regression coefficients, β_1 and β_2.
c. Interpret the coefficient of determination and the result of each significance test done by the computer.
d. Analyze the residuals. Does the analysis support the applicability of the multiple regression model to these data?

16.45 Of all U.S. credit unions, 67% have a "courtesy overdraft" policy that automatically covers overdrafts, with the median fee for each overdraft being approximately \$25. For 14 federal credit unions with a courtesy overdraft policy, the accompanying table shows their fee income, total assets, and average fee income per member for 2008. Source: Kathy Chu, "Credit Unions Hit Customers with Fees, Too," *USA Today*, August 4, 2009, p. 2B.

Credit Union	Fee Income, y	Total Assets, x_1	Average Fee Income per Member, x_2
1	\$0.5 million	\$19.0 million	\$140.7
2	2.8	128.2	140.6
3	1.1	53.0	165.1
4	0.6	42.1	107.4
5	0.3	29.0	93.9
6	1.8	147.7	69.4
7	2.2	211.2	97.9
8	0.6	58.4	90.8
9	4.0	443.3	86.3
10	3.4	395.4	85.6
11	9.0	1249.8	54.6
12	6.8	986.0	93.3
13	2.5	417.5	76.6
14	0.2	178.4	18.1

a. Determine the multiple regression equation and interpret the partial regression coefficients.
b. Determine the 95% confidence interval for the population partial regression coefficients, β_1 and β_2.
c. Interpret the coefficient of determination and the result of each significance test done by the computer.
d. Analyze the residuals. Does the analysis support the applicability of the multiple regression model to these data?

16.7 ADDITIONAL TOPICS IN MULTIPLE REGRESSION AND CORRELATION

Although we can't cover all facets of multiple regression and correlation analysis, two additional topics are deserving of our attention. These include (1) using a dummy variable to incorporate qualitative data into the analysis, and (2) a discussion of the potential problem known as multicollinearity.

Including Qualitative Data: The Dummy Variable

There may be situations where some of our data do not meet the requirements of either the interval or the ratio scales of measurement. In such cases, we can still include the data in our analysis by using one or more **dummy variables.** A dummy variable will have a value of either one or zero, depending on whether a given characteristic is present or absent.

For example, if some of the 10 franchise managers were college graduates and others were not, we could include an x_3 variable reflecting this information. Depending on the presence or absence of a college degree, x_3 could be recorded as either one or zero:

$$x_3 = 1 \text{ if the franchise is managed by a college graduate}$$

and

$$x_3 = 0 \text{ if the manager is not a college graduate}$$

The dummy variable has a partial regression coefficient, in this case b_3. As an example of its interpretation, if $b_3 = 0.300$, the estimated profit will be \$0.300 million higher if a franchise is managed by a college graduate.

Dummy variables can represent information such as marital status, gender, city versus urban location, trained versus untrained employees, or even whether someone wears contact lenses. As with other independent variables, computer analysis tests the significance of the partial regression coefficient for a dummy variable. For example, if b_3 were to differ significantly from zero, we would conclude that whether a manager is a college graduate is an important predictor for the net profit of a franchise. If b_3 were positive *and* significant, this would indicate that franchises managed by college graduates are significantly more profitable than those managed by nongraduates.

To further demonstrate how dummy variables can be employed, we will use actual data for a sample of 20 personal printers.[2] Some of the printers can print in color, while others are monochrome, and the models vary in how fast they can generate text-only materials. The data are in file **CX16PRIN** and represent the following variables:

PRICE = The list price for the printer
SPEED = Printing speed for text, pages per minute
COLOR = A dummy variable reflecting color versus monochrome printer (1 = color, 0 = monochrome)

[2]Source: "Top 10 Monochrome Laser Printers," *PC World,* August 2009, p. 70; and "Top 10 Color Laser Printers," *PC World,* March 2009, p. 48.

Using PRICE as the dependent variable, the estimation equation is:

Estimated list price of the printer (dollars)	For each additional page that can be printed per minute, the estimated price goes up by \$19.62.	If the printer is color instead of monochrome, the estimated price goes up by \$199.01.

$$\text{PRICE} = -184.29 + 19.62 \text{ SPEED} + 199.01 \text{ COLOR}$$

Point and interval price estimates, confidence intervals, significance testing, and the coefficient of multiple determination are treated the same way as they were in earlier portions of the chapter. For example, a machine that prints 20 text pages per minute (SPEED = 20) and is capable of color printing (COLOR = 1) would have an estimated price of $-184.29 + 19.62(20) + 199.01(1)$, or \$407.12.

Multicollinearity

Multicollinearity is a situation in which two or more of the independent variables are highly correlated with each other. When this happens, the partial regression coefficients become both statistically unreliable and difficult to interpret. This is because they are not really "saying" entirely different things about y. If our fast-food example had included x_3 = number of vehicles using the drive-through, this would be very highly correlated with x_2 = drive-through sales. As a result, these two variables would be the mathematical equivalent of the old-time comedians Laurel and Hardy carrying a piano. They would be very unsure of how much of the "load" each should carry, and the piano would tend to be quite unstable.

Multicollinearity is *not* a problem when we simply substitute x values into the regression equation in order to obtain an estimated value for y. However, it *is* a problem when we're trying to interpret the partial regression coefficients for the x variables that are highly correlated.

When multicollinearity is present, strange things can happen. Along with a high correlation between two or more independent variables, the following are clues to its presence:

1. An independent variable known to be an important predictor ends up having a partial regression coefficient that is not significant.
2. A partial regression coefficient that should be positive turns out to be negative, or vice versa.
3. When an independent variable is added or removed, the partial regression coefficients for the other independent variables change drastically.

One way to deal with multicollinearity is to avoid having a set of independent variables in which two or more variables are highly correlated with each other. In the fast-food example used throughout this chapter, the inclusion of a third independent variable representing the number of vehicles using the drive-through would likely cause multicollinearity problems, because x_2 = drive-through sales would probably be highly correlated with x_3 = drive-through traffic count—each variable is essentially describing the "bigness" of the drive-through aspect of the restaurant.

In Chapter 17, we will discuss **stepwise regression,** a regression model in which the independent variables enter the regression analysis one at a time. The first x variable to enter is the one that explains the greatest amount of variation in y. The second x variable to enter will be the one that explains the greatest amount of the *remaining* variation in y, and so on. At each "step," the computer decides which, if any, of the remaining x variables should be brought in. Every step results in a new regression equation and an updated printout of this equation and its analysis. The general idea of stepwise regression is the balancing act of trying to (1) explain the most possible variation in y, while (2) using the fewest possible x variables.

EXERCISES

16.46 What is a dummy variable, and how is it useful to multiple regression? Give an example of three dummy variables that could be used in describing your home town.

16.47 A multiple regression equation has been developed for y = daily attendance at a community swimming pool, x_1 = temperature (degrees Fahrenheit), and x_2 = weekend versus weekday ($x_2 = 1$ for Saturday and Sunday, and 0 for the other days of the week). For the regression equation shown below, interpret each partial regression coefficient.

$$\hat{y} = 100 + 8x_1 + 150x_2$$

16.48 For the regression equation in Exercise 16.47, the estimated number of persons swimming on a zero-degree weekday would be 100 persons. Since this level of pool attendance is unlikely on such a day, does this mean that an error was made in constructing the regression equation? Explain.

16.49 What is multicollinearity, and how can it adversely affect multiple regression analysis? How can we tell whether multicollinearity is present?

(DATA SET) *Note:* Exercises 16.50 and 16.51 require a computer and statistical software.

16.50 For 12 recent clients, a weight-loss clinic has collected the following data. (Session is coded as 1 for day, 0 for evening. Gender is coded as 1 for male, 0 for female.) Using a computer statistical package, carry out a multiple regression analysis, interpret the partial regression coefficients, and discuss the results of each significance test.

Client	Pounds Lost	Months as a Client	Session	Gender
1	31	5	1	1
2	49	8	1	1
3	12	3	1	0
4	26	9	0	0
5	34	8	0	1
6	11	2	0	0
7	4	1	0	1
8	27	8	0	1
9	12	6	1	1
10	28	9	1	0
11	41	6	0	0
12	16	6	0	0

16.51 A safety researcher has measured the speed of automobiles passing his vantage point near an interstate highway. He has also recorded the number of occupants and whether the driver was wearing a seat belt. The data are shown in the following table with seat belt users coded as 1, nonusers as 0. Using a computer statistical package, carry out a multiple regression analysis on these data, interpret the partial regression coefficients, and discuss the results of each significance test.

y = speed (mph):	61	63	55	59	59	52	61
x_1 = occupants:	2	1	2	1	3	1	2
x_2 = belt usage:	1	1	1	0	0	1	0
y = speed (mph):	51	57	55	54	49	61	73
x_1 = occupants:	2	2	2	3	1	1	1
x_2 = belt usage:	1	1	1	0	1	1	0

16.8 SUMMARY

• Multiple regression model

Unlike its counterpart in Chapter 15, multiple regression and correlation analysis considers two or more independent (x) variables. The model is of the form $y_i = \beta_0 + \beta_1 x_{1i} + \beta_2 x_{2i} + \cdots + \beta_k x_{ki} + \epsilon_i$ and, for a given set of x values, the expected value of y is given by the regression equation, $E(y) = \beta_0 + \beta_1 x_{1i} + \beta_2 x_{2i} + \cdots + \beta_k x_{ki}$. For each x term in the estimation equation, the corresponding β is referred to as the *partial regression coefficient*. Each β_i ($i = 1, 2, \ldots, k$) is a slope relating changes in $E(y)$ to changes in one x variable when all other x's are held constant. The multiple standard error of estimate expresses the amount of dispersion of data points about the regression equation.

Estimation and computer assistance

Calculations involving multiple regression and correlation analysis are tedious and best left to a computer package such as Excel or Minitab. The chapter includes formulas for determining the approximate confidence interval for the conditional mean of y as well as the approximate prediction interval for an individual y value, given a set of x values. The exact intervals can be obtained using Excel or Minitab.

Coefficient of multiple determination

Multiple correlation measures the strength of the relationship between the dependent variable and the set of independent variables. The coefficient of multiple determination (R^2) is the proportion of the variation in y that is explained by the regression equation.

Significance testing and residual analysis

The overall significance of the regression equation can be tested through the use of analysis of variance, while a t-test is used in testing the significance of each partial regression coefficient. The results of these tests are typically included when data have been analyzed using a computer statistical package. Most computer packages can also provide a summary of the residuals, which may be analyzed further in testing the regression model and identifying data points especially distant from the regression equation.

Dummy variables and multicollinearity

A dummy variable has a value of either one or zero, depending on whether a given characteristic is present, and it allows qualitative data to be included in the analysis. If two or more independent variables are highly correlated with each other, multicollinearity is present, and the partial regression coefficients may be unreliable. Multicollinearity is not a problem when the only purpose of the equation is to predict the value of y.

Stepwise regression

In stepwise regression, independent variables enter the equation one at a time, with the first one entered being the x variable that explains the greatest amount of the variation in y. At each step, the variable introduced is the one that explains the greatest amount of the remaining variation in y. Stepwise regression will be discussed more completely in Chapter 17.

EQUATIONS

The Multiple Regression Model

$$y_i = \beta_0 + \beta_1 x_{1i} + \beta_2 x_{2i} + \cdots + \beta_k x_{ki} + \epsilon_i$$

where y_i = a value of the dependent variable, y

β_0 = a constant

$x_{1i}, x_{2i}, \ldots, x_{ki}$ = values of the independent variables, $x_1, x_2, \ldots, x_k$

$\beta_1, \beta_2, \ldots, \beta_k$ = partial regression coefficients for the independent variables, $x_1, x_2, \ldots, x_k$

ϵ_i = random error, or residual

The Sample Multiple Regression Equation

$$\hat{y} = b_0 + b_1x_1 + b_2x_2 + \cdots + b_kx_k$$

where $\hat{y}$ = the estimated value of the dependent variable, y for a given set of values for $x_1, x_2, \ldots, x_k$
k = the number of independent variables
b_0 = the y-intercept; this is the estimated value of y when all of the independent variables are equal to 0
b_i = the partial regression coefficient for x_i

The Multiple Standard Error of Estimate

$$s_e = \sqrt{\frac{\Sigma(y_i - \hat{y}_i)^2}{n - k - 1}}$$

where y_i = each observed value of y in the data
$\hat{y}_i$ = the value of y that would have been estimated from the regression equation
n = the number of data points
k = the number of independent (x) variables

Approximate Confidence Interval for the Conditional Mean of *y*

$$\hat{y} \pm t\frac{s_e}{\sqrt{n}}$$

where $\hat{y}$ = the estimated value of y based on the set of x values provided
t = t value from the t distribution table for the desired confidence level and $df = n - k - 1$ (n = number of data points, k = number of x variables)
s_e = the multiple standard error of estimate

Approximate Prediction Interval for an Individual *y* Observation

$$\hat{y} \pm ts_e$$

where $\hat{y}$ = the estimated value of y based on the set of x values provided
t = t value from the t distribution table for the desired confidence level and $df = n - k - 1$ (n = number of data points, k = number of x variables)
s_e = the multiple standard error of estimate

Coefficient of Multiple Determination

$$R^2 = 1 - \frac{\textit{Error}\text{ variation, which is }\textit{not}\text{ explained by the regression equation}}{\textit{Total}\text{ variation in the } y \text{ values}}$$

$$= 1 - \frac{\Sigma(y_i - \hat{y}_i)^2}{\Sigma(y_i - \bar{y})^2} \quad \text{or} \quad 1 - \frac{SSE}{SST}$$

As in Chapter 15, the total variation in y is the sum of the explained variation plus the unexplained (error) variation, or

Total variation in y values (SST)	=	Variation explained by the regression equation (SSR)	+	Variation not explained by the regression equation (SSE)

ANOVA Test for the Overall Significance of the Multiple Regression Equation

- Null hypothesis

H_0: $\beta_1 = \beta_2 = \cdots = \beta_k = 0$ The regression equation is not significant.

- Alternative hypothesis

H_1: One or more of the β_i $(i = 1, 2, \ldots, k) \neq 0$ — The regression equation is significant.

- Test statistic

$$F = \frac{SSR/k}{SSE/(n - k - 1)}$$

where SSR = regression sum of squares
SSE = error sum of squares
n = number of data points
k = number of independent (x) variables

- Critical value of the F statistic

F for significance level specified and df (numerator) $= k$, and df (denominator) $= n - k - 1$

t-Test for the Significance of a Partial Regression Coefficient, b_i

- Null hypothesis

H_0: $\beta_i = 0$ — The population coefficient is 0.

- Alternative hypothesis

H_1: $\beta_i \neq 0$ — The population coefficient is not 0.

- Test statistic

$$t = \frac{b_i - 0}{s_{b_i}}$$

where b_i = the observed value of the regression coefficient
s_{b_i} = the estimated standard deviation of b_i

NOTES

1. We can also test H_0: $\beta_i = \beta_{i0}$ versus H_1: $\beta_i \neq \beta_{i0}$, where β_{i0} is any value of interest. With the exception of β_{i0} replacing zero, the test is the same as described here.
2. Critical values of t are $\pm t$ for the level of significance desired and $df = n - k - 1$ (n = number of data points, k = number of independent variables).

Confidence Interval for a Partial Regression Coefficient, β_i

The interval is $b_i \pm t(s_{b_i})$

where b_i = partial regression coefficient in the sample regression equation
t = t value for the confidence level desired, and $df = n - k - 1$
s_{b_i} = estimated standard deviation of b_i

NOTE

Excel provides these confidence intervals as part of the regression analysis printout.

CHAPTER EXERCISES

(**DATA SET**) *Note:* Exercises 16.52–16.58 require a computer and statistical software.

16.52 Interested in the possible relationship between the size of his tip versus the size of the check and the number of diners in the party, a food server has recorded the following for a sample of 8 checks:

Observation Number	y = Tip	x_1 = Check	x_2 = Diners
1	$7.5	$40	2
2	0.5	15	1
3	2.0	30	3
4	3.5	25	4
5	9.5	50	4
6	2.5	20	5
7	3.5	35	5
8	1.0	10	2

a. Determine the multiple regression equation and interpret the partial regression coefficients.
b. What is the estimated tip amount for 3 diners who have a $40 check?
c. Determine the 95% prediction interval for the tip left by a dining party like the one in part (b).
d. Determine the 95% confidence interval for the mean tip left by all dining parties like the one in part (b).
e. Determine the 95% confidence interval for the partial regression coefficients, β_1 and β_2.
f. Interpret the significance tests in the computer printout.
g. Analyze the residuals. Does the analysis support the applicability of the multiple regression model to these data?

16.53 Annual per-capita consumption of all fresh fruits versus that of apples and grapes from 1998 through 2003 was as shown in the table. Source: Bureau of the Census, *Statistical Abstract of the United States 2006*, p. 137.

Year	y = All Fresh Fruits	x_1 = Apples	x_2 = Grapes
1998	128.9 lb/person	19.0 lb/person	7.1 lb/person
1999	129.8	18.7	8.1
2000	128.0	17.6	7.4
2001	125.7	15.8	7.7
2002	126.9	16.2	8.7
2003	126.7	16.7	7.5

a. Determine the multiple regression equation and interpret the partial regression coefficients.
b. What is the estimated per-capita consumption of all fresh fruits during a year when 17 pounds of apples and 6 pounds of grapes are consumed per person?
c. Determine the 95% prediction interval for per-capita consumption of fresh fruits during a year like the one described in part (b).
d. Determine the 95% confidence interval for mean per-capita consumption of fresh fruits during years like the one described in part (b).
e. Determine the 95% confidence interval for the population partial regression coefficients, β_1 and β_2.
f. Interpret the significance tests in the computer printout.
g. Analyze the residuals. Does the analysis support the applicability of the multiple regression model to these data?

16.54 A university placement director is interested in the effect that grade point average (GPA) and the number of university activities listed on the résumé might have on the starting salaries of this year's graduating class. He has collected these data for a sample of 10 graduates:

Graduate	y = Starting Salary (Thousands)	x_1 = Grade Point Average	x_2 = Number of Activities
1	$40	3.2	2
2	46	3.6	5
3	38	2.8	3
4	39	2.4	4
5	37	2.5	2
6	38	2.1	3
7	42	2.7	3
8	37	2.6	2
9	44	3.0	4
10	41	2.9	3

a. Determine the multiple regression equation and interpret the partial regression coefficients.
b. Dave has a 3.6 grade point average and 3 university activities listed on his résumé. What would be his estimated starting salary?
c. Determine the 95% prediction interval for the starting salary of the student described in part (b).
d. Determine the 95% confidence interval for the mean starting salary for all students like the one described in part (b).
e. Determine the 95% confidence interval for the population partial regression coefficients, β_1 and β_2.
f. Interpret the significance tests in the computer printout.
g. Analyze the residuals. Does the analysis support the applicability of the multiple regression model to these data?

16.55 An admissions counselor, examining the usefulness of x_1 = total SAT score and x_2 = high school class rank in predicting y = freshman grade point average (GPA) at her university, has collected the sample data shown here and listed in file **XR16055**. Rank in high school class is expressed as a cumulative percentile, i.e., 100.0% reflects the top of the class.

Freshman GPA	SAT, Total	High School Rank	Freshman GPA	SAT, Total	High School Rank
2.66	1153	61.5	2.45	1136	45.5
2.10	1086	84.5	2.50	966	60.2
3.33	1141	92.0	2.29	1023	74.0
3.85	1237	94.4	2.24	976	86.4
2.51	1205	89.5	1.81	1066	73.0
3.22	1205	97.0	2.99	1076	55.2
2.92	1163	95.9	3.14	1152	72.1
1.95	1121	64.1	1.86	955	51.0

a. Determine the multiple regression equation and interpret the partial regression coefficients.
b. What is the estimated freshman GPA for a student who scored 1100 on the SAT exam and had a cumulative class rank of 80%?
c. Determine the 95% prediction interval for the GPA of the student described in part (b).
d. Determine the 95% confidence interval for the mean GPA of all students similar to the one described in part (b).
e. Determine the 95% confidence interval for the population partial regression coefficients, β_1 and β_2.
f. Interpret the significance tests in the computer printout.
g. Analyze the residuals. Does the analysis support the applicability of the multiple regression model to these data?

16.56 Data file **XR16056** lists the following information for a sample of local homes sold recently: selling price (dollars), lot size (acres), living area (square feet), and whether the home is air conditioned (1 = A/C, 0 = no A/C).
a. Determine the multiple regression equation and interpret the partial regression coefficients.
b. What is the estimated selling price for a house sitting on a 0.9-acre lot, with 1800 square feet of living area and central air conditioning?
c. Determine the 95% prediction interval for the selling price of the house described in part (b).
d. Determine the 95% confidence interval for the mean selling price of all houses like the one described in part (b).
e. Determine the 95% confidence interval for the population partial regression coefficients, β_1, β_2, and β_3.
f. Interpret the significance tests in the computer printout.
g. Analyze the residuals. Does the analysis support the applicability of the multiple regression model to these data?

16.57 In Exercise 16.56, what would be the estimated selling price of a house occupying a 0.01-acre lot, with 100 square feet of living area and no central air conditioning? Considering the nature of this "house," does this selling price seem reasonable? If not, might the computer have made an error in obtaining the regression equation? Explain.

16.58 Data file **XR16058** lists the following data for a sample of automatic teller machine (ATM) customers: time (seconds) to complete their transaction, estimated age, and gender (1 = male, 0 = female).
a. Determine the multiple regression equation and interpret the partial regression coefficients.
b. What is the estimated time required by a female customer who is 45 years of age?
c. Determine the 95% prediction interval for the time required by the customer described in part (b).
d. Determine the 95% confidence interval for the mean time required by customers similar to the one described in part (b).
e. Determine the 95% confidence interval for the population partial regression coefficients, β_1 and β_2.
f. Interpret the significance tests in the computer printout.
g. Analyze the residuals. Does the analysis support the applicability of the multiple regression model to these data?

INTEGRATED CASES

Thorndike Sports Equipment

For several years, Thorndike Sports Equipment has been a minority stockholder in the Snow Kingdom Ski Resort and Conference Center. The Thorndikes visit Snow Kingdom several times each winter to meet with management and find out how business is going. In addition, the visits give them a chance for informal discussions with customers and potential customers of Thorndike ski clothing and equipment. Luke claims that many good product ideas have been inspired by a warm drink in the Snow Kingdom lodge.

On their current visit, Ted and Luke are asked by Snow Kingdom management to lend a hand in analyzing some data that might help in predicting how many customers to expect on any given day. Overall business has been rather steady over the past several years, but the daily customer count seems to have ups and downs. The people at Snow Kingdom are curious as to what factors might be causing the seemingly random levels of patronage from one day to the next.

In response to Ted's request, management supplies data for a random sample of 30 days over the past two seasons. The information includes the number of skiers, the high temperature (degrees Fahrenheit), the number of inches of snow on the ground at noon, and whether the day fell on a weekend (1 = weekend, 0 = weekday). Data are shown here and are also in file **THORN16**.

Using multiple regression and correlation analysis, do you think this information appears to be helpful in explaining the level of daily patronage at Snow Kingdom? Help Ted in interpreting the associated computer printout for an upcoming presentation to the Snow Kingdom management.

Day	Skiers	Weekend	Snow (Inches)	Temperature (Degrees)	Day	Skiers	Weekend	Snow (Inches)	Temperature (Degrees)
1	402	0	22	24	16	648	0	14	8
2	337	0	17	25	17	540	0	17	11
3	471	0	17	28	18	614	1	36	6
4	610	0	39	21	19	796	1	25	29
5	620	0	11	18	20	477	0	13	5
6	545	1	24	17	21	532	0	35	30
7	523	0	25	29	22	732	0	36	15
8	563	0	34	17	23	618	0	18	11
9	873	1	18	11	24	728	1	19	28
10	358	0	28	6	25	620	0	29	8
11	568	0	14	9	26	551	0	18	6
12	453	0	12	18	27	816	0	31	13
13	485	0	27	27	28	765	0	24	19
14	767	1	37	16	29	650	1	22	27
15	735	1	12	6	30	732	0	11	24

Springdale Shopping Survey

This case involves the variables shown here and described in data file **SHOPPING**. Information like that gained from this case could provide management with useful insights into how two or more variables (including dummy variables) can help describe consumers' overall attitude toward a shopping area.

- **Variables 7–9** Overall attitude toward the shopping area. The highest possible rating is 5. In each analysis, one of these will be the dependent variable.
- **Variables 18–25** The importance (highest rating = 7) placed on each of 8 attributes that a shopping area might possess. In each analysis, these will be among the independent variables.
- **Variables 26 and 28** The respondent's gender and marital status. These independent variables will be dummy variables. Recode them so that (1 = male, 0 = female) and (1 = married, 0 = single or other).

The necessary commands will vary, depending on your computer statistical package. Be careful not to save the recoded database, because the original values will be overwritten if you do.

1. With variable 7 (attitude toward Springdale Mall) as the dependent variable, perform a multiple regression analysis using variables 21 (good variety of sizes/styles), 22 (sales staff helpful/friendly), 26 (gender), and 28 (marital status) as the four independent variables. If possible, have the residuals and the predicted *y* values retained for later analysis.
   a. Interpret the partial regression coefficient for variable 26 (gender). At the 0.05 level, is it significantly different from zero? If so, what does this say about the respective attitudes of males versus females toward Springdale Mall? Interpret the other partial regression coefficients and the results of the significance test for each.
   b. At the 0.05 level, is the overall regression equation significant? At exactly what *p*-value is the equation significant?
   c. What percentage of the variation in *y* is explained by the regression equation? Explain this percentage in terms of the analysis of variance table that accompanies the printout.

d. If possible with your computer package, generate a plot of the residuals (vertical axis) versus each of the independent variables (horizontal axis). Evaluate each plot in terms of whether patterns exist that could weaken the validity of the regression model.

e. If possible with your computer statistical package, use the normal probability plot to examine the residuals.

2. Repeat question 1, but with variable 8 (attitude toward Downtown) as the dependent variable.
3. Repeat question 1, but with variable 9 (attitude toward West Mall) as the dependent variable.
4. Compare the regression equations obtained in questions 1, 2, and 3. For which one of the shopping areas does this set of independent variables seem to do the best job of predicting shopper attitude? Explain your reasoning.

BUSINESS CASE

Easton Realty Company (A)

Digital Vision/Getty Images

Sam Easton started out as a real estate agent in Atlanta ten years ago. After working two years for a national real estate firm, he transferred to Dallas, Texas, and worked for another realty agency. His friends and relatives convinced him that with his experience and knowledge of the real estate business, he should open his own agency. He eventually acquired his broker's license and before long started his own company, Easton Realty Company, in Fort Worth, Texas.

Two salespeople at the previous company agreed to follow him to the new company. Easton currently has eight real estate agents working for him. Before the real estate slump, the combined residential sales for Easton Realty amounted to approximately $15 million annually.

Recently, the Dallas–Fort Worth (DFW) metroplex and the state of Texas have suffered economic problems from several sources. Much of the wealth in Texas was generated by the oil industry, but the oil industry has fallen on hard times in recent years. Many savings and loan (S&L) institutions loaned large amounts of money to the oil industry and to commercial and residential construction. As the oil industry fell off and the economy weakened, many S&Ls found themselves in difficulty as a result of poor real estate investments and the soft real estate market that was getting worse with each passing month. With the lessening of the Cold War, the federal government closed several military bases across the country, including two in the DFW area. Large government contractors, such as General Dynamics, had to trim down their operations and lay off many workers. This added more pressure to the real estate market by putting more houses on an already saturated market. Real estate agencies found it increasingly difficult to sell houses.

Two days ago, Sam Easton received a special delivery letter from the president of the local Board of Realtors. The board had received complaints from two people who had listed and sold their homes through Easton Realty in the past month. The president of the Board of Realtors was informing Sam of these complaints and giving him the opportunity to respond. Both complaints were triggered by a recent article on home sales that appeared in one of the local newspapers. The article contained the table shown below.

Typical Home Sale, DFW Metroplex

Average Sales Price	$104,250
Average Size	1860 sq. ft.

Note: Includes all homes sold in the Dallas–Fort Worth metroplex over the past 12 months.

The two sellers charged that Easton Realty Company had underpriced their homes in order to accelerate the sales. The first house is located outside of the DFW area, is four years old, has 2190 square feet, and sold for $88,500. The second house is located in Fort Worth, is nine years old, has 1848 square feet, and sold for $79,500. Both houses in question are three-bedroom houses. Both sellers believe that they would have received more money for their houses if Easton Realty had priced them at their true market value.

Sam knew from experience that people selling their homes invariably overestimate the value. Most sellers believe they could have gotten more money from the sale of their homes. But Sam also knew that his agents would not intentionally underprice houses. However, in these bad economic times, many real estate companies, including Easton Realty, had large inventories of houses for sale and needed to make sales. One quick way to unload these houses is to underprice them. On a residential sale, an agent working under a real estate broker typically makes about 3% of the sales price if he originally listed the property. Dropping the sales price of a $100,000 home down to $90,000 would speed up the sale and the agent's commission would fall only from $3,000 to $2,700. Some real estate agents might consider sacrificing $300 in order to get their commission sooner, but it is unethical because the agent is supposed to be representing the seller and acting in the seller's best interests. Sam had to convince the two sellers and the Board of Realtors that there was no substance to the complaints. The question was, How was he going to do it?

First, he needed to obtain recent residential sales data. Unfortunately, the local multiple listing service (MLS) did not contain actual sales prices of homes. However, Pat McCloskey, a local real estate appraiser, did maintain a database that had the sales information Sam needed. Phoning Pat, Sam found that she indeed had the data he required, but she would have to merge her personal database with data downloaded from the MLS in order to give Sam the necessary information. Fortunately, this was a relatively simple task, and Pat could get the data to Sam the next day.

Sam had asked Pat to give him all the data she had on home sales that had taken place in the DFW area over the previous three months. Although Pat's database did not contain all home sales in the DFW metroplex over that period of time, she felt the data she had were representative of the entire population. The data for each home sold included the sale month, the sale price, the size of the home (in square feet of heated floor space), the number of bedrooms, the age of the house, whether the house was located in Dallas versus Fort Worth or elsewhere within the metroplex, and the real estate company that made the sale.

Assignment

The real estate data compiled by Pat McCloskey for Sam Easton are contained in the file **EASTON**. The Data Description section provides a partial listing of this data file along with definitions of the variables. Using this data set and other information given in the case, help Sam Easton respond to the underpricing claims of his former clients. The Case Questions will assist you in your analysis of the data. Use important details from your analysis of the data to support your recommendation.

Data Description

The **EASTON** file contains data on home sales over the past three months in the DFW metroplex. A partial listing of the data is shown at the top of the next page.

The variables are defined as follows:

MONTH: Month in which the sale took place:
- 4, if April
- 5, if May
- 6, if June

PRICE: Sale price of the house in dollars

SQFEET: Square feet of heated floor space

BEDROOMS: Number of bedrooms in the house

AGE: Age of the house in years

DALLAS: Area within the DFW metroplex where the house is located:
- 1, if in Dallas
- 0, if in Fort Worth or elsewhere in the metroplex

EASTON:
- 1, if Easton Realty Company sold the house
- 0, otherwise

Case Questions

1. Considering the claim that the two houses in question did not sell for their fair market value:
   a. Compare the selling prices of the two houses to the average selling price of all houses sold in the most recent three-month period. Does the difference appear to be substantial?
   b. Provide at least two reasons why the comparison in part (a) is not fair—i.e., describe at least two pricing factors that are not being considered in drawing the comparison in part (a).
   c. In making their argument, the complaining sellers are relying heavily on the average selling price ($104,250) stated in the article for all homes sold in the area during the previous *twelve* months during a weakening housing market. Compare the $104,250 average with that for all houses sold in the area during the most recent three months, then comment on how this comparison might affect the validity of their argument.
2. Use a multiple regression model to estimate PRICE as a function of SQFEET, BEDROOMS, AGE, DALLAS, and EASTON.
   a. Interpret the partial regression coefficients in the equation. Specifically, comment on the sign and numerical value of the partial regression coefficient for EASTON in terms of the claim that Easton Realty Company has been underpricing its residential properties relative to other real estate companies.
   b. For each of the two houses that are the subject of complaints, construct and interpret the 95% prediction interval for a comparable house that is being sold by a realtor other than Easton.
   c. On the basis of your analyses in question 1 and in parts (a) and (b) of this question, prepare a brief, convincing response to the claims of underpricing.

Month	Price	SqFeet	Bedrooms	Age	Dallas	Easton
4	$73,800	1284	2	6	1	0
4	69,200	1919	3	8	0	0
4	98,500	2316	3	7	0	0
4	82,200	1821	4	4	0	0
4	100,300	1703	3	4	1	0
⋮	⋮	⋮	⋮	⋮	⋮	⋮

BUSINESS CASE

Circuit Systems, Inc. (C)

In Chapters 11 and 14, we visited Circuit Systems, Inc., a company that was concerned about the effectiveness of its new program for reducing the cost of absenteeism among hourly workers. Besides allowing employees to be paid for unused sick leave, the program also pays $200 toward membership in a local health club for those employees who wish to join. Thomas Nelson, who introduced the new program, has collected data describing the 233 employees who have been with the company for both the year preceding the new program and the year following its implementation. The information Mr. Nelson has collected is in the **CIRCUIT** data file, and the variables and description of his program are as described in Circuit Systems, Inc. (A), in Chapter 11.

1. Use a multiple regression model to estimate the number of days of sick leave this year as a function of two variables: days of sick leave taken last year and whether the employee is a participant in the exercise program. Interpret the partial regression coefficients in the equation. Specifically, comment on the sign and numerical value of the partial regression coefficient for the dummy variable representing participation in the exercise program. On this basis, does the exercise program seem worthy of continuation?
2. What percentage of the variation in days of sick leave this year is explained by the regression equation you generated for question 1? To the extent that there is unexplained variation in the number of sick days taken this year, what other variables—including possible variables not in the database—might influence the number of days of sick leave taken this year?

Model Building

THE NFL PASSER RATING

To many fans, one of the more mysterious facets of the National Football League (NFL) is its numerical rating of quarterbacks. It's something the sports announcers and "Monday morning quarterbacks" like to talk about, but they spend little or no time talking about how this rating is determined. Media listings of passer ratings are usually accompanied by other performance statistics, such as passing yards, completion percentage, touchdowns, interceptions, and so on.

In this chapter, we learn about stepwise regression, a method by which the candidate predictor variables are mathematically evaluated before being invited into the equation. As in the NFL, not all of them make the team. At each round of the evaluations, just one is invited in, and the invitations stop when we run out of variables that are able to make a meaningful contribution to estimating the dependent variable.

Later in the chapter, you can use Exercise 17.73 and stepwise regression in examining the NFL passer rating. The idea is to find out which variables are actually used, and how they might be combined. Although the passer rating is an interesting and useful measure, the official NFL site is quick to add this caveat: "Judging a quarterback based solely on statistical success overlooks critical intangibles such as leadership and decision making. While an elite quarterback routinely demonstrates his ability to put up big numbers, the great ones deliver those performances while leading their teams to victories."

Source: Bucky Brooks, "Game's Best Quarterbacks Produce Stats as Well as Victories," NFL.com, August 18, 2009.

Models represent the real thing, but are a lot easier to work with.

- Build polynomial regression models to describe curvilinear relationships.
- Apply qualitative variables representing two or more categories.
- Use logarithmic transformations in constructing exponential and multiplicative models.
- Identify and compensate for multicollinearity in the data.
- Apply stepwise regression in selecting which variables to use in a model.
- Determine which of several competing models might be most suitable for the data and the situation

LEARNING OBJECTIVES

After reading this chapter, you should be able to:

INTRODUCTION (17.1)

In Chapter 15, we applied the simple linear regression model in estimating the value of a dependent variable on the basis of a single independent variable. In Chapter 16, we expanded this model to include two or more independent variables. Both models assume a linear relationship, and the multiple regression model assumes that the independent variables are relatively independent from each other. In this chapter, we will examine more sophisticated models for the estimation of y. These will include curvilinear models as well as models in which qualitative variables are considered. We will also consider models in which the independent variables are related to each other. We will introduce a regression method that allows us to strategically choose a subset of the independent variables that has nearly the predictive power of the original, larger set.

The final section focuses on an important concept that is considered throughout the chapter: the importance of selecting a model that is appropriate to both the data and our analysis objectives. As an introductory example, note the scatter diagrams in parts (a) and (b) of Figure 17.1. As x increases, y initially increases, then it decreases, and then it increases again. As part (a) indicates, a linear model is not a very good fit to the data. In part (b), we use one of the models that will be presented in Section 17.2, and it does a much better job of estimating y for a given value of x.

POLYNOMIAL MODELS WITH ONE QUANTITATIVE PREDICTOR VARIABLE (17.2)

By extending the simple linear regression model of Chapter 15 to include terms that include the independent (or *predictor*) variable x to a power greater than 1, we can construct models to fit data that are decidedly nonlinear. Such models can fit even data like those shown in Figure 17.1, where y responds to increases in x by initially increasing, then decreasing, then increasing again. The equation shown in part (b) of Figure 17.1 includes both an x^2 term and an x^3 term, and is a member of the family of polynomial models discussed in this section.

The General Polynomial Model

The polynomial model includes p terms that contain the independent variable x, with the size of the exponent increasing in consecutive terms from 1 to a maximum of p. There is only one independent, or predictor variable (x), but each x

FIGURE 17.1

It is important to select an appropriate model for a given set of data. In part (a), the simple regression model of Chapter 15 is applied to the data. The model in part (b), to be introduced in Section 17.2, is a much better fit.

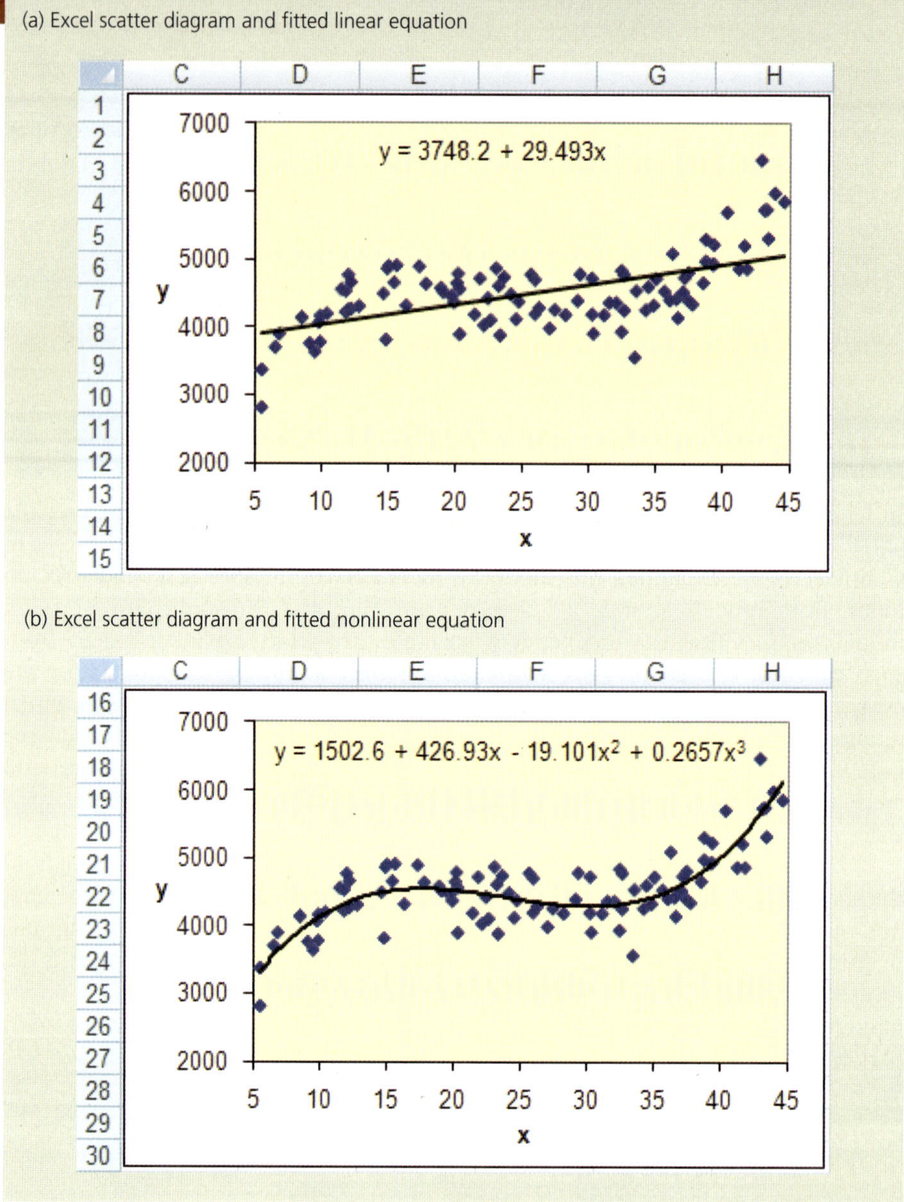

term in the equation is actually being treated as if it were a separate independent variable. The general form of the model is shown here:

The general polynomial model:

$$y_i = \beta_0 + \beta_1 x_i + \beta_2 x_i^2 + \beta_3 x_i^3 + \cdots + \beta_p x_i^p + \epsilon_i$$

where y_i = a value of the dependent variable, y

β_0 = a constant

x_i = a value of the independent (predictor) variable, x

$\beta_1, \beta_2, \ldots, \beta_p$ = partial regression coefficients

ϵ_i = random error, or residual

In the general polynomial model, $y_i = \beta_0 + \beta_1 x_i + \beta_2 x_i^2 + \beta_3 x_i^3 + \cdots + \beta_p x_i^p + \epsilon_i$, the *order* of the polynomial is the integer p. As in simple and multiple regression, the constant (β_0) and the partial regression coefficients ($\beta_1, \beta_2, \ldots, \beta_p$) must be estimated on the basis of sample data. Our discussion in the remainder of this section pertains to polynomial models in which $p = 1, 2,$ or 3, because practical situations rarely call for a model that is higher than the third order. The first-, second-, and third-order polynomial models will be discussed individually.

First-Order Polynomial Model

The first-order polynomial model is the same as the linear regression model presented in Chapter 15, and is appropriate when there is a linear relationship between y and x. With $E(y)$ as the expected value of y for a given value of the predictor variable x, the format is

First-order polynomial model:

$$E(y) = \beta_0 + \beta_1 x$$

As shown in part (a) of Figure 17.1, this model is not suitable when the relationship between y and x is nonlinear. When the relationship does happen to be linear, the equation can have either a positive slope ($\beta_1 > 0$) or a negative slope ($\beta_1 < 0$), as shown in Figure 17.2. If you were driving along the plotted line in a car, you would always be driving straight.

Second-Order Polynomial Model

The second-order polynomial model has a parabolic shape in that the curve can resemble a "U" that is either upright or upside-down. The format for this model is

Second-order polynomial model:

$$E(y) = \beta_0 + \beta_1 x + \beta_2 x^2$$

In this model, the $\beta_1 x$ term provides a linear component while the $\beta_2 x^2$ term provides a nonlinear component that allows the line to curve. The shape of the curve will be determined by the relative sizes of β_1 and β_2 and on their signs, and typical curves for this model are shown in Figure 17.3. To the extent that

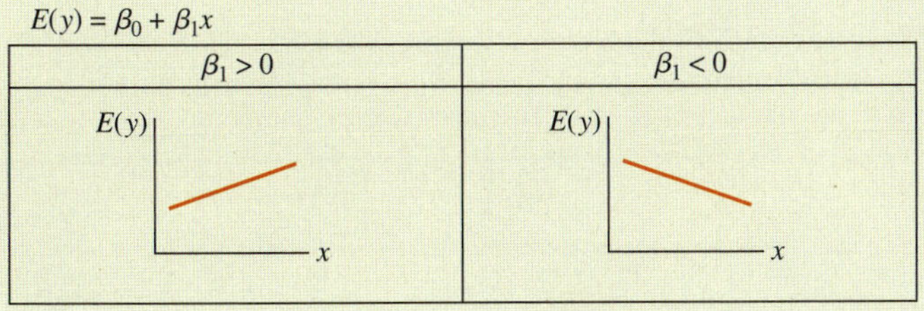

FIGURE 17.2

In the first-order polynomial model, the expected value of y is $E(y) = \beta_0 + \beta_1 x$. This is the same as the simple regression model in Chapter 15, and the slope can be either positive ($\beta_1 > 0$) or negative ($\beta_1 < 0$).

FIGURE 17.3

For the second-order polynomial model, the expected value of y will be $E(y) = \beta_0 + \beta_1 x + \beta_2 x^2$, and the shape of the curve will depend on both the signs and the relative sizes of the coefficients β_1 and β_2. With $x > 0$, combinations of positive and negative values of these coefficients can result in curves like these.

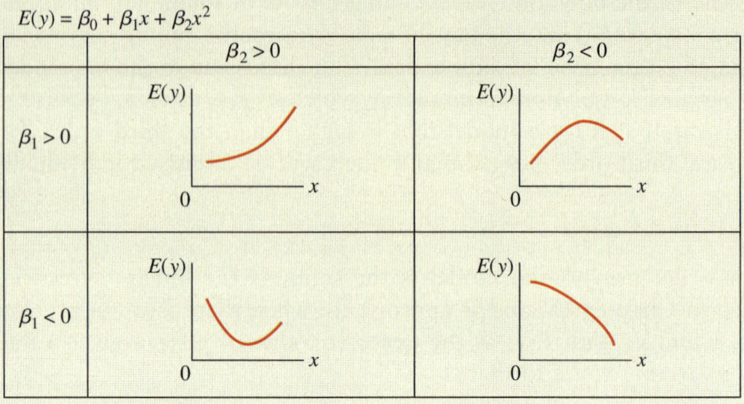

any given second-order polynomial exhibits a curve, it will always curve in the same direction.

Third-Order Polynomial Model

As with the second-order model, the third-order polynomial model allows the plotted line to curve. However, it makes it possible for the line *to reverse the direction in which it is curving*. The format for this model is

Third-order polynomial model:

$$E(y) = \beta_0 + \beta_1 x + \beta_2 x^2 + \beta_3 x^3$$

In this model, the $\beta_1 x$ term provides a linear component, the $\beta_2 x^2$ term provides a nonlinear component allowing the line to curve, and the $\beta_3 x^3$ term provides a nonlinear component enabling the line to reverse the direction of curvature. The sign of β_1 determines the initial slope of the line for small positive values of x. If β_2 and β_3 have opposite signs, the line will be able to reverse the direction it is curving. If β_2 and β_3 have the same sign, the effect of the $\beta_3 x^3$ term will be to make the line curve more strongly in the direction it is already curving. Figure 17.4 shows example plots of the third-order model when β_2 and β_3 have opposite signs.

FIGURE 17.4

In the third-order polynomial model, the expected value of y is $E(y) = \beta_0 + \beta_1 x + \beta_2 x^2 + \beta_3 x^3$. When β_2 and β_3 have opposite signs, the fitted line can reverse its direction of curvature. For positive values of x and β_1, these are examples of curves when $\beta_2 > 0$ and $\beta_3 < 0$, and vice versa.

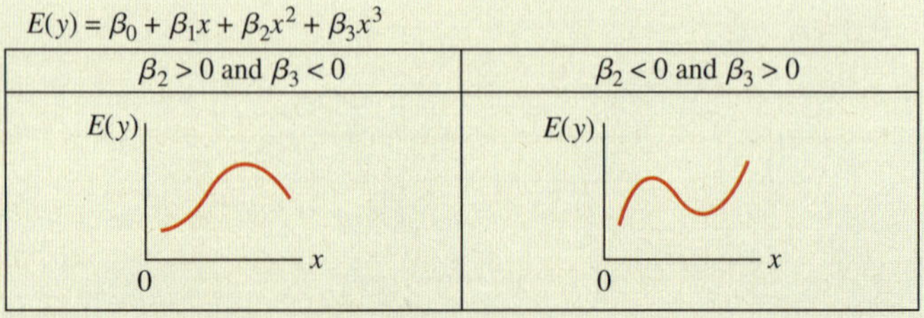

EXAMPLE

Fitting a Polynomial Model

U.S. News & World Report compiles information each year regarding leading U.S. graduate schools in a variety of fields. Data file **CX17MBA** includes the magazine's overall score for each of 25 leading MBA programs along with the average starting salary and bonus for the program's most recent graduating class.[1] In our analysis, we will use overall score as a predictor variable to estimate starting salary and bonus.

SOLUTION

Computer Solutions 17.1 describes the procedures and shows the results when we use Excel and Minitab in fitting a polynomial model to a set of data. With either software package, it is very easy to select whether the model will be first, second, or third order. Note that Minitab refers to these as linear, quadratic, or cubic, respectively. In Computer Solutions 17.1, we fit a second-order polynomial model to the data. With x = overall score and y = average starting salary and bonus (thousands of dollars), the estimation equation and coefficient of determination, R^2, are

$$\hat{y} = -39.678 + 2.887x - 0.0109x^2, \quad \text{with } R^2 = 0.885$$

COMPUTER 17.1 SOLUTIONS

Fitting a Polynomial Regression Equation, One Predictor Variable

EXCEL

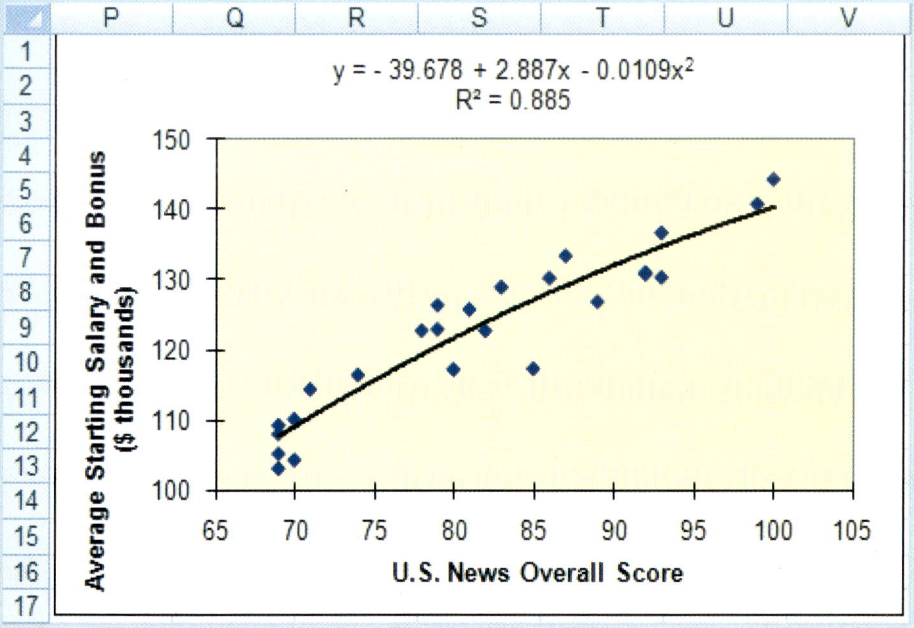

(continued)

[1]Source: *U.S. News & World Report: America's Best Graduate Schools 2009*, p. 14.

Fitting a Polynomial Regression Equation with Excel

1. Using Excel file **CX17MBA**, with MBA average starting salary and bonus data (in thousands of dollars) in A2:A26, *U.S. News & World Report* overall scores in F2:F26 and labels in A1 and F1: From the **Data** ribbon, click **Data Analysis.** Select **Regression.** Click **OK.**
2. Enter **A1:A26** into the **Input Y Range** box. Enter **F1:F26** into the **Input X Range** box. Select **Labels.** Enter **H1** into the **Output Range** Box. Select **Line Fit Plots.** Click **OK.**
3. Right-click within the legend descriptions to the right of the chart and click **Delete.** Right-click on any one of the linearly-arranged squares describing predicted values for the data points and click **Delete.**
4. Right-click on any one of the data points in the scatter diagram and click **Add Trendline.** In the **Trendline Options** menu, select **Polynomial** and enter **2** into the **Order** box. Click to select the **Display Equation on chart** and **Display R-squared value on chart** options. Click **Close.** Make formatting adjustments to the chart.

Note: To obtain the Excel printout in Table 17.1: In step 2, above, enter **F1:G26** into the **Input X Range** box and do not select **Line Fit Plots.** Omit steps 3 and 4. (The squares of the overall scores must be in column G.)

MINITAB

Fitting a Polynomial Regression Equation with Minitab

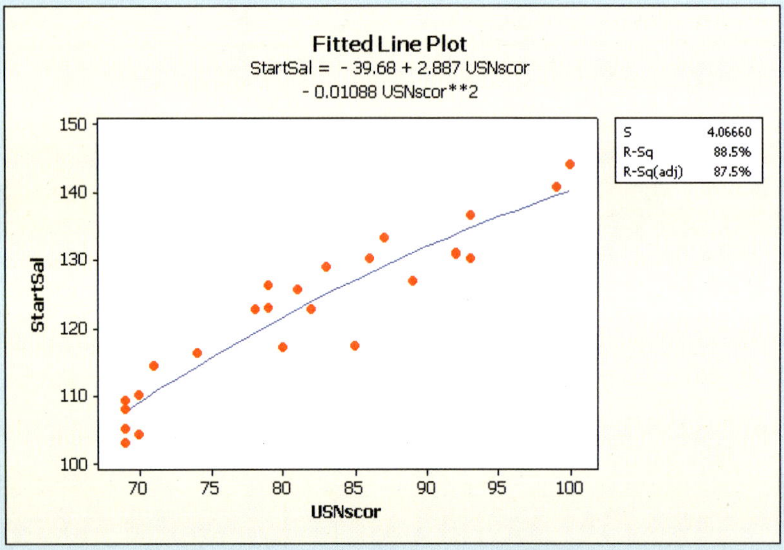

Using Minitab file **CX17MBA**, with MBA average starting salary and bonus data (in thousands of dollars) in C1 and *U.S. News & World Report* overall scores in C6: Click **Stat.** Select **Regression.** Click **Fitted Line Plot.** Enter **C1** into the **Response (Y)** box. Enter **C6** into the **Predictor (X)** box. We will be using the second-order polynomial model in this example. Within the **Type of regression model** box, select **Quadratic.** Click **OK.** The Minitab plot, fitted equation, and R^2 items appear as shown here.

Note: The Minitab printout in Table 17.1 is automatically generated when we carry out the Fitted Line Plot procedure described above.

EXAMPLE

Interpreting $R^2 = 0.885$, we can say that 88.5% of the variation in average starting salary and bonus is explained by the predictor variable, overall score, through the fitted equation. Table 17.1 shows the printouts obtained from performing a multiple regression analysis with x and x^2 treated as separate variables. These printouts provide greater detail, and we can see that the regression equation is significant at the 0.000 level (rounded). The procedures for obtaining the printouts in Table 17.1 are described in the Excel and Minitab portions of Computer Solutions 17.1.

TABLE 17.1

Regression printouts corresponding to the plots in Computer Solutions 17.1. In each case, we have created a new variable: the square of the *U.S. News & World Report* overall score. The *p*-value for the regression equation is 0.000 (rounded), and Excel reports it in scientific notation (4.64×10^{-11}).

EXCEL

	H	I	J	K	L	M	N
1	SUMMARY OUTPUT						
2							
3	*Regression Statistics*						
4	Multiple R	0.941					
5	R Square	0.885					
6	Adjusted R Square	0.875					
7	Standard Error	4.067					
8	Observations	25					
9							
10	ANOVA						
11		*df*	*SS*	*MS*	*F*	*Significance F*	
12	Regression	2	2800.90	1400.45	84.68	4.64E-11	
13	Residual	22	363.82	16.54			
14	Total	24	3164.71				
15							
16		*Coefficients*	*Standard Error*	*t Stat*	*P-value*	*Lower 95%*	*Upper 95%*
17	Intercept	-39.6777	62.6340	-0.6335	0.5329	-169.5726	90.2172
18	USNscor	2.8870	1.5309	1.8858	0.0726	-0.2879	6.0619
19	USNscor_sq	-0.0109	0.0093	-1.1756	0.2523	-0.0301	0.0083

MINITAB

```
Polynomial Regression Analysis: StartSal versus USNscor

The regression equation is
StartSal = - 39.68 + 2.887 USNscor - 0.01088 USNscor**2

S = 4.06660    R-Sq = 88.5%    R-Sq(adj) = 87.5%

Analysis of Variance

Source      DF       SS       MS      F      P
Regression   2  2800.90  1400.45  84.68  0.000
Error       22   363.82    16.54
Total       24  3164.71
```

If we use the procedures in Computer Solutions 17.1 to fit first-order and third-order polynomial models, the resulting equations and their R^2 values are as shown below.

First-order polynomial:

$$\hat{y} = 33.505 + 1.09x, \quad \text{with } R^2 = 0.878$$

Third-order polynomial:

$$\hat{y} = -1552.294 + 58.031x - 0.675x^2 + 0.0026x^3, \quad \text{with } R^2 = 0.910$$

The third-order polynomial, with $R^2 = 0.910$, explains a greater percentage of the variation in average starting salary and bonus than either of the other models. This is not surprising, because it includes two more terms than the first-order model and one more term than the second-order model. However, its R^2 advantage over the second-order model is relatively slight ($R^2 = 0.910$ versus $R^2 = 0.885$, respectively), and the greater simplicity of the second-order model may make it the most preferable of the polynomial models. In Section 17.8, we will further discuss the process of selecting the appropriate model for a set of data.

EXERCISES

17.1 The models in this section include the first-, second-, and third-order polynomials. What would be the form of a fifth-order polynomial? Would such a polynomial model tend to be practical for use with business data?

17.2 As x increases, y increases, but at a decreasing rate. If a second-order polynomial model were fitted to the scatterplot of the data, what would be the signs of the partial regression coefficients in the model?

17.3 As x increases, y decreases, and at an increasing rate. If a second-order polynomial model were fitted to the scatterplot of the data, what would be the signs of the partial regression coefficients in the model?

17.4 As the temperature increases, an experimental lubricant initially decreases in viscosity (resistance to flow), then increases, then decreases again. Of the polynomial models discussed in this section, which one would tend to be most applicable for such data? What would be the signs of the partial regression coefficient(s) in the model?

17.5 A communications firm employs a large number of technical sales representatives who make personal visits to potential customers. For data collected on y = average amount of sales per call and x = number of years of experience, the accompanying scatter diagram has been constructed. If a third-order polynomial model were fitted to these data, what would be the signs of the partial regression coefficients in the model?

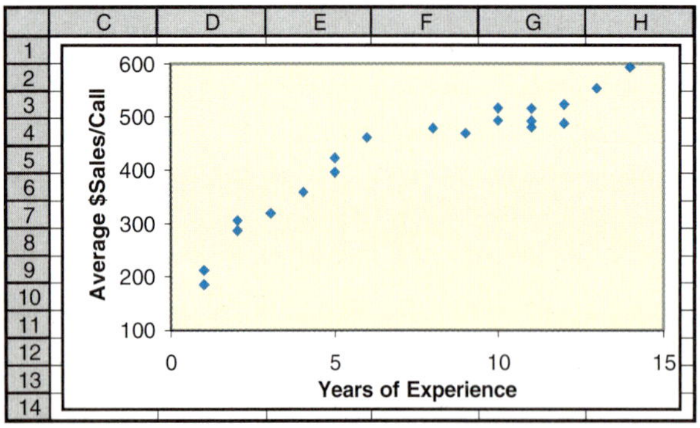

17.6 As the number of cable television subscribers has increased, the number of subscribers to satellite television services has increased as well. For y = number of satellite subscribers and x = number of cable subscribers, the scatter diagram shown here has been constructed. Of the polynomial models discussed in this section, which one would tend to be most applicable for such data, and what would be the signs of the partial regression coefficient(s) in the model? Source: *USA Today*, May 3, 2000, p. 26A.

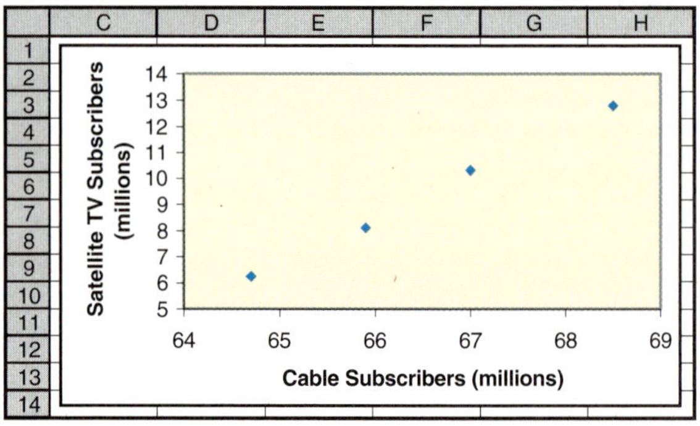

(DATA SET) *Note:* Exercises 17.7–17.10 require a computer and statistical software.

17.7 A travel research firm has reported average occupancy and room rates for hotels in 26 U.S. hotel markets. The data are also in file **XR17007**. For y = average room rate (dollars) and x = average occupancy rate (percent), fit a second-order polynomial model to the data and determine the percentage of the variation in average room rates that is explained by the model. Given the nature of these variables, comment on causation possibilities that might be present.

17.8 During the years 2005–2009, the world record for the 100-meter dash was repeatedly broken as never before. The years and corresponding record times are in data file **XR17008** and shown in the table. Using $x = 1$ through 5 to represent years 2005–2009, fit a second-order polynomial model to the data and estimate the world record time for the 100-meter dash in 2014. Source: Karen Rosen, "9.58: Bolt Strikes Again at Worlds," *USA Today*, August 17, 2009, p. 1C.

Year	x = Year code	World Record (seconds)
2005	1	9.77
2006	2	9.77
2007	3	9.74
2008	4	9.69
2009	5	9.58

17.9 An early edition of *Car and Driver* magazine reported the engine horsepower, top speed (mph), curb weight (pounds), and 0–60 mph acceleration time (seconds) for a selection of automobiles that ranged from large American "muscle cars" to small economy imports. The data are in file **XR17009**. Using y = 0–60 time and x = horsepower, fit a second-order polynomial model and predict the 0–60 time for a car with 300 horsepower. At the 0.05 level of significance, is the regression equation significant? Source: *1964 Car and Driver Yearbook*, pp. 14–63.

17.10 Repeat Exercise 17.9, but for a third-order model and using y = top speed (mph) as the dependent variable. During the 1960s, manufacturers of the so-called "muscle cars" were accused of deliberately understating the massive amount of horsepower these engines produced. Examining the curve for the third-order model, would it appear that these accusations might have been justified?

17.11 For the 20 top-grossing U.S. films of the 1990s, data have been reported for y = foreign gross ticket sales and x = domestic gross ticket sales, both in millions of dollars. The printout here shows the results for a second-order polynomial model fitted to the data. What amount of foreign gross ticket sales would be predicted for a film that generated $400 million in domestic gross ticket sales? At the 0.05 level of significance, is this model significant? Source: "Overseas Grosses Outdoing North America," *USA Today,* December 13, 2000.

	D	E	F	G	H	I	J
1	SUMMARY OUTPUT						
2							
3	*Regression Statistics*						
4	Multiple R	0.874					
5	R Square	0.764					
6	Adjusted R Square	0.736					
7	Standard Error	112.603					
8	Observations	20					
9							
10	ANOVA						
11		*df*	*SS*	*MS*	*F*	*Significance F*	
12	Regression	2	697099.23	348549.61	27.49	4.71E-06	
13	Residual	17	215551.88	12679.52			
14	Total	19	912651.11				
15							
16		*Coefficients*	*Standard Error*	*t Stat*	*P-value*	*Lower 95%*	*Upper 95%*
17	Intercept	860.8444	244.9641	3.5142	0.0027	344.0146	1377.6743
18	DomGross	-4.1517	1.4392	-2.8848	0.0103	-7.1882	-1.1153
19	DomGross_sq	0.0077	0.0019	4.0038	0.0009	0.0036	0.0117

POLYNOMIAL MODELS WITH TWO QUANTITATIVE PREDICTOR VARIABLES (17.3)

In this section, we consider polynomial models in which two quantitative predictor variables are used in estimating the value of the dependent variable. Our discussion will include three such models.

First-Order Model with Two Variables

The first-order polynomial model with two quantitative predictor variables is the same as the two-variable multiple regression model of Chapter 16. Accordingly, the format is simply

First-order model with two predictor variables:

$$E(y) = \beta_0 + \beta_1 x_1 + \beta_2 x_2$$

In this model, the expected value of y is based on x_1 and x_2, as in the multiple regression models of Chapter 16. The model is applicable when there is *no interaction* between x_1 and x_2. Interaction between x_1 and x_2 is absent whenever the following conditions are met:

1. For a given value of x_1, a 1-unit increase in x_2 will cause $E(y)$ to increase by β_2.
2. For a given value of x_2, a 1-unit increase in x_1 will cause $E(y)$ to increase by β_1.

In other words, interaction between the two predictor variables is absent when the effect on $E(y)$ of a 1-unit increase in x_2 does not depend on the value of x_1, and vice versa. Figure 17.5 shows a situation in which interaction is absent. For the estimation equation $\hat{y} = 10 + 5x_1 + 3x_2$, we have plotted three separate linear relationships between $\hat{y}$ and x_1: one for each of the assumed values of x_2. It is very important to note that the lines are *parallel*—regardless of the value of x_2, the slope is the same.

FIGURE 17.5

In this first-order polynomial model with two quantitative predictor variables, there is no interaction between x_1 and x_2. Regardless of the value of x_2, the slope of the line relating $\hat{y}$ to x_1 remains the same.

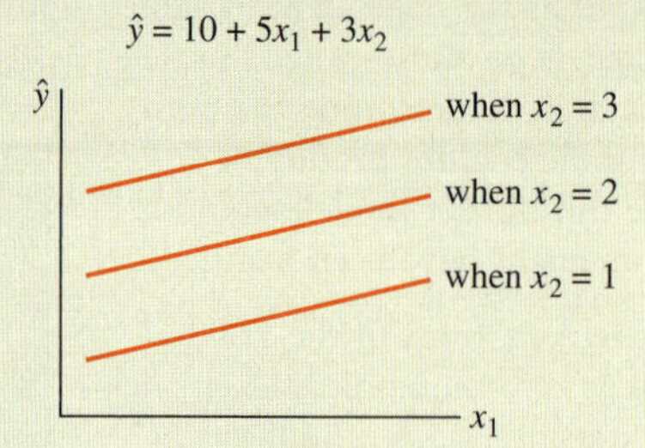

FIGURE 17.6

In this first-order polynomial model with an interaction term included, there is interaction between x_1 and x_2. The slope of the relationship between $\hat{y}$ and x_1 depends on the value of x_2.

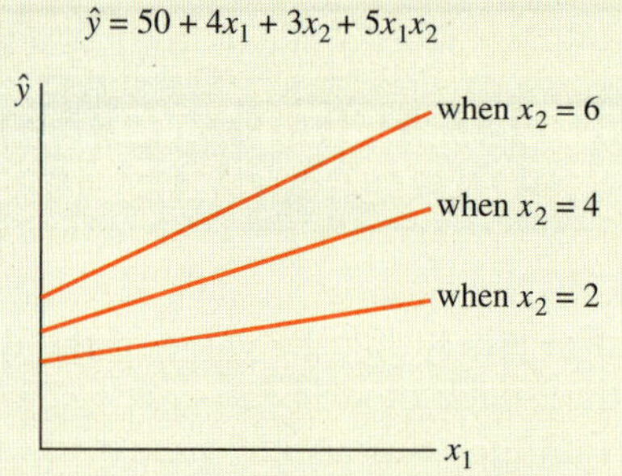

First-Order Model with an Interaction Term

When we believe there may be interaction between the two predictor variables, we can enhance the first-order model just discussed by adding an interaction term. The format for this model is

First-order model with two predictor variables and interaction:

$$E(y) = \beta_0 + \beta_1 x_1 + \beta_2 x_2 + \beta_3 x_1 x_2$$

In this model, the interaction term is $\beta_3 x_1 x_2$. The term includes the partial regression coefficient β_3 and the cross product $x_1 x_2$. Because the sum of the exponents in the cross product is 2, this model can technically be considered a second-order model. However, to avoid confusion with what is typically referred to as the "full" second-order model discussed next, we will consider this to be a first-order model that has had an interaction term added to it.

In this model, interaction is present because the effect on $E(y)$ of a 1-unit increase in x_2 will depend on the value of x_1, and vice versa. Figure 17.6 shows a situation in which interaction is present. For the estimation equation $\hat{y} = 50 + 4x_1 + 3x_2 + 5x_1x_2$, we have plotted three separate linear relationships between $\hat{y}$ and x_1: one for each of the assumed values of x_2. Note that the lines are not parallel and that the slope of the relationship between $\hat{y}$ and x_1 depends on the value of x_2.

Second-Order Model with Interaction

When, in addition to the possibility of interaction between the predictor variables, we believe there could be a nonlinear relationship between y and each of the predictor variables, we can use the second-order model with interaction. The format for this model is

Second-order model with interaction:

$$E(y) = \beta_0 + \beta_1 x_1 + \beta_2 x_2 + \beta_3 x_1^2 + \beta_4 x_2^2 + \beta_5 x_1 x_2$$

In addition to the interaction term $\beta_5 x_1 x_2$, this model includes the quadratic terms $\beta_3 x_1^2$ and $\beta_4 x_2^2$. As mentioned previously, this model is typically referred to as the "full" second-order model—not only do the exponents in the interaction term sum to 2, but each predictor variable is in a term where it is squared.

EXAMPLE

Polynomial Model with Two Predictor Variables

Residents of warm climates are familiar with the heat index that estimates how hot it feels outside. This index is based on two variables: air temperature and relative humidity. File **CX17HEAT** contains data showing the heat index for various combinations of air temperature and relative humidity.[2] The file also includes, as separate variables, the temperature-squared, relative humidity-squared, and (temperature × relative humidity) values. In Computer Solutions 17.2, we use Excel and Minitab to obtain a second-order polynomial model with two predictor variables and an interaction term. The Minitab equation is

$$\text{FeelsLike} = 127.74 - 1.76\text{AirTemp} - 1.95\text{RelHumid} + 0.0134\text{AirTemp\_sq} + 0.0011\text{RelHumid\_sq} + 0.0259(\text{AirTemp} \times \text{RelHumid})$$

SOLUTION

Interpreting the R^2 portion of the printouts, we can say that 98.0% of the variation in heat index is explained by this model. However, if we apply either of the two simpler models discussed in this section, we notice that one of them ends up obtaining nearly the same R^2 value for these data. The procedures for applying these models are similar to the steps described in Computer Solutions 17.2, and the resulting Minitab equations and R^2 values are

First-order polynomial:

$$\text{FeelsLike} = -55.13 + 1.52\text{AirTemp} + 0.341\text{RelHumid}, \quad \text{with } R^2 = 0.855$$

First-order polynomial, with interaction term added:

$$\text{FeelsLike} = 2.54 + 0.867\text{AirTemp} - 1.467\text{RelHumid} + 0.0212(\text{AirTemp} \times \text{RelHumid}), \quad \text{with } R^2 = 0.968$$

[2] *Time Almanac 2009*, p. 176.

Compared to the first-order model with no interaction, including the interactive term improves R^2 quite substantially, from 0.855 to 0.968. However, if we also include the square terms, R^2 goes up just slightly, to 0.980. Because the simpler interactive model does practically as well as the more complex one in its ability to explain variation in the heat index values, it would be the preferred model to fit to these data.

COMPUTER 17.2 SOLUTIONS

Fitting a Polynomial Regression Equation, Two Predictor Variables

EXCEL

	G	H	I	J	K	L	M	N
1	SUMMARY OUTPUT							
2								
3	*Regression Statistics*							
4	Multiple R	0.990						
5	R Square	0.980						
6	Adjusted R Square	0.979						
7	Standard Error	3.317						
8	Observations	93						
9								
10	ANOVA							
11		*df*	*SS*	*MS*	*F*	*Significance F*		
12	Regression	5	46350.120	9270.024	842.617	0.00000		
13	Residual	87	957.127	11.001				
14	Total	92	47307.247					
15								
16		*Coefficients*	*Standard Error*	*t Stat*	*P-value*	*Lower 95%*	*Upper 95%*	*Lower 95.0%*
17	Intercept	127.74311	17.8207	7.1683	0.00000	92.32260	163.16362	92.3226038
18	AirTemp	-1.75589	0.3701	-4.7449	0.00001	-2.49142	-1.02037	-2.4914155
19	RelHumid	-1.95291	0.1301	-15.0099	0.00000	-2.21151	-1.69430	-2.2115116
20	AirTemp_sq	0.01341	0.0019	7.0907	0.00000	0.00965	0.01717	0.009654
21	RelHumid_sq	0.00108	0.0005	2.2802	0.02504	0.00014	0.00202	0.00013838
22	AirTempxRelHumid	0.02586	0.0012	21.1306	0.00000	0.02342	0.02829	0.02342466

Fitting a Polynomial Regression Equation, Two Predictors, with Excel

1. The heat index data are in Excel file **CX17HEAT**, with heat index values in A2:A94, temperatures (degrees Fahrenheit) in B2:B94, relative humidity values in C2:C94, and their labels in A1:C1. The squares of the temperature and relative humidity values are in columns D and E, respectively, and the cross-products (temperature × relative humidity) are in column F.
2. We will generate the regression equation for the second-order model with interaction: From the **Data** ribbon, click **Data Analysis.** Select **Regression.** Click **OK.**
3. Enter **A1:A94** into the **Input Y Range** box. Enter **B1:F94** into the **Input X Range** box. Select **Labels.** Enter **G1** into the **Output Range** box. Click **OK.** After slight adjustments to the column widths and the number of decimal places, the results are as shown in the printout.

MINITAB

Fitting a Polynomial Regression Equation, Two Predictors, with Minitab

```
Regression Analysis: FeelsLike versus AirTemp, RelHumid, ...

The regression equation is
FeelsLike = 128 - 1.76 AirTemp - 1.95 RelHumid + 0.0134 AirTemp_sq
            + 0.00108 RelHumid_sq + 0.0259 AirTempxRelHumid

Predictor               Coef     SE Coef        T      P
Constant              127.74       17.82     7.17  0.000
AirTemp              -1.7559      0.3701    -4.74  0.000
RelHumid             -1.9529      0.1301   -15.01  0.000
AirTemp_sq          0.013414    0.001892     7.09  0.000
RelHumid_sq        0.0010784   0.0004730     2.28  0.025
AirTempxRelHumid    0.025857    0.001224    21.13  0.000

S = 3.31685   R-Sq = 98.0%   R-Sq(adj) = 97.9%

Analysis of Variance

Source          DF       SS      MS       F      P
Regression       5  46350.1  9270.0  842.62  0.000
Residual Error  87    957.1    11.0
Total           92  47307.2
```

The heat index data are in Minitab file **CX17HEAT**, with heat index values in C1, temperatures (degrees Fahrenheit) in C2, relative humidity values in C3, temperature-squared values in C4, relative humidity-squared values in C5, and the cross-products (temperature × relative humidity) in C6: Click **Stat.** Select **Regression.** Click **Regression.** Enter **C1** into the **Response** box. Enter **C2-C6** into the **Predictors** box. Click **OK.**

EXERCISES

17.12 A researcher suspects that the dependent variable y is linearly related to both x_1 and x_2, but believes there is little or no interaction between the two predictor variables. Of the models in this section, which one would likely be most suitable for this situation?

17.13 In Exercise 17.12, suppose the researcher wishes to use one of the models in this section, but is uncertain as to whether y might be nonlinearly related to x_1 and/or x_2. Also, the researcher doesn't have any idea as to whether interaction might exist between x_1 and x_2. Of the models in this section, which one would result in the highest R^2 value while at the same time reducing the possibility that the researcher might "miss" an important aspect of the relationship between the dependent and predictor variables?

17.14 The regression model $\hat{y} = 200 + 5x_1 - 2x_2$ has been fitted to a set of data. If $x_2 = 10$, what will be the effect on $\hat{y}$ if x_1 increases by 1 unit? Will your answer change if $x_2 = 20$? Given the preceding, comment on whether interaction appears to exist between x_1 and x_2.

(DATA SET) *Note:* Exercises 17.15–17.20 require a computer and statistical software.

17.15 The personnel director of the communications firm in Exercise 17.5 has just provided another variable describing the company's technical sales representatives: their scores on a sales aptitude test they were recently given. Using data file **XR17015**, with y = average dollar sales per call, x_1 = years of experience, and x_2 = score on the sales aptitude test, fit a second-order model with interaction to the data. Interpret the R^2 value for the model and indicate whether, at the 0.05 level, the model is significant.

17.16 *Auto Rental News* has reported U.S. operations data for 15 top rental-car companies. Using data file **XR17016**, with y = rental revenue (millions of dollars), x_1 = thousands of cars in service, and x_2 = number of rental locations: Source: *Wall Street Journal Almanac 1999*, p. 649.

a. Fit a first-order model of the form $E(y) = \beta_0 + \beta_1x_1 + \beta_2x_2$. Interpret the R^2 value for the model and indicate whether, at the 0.01 level, the model is significant.
b. Repeat part (a), but with an interaction term included. Interpreting the R^2 value, to what extent has inclusion of the interaction term improved the model's explanatory power?

17.17 The Air Transport Association has reported the following kinds of information describing 23 individual U.S. airlines: total operating revenue, number of aircraft, number of employees, and number of aircraft departures. Using data file **XR17017**, with y = total operating revenue (millions of dollars), x_1 = number of employees, and x_2 = number of aircraft departures: Source: *Wall Street Journal Almanac 1999*, p. 644.

a. Fit a first-order model of the form $E(y) = \beta_0 + \beta_1x_1 + \beta_2x_2$. Interpret the R^2 value for the model and indicate whether, at the 0.01 level, the model is significant.
b. Repeat part (a), but with an interaction term included. Interpreting the R^2 value, to what extent has inclusion of the interaction term improved the model's explanatory power?

17.18 Repeat Exercise 17.17, using the data in file **XR17017**, but with y = total operating revenue (millions of dollars), x_1 = number of aircraft, and x_2 = number of employees.

17.19 For the *Car and Driver* information described in Exercise 17.9, it would seem reasonable that the number of seconds required to accelerate to 60 mph would depend on both a car's weight and its horsepower. Further, it would be logical to assume that horsepower and weight might interact. Using data file **XR17009**, with y = 0–60 mph acceleration time (seconds), x_1 = horsepower, and x_2 = curb weight (pounds), fit a model of the form $E(y) = \beta_0 + \beta_1x_1 + \beta_2x_2 + \beta_3x_1x_2$. Interpret the R^2 value for the model and indicate whether, at the 0.02 level, the model is significant.

17.20 Repeat Exercise 17.19, but using a second-order model with interaction. Interpreting the R^2 value, to what extent has inclusion of the squared terms improved the model's explanatory power?

(17.4) QUALITATIVE VARIABLES

In Chapter 16, we briefly discussed the idea of including a binary qualitative variable in the regression analysis. Such a variable has two possible values, typically coded as 1 and 0. In examining home prices, whether a home has central air conditioning can be an informative variable. In studying employee salaries and wages, using gender as a qualitative variable can help in determining whether discrimination might exist within a company's workforce. In this section, we begin by reviewing the inclusion of a qualitative variable representing two possible categories, then go on to consider qualitative variables that happen to have more than two possible values. In regression analysis, qualitative variables are sometimes referred to as *dummy variables.*

Qualitative Variable Representing Two Possible Categories

One of the most important variables in determining a vehicle's highway fuel economy is weight. However, body style could also be important, since some types of vehicles are not very aerodynamic and would tend to get fewer miles per gallon at highway speeds. In a multiple regression of highway fuel economy on

(1) weight and (2) whether the vehicle is a sports-utility vehicle, the format of the fitted equation is

$$\text{HwyMpg} = b_0 + b_1\text{Weight} + b_2\text{SUV?},$$

where HwyMpg = highway miles per gallon
Weight = curb weight (pounds)
SUV? = sports-utility vehicle dummy variable (1 if vehicle is an SUV, 0 if not)

EXAMPLE

Qualitative Variable with Two Categories

For a sample of vehicles described in *Consumer Reports' New Car Preview 2001,* data include highway fuel economy test results, vehicle weights, and vehicle types.[3] Data file **CX17FUEL** describes 53 sports-utility vehicles, large or family sedans, and small cars in terms of their highway fuel economy and their curb weight. With the variable SUV? coded as 1 for sports-utility vehicles and 0 for other vehicles, we obtain the following fitted regression equation and R^2:

$$\text{HwyMpg} = 52.3 - 0.00534\text{Weight} - 6.01\text{SUV?}, \quad \text{with } R^2 = 0.784$$

SOLUTION

Interpreting this equation, we can say that the estimated highway fuel economy will decrease by 0.00534 mpg for each additional pound of curb weight. The coefficient for the SUV? variable indicates that, for a given curb weight, the estimated highway fuel economy will be 6.01 mpg lower if the vehicle is an SUV. The regression equation explains 78.4% of the variation in highway fuel economy for the vehicles in the sample.

EXAMPLEEXAMPLEEXAMPLEEXAMPLE

Qualitative Variables Representing More Than Two Possible Categories

With slight modification, the strategy we've just employed can also be used when a qualitative variable happens to represent more than two possible categories. For the new-car data described previously, we have $n = 3$ types of vehicles in data file **CX17FUEL**, but we will need only $(n - 1) = 2$ qualitative variables to represent

[3]Consumer Reports, *New Car Preview 2001.*

them. Using SUV? and FamSedan? as the two qualitative variables, our coding scheme is as follows:

SUV? =	FamSedan? =	Represents:
1	0	sports-utility vehicle
0	1	large/family sedan
0	0	small car

A portion of the data from file **CX17FUEL** is shown here; it represents the first five vehicles in the listing.

HwyMpg	Weight	SUV?	FamSedan?	
26	3225	1	0	← sports-utility vehicle
33	3455	0	1	← large/family sedan
30	4070	0	1	
38	3325	0	1	
39	2780	0	0	← small car

EXAMPLE

Qualitative Variable with More Than Two Categories

For the sample of vehicles in data file **CX17FUEL**, and using the coding scheme described above, we carry out a multiple regression with HwyMpg as the dependent variable, and Weight, SUV?, and FamSedan? as the predictor variables. The procedures and results for Excel and Minitab are shown in Computer Solutions 17.3. The equation can be expressed as

$$\text{HwyMpg} = 50.54 - 0.00428\text{Weight} - 8.62\text{SUV?} - 2.58\text{FamSedan?},$$
$$\text{with } R^2 = 0.799$$

SOLUTION

Interpreting this equation, we can say that the estimated highway fuel economy will decrease by 0.00428 mpg for each additional pound of curb weight. The estimated highway fuel economy will be 8.62 mpg lower if the vehicle is an SUV and 2.58 mpg lower if it is a large or family sedan. If the vehicle is a small car (i.e., SUV? = 0 and FamSedan? = 0), the estimated highway fuel economy will be 50.54 − 0.00428Weight − 8.62(0) − 2.58(0), or simply 50.54 − 0.00428Weight. The equation explains 79.9% of the variation in highway fuel economy.

Figure 17.7 describes the linear estimation equations when we consider the three types of vehicles individually. For the equation estimating highway fuel economy only for small cars, SUV? = 0, FamSedan? = 0, and the constant in the equation is 50.54. The constants in the corresponding sports-utility vehicle and large/family sedan equations will be smaller. As shown in Figure 17.7, the

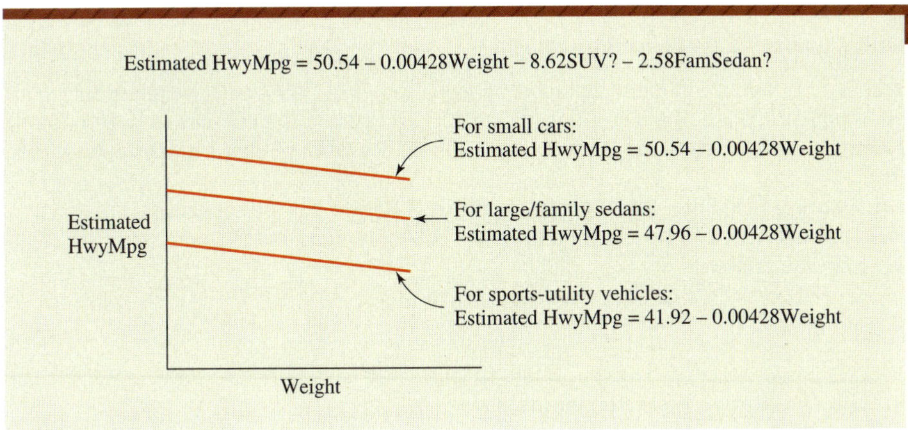

FIGURE 17.7

Given the multiple regression equation developed in Computer Solutions 17.3, these are the highway fuel-economy estimation equations for the individual types of vehicles in the analysis.

constant in the estimation equation for sports-utility vehicles is 50.54 − 8.62, or 41.92, and the constant in the equation for large/family sedans is 50.54 − 2.58, or 47.96.

EXAMPLE

COMPUTER 17.3 SOLUTIONS

Multiple Regression with Qualitative Predictor Variables

EXCEL

	G	H	I	J	K	L	M
1	SUMMARY OUTPUT						
2							
3	*Regression Statistics*						
4	Multiple R	0.894					
5	R Square	0.799					
6	Adjusted R Square	0.787					
7	Standard Error	3.089					
8	Observations	53					
9							
10	ANOVA						
11		*df*	*SS*	*MS*	*F*	*Significance F*	
12	Regression	3	1856.24	618.75	64.86	0.0000	
13	Residual	49	467.46	9.54			
14	Total	52	2323.70				
15							
16		*Coefficients*	*Standard Error*	*t Stat*	*P-value*	*Lower 95%*	*Upper 95%*
17	Intercept	50.5445	2.506	20.168	0.0000	45.508	55.581
18	Weight	-0.0043	0.001	-4.738	0.0000	-0.006	-0.002
19	SUV?	-8.6198	1.764	-4.885	0.0000	-12.166	-5.074
20	FamSedan?	-2.5828	1.348	-1.915	0.0613	-5.292	0.127

(continued)

Excel Multiple Regression with Qualitative Predictor Variables

1. The new-car data are in Excel file **CX17FUEL** with highway fuel economy (mpg) in A2:A54, curb weight (pounds) in B2:B54, whether the vehicle is an SUV (coded as 1 if yes, 0 if no) in C2:C54, whether the vehicle is a large or family sedan (coded as 1 if yes, 0 if no) in D2:D54, and their labels in A1:D1. From the **Data** ribbon, click **Data Analysis.** Select **Regression.** Click **OK.**
2. Enter **A1:A54** into the **Input Y Range** box. Enter **B1:D54** into the **Input X Range** box. Select **Labels.** Enter **G1** into the **Output Range** box. Click **OK.** After slight adjustments to the column widths and the number of decimal places, the results are as shown in the printout.

MINITAB

Minitab Multiple Regression with Qualitative Predictor Variables

```
Regression Analysis: HwyMpg versus Weight, SUV?, FamSedan?

The regression equation is
HwyMpg = 50.5 - 0.00428 Weight - 8.62 SUV? - 2.58 FamSedan?

Predictor          Coef     SE Coef       T       P
Constant         50.545       2.506   20.17   0.000
Weight       -0.0042766   0.0009026   -4.74   0.000
SUV?             -8.620       1.764   -4.89   0.000
FamSedan?        -2.583       1.348   -1.92   0.061

S = 3.08868    R-Sq = 79.9%    R-Sq(adj) = 78.7%

Analysis of Variance

Source           DF       SS      MS      F       P
Regression        3  1856.24  618.75  64.86   0.000
Residual Error   49   467.46    9.54
Total            52  2323.70
```

Using Minitab file **CX17FUEL**, with highway fuel economy (mpg) in C1, curb weight (pounds) in C2, whether the vehicle is an SUV (coded as 1 if yes, 0 if no) in C3, and whether the vehicle is a large or family sedan (coded as 1 if yes, 0 if no) in C4: Click **Stat.** Select **Regression.** Click **Regression.** Enter **C1** into the **Response** box. Enter **C2–C4** into the **Predictors** box. Click **OK.**

EXERCISES

17.21 For a sample of 100 recent credit-card customers of a company's on-line store, the following data are available: the dollar amount of merchandise ordered during the visit, how much time (minutes) the customer spent at the site during the purchase visit, and whether the customer used a Visa credit card. Describe the format of a linear regression equation for estimating the dollar amount of the order on the basis of the other variables described above. If any qualitative variables are used, indicate the strategy for coding them.

17.22 Repeat Exercise 17.21, but assume that more information is available regarding the specific type of credit card that was used for the purchase, that is, Visa, MasterCard, or other.

17.23 Data to be used in a linear regression analysis involving clients of an investment counselor include information regarding the risk level each client has specified: low, medium, or high. Determine the number of qualitative independent variables that will be required to represent this information and describe the coding scheme for each.

17.24 Repeat Exercise 17.23, but for clients whose risk levels have been specified as low, medium, high, or very high.

17.25 The manager of an ice cream store has developed a regression equation for estimating the number of customers on the basis of two independent variables: x_1 = the high temperature (degrees Fahrenheit) for the day, and x_2 = whether the day is during the weekend (coded as 1) or is a weekday (coded as 0). For the estimation equation $\hat{y} = 150 + 5x_1 + 50x_2$, interpret the partial regression coefficients, then estimate the number of customers on an 80-degree Saturday.

17.26 The dean of a law school has developed a regression equation for estimating the starting salary (thousands of dollars) of a new graduate on the basis of two independent variables: x_1 = the score achieved on the Law School Admission Test (LSAT) at the time of application, and x_2 = whether the new graduate's position is in the private sector (coded as 1) or the public sector (coded as 0). For the estimation equation $\hat{y} = 50 + 0.1x_1 + 30x_2$, interpret the partial regression coefficients, then estimate the starting salary for a new graduate with an LSAT score of 160 who is entering the private sector.

(DATA SET) *Note:* Exercises 17.27–17.30 require a computer and statistical software.

17.27 In a *PC World* comparison of computer hard drives, data describing the drives included the following variables: price (dollars), disk capacity (gigabytes), and rotational speed (RPM). The data are in file **XR17027**. With y = price, x_1 = disk capacity, and x_2 = rotational speed (coded as 1 = 7200 RPM and 0 = 5400 RPM), determine the linear regression equation for estimating price as a function of the two predictor variables and interpret the partial regression coefficients. For a given storage capacity, how much extra would one expect to pay for a hard drive with the higher rotational speed? Source: Stan Miastkowski, "Livin' Large," *PC World*, March 2001, p. 109.

17.28 A researcher has observed 50 automatic teller machine (ATM) users outside the gift shop of a museum, and the data are in file **XR17028**. With y = amount withdrawn (dollars), x_1 = time required for the transaction (seconds), and x_2 = gender (1 = female, 0 = male), determine the linear regression equation for estimating the withdrawal amount as a function of the two predictor variables and interpret the partial regression coefficients. According to the equation, how much difference does gender tend to make in the amount withdrawn?

17.29 An efficiency expert has studied 12 employees who perform similar assembly tasks, recording productivity (units per hour), number of years of experience, and which one of three popular assembly methods the individual has chosen to use in performing the task. Given the data shown here, determine the linear regression equation for estimating productivity based on the other variables. For any qualitative variables that are used, be sure to specify the coding strategy employed for each. Comment on the explanatory power of the regression equation and interpret the partial regression coefficients.

Worker	Productivity units/hour	Years of Experience	Assembly Method
1	75	7	A
2	88	10	C
3	91	4	B
4	93	5	B
5	95	11	C
6	77	3	A
7	97	12	B
8	85	10	C
9	102	12	C
10	93	13	A
11	112	12	B
12	86	14	A

17.30 For a small sample of suburban homes recently sold, the selling price ($), living space (square feet), and suburban area (Sunny Haven, Easy Acres, or Rolling Hills) have been recorded. Given the data shown here, determine the linear regression equation for estimating selling price based on the other variables. For any qualitative variables that are used, be sure to specify the coding strategy employed for each. Comment on the explanatory power of the regression equation and interpret the partial regression coefficients.

Home	Price	Space	Area
1	167,900	2860	Easy
2	184,800	2050	Sunny
3	161,000	2110	Rolling
4	160,800	2950	Easy
5	148,200	1570	Sunny
6	128,600	1860	Rolling
7	145,700	2620	Rolling
8	178,500	2010	Sunny
9	146,100	2710	Easy
10	146,300	2200	Rolling

DATA TRANSFORMATIONS 17.5

When the relationship between the dependent variable and one or more predictor variables is not linear, one of the polynomial models discussed in Sections 17.2 and 17.3 will often provide satisfactory results. However, we sometimes encounter variables for which the underlying relationship is better described by other models that can be converted to linear form. This is possible through *data transformation,* which is what we did in Section 17.2 when we complemented x by forming the x^2 and x^3 terms in the third-order model. Many other types of transformations are possible, such as converting x to its square root, its inverse, or its logarithm. We will examine two nonlinear models that rely on logarithmic transformations to achieve linearity.

For the **exponential model** with estimation equation $\hat{y} = b_0 b_1^x$, conversion to a linear form involves taking the logarithms of both sides of the equation. Either natural logarithms (base $e = 2.71828\ldots$) or common logarithms (base 10) can be used, as long as we are consistent. We will be using common logarithms. The exponential estimation equation is converted to linear form as follows:

Conversion of the exponential equation, $\hat{y} = b_0 b_1^x$, to linear form:

$$\log \hat{y} = \log b_0 + x \log b_1, \text{ or } \log \hat{y} = \log b_0 + (\log b_1)x$$

As a brief review of logarithms, recall that a number is converted to a common logarithm by expressing it in terms of the power to which 10 must be raised. Also, multiplying two numbers is equivalent to adding their logarithms, and raising a quantity to a power is the same as multiplying its logarithm by that power. For example,

- **Logarithms and multiplication**

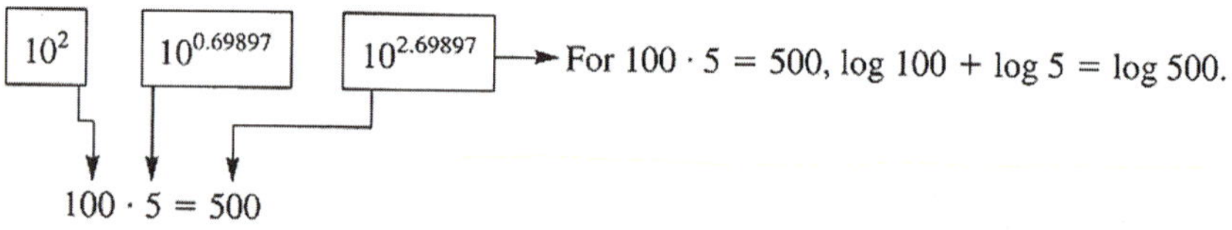

- **Logarithms and exponentiation**

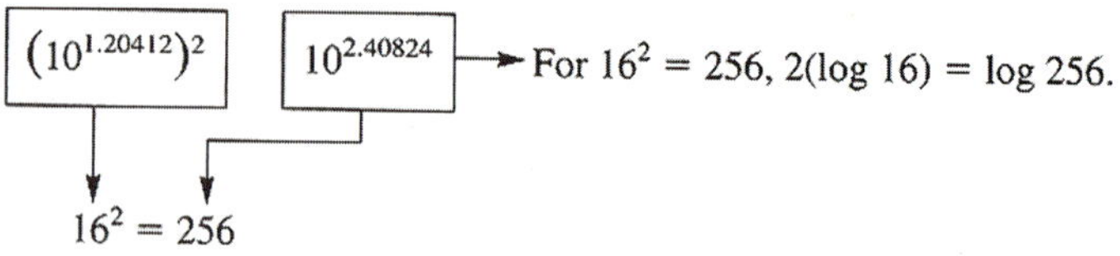

The second model we will consider is the **multiplicative model.** If there are two predictor variables, the estimated value of the dependent variable will be $\hat{y} = b_0 x_1^{b_1} x_2^{b_2}$. The multiplicative equation is converted to linear form as shown next.

Conversion of the multiplicative equation, $\hat{y} = b_0 x_1^{b_1} x_2^{b_2}$, to linear form:

$$\log \hat{y} = \log b_0 + b_1 \log x_1 + b_2 \log x_2$$

As an example of the logarithmic transformation of data, we will use data file **CX17ACC**, which lists the following variables for a sample consisting of sports cars and sedans described in an early issue of the *Car and Driver Yearbook*:[4]

y = 0–60 mph acceleration time (seconds)
x_1 = engine horsepower
x_2 = curb weight (pounds)

Considering the possibility that the time to reach 60 mph might be directly proportional to horsepower and inversely proportional to curb weight (in essence, a multiplicative model with estimation equation $\hat{y} = b_0 x_1^{b_1} x_2^{b_2}$), we can use logarithms in converting this model to linear form. This can be done with either Excel or Minitab, and the procedures and results are shown in Computer Solutions 17.4. The equation and the result when it is converted back to multiplicative form are shown here:

Logarithmic form: Log_0–60 = −0.0528 − 0.8607 log_hp + 0.8691 log_wt

Multiplicative form: 0–60 = $0.8855 \times \text{hp}^{-0.8607} \times \text{curbwt}^{0.8691}$

In this conversion, the antilog of −0.0528 is $10^{-0.0528} = 0.8855$. The estimated 0–60 mph time for a 100-horsepower car weighing 2000 pounds is calculated as $0.8855 \times 100^{-0.8607} \times 2000^{0.8691} = 12.44$ seconds.

In applying the models in this section, all of the original data values must be greater than zero. Also, because the dependent variable is log y instead of y, a comparison of R^2 values with those from regression models having y as the dependent variable may not be appropriate because of the differing units of measurement.[5]

COMPUTER 17.4 SOLUTIONS

Transformation of the Multiplicative Model

EXCEL

	G	H	I	J	K	L	M
1	SUMMARY OUTPUT						
2							
3	*Regression Statistics*						
4	Multiple R	0.923					
5	R Square	0.852					
6	Adjusted R Square	0.836					
7	Standard Error	0.066					
8	Observations	21					
9							
10	ANOVA						
11		*df*	*SS*	*MS*	*F*	*Significance F*	
12	Regression	2	0.455	0.228	52.014	0.000	
13	Residual	18	0.079	0.004			
14	Total	20	0.534				
15							
16		*Coefficients*	*Standard Error*	*t Stat*	*P-value*	*Lower 95%*	*Upper 95%*
17	Intercept	-0.0528	0.5531	-0.0954	0.925	-1.2148	1.1092
18	Log_hp	-0.8607	0.1062	-8.1022	0.000	-1.0839	-0.6375
19	Log_wt	0.8691	0.2170	4.0042	0.001	0.4131	1.3251

(continued)

[4]*1964 Car and Driver Yearbook*, pp. 14–63.
[5]T. Dielman, *Applied Regression Analysis*, Pacific Grove, CA: Duxbury Press, 1996, p. 240.

Excel Transformation of the Multiplicative Model

1. The data are in Excel file **CX17ACC**, with 0–60 mph times (seconds) in A2:A22, horsepower ratings in B2:B22, curb weights (pounds) in C2:C22, and their labels in A1:C1.
2. Click to select cell D2. From the **Formulas** ribbon, click the f_x **(Insert Function)** symbol. In the **Category** box, select **Math & Trig.** Within the **Function** box, select **LOG10.** Click **OK.** Enter **A2** into the **Number** box. Click **OK.** The logarithm of the first 0–60-mph time will now be in cell D2.
3. Click the lower right corner of cell D2 (cursor turns to "+" sign), then drag and fill to F2. Similarly, click the lower right corner of F2, then drag and fill to F22. Enter the column labels (**Log_0-60, Log_hp,** and **Log_wt,** respectively) into D1:F1.
4. From the **Data** ribbon, click **Data Analysis.** Select **Regression.** Click **OK.**
5. Enter **D1:D22** into the **Input Y Range** box. Enter **E1:F22** into the **Input X Range** box. Select **Labels.** Enter **G1** into the **Output Range** box. Click **OK.** After slight adjustments to the column widths and the number of decimal places, the results are as shown above.

MINITAB

Minitab Transformation of the Multiplicative Model

```
Regression Analysis: Log_60 versus Log_hp, Log_wt

The regression equation is
Log_60 = - 0.053 - 0.861 Log_hp + 0.869 Log_wt

Predictor        Coef    SE Coef       T       P
Constant      -0.0528     0.5531   -0.10   0.925
Log_hp        -0.8607     0.1062   -8.10   0.000
Log_wt         0.8691     0.2170    4.00   0.001

S = 0.0661493    R-Sq = 85.2%    R-Sq(adj) = 83.6%

Analysis of Variance

Source           DF        SS        MS       F       P
Regression        2   0.45520   0.22760   52.01   0.000
Residual Error   18   0.07876   0.00438
Total            20   0.53396
```

1. Using Minitab file **CX17ACC**, 0–60 times in C1, horsepower ratings in C2, and curb weights in C3: Click **Calc.** Click **Calculator.** In the **Functions** box, select **Log base 10.** Within the parentheses in the **Expression** box, enter **C1.** Enter **C4** into the **Store result in variable** box. Click **OK.**
2. Repeat this procedure two times, placing the logarithms for C2 into column C5 and placing the logs for C3 into C6.
3. Enter the **Log_0-60, Log_hp,** and **Log_wt** labels at the top of columns **C4–C6,** respectively. Click **Stat.** Select **Regression.** Click **Regression.** Enter **C4** into the **Response** box. Enter **C5–C6** into the **Predictors** box. Click **OK.**

The transformations in this section are but two of the many that can be done. For a more complete discussion of the possibilities and their applications, the interested reader may refer to one of the more advanced treatments cited in the footnote.[6]

[6]See Dielman or J. Neter, M. Kutner, C. Nachtsheim, and W. Wasserman, *Applied Linear Statistical Models,* Homewood, IL: Irwin, 1996.

EXERCISES

17.31 Following data transformation, regression analysis results in the estimation equation $\log \hat{y} = -0.179 + 0.140x$. Transform this equation to the equivalent exponential model with estimation equation $\hat{y} = b_0 b_1^x$. (The base of the logarithms is 10.)

17.32 Following data transformation, regression analysis results in the estimation equation $\log \hat{y} = 3.15 + 0.473 \log x_1 + 0.354 \log x_2$. Transform this equation to the equivalent multiplicative model with estimation equation $\hat{y} = b_0 x_1^{b_1} x_2^{b_2}$. (The base of the logarithms is 10.)

(DATA SET) *Note:* Exercises 17.33–17.36 require a computer and statistical software. Use logarithms with base 10.

17.33 For the situation and variables in Exercise 17.7, fit an exponential model and use it to estimate the average room rate for a market in which the average occupancy rate is 70%. The underlying data are in file **XR17007**.

17.34 For the situation and variables in Exercise 17.8, fit an exponential model and use it to estimate the world record time for the 100-meter dash in the year 2014. The underlying data are in file **XR17008**.

17.35 For the situation and variables in Exercise 17.17, fit a multiplicative model and use it to estimate the total operating revenue for an airline with 2000 employees and 10,000 departures. The underlying data are in file **XR17017**.

17.36 For the situation and variables in Exercise 17.16, fit a multiplicative model and use it to estimate the rental revenue for a company with 80,000 cars in service and 2500 rental locations. The underlying data are in file **XR17016**.

MULTICOLLINEARITY (17.6)

In Chapter 16, we briefly discussed *multicollinearity*, a condition in which two or more of the independent variables are highly correlated with each other. When this happens, partial regression coefficients can become both statistically unreliable and difficult to interpret. This is because they are not really "saying" different things about *y*. We likened multicollinearity to a situation where old-time comedians Laurel and Hardy are carrying a piano, and each comedian is unsure as to how much of the load he should carry. As a result, the piano will tend to be very unstable. Multicollinearity is *not* a problem if we are simply substituting *x* values into the equation and attempting to estimate *y*. However, it *is* a problem when we're trying to interpret the partial regression coefficients for the *x* variables that are highly correlated.

Besides the high correlation between two or more independent variables, there can be other clues to the presence of multicollinearity: (1) an independent variable known to be an important predictor ends up having a partial regression coefficient that is not significant; (2) a partial regression coefficient that should be positive turns out to be negative, or vice versa; or (3) when an independent variable is added or removed, the partial regression coefficients for the other independent variables change drastically.

A practical way to identify multicollinearity is to generate a **correlation matrix.** This is a matrix showing the correlation of each variable with each of the other variables. Computer Solutions 17.5 shows the correlation matrix for the following *y* and *x* variables for a sample consisting of 34 large automobile dealerships.[7] The data are in file **CX17DEAL**.

y = TOTSALES, total sales generated during the year (thousands of dollars)
x_1 = PARTSERV, parts and service sales for the year (thousands of dollars)
x_2 = RETUSED, number of used cars sold at retail

[7]Crain Communications, Inc., *Automotive News 1988 Market Data Book,* pp. 95–108.

x_3 = RETNEW, number of new cars sold at retail
x_4 = FLEET, number of cars sold to fleet customers
x_5 = YRSINBUS, number of years the dealership has been in business

COMPUTER 17.5 SOLUTIONS

The Correlation Matrix

EXCEL

	G	H	I	J	K	L	M
1		*TOTSALES*	*PARTSERV*	*RETUSED*	*RETNEW*	*FLEET*	*YRSINBUS*
2	TOTSALES	1.000					
3	PARTSERV	0.794	1.000				
4	RETUSED	0.780	0.386	1.000			
5	RETNEW	0.759	0.333	0.835	1.000		
6	FLEET	0.358	0.322	0.238	0.201	1.000	
7	YRSINBUS	0.190	0.275	-0.085	-0.130	0.336	1.000

Excel Correlation Matrix

1. The dealership data are in Excel file **CX17DEAL**, with total sales in A2:A35, parts and service sales in B2:B35, used retail units in C2:C35, new retail units in D2:D35, fleet units in E2:E35, years in business in F2:F35, and their labels in A1:F1. From the **Data** ribbon, click **Data Analysis.** Select **Correlation.** Click **OK.**
2. Enter **A1:F35** into the **Input Range** box. In the **Grouped By** menu, select **Columns.** Select **Labels in First Row.** Enter **G1** into the **Output Range** box. Click **OK.** After adjusting the column widths and the number of decimal places, the results are as shown here.

MINITAB

Minitab Correlation Matrix

Correlations: TOTSALES, PARTSERV, RETUSED, RETNEW, FLEET, YRSINBUS

	TOTSALES	PARTSERV	RETUSED	RETNEW	FLEET
PARTSERV	0.794				
RETUSED	0.780	0.386			
RETNEW	0.759	0.333	0.835		
FLEET	0.358	0.322	0.238	0.201	
YRSINBUS	0.190	0.275	-0.085	-0.130	0.336

Cell Contents: Pearson correlation

Using Minitab file **CX17DEAL**, with total sales in C1, parts and service sales in C2, used retail units in C3, new retail units in C4, fleet units in C5, and years in business in C6: Click **Stat.** Select **Basic Statistics.** Click **Correlation.** Enter **C1–C6** into the **Variables** box. Click **OK.**

With a total of six variables, there are $(6 \cdot 5)/2 = 15$ possible pairs of variables. As shown in the first column of the correlation matrix, several of the independent variables are highly correlated with the dependent variable, TOTSALES. However, there is a very high correlation (0.835) between the number of used (RETUSED) and new (RETNEW) cars sold at retail. This high correlation between two of the independent variables is an example of multicollinearity.

We can reduce the impact of multicollinearity by breaking up "partners" like the RETUSED and RETNEW variables and not including both of them in the equation at the same time. For example, we could decide to include only the "partner" with the higher correlation with y. An alternative approach involves the stepwise regression method described in the next section.

EXERCISES

17.37 What is a correlation matrix, and what role does it play in multiple regression?

17.38 One of the indicators that multicollinearity is present is when a variable that is known to be an important predictor ends up having a partial regression coefficient that is not significant. What are some of the other signals for the presence of multicollinearity?

17.39 In linear regression models, for what types of uses or applications would multicollinearity tend to be a problem? Are there any applications or uses for which multicollinearity would *not* be a problem?

17.40 A linear regression analysis was performed using x_1 as a predictor variable. The p-value associated with the regression coefficient for x_1 was 0.02, and the sign of the coefficient was negative. Another analysis was performed, this time using both x_1 and x_2 as predictor variables. In the second analysis, the partial regression coefficient for x_1 had a positive sign and the p-value associated with it was 0.39. Given these results, would multicollinearity appear to be present?

17.41 Four predictor variables are being considered for use in a linear regression model. Given the accompanying correlation matrix, does it appear that multicollinearity could be a problem?

	E	F	G	H	I
1		x1	x2	x3	x4
2	x1	1.000			
3	x2	0.483	1.000		
4	x3	0.952	0.516	1.000	
5	x4	-0.164	-0.202	-0.113	1.000

(DATA SET) *Note:* A computer and statistical software are required for Exercises 17.42–17.45.

17.42 For the situation and variables in Exercise 17.16, use data file **XR17016** and generate a correlation matrix for the independent variables. Examine the correlation matrix and comment on the possible presence of multicollinearity.

17.43 For the situation and variables in Exercise 17.17, use data file **XR17017** and generate a correlation matrix for these independent variables: number of aircraft, number of employees, and number of aircraft departures. Examine the correlation matrix and comment on the possible presence of multicollinearity.

17.44 Circulation (millions) for Sunday newspapers versus morning and evening daily papers for 1999–2007 are shown here. The data are also in file **XR17044**. With Sunday circulation as the dependent variable, generate the correlation matrix for the independent variables and comment on the possible presence of multicollinearity.

Source: Bureau of the Census, *Statistical Abstract of the United States 2006*, p. 738 and *Statistical Abstract of the United States 2009*, p. 697.

Year	Sunday	Daily, Morning	Daily, Evening
1999	59.9	46.0	10.0
2000	59.4	46.8	9.0
2001	59.1	46.8	8.8
2002	58.8	46.6	8.6
2003	58.5	46.9	8.3
2004	57.8	46.9	7.7
2005	55.3	46.1	7.2
2006	53.2	45.4	6.9
2007	51.2	44.5	6.2

17.45 For the years 1998–2008, production in the U.S. energy sector included the quantities shown here. The variables represent total energy production as well as energy production from nonrenewable fossil fuels. The units are in quadrillion Btu and the data are also in file **XR17045**. For these data, and with y = total energy production, generate the correlation matrix for the independent variables and comment on the possible presence of multicollinearity. Source: U.S. Energy Information Administration, *Annual Energy Review 2008*, pp. 5, 7.

Year	Total Energy Production	Crude Oil	Natural Gas	Coal
1998	66.39	13.24	22.03	24.05
1999	65.23	12.45	21.87	23.30
2000	65.23	12.36	22.27	22.74
2001	66.57	12.28	22.71	23.55
2002	65.03	12.16	22.00	22.73
2003	64.12	12.03	22.04	22.09
2004	64.13	11.50	21.56	22.85
2005	63.22	10.96	20.91	23.19
2006	64.18	10.80	21.38	23.79
2007	64.70	10.72	22.03	23.49
2008	66.41	10.52	23.57	23.86

(17.7) STEPWISE REGRESSION

In **stepwise regression,** the variables enter the regression analysis one at a time. The first x variable to enter is the one that explains the greatest amount of variation in y. The second x variable to enter will be the one that explains the greatest amount of the *remaining* variation in y, and so on. At each "step," the computer decides which, if any, of the remaining x variables should be brought in. Every step results in a new regression equation and (if you choose) an updated printout of this equation and its analysis.

Stepwise regression is especially useful when there are a great many x variables. Under these conditions, a lot of independent variables may be highly correlated, and it can be difficult for us humans to sort out the confusion they represent. The general idea of stepwise regression is the balancing act of trying to (1) explain the most possible variation in y, while (2) using the fewest possible x variables.

In general, it's advisable to test a model that has already been hypothesized (e.g., a multiple linear regression between y and x_1, x_3, and x_6). If you don't have a specific model in mind, however, stepwise regression is very useful when you have a great many independent variables and wish to determine which linear model might best apply.

EXAMPLE

Stepwise Regression

As shown in Computer Solutions 17.5, two of the five independent variables (RETUSED and RETNEW) are highly correlated with each other. In viewing the application of stepwise regression to these data, we'll observe how the analysis is influenced by this situation in which multicollinearity is obviously present.

SOLUTION

Computer Solutions 17.6 shows the stepwise regression analysis of the underlying data for the six dealership variables. Although there are five independent variables, the procedure has gone through just four steps. The number of vehicles sold to fleet customers (FLEET) is denied entrance to the regression equation. Here is a brief discussion of the steps in the Computer Solutions 17.6 printout.

Step 1 The first variable invited into the equation is PARTSERV, sales revenue from parts and service. This is no surprise, since the correlation matrix shows this to be the independent variable most highly correlated with TOTSALES. At this step, the regression equation is

$$\text{TOTSALES} = -347.6 + 9.55\text{PARTSERV}$$

COMPUTER 17.6 SOLUTIONS

Stepwise Regression

EXCEL

	H	I	J	K	L	M	N
1	**Results of stepwise regression**						
2							
3	**Step 1 - Entering variable: PARTSERV**						
4							
5	Summary measures						
6		Multiple R	0.7940				
7		R-Square	0.6304				
8		Adj R-Square	0.6189				
9		StErr of Est	23143				
10							
11	ANOVA Table						
12		Source	df	SS	MS	F	p-value
13		Explained	1	29236853124	29236853124	54.59	0.000
14		Unexplained	32	17138622464	535581952		
15							
16	Regression coefficients						
17			Coefficient	Std Err	t-value	p-value	
18		Constant	-347.57	7685.61	-0.045	0.964	
19		PARTSERV	9.55	1.29	7.388	0.000	
20							

	H	I	J	K	L	M	N
21	**Step 2 - Entering variable: RETNEW**						
22							
23	Summary measures			Change	% Change		
24		Multiple R	0.9513	0.1573	%19.8		
25		R-Square	0.9050	0.2746	%43.6		
26		Adj R-Square	0.8989	0.2800	%45.2		
27		StErr of Est	11920	-11223	-%48.5		
28							
29	ANOVA Table						
30		Source	df	SS	MS	F	p-value
31		Explained	2	41971102084	20985551042	147.71	0.000
32		Unexplained	31	4404373504	142076565		
33							
34	Regression coefficients						
35			Coefficient	Std Err	t-value	p-value	
36		Constant	-6156.03	4005.73	-1.537	0.134	
37		PARTSERV	7.32	0.71	10.372	0.000	
38		RETNEW	10.81	1.14	9.467	0.000	
39							

(continued)

	H	I	J	K	L	M	N
40	**Step 3 - Entering variable: RETUSED**						
41							
42	Summary measures			Change	% Change		
43		Multiple R	0.9632	0.0119	%1.3		
44		R-Square	0.9279	0.0228	%2.5		
45		Adj R-Square	0.9206	0.0217	%2.4		
46		StErr of Est	10561	-1359	-%11.4		
47							
48	ANOVA Table						
49		Source	df	SS	MS	F	p-value
50		Explained	3	43029497732	14343165911	128.60	0.000
51		Unexplained	30	3345977856	111532595		
52							
53	Regression coefficients						
54			Coefficient	Std Err	t-value	p-value	
55		Constant	-5874.83	3550.30	-1.655	0.108	
56		PARTSERV	6.91	0.64	10.812	0.000	
57		RETNEW	6.46	1.74	3.724	0.001	
58		RETUSED	10.20	3.31	3.081	0.004	
59							

	H	I	J	K	L	M	N
60	**Step 4 - Entering variable: YRSINBUS**						
61							
62	Summary measures			Change	% Change		
63		Multiple R	0.9691	0.0058	%0.6		
64		R-Square	0.9391	0.0112	%1.2		
65		Adj R-Square	0.9307	0.0100	%1.1		
66		StErr of Est	9871	-690	-%6.5		
67							
68	ANOVA Table						
69		Source	df	SS	MS	F	p-value
70		Explained	4	43549942148	10887485537	111.74	0.000
71		Unexplained	29	2825533440	97432188		
72							
73	Regression coefficients						
74			Coefficient	Std Err	t-value	p-value	
75		Constant	-10235.89	3817.27	-2.681	0.012	
76		PARTSERV	6.41	0.64	10.100	0.000	
77		RETNEW	6.92	1.63	4.237	0.000	
78		RETUSED	10.42	3.10	3.362	0.002	
79		YRSINBUS	388.30	168.01	2.311	0.028	

Excel Stepwise Regression

1. Using Excel file **CX17DEAL**, with total sales in A2:A35, parts and service sales in B2:B35, used retail units in C2:C35, new retail units in D2:D35, fleet units in E2:E35, years in business in F2:F35, and their labels in A1:F1: Click on any cell within the data field. From the **Add-Ins** ribbon, click **Data Analysis Plus.** Select **Stepwise Regression.** Click **OK.**
2. Check to verify that the cursor is on a cell within the data field, then click **Continue.** If there are no missing values within the data, click **OK.** Click on **TOTSALES** as the response variable and click **OK.** Holding down the **Ctrl** key, click on each of the five explanatory variables to be used: **PARTSERV, RETUSED, RETNEW, FLEET,** and **YRSINBUS.** Click **OK.**
3. In the **Stepwise regression parameters** menu, click **p-values** and click **OK.** Accept the default **p-to-enter** and **p-to-leave** parameters (0.05 and 0.10, respectively) and click **OK.** Select any scatterplots that are desired, and click **OK.** In the **Location of results** menu, select **Same worksheet as the data.** Click **OK.** After adjusting the column widths and the number of decimal places, the results are as shown here.

MINITAB

Minitab Stepwise Regression

```
Stepwise Regression: TOTSALES versus PARTSERV, RETUSED, ...

  Alpha-to-Enter: 0.05  Alpha-to-Remove: 0.1

Response is TOTSALES on 5 predictors, with N = 34

Step              1        2        3         4
Constant     -347.6  -6156.0  -5874.8  -10235.9

PARTSERV       9.55     7.32     6.91      6.41
T-Value        7.39    10.37    10.81     10.10
P-Value       0.000    0.000    0.000     0.000

RETNEW                  10.8      6.5       6.9
T-Value                 9.47     3.72      4.24
P-Value                0.000    0.001     0.000

RETUSED                          10.2      10.4
T-Value                          3.08      3.36
P-Value                         0.004     0.002

YRSINBUS                                    388
T-Value                                    2.31
P-Value                                   0.028

S             23143    11920    10561      9871
R-Sq          63.04    90.50    92.79     93.91
R-Sq(adj)     61.89    89.89    92.06     93.07
Mallows Cp    140.1     15.7      7.2       4.0
```

1. Using Minitab file **CX17DEAL**, with total sales in C1, parts and service sales in C2, used retail units in C3, new retail units in C4, fleet units in C5, and years in business in C6: Click **Stat.** Select **Regression.** Click **Stepwise.** Enter **C1** into the **Response** box. Enter **C2–C6** into the **Predictors** box.
2. For consistency, we will use the same entry/removal parameters as the defaults in the Excel procedure: Click **Methods.** Select **Use alpha values.** Select **Stepwise (forward and backward).** Enter **0.05** into the **Alpha to enter** box. Enter **0.10** into the **Alpha to remove** box. Click **OK.** Click **OK.**

This regression equation explains 63.04% of the variation in TOTSALES, and the multiple standard error of estimate is listed as 23143, or \$23.143 million.

Step 2 The next variable to enter is RETNEW, the number of new cars sold at retail. The regression equation now becomes

$$\text{TOTSALES} = -6156.0 + 7.32\text{PARTSERV} + 10.8\text{RETNEW}$$

The inclusion of RETNEW has shaken things up a bit by changing the constant from −347.6 to −6156.0. This is not too unusual, since improving the equation means changing the equation. The percentage of variation in TOTSALES that is explained rises from 63.04% to 90.50%, a hefty increase. The estimating ability of the equation has also improved: The multiple standard error of estimate is reduced to \$11.920 million.

Step 3 Here comes the Laurel and Hardy piano act. Recall that RETNEW and RETUSED are highly correlated ($r = 0.835$). However, of the variables not yet included, RETUSED explains the greatest amount of remaining variability in TOTSALES, so it comes into the regression equation:

$$\text{TOTSALES} = -5874.8 + 6.91\text{PARTSERV} + 6.5\text{RETNEW} + 10.2\text{RETUSED}$$

The constant term and the partial regression coefficient for PARTSERV have changed slightly, but the partial regression coefficient for RETNEW has decreased dramatically. (In the Laurel and Hardy analogy, RETNEW was carrying most of the piano in step 2, but now RETUSED has taken up part of the load.) Notice how the t-ratio for RETNEW has gone down from 9.47 to 3.72, another reflection of the new sharing of the load. Although RETUSED has a high correlation with TOTSALES ($r = 0.780$), it has not added much to the predictive power of the regression equation. The percentage of variation explained has increased only slightly, from 90.50% to 92.79%. Likewise, the multiple standard error of estimate is reduced very little and is now $10.561 million.

If our sole purpose were the prediction of TOTSALES, there would be no problem in using the equations in this step and the next. On the other hand, if our purpose is to be able to interpret the partial regression coefficients, we should use only the equation that resulted from step 2. At step 3, multicollinearity has set in, and we have lost the ability to make meaningful interpretations of the partial regression coefficients.

Step 4 The next entrant is YRSINBUS, the number of years the dealership has been in business. The regression equation is now

$$\begin{aligned}\text{TOTSALES} = &-10{,}235.9 + 6.41\text{PARTSERV} + 6.9\text{RETNEW} \\ &+ 10.4\text{RETUSED} + 388.0\text{YRSINBUS}\end{aligned}$$

Now that RETNEW and RETUSED have the piano under control, their partial regression coefficients have not changed much from the previous step. However, the constant term is now −10,235.9. Bringing YRSINBUS into the equation increases the explained percentage of variation in TOTSALES to 93.91%, and the multiple standard error of estimate decreases to $9.871 million.

Once stepwise regression identifies a subset of the original independent variables and lists them in a multiple regression equation, it is useful to include only these independent variables in a conventional multiple regression analysis. The procedure described in this section is known as *forward selection stepwise regression,* since we begin with no predictor variables and add one predictor at each iteration. There is also a *backward elimination stepwise regression,* in which we begin with the entire set of predictor variables and eliminate one variable at each iteration.

EXERCISES

17.46 What is stepwise regression, and when is it desirable to make use of this multiple regression technique?

17.47 In general, on what basis are independent variables selected for entry into the equation during stepwise regression?

17.48 The two largest values in a correlation matrix are the 0.850 correlation between y and x_4, and the 0.790 correlation between y and x_9. During a stepwise regression analysis, x_4 is the first independent variable brought into the equation. Will x_9 necessarily be next? If not, why not?

17.49 The following Minitab printout is for a stepwise regression with 10 independent variables and 50 observations:

Stepwise Regression: y versus x1, x2, x3, x4, x5, x6, x7, x8, x9, x10

```
 Alpha-to-Enter: 0.05  Alpha-to-Remove: 0.1
Response is     y    on 10 predictors, with N =  50
     Step          1          2          3
Constant       83.91     106.85     124.76

x5             -0.33      -0.35      -0.30
T-Value        -2.39      -2.65      -2.30
P-Value        0.021      0.011      0.026

x2                        -0.33      -0.38
T-Value                   -2.30      -2.70
P-Value                   0.026      0.010

x9                                   -0.27
T-Value                              -2.07
P-Value                              0.044

S               13.8       13.2       12.8
R-Sq           10.63      19.66      26.52
R-Sq(adj)       8.76      16.25      21.73
C-p              3.9        0.9       -0.9
```

a. List the order in which the independent variables have been introduced.
b. What is the multiple regression equation after two independent variables have been introduced?
c. As of step 3, three variables have been introduced. Which, if any, partial regression coefficients differ significantly from zero at the 0.05 level? At the 0.02 level?

(DATA SET) *Note:* Exercises 17.50–17.57 require a computer and statistical software. We suggest that Minitab users specify alpha-to-enter as 0.05 and alpha-to-remove as 0.10. (See Computer Solutions 17.6.)

17.50 For the data in Exercise 17.44, and using Sunday circulation as the dependent variable, perform conventional multiple regression and stepwise regression analyses on the data, then compare the results of the two approaches. The data are in file **XR17044**.

17.51 For the data in Exercise 17.45, and using total nonrenewable energy production as the dependent variable, perform conventional multiple regression and stepwise regression analyses on the data, then compare the results of the two approaches. The data are in file **XR17045**.

17.52 For the situation, data, and variables described in Exercise 17.16, perform conventional multiple regression and stepwise regression analyses on the data, then compare the results of the two approaches. The data are in file **XR17016**.

17.53 For the situation, data, and variables described in Exercise 17.15, perform conventional multiple regression and stepwise regression analyses on the data, then compare the results of the two approaches. The data are in file **XR17015**.

17.54 For the situation and data described in Exercise 17.17, perform conventional multiple regression and stepwise regression analyses on the data, then compare the results of the two approaches. Use total operating revenue as the dependent variable and number of aircraft, number of employees, and number of aircraft departures as the three independent variables. The data are in file **XR17017**.

17.55 An officer training program has 9 board members who provide ratings for the candidates. The high and low scores are dropped before the overall rating is computed. For a sample of 28 persons from the most recent class of candidates, the overall ratings and ratings by the individual board members are provided in data file **XR17055**.

a. Construct the correlation matrix for all 10 variables.
b. With overall rating as the dependent variable, perform a stepwise regression and interpret the results.
c. From the results in part (b), does it appear that the size of the board could be reduced without losing very much explanatory power? Explain.
d. Using only the data provided by the first two board members introduced in part (b), perform a multiple regression analysis. Compare the results with the stepwise regression as it appeared following the introduction of these board members.

17.56 Using the data file for Exercise 17.55:

a. Apply stepwise regression with the rating by board member 1 as the dependent variable and the ratings by the other 8 members as the set of independent variables from which to select. Interpret the results.

b. Using only the independent variables introduced during the stepwise regression, perform a multiple regression analysis.
c. Compare the results in parts (a) and (b).

17.57 Using the data file for Exercise 17.55:
a. Apply stepwise regression with the rating by board member 9 as the dependent variable and the ratings by the other 8 members as the set of independent variables from which to select, then interpret the results.
b. Including only the independent variables introduced during the stepwise regression, perform a multiple regression analysis.
c. Compare the results in parts (a) and (b).

17.8 SELECTING A MODEL

Given the variety of models discussed in this chapter, and in Chapters 15 and 16, it is natural to wonder which model is best for a given situation. However, the selection of which model to use is not quite so straightforward as the workings of the individual models themselves. Rather than feature a complex flow diagram and an elaborate discourse on each of its components, in this section we will consider a general sequence of steps and some of the more important considerations involved with each. For a more advanced discussion of the process of selecting the best model, the interested reader may wish to refer to a more advanced text, such as the excellent treatment given this topic by Kleinbaum, Kupper, Muller, and Nizam.[8]

1. **Specify the dependent variable and determine the goal for the analysis.** If the goal is simply to predict the value of the dependent variable based on a set of predictor variables, we need not worry about multicollinearity. Accordingly, we can use a large number of predictor variables without worrying about whether some of them might be highly correlated with each other. On the other hand, if we are concerned with interpreting partial regression coefficients in an attempt to better understand the relationship between the dependent variable and one or more of the predictor variables, it will be desirable to build a smaller model with predictor variables that are more independent from each other.
2. **Specify possible predictor variables.** Based on a combination of experience and knowledge about the dependent variable and the environment in which it exists, develop a list of predictor variables that could be related to the dependent variable. In doing so, keep in mind the data-collection step that comes next. It may not be practical or economically feasible to gather relevant data for all of the predictor variables that could be candidates for the model.
3. **Collect the data.** This may involve primary data, secondary data, or both. The kinds of methods and data sources described in Chapter 4 could be utilized. Consider how large the eventual set of predictor variables might be, and collect data accordingly. As a rule of thumb, the number of observations should be at least five times the number of predictor variables in the model.[9] This is due to degrees-of-freedom considerations. The larger the number of predictor variables, the easier it is for the model to fit any given set of data, even data that are completely random.
4. **Specify the candidate models.** The models under consideration could include any of those discussed in this chapter and the two that preceded, as well as models that are larger or more complex. In general, we should attempt to use the simplest possible model that satisfies the goal for which the model is being developed.

[8]Kleinbaum, D.G., L.L. Kupper, K.E. Muller, and A. Nizam, *Applied Regression Analysis and Multivariable Methods,* Pacific Grove, CA: Duxbury Press, 1998.
[9]Ibid., p. 389.

5. **For each candidate model, decide which of the available predictor variables will be included in the model.** Deciding which variable(s) to include can be done one variable at a time, as in the stepwise regression method presented in Section 17.7, and the approach can involve either forward selection or reverse elimination. However, with increasingly powerful computers and the greater sophistication of statistical software, more "brute force" methods are available. In an approach called *best subsets regression,* the computer considers all of the possible combinations or subsets of the available predictor variables, then reports which models are the best performers for each given number of predictors. Implicit in this process is the criterion by which one model is judged to perform better than another. In this chapter, our focus has been on the coefficient of determination (R^2) that indicates the proportion of variation in the dependent variable that is "explained" by the set of predictor variables, but many other criteria can be applied.
6. **Select a model and evaluate its suitability.** Once a model has been selected, we can evaluate whether it satisfies the assumptions underlying its use. This evaluation should include a residual analysis and related diagnostics.
7. **Test the reliability of the model by applying it to new or withheld data.** The model can be tested by applying it to new data as time goes by. Another possibility is to build the model on the basis of a portion of the available data, then test its performance on the "fresh" data that were initially withheld from the model.

SUMMARY (17.9)

Models, variables, and transformations

In this chapter, we go beyond the simple and multiple regression estimation models in Chapters 15 and 16 to include curvilinear models as well as models where a variable is qualitative and can represent two or more categories. For underlying relationships that are not linear, logarithmic transformation methods can be employed in constructing exponential and multiplicative models.

Multicollinearity

If two or more independent variables are highly correlated with each other, multicollinearity is present, and the partial regression coefficients may be unreliable. One way to identify multicollinearity is to generate a correlation matrix that shows the correlation of each variable with each of the other variables. Multicollinearity is not a problem when the only purpose of the equation is to predict the value of the dependent variable, but it can be a problem when we are trying to interpret the meaning of the partial regression coefficients.

Stepwise regression

Stepwise regression is introduced and the forward selection method is presented. In this approach, independent variables enter the equation one at a time, with the first one entering being the x variable that explains the greatest amount of the variation in y. At each step, the variable is introduced that explains the greatest amount of the remaining variation in y.

Model selection

Determining the best model for a given situation can be a complex process, and a sequence of general steps is presented for model selection.

EQUATIONS

Polynomial models with one quantitative predictor variable

General polynomial model: $y_i = \beta_0 + \beta_1 x_i + \beta_2 x_i^2 + \beta_3 x_i^3 + \cdots + \beta_p x_i^p + \epsilon_i$

First-order polynomial model: $E(y) = \beta_0 + \beta_1 x$

Second-order polynomial model: $E(y) = \beta_0 + \beta_1 x + \beta_2 x^2$

Third-order polynomial model: $E(y) = \beta_0 + \beta_1 x + \beta_2 x^2 + \beta_3 x^3$

Polynomial models with two quantitative predictor variables

First-order model with two predictor variables: $E(y) = \beta_0 + \beta_1 x_1 + \beta_2 x_2$

First-order model with two predictor variables and interaction:

$$E(y) = \beta_0 + \beta_1 x_1 + \beta_2 x_2 + \beta_3 x_1 x_2$$

Second-order model with interaction:

$$E(y) = \beta_0 + \beta_1 x_1 + \beta_2 x_2 + \beta_3 x_1^2 + \beta_4 x_2^2 + \beta_5 x_1 x_2$$

Conversion of the exponential equation, $\hat{y} = b_0 b_1^x$, to linear form

$$\log \hat{y} = \log b_0 + x \log b_1, \quad \text{or} \quad \log \hat{y} = \log b_0 + (\log b_1)x$$

Conversion of the multiplicative equation, $\hat{y} = b_0 x_1^{b_1} x_2^{b_2}$, to linear form

$$\log \hat{y} = \log b_0 + b_1 \log x_1 + b_2 \log x_2$$

CHAPTER EXERCISES

(DATA SET) *Note:* Exercises 17.58–17.73 require a computer and statistical software.

17.58 As reported by the Recording Industry Association of America, the percentage of recorded music dollar sales in CD format was as shown here for the years 2000 through 2007. The data are also in file **XR17058**. Using $x = 1$ through 8 to represent years 2000–2007, fit a second-order polynomial model to the data and estimate the percentage of music sales in CD format for the year 2015. Source: *The World Almanac and Book of Facts 2009*, p. 288.

Year	x = Year Code	CD Music Percentage
2000	1	89.3
2001	2	89.2
2002	3	90.5
2003	4	87.8
2004	5	90.3
2005	6	87.0
2006	7	85.6
2007	8	82.6

17.59 J.K. Rowling's popular *Harry Potter* books have tended to become thicker with each new edition. The publication year and page count for U.S. editions for the first seven books were as shown on the next page. The data are also in file **XR17059**. Using year codes so that $x = 1$ corresponds to 1998, the year the first book was published, fit a third-order polynomial model to the data and estimate the page count for a *Harry Potter* book published in 2015. Source: Barnes & Noble, bn.com, August 20, 2009.

17.60 For 1995 through the first half of 2009, PricewaterhouseCoopers and the National Venture Capital Association reported quarterly venture capital investments (trillions of dollars) and the number of investment deals as shown in data file **XR17060**. Generate a scatter diagram with the number of deals as the independent variable, then apply an appropriate polynomial model to fit the data. Does there appear to be any relationship between the variables? How much of the variation in quarterly venture capital investments is explained by the model? Source: PricewaterhouseCoopers, pwcmoneytree.com, August 20, 2009.

Data for Exercise 17.59

Title	Publication Year	x = Year Code	Page Count, U.S. Edition
The Sorcerer's Stone	1998	1	309
The Chamber of Secrets	1999	2	341
The Prisoner of Azkaban	1999	2	435
The Goblet of Fire	2000	3	734
The Order of the Phoenix	2003	6	870
The Half-Blood Prince	2005	8	652
The Deathly Hallows	2007	10	759

17.61 Each workstation in the final assembly area of a large manufacturing facility has a "backlog" area to hold units that are to be processed. A production engineer believes there could be a relationship between the number of units in the backlog area and the productivity of the worker. Specifically, the engineer thinks increasing the number of units in the backlog area might lead to workers being more in a hurry to "catch up," hence increasing their productivity. The data in file **XR17061** show the results of a test in which the engineer tried various backlog levels and measured productivity at 50 different workstations. Generate a scatter diagram relating y = productivity (units/hour) to x = number of units in the backlog area, then apply an appropriate polynomial model to fit the data. Does there appear to be any relationship between the variables? How much of the variation in productivity is explained by the model?

17.62 The data in the table show the number of years it took for selected technological innovations to spread to 25% of the U.S. population. The data are also in file **XR17062**. For y = years to spread and x = years past 1800 (e.g., the code for 1975 is 175), generate a scatter diagram and apply an appropriate polynomial model to fit the data. Does there appear to be any relationship between the variables? How much of the variation in dissemination time is explained by the model? Using your model, predict how long it will take for a major technological invention that is invented in 2015 to spread to 25% of the U.S. population. Source: *Wall Street Journal Almanac 1999*, p. 350.

Invention	Year Code	Years to Spread
Home electricity (1873)	73	46
Telephone (1875)	75	35
Automobile (1885)	85	55
Radio (1906)	106	22
Television (1925)	125	26
VCR (1952)	152	34
Microwave oven (1953)	153	30
Personal computer (1975)	175	15
Cellular phone (1983)	183	13

17.63 The data in file **XR17063** include the wind chill index (how cold it "feels") at different combinations of outside temperature (degrees Fahrenheit) and wind speed (miles per hour). For y = wind chill index, x_1 = actual outside temperature, and x_2 = wind speed, fit a first-order polynomial model to the data and interpret the partial regression coefficients and the R^2 value. Add an interaction term to the model. To what extent has the addition of the interaction term improved the explanatory power of the model? Source: *New York Times Almanac 2009*, p. 478.

17.64 For loans at insured commercial banks, the Federal Financial Institutions Examination Council has reported the percentage of various kinds of loans that were at least 30 days in delinquency during each of the years shown in the table. The data are also listed in file **XR17064**. For y = percent of all insured commercial bank loans overdue, x_1 = percent of consumer credit cards overdue, and x_2 = percent of residential loans overdue, fit a first-order polynomial model to the data and interpret the partial regression coefficients and the R^2 value. Add an interaction term to the model. To what extent has the addition of the interaction term improved the explanatory power of the model? Source: Bureau of the Census, *Statistical Abstract of the United States 2009*, p. 727.

Year	All Loans	Credit Cards	Residential Loans
2000	2.18%	4.50%	2.11%
2001	2.61	4.86	2.29
2002	2.69	4.87	2.11
2003	2.33	4.47	1.83
2004	1.80	4.11	1.55
2005	1.57	3.70	1.55
2006	1.57	4.01	1.73
2007	2.06	4.25	2.57

17.65 The production engineer for the manufacturing plant in Exercise 17.61 has just included another variable: the worker's gender, coded as 1 = female and 0 = male. File **XR17065** contains the expanded set of data. With y = productivity (units/hour), x_1 = number of units in the backlog area, and x_2 = gender, fit a first-order model with no interaction term. Interpret the partial regression coefficients and the R^2 value. At the 0.02 level, is the model significant?

17.66 The sales director for an industrial supplies firm has collected information describing the performance and personal characteristics of 80 members of her sales force. The data in file **XR17066** include the percent of sales quota the salesperson achieved last year, age, number of years with the firm, and personal appearance (very neat, about average, or a little sloppy). In the data file, a very neat person is coded as neat = 1, average = 0, a person of average neatness is coded as neat = 0, average = 1, and a sloppy person is coded as neat = 0, average = 0. With y = percent of sales quota, perform a linear regression using the other variables as predictors. Interpret the partial regression coefficients and the R^2 value for the model. At the 0.05 level, is the model significant?

17.67 For the loan delinquency data in Exercise 17.64 and listed in file **XR17064**, fit a multiplicative model and use it to estimate the delinquency rate for all insured commercial bank loans when the delinquency rates for credit cards and residential loans are 4.0% and 2.0%, respectively.

17.68 The Dow Jones Global Indexes describe over 3000 companies worldwide that represent more than 80% of the equity capital in the world's stock markets. Data file **XR17068** shows the quarterly indexes for the world as a whole, and indexes for the United States, the Americas, Europe, and the Asia/Pacific region over 23 consecutive quarters. Given that the dependent variable will be y = world index, construct a correlation matrix for the four possible predictor variables. Does multicollinearity seem to be present in the data? If so, would this present a problem if we were using the predictor variables to estimate the world index? If there is multicollinearity, would this present a problem if we were interpreting the partial regression coefficients in a linear model applied to the data? Source: *Wall Street Journal Almanac 1999*, p. 264.

Note: For Exercises 17.69–17.73, we suggest that Minitab users specify alpha-to-enter as 0.05 and alpha-to-remove as 0.10. (See Computer Solutions 17.6.)

17.69 Data file **XR17069** lists a variety of information for a selection of aircraft. The variables and their units are: operating cost (dollars per hour), number of seats, speed (mph), flight range (miles), and fuel consumption (gallons per hour). Considering operating cost per hour as the dependent variable, construct and interpret the correlation matrix for the possible predictor variables. Does it appear that multicollinearity could be present? For y = operating cost per hour, perform a stepwise regression analysis using the other variables as predictors and interpret the results. How much of the variation in operating costs is explained by the model? Source: *World Almanac and Book of Facts 2009*, p. 365.

17.70 Given the situation and variables described in Exercise 17.66 and data file **XR17066**, use y = percent of sales quota in performing a stepwise regression using the other variables as predictors. Interpret the results, including a commentary on which variables seem to be of greater importance in predicting sales performance. How much of the variation in sales performance is explained by the model?

17.71 The students in a training program have taken four tests along with a final exam, with the scores shown in the table. The data are also in file **XR17071**.

Student	Test 1	Test 2	Test 3	Test 4	Final Exam
1	59	65	53	65	62
2	69	77	57	76	78
3	79	74	81	86	82
4	87	88	83	90	96
5	92	89	97	92	94
6	65	69	60	66	68
7	71	70	68	81	78
8	93	87	85	95	94
9	83	92	89	90	91
10	71	68	69	74	77
11	64	61	74	74	73
12	76	82	79	78	86
13	65	75	67	77	79
14	90	88	75	94	89
15	69	68	74	76	72
16	85	81	99	79	96

a. Construct the correlation matrix for these variables. Does it appear that multicollinearity could be a problem if all of the independent variables are used in estimating y = final exam score? In a stepwise regression, which independent variable would be the first one introduced into the equation, and why?
b. With y = final exam score and the independent variables as shown above, apply stepwise regression analysis to these data.
c. Describe each step of the stepwise regression in terms of the variable introduced, the regression equation after its introduction, and the proportion of variation in y that has been explained.
d. Using only the independent variables introduced during the stepwise regression, carry out a conventional multiple regression analysis on the data. What, if any, additional information does your computer statistical package provide that was not given in the printout for the stepwise regression?

17.72 Repeat parts (b)–(d) of Exercise 17.71, with Test 1 as the dependent variable, and Test 2, Test 3, Test 4, and Final Exam as the independent variables.

17.73 As indicated in the chapter-opening vignette, the National Football League quarterback passer rating is somewhat of a mystery to most observers. Data file **XR17073** includes the passer rating for 32 NFL quarterbacks during the 2008 season, along with these ten performance measurements for each quarterback: passing attempts, completions, completion percentage, yards gained, average yards per attempt, number of touchdowns, percentage of passes that were touchdowns, longest pass completed, number of interceptions, and percentage of passes that were intercepted. With passer rating as the dependent variable, use stepwise regression to find out which predictor variables are especially important in determining the passer rating. Comment on the results, including the percentage of the variation in passer ratings that is explained by the model. Source: NFL.com, August 21, 2009.

INTEGRATED CASES

Thorndike Sports Equipment

Ted Thorndike has been asked to pay another visit to the statistics class taught by his friend, Mary Stuart. One of the university's track athletes has an excellent chance to make the Olympics in the 1500 meters, known as the "metric mile," and Ted has brought along his laptop computer, a statistical package, and data on the world record for the one-mile run.

In 1954, Great Britain's Roger Bannister became the first person to run a mile in less than four minutes (3:59.4). Since then, the record has been broken 18 times, most recently by Morocco's Hicham El Guerrouj, at 3:43.13. The following data describe the record runs and the years in which they occurred. The data are also in file **THORN17**. Source: *Sports Illustrated 2009 Almanac*, p. 526.

Note These times are not recorded to the same number of decimal places for the seconds, because officiating policies and timing technologies may have differed from one record run to the next. For convenience, you may wish to (1) code the times so that each is the number of seconds over 3 minutes and (2) code the years so that the year 1901 = 1.

Person	Time	Year
Roger Bannister	3:59.4	1954
John Landy	3:58	1954
Derek Ibbotson	3:57.2	1957
Herb Elliott	3:54.5	1958
Peter Snell	3:54.4	1962
Peter Snell	3:54.1	1964
Michael Jazy	3:53.6	1965
Jim Ryun	3:51.3	1966
Jim Ryun	3:51.1	1967
Filbert Bayi	3:51	1975
John Walker	3:49.4	1975
Sebastian Coe	3:49	1979
Steve Ovett	3:48.8	1980
Sebastian Coe	3:48.53	1981
Steve Ovett	3:48.4	1981
Sebastian Coe	3:47.33	1981
Steve Cram	3:46.32	1985
Noureddine Morceli	3:44.39	1993
Hicham El Guerrouj	3:43.13	1999

1. Fit a simple linear regression ($\hat{y} = b_0 + b_1x$) to the time series.
2. Fit a polynomial regression equation ($\hat{y} = b_0 + b_1x + b_2x^2$) to the time series.
3. Comparing the R^2 values for the equations from parts 1 and 2, which equation is the better fit to the data?
4. Using the equation identified in question 3, what is the predicted world record for the one-mile run in the year 2015?

Fast-Growing Companies

Computer database **GROWCO** describes the characteristics of 100 companies identified by *Fortune* as among the fastest growing. Source: "*Fortune's* 100 Fastest-Growing Companies," *Fortune*, September 4, 2000, pp. 142–158. The variables include:

1. Name
2. Earnings-per-share growth, 3-year annual rate (percent)
3. Net income, past four quarters (millions of dollars)
4. Revenue growth, 3-year annual rate (percent)
5. Revenue, past four quarters (millions of dollars)
6. Total return, 3-year annual rate (percent)
7. Stock price change in past year (percent)
8. Stock price (dollars)

9. Estimated price/earnings ratio a year from now
10. Category (1 = technology, 2 = industrial, 3 = retail, 4 = health care, 5 = financial services, 6 = telecommunications, 7 = other)

Variables 11–16 are dummy variables representing the categories of variable 10. They are coded according to the scheme described in Section 17.4. For a technology company, technology = 1 and the other dummy variables will be 0. For a financial services company, financial = 1 and the others will be 0. If a company is "other," all six values will be 0. Thus, we need only six variables to describe the seven possible categories.

1. Generate a correlations matrix that includes variables 2–9 and 11–16. (Variable 10 would be meaningless here because its values are 1–7 for the 7 company categories.) Do the correlations "make sense" in terms of which correlations are positive and which are negative?
2. Using y = stock price (variable 8) as the dependent variable, carry out a conventional multiple regression analysis using variables 2–7, 9, and 11–16. Excel users, see note 1, which follows. Examine the partial regression coefficients to see whether their signs match the signs of the corresponding correlation coefficients associated with stock price. What percentage of the variation in stock prices is explained by the set of predictor variables?
3. Repeat step 2, but this time perform a stepwise regression analysis. Minitab and Excel users should refer to the following notes. Identify the variables that were introduced and interpret the printout. What percentage of the variation in stock prices is explained by the reduced set of predictor variables?

Minitab users, note: In the solutions materials, we use alpha-to-enter = 0.05 and alpha-to-remove = 0.10 for stepwise regression analyses. This is to provide consistency with Excel and the Data Analysis Plus add-in used with the text. See Computer Solutions 17.6.

Excel users, note 1: Excel requires that the predictor variables be arranged in adjacent columns when we carry out a multiple regression analysis. The same is true for all variables for which a correlation matrix is being generated. Depending on the predictor variables being used in a given analysis, you may need to rearrange some of the columns in the database. If you save the revised database, it is advisable to change the name of the file when doing so.

Excel users, note 2: The estimated price/earnings ratio a year from now was not available for four of the companies, so these are "missing data." If you are using the Data Analysis Plus add-in for the stepwise regression, you may wish to either omit this variable or omit these four companies from the analysis.

BUSINESS CASE

Westmore MBA Program

The MBA program at Westmore University has undergone several dramatic changes over the past five years. During this time, the goal of the business school was to recruit as many students as possible into the MBA program in order to build up their student base and credit hour production. A massive campaign was launched five years ago in order to attract more applicants to the program. Special brochures containing information about the program were printed and mailed to prospective students as well as to other colleges and universities that were likely to have undergraduate students who might be interested in coming to Westmore. Mailings were also sent to students who indicated an interest in Westmore on their GMAT exam. (The GMAT exam is a national standardized test used by most business schools in making admissions decisions for applicants to their graduate programs.) Representatives from the Westmore School of Business began attending regional "MBA fairs," where MBA program representatives meet with prospective MBA students and share information.

In the beginning, the number of students applying to the Westmore MBA program was small, but eventually the advertising campaign began to work and the number of qualified applicants each year increased to the target value of 150 initially set by the dean of the business school and the director of the MBA program. The yield—the percentage of admitted applicants who actually enroll and attend Westmore—is typically around 70%. Admitted students who do not enroll either attend other MBA programs or accept job offers. The accompanying table shows the admissions and enrollment figures for the five years of the plan to build the MBA student base at Westmore.

Year	1	2	3	4	5
Admissions	86	108	134	141	154
Enrollment	54	77	91	96	106

Wayne McDonald, the director of the program, is currently putting the second phase of the plan into action. He knows that in order for the MBA program at Westmore to attain national recognition, administrators must become more selective in the admissions process. The number of applicants is now large enough to do this without falling below a critical mass of 60 enrolled students each year.

The major issue facing Wayne and the MBA program is how to go about selecting the best students for the program. Wayne recently met with the MBA Admissions Committee, which consists of himself and two faculty members, Dr. Susan Thompson, who is a finance professor, and Dr. Hector Gonzalez, who is a marketing professor.

Wayne: "Thanks for coming to the meeting today. As you both know, our recruiting effort over the past five years has been extremely successful. We were able to exceed our original enrollment goal last year. While many of our students have been outstanding and have given our program visibility in the business community, we have had a number of weak performers. Some professors have had to water down their courses to keep these people afloat in the program. If we are to have a nationally recognized, quality MBA program, we must become stricter in our admission policies. Fortunately, we are now at the point where we can be much more selective and still have our minimum critical mass of 60 enrolled students each year."

Susan: "Wayne is right. Our current admission standards require a minimum score of 400 on the GMAT and a minimum undergraduate grade point average of 2.0. Obviously, this is not much of a hurdle. Personally, I would like to see the minimum requirements set at 580 on the GMAT and 2.75 for the undergraduate grade point average."

Wayne: "Well, raising the minimums is one way of going about it, but there are many other factors that determine the degree of success a student has in our MBA program. We should consider including these factors in our decision making process."

Hector: "Too bad we don't know in advance which students are going to excel. Wayne, do you know what other schools are doing?"

Wayne: "From conferences that I've attended, I have found that many MBA programs put a lot of emphasis on the GMAT score and the undergraduate grade point average of the student. While some schools set a minimum entrance requirement on each of these criteria as we do currently, other schools combine these measures into a single overall score. For instance, there is a 'formula score' that many schools use which multiplies the undergraduate GPA by 200 and adds the result to the GMAT score. If the formula score is above a certain figure, say, 1000, then the student is considered to be admittable."

Susan: "But there are so many other factors to consider. Surely we don't want our admissions decisions to be based solely on a formula. There are many students who attend colleges with high grade inflation. Those applicants would have an unfair advantage over applicants from stronger schools with regard to undergraduate grade point average."

Hector: "Yes, I agree. There are also studies that have indicated the GMAT is not a strong predictor of success in graduate school for many reasons, including cultural bias."

Wayne: "I am not advocating that we go to a strictly mathematical basis for making our decisions. However, higher minimum standards than we currently have or some sort of formula involving GMAT and undergraduate grade point average might be a useful screening device for sifting through applicants."

Susan: "I'm not opposed to your suggestion. Such an approach could be used to identify those students with high potential for succeeding in our program. In a sense, many of these decisions could be automated."

Wayne: "That would certainly be a great timesaver. Our admissions committee would only have to meet to discuss those applicants with other strong characteristics or extenuating circumstances."

Susan: "Excellent idea! Now, if we go with raising the minimum requirements for GMAT and undergraduate GPA, how much should we raise them? Or if we go with a combined score approach, what formula should we use and how should we set its cutoff values?"

Wayne: "We could go with your earlier suggestion of a 580/2.75 minimum requirement or with the formula score I described earlier. I could talk with directors of other MBA programs to get some feel for how they set their cutoff criteria. After we gain some experience with future admissions, we could adjust the cutoff criteria."

Hector: "Why wait until then? We have five years worth of data already! We should be able to develop our own criteria based on our own past experience."

Susan: "We might even want to consider developing our own formula."

Wayne: "Great ideas! That's why I like working with the two of you on this committee. However, I would limit the data to the last two years because of several changes we made in the program a few years back. The data for the last two years are more reflective of our current program."

Hector: "In looking at these data, how are we going to measure the degree to which a student is successful in our program? Whether they graduate or not?"

Wayne: "Fortunately or unfortunately, depending on how you look at it, practically all of our students have eventually graduated. One of the things we are trying to accomplish is to make our program more rigorous and demanding, to raise the level of quality. If this is going to happen, we have to be more selective with the students we admit."

Susan: "Why not consider the grade point average of the student at the end of the program?"

Wayne: "The major problem there is that the students in our program do not take the same set of courses in their second year because they select different areas of concentration. Some go into marketing, some into finance, others into either accounting or management. There is a real lack of comparability in those final GPA figures. But what we might do is look at the first-year GPA in the MBA program. The courses taken by students in the first year are essentially the same because they are required core courses. It is not until their second year that they began taking elective courses in the different concentration areas. What first-year MBA grade point average would the two of you, as faculty members, define as indicating a successful first year?"

Hector: "Given the breadth of those first-year core courses, their level of difficulty, and our mild degree of grade inflation, I would say that any of our students in the past two years with at least a 3.2 average would be considered successful in the first year. Would you agree, Susan?"

Susan: "I believe most of the faculty would go along with that."

Wayne: "Don't set your goals too high! Remember, we need at least 60 students per year to even have a program. We probably need to look at the data to see what's possible."

Hector: "When can we get access to the past data? I really would like to get started."

Wayne: "I'll have one of the staff members write a database program to pull up the relevant information on all students who have completed the first year of the program in the past two years. I'll get the data to you as soon as I can. Let's plan to meet again in two weeks to see what we have discovered by then. I look forward to hearing your ideas."

Assignment

Wayne McDonald's assistant gathered the data for the past two years of experience with the Westmore MBA program and the information is stored in the **WESTMORE** data file. The Data Description section provides a partial listing of the data along with definitions of the variables. Using this data set and other information given in the case, help Wayne McDonald and the MBA Admissions Committee in their quest to develop better admissions guidelines for the Westmore MBA program. The case questions will assist you in your analysis of the data. Use important details from your analysis to support your recommendations.

Data Description

The data for the Westmore MBA admissions case are contained in the **WESTMORE** data file. The file contains data for the 202 students who completed their first year in the MBA program over the past two years. A partial listing of the data is shown below.

The variables are defined as follows:

- ID_No: Student identification number.
- GPA_Yr1: Grade point average for the first year of courses in the Westmore MBA program.
- GMAT: Score on the GMAT test.
- UG_GPA: Undergraduate grade point average.

Undergraduate Major is a qualitative measure broken down into two variables:

- UG_Bus = 1 for undergraduate business major, otherwise = 0.
- UG_Tech = 1 for undergrad science, engineering, or other technical field, otherwise = 0. (*Note:* A student with an undergraduate degree in a nonbusiness and nontechnical field is the "baseline" and will have both UG_Bus = 0 and UG_Tech = 0.)

Undergraduate School Rating is a qualitative measure broken down into four variables:

- UG_1st5th = 1 if student is from the top 20% of undergraduate schools, otherwise = 0.
- UG_2nd5th = 1 if student is from the second 20% of undergraduate schools, otherwise = 0.
- UG_3rd5th = 1 if student is from the third 20% of undergraduate schools, otherwise = 0.
- UG_4th5th = 1 if student is from the fourth 20% of undergraduate schools, otherwise = 0. (*Note:* A student with a degree from a school in the bottom fifth of undergraduate schools is the "baseline" and will have a 0 value for all four of the variables shown above.)
- Age: Age of the student, in years.
- US_Cit: 1, if student is a U.S. citizen, 0 if otherwise.

ID_ No	GPA_ Yr1	GMAT	UG_ GPA	UG_ Bus	UG_ Tech	UG_ 1st 5th	UG_ 2nd 5th	UG_ 3rd 5th	UG_ 4th 5th	Age	US_ Cit
4001	3.201	540	2.974	1	0	0	0	1	0	25	1
4002	2.964	540	2.529	1	0	0	1	0	0	23	1
4003	3.745	510	3.727	0	0	0	1	0	0	25	1
4004	3.290	570	2.905	1	0	0	0	1	0	23	1
4005	3.028	640	2.641	1	0	1	0	0	0	23	1
⋮	⋮	⋮	⋮	⋮	⋮	⋮	⋮	⋮	⋮	⋮	⋮

Case Questions

1. The committee wishes to use first-year MBA grade point average as its measure of the quality of a student. As a first step in assisting committee members, construct a correlation matrix and briefly discuss the information it contains that could be of help to the committee members. If they were to choose only one variable in predicting the first-year MBA GPA, which variable would you recommend? Do you think it would be a good idea to evaluate applicants on the basis of only this one variable?
2. Because the committee is very interested in using predicted first-year MBA grade point average in measuring the quality of applicants to its program, assist the members further by carrying out and interpreting the results of an appropriate stepwise regression of the data. (*Note to Minitab users:* For consistency with Excel results, you may wish to select the alpha-to-enter and alpha-to-remove values listed in the Minitab portion of Computer Solutions 17.6.) Discuss the results, including making and supporting any recommendations that you think the committee would find useful.
3. Based on your analysis in Case Question 2, and even considering the results from a complete regression that uses *all* of the possible predictive variables, does there seem to be very much variation in the first-year MBA grade point average that is *not* explained by the independent variables provided in the data? If so, provide the committee with some examples of other quantitative or nonquantitative applicant variables that you think could help them in evaluating the potential for academic success of persons applying for admission to the Westmore MBA program.

BUSINESS CASE

Easton Realty Company (B)

When we visited Easton Realty Company (A), in Chapter 16, Sam Easton was concerned about two former clients who complained that Easton Realty had underpriced their houses in order to get a quick sale and collect their realty fee. In defending his company against these charges, Mr. Easton obtained a set of residential sales data on recent home sales in the Dallas–Fort Worth metroplex area. The data included the sale month, the sale price, the size of the home in square feet, the number of bedrooms, the age of the house, whether the house was located in Dallas versus somewhere else in the Dallas–Fort Worth metroplex, and whether Easton was the real estate company that handled the sale. In this chapter, you will be asked to help Mr. Easton gain a better understanding of what determines the selling price of a house, and you will be using the same variables that were introduced earlier, in Easton Realty (A). The data are in the **EASTON** data file.

Case Questions

1. As a first step in helping Sam Easton better understand how the sales price of a residential property is determined, construct a correlation matrix for the variables PRICE, SQFEET, BEDROOMS, AGE, DALLAS, and EASTON. Examine the matrix and advise Mr. Easton as to which variable seems to be the most important in explaining the variation in residential housing prices. You can expect Mr. Easton to be a little skeptical, since he realizes that the sales price of a house depends on many other factors besides the one you have identified.

2. Provide further assistance to Sam Easton by carrying out and interpreting the results of an appropriate stepwise regression designed to indicate which of the predictive variables mentioned in Case Question 1 are most important in determining the sales price of a house. (*Note to Minitab users:* For consistency with Excel results, you may wish to select the alpha-to-enter and alpha-to-remove values listed in the Minitab portion of Computer Solutions 17.6.) Briefly discuss the results, including making and supporting any recommendations you believe could benefit Mr. Easton.

3. Based on your analysis in Case Question 2, and even considering the results from a complete regression that uses *all* of the possible predictive variables identified in Case Question 1, does there seem to be very much variation in sales price that is *not* explained by the independent variables provided in the data? If so, what other quantitative or nonquantitative variables associated with a house might cause its sales price to be especially high or low compared to the estimated value based on models such as the ones you applied here and in Case Question 2?

CHAPTER 18

Models for Time Series and Forecasting

TIME SERIES-BASED FORECASTING AND THE ZÜNDAPP

During the late 1950s, some very unusual, miniature-size economy cars were imported to the United States. One of these was the German Zündapp, a 9.5-foot, 14-horsepower four-seater with two doors: one at the front of the car, the other at the rear. The model designation was *Janus,* for the two-faced Roman god who guarded the gates with the help of peripheral vision that athletes and city pedestrians could only dream of. The Zündapp's driver and front passenger faced forward, while the two rear passengers faced (where else?) . . . to the rear.

Being a rear-seat passenger in the Zündapp was a lot like engaging in the time series-based forecasting approaches discussed in this chapter. The only thing you could see was where you'd been, and you often had little idea where you were going.

Imagine sitting in the rear of the Zündapp and having to give driving instructions to a blindfolded driver. Giving such directions would be no problem as long as no bridges were out, no busy intersections loomed ahead, and no freeway ramps were coming up. In other words, you could do a good job if the future were not much different from the past.

While forecasting isn't bothered by fallen bridges, intersections, and freeway ramps, unpredictable events such as technological advances and changes in either the economy or consumer demands can play havoc with the accuracy of a forecast. The moral of the story: When engaged in time series-based forecasting, remember the Zündapp.

The 1958 Zündapp Janus

LEARNING OBJECTIVES

After reading this chapter, you should be able to:

- Describe the trend, cyclical, seasonal, and irregular components of the classical time series model.
- Fit a linear or a polynomial trend equation to a time series.
- Smooth a time series with the centered moving average and exponential smoothing techniques.
- Determine seasonal indexes and use them to compensate for the seasonal effects in a time series.
- Use the trend extrapolation and the exponential smoothing forecast methods to estimate a future value.
- Use the mean absolute deviation (*MAD*) and mean squared error (*MSE*) criteria to compare how well fitted equations or curves fit a time series.
- Use the Durbin-Watson test to determine whether regression residuals are autocorrelated.
- Use index numbers to compare business or economic measurements from one period to the next.

18.1 INTRODUCTION

When data are a sequence of observations over regular intervals of time, they are referred to as a **time series.** A time series may consist of weekly production output, monthly sales, annual rainfall, or any other variable for which the value is observed or reported at regular intervals.

Time series are examined for two reasons: (1) to understand the past and (2) to predict the future. By analyzing a time series, we can identify patterns and tendencies that help explain variation in past sales, shipments, rainfall, or any other variable of interest. Likewise, this understanding contributes to our ability to forecast future values of the variable.

18.2 TIME SERIES

A time series can be visually portrayed as a graph in which the variable of interest (e.g., sales, costs, shipments) is the vertical (y) axis and time is the horizontal (x) axis. An example is shown in Figure 18.1, where U.S. hybrid vehicle sales are plotted against time for the years 2000–2005.

Components of a Time Series

Analysis of a time series involves identifying the components that have led to the fluctuations in the data. In what is known as the "classical" approach to time series, these components are as follows:

TREND (T) This is an overall upward or downward tendency. As Figure 18.1 shows, U.S. sales of hybrid vehicles tended to increase during the 2000–2005 period. To the extent that the trend component is present, a regression line fitted to the points on the time series plot will have either a positive or negative slope. The line fitted in Figure 18.1 has a positive slope.

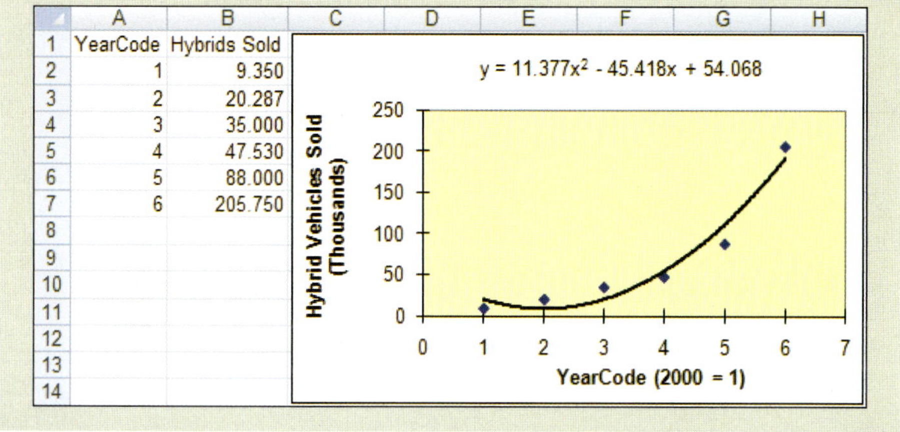

FIGURE 18.1

Excel plot with superimposed quadratic trendline for U.S. hybrid vehicle sales from 2000 through 2005. It was generated using the procedures in Computer Solutions 18.1, shown later in this section. The years are coded so that 2000 = 1.
Data source: HybridCars.com, August 25, 2006.

CYCLICAL (*C*) These are fluctuations that repeat over time, with the period being more than a year from one peak to the next. The sunspot and business cycles are examples of this type of fluctuation.

SEASONAL (*S*) These are also periodic fluctuations, but they repeat over a period of one year or less. For example, retail sales of bathing suits tend to be higher during the spring months and lower during the winter. Likewise, department stores typically do a great deal of their business during the Christmas period, vacation travel peaks during the summer months when children are out of school, and grocery stores may experience greater sales every other week, when employees at a local plant are paid.

IRREGULAR (*I*) This component represents random, or "noise" fluctuations that are the result of chance events, such as work stoppages, oil embargoes, equipment malfunction, or other happenings that either favorably or unfavorably influence the value of the variable of interest. Random variation can make it difficult to identify the effect of the other components.

In the classical time series model, the preceding components may be combined in various ways. In our discussion, we will use the *multiplicative* approach, described below and illustrated in Figure 18.2.

Classical time series model:

$y = T \cdot C \cdot S \cdot I$ where y = observed value of the time series variable
T = trend component
C = cyclical component
S = seasonal component
I = irregular component

The model assumes that any observed value of y is the result of influences exerted by the four components, and that these influences are multiplicative. The observed value of y is assumed to be the trend value, T, adjusted upward or downward by the combined influence of the cyclical, seasonal, and irregular components.

Several alternative models exist, including an additive model, with $y = T + C + S + I$. It can also be assumed that the value of y depends on the value of y

FIGURE 18.2

The components of the classical multiplicative time series model.

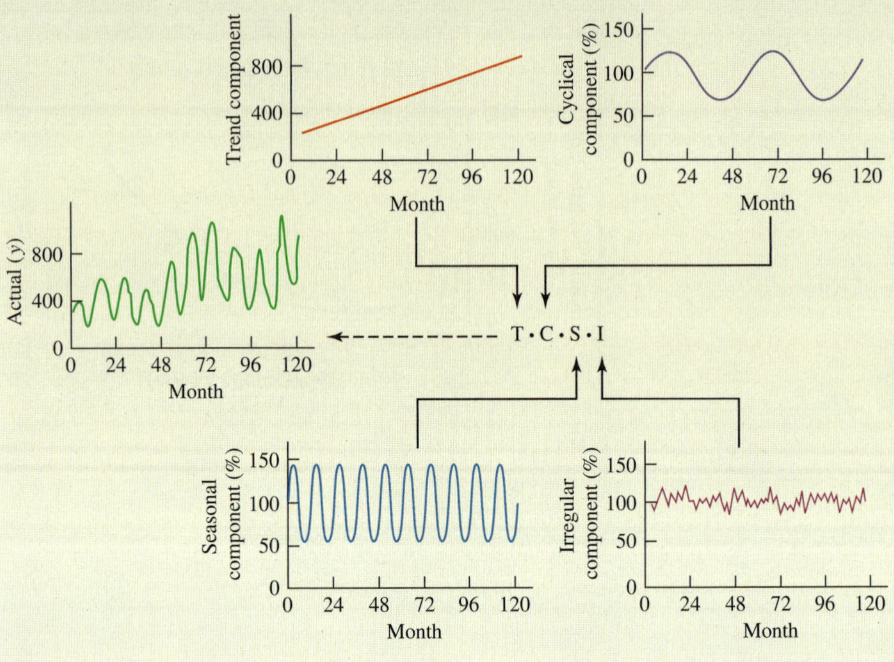

Example: If trend = 340 units, cyclical = 120%, seasonal = 85%, and irregular = 110%, the observed value will be 340 • 1.20 • 0.85 • 1.10 = 381.5 units.

for the preceding period. In other words, y_t (the observed value for time period t) may be expressed as a function of T, C, S, I, and y_{t-1} (the value for the preceding time period). Also, the multiplicative and additive models may be combined, as in $y = (T + C) \cdot S \cdot I$.

In general, the most important component of most time series is the trend, which can be examined (1) by using regression techniques to fit a trend line to the data, or (2) by using the smoothing techniques described in the next section to moderate the peaks and valleys within the series.

In fitting a trend line with regression techniques, the line may be linear (e.g., $\hat{y} = b_0 + b_1x$) or nonlinear (e.g., $\hat{y} = b_0 + b_1x + b_2x^2$), where $\hat{y}$ is the trend line estimate for y and x = time. The nonlinear example is a quadratic equation of the type we discussed in Chapter 17.

Fitting a Linear Trend Equation

In this method, a least-squares linear equation is fitted to the set of observed data points. The approach is the same as in Chapter 15 and can be summarized as follows:

The linear trend equation:

$$\hat{y} = b_0 + b_1x \qquad \text{where } \hat{y} = \text{the trend line estimate of } y$$
$$x = \text{time period}$$

In determining this and other trend equations, we will code the time periods so that the first one has a value of 1, the second has a value of 2, and so on—this is the same approach taken by Excel and Minitab. Also, if you are not using the computer, this offers the advantage of having easier calculator computations because the x values have fewer digits than the four-digit years of the original data.

Using this coding technique and the procedures in Computer Solutions 18.1, we could have obtained the least-squares regression line ($\hat{y} = -52.114 + 34.219x$).

The slope of the equation is positive because vehicle sales tended to increase from 2000 to 2005. The slope, $b_1 = +34.219$, indicates that sales tended to increase by 34.219 thousand vehicles per year during the period. For any given year, we can substitute the appropriate value of x into the trend line equation to find the estimated value of y. For example, 2013 is coded as $x = 14$, and the trend estimate for that year is $\hat{y} = -52.114 + 34.219(14)$, or 426.95 thousand vehicles.

COMPUTER 18.1 SOLUTIONS

Fitting a Linear or Quadratic Trend Equation

EXCEL

Fitting a Linear or Quadratic Trend Equation with Excel (see Figure 18.1)

1. The data in Figure 18.1 are also in Excel file **CX18HYBRID**, with year codes in A2:A7, sales data (thousands of vehicles) in B2:B7, and labels in A1 and B1. From the **Data** ribbon, click **Data Analysis.** Select **Regression.** Click **OK.**
2. Enter **B1:B7** into the **Input Y Range** box. Enter **A1:A7** into the **Input X Range** box. Select **Labels.** Enter **J1** into the **Output Range** box. Select **Line Fit Plots.** Click **OK.**
3. Right-click within the legend descriptions to the right of the chart and click **Delete.** Right-click on any one of the linearly-arranged squares that describe predicted values for the data points and click **Delete.**
4. Right-click on any one of the data points in the scatter diagram and click **Add Trendline.** In the **Trendline Options** menu, select the **Polynomial** option and enter **2** into the **Order** box. Click to select **Display Equation on chart.** Click **Close.** Make formatting adjustments to the chart.

MINITAB

Fitting a Linear or Quadratic Trend Equation with Minitab

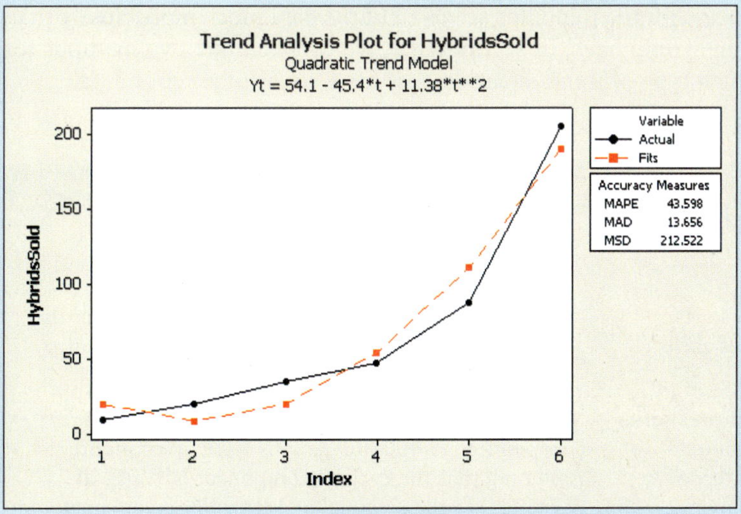

Using Minitab file **CX18HYBRID**, with the year codes in C1 and the vehicle sales data in C2: Click **Stat.** Select **Time Series.** Click **Trend Analysis.** Enter **C2** into the **Variable** box. Within the **Model Type** menu, select either **Linear** or **Quadratic.** We selected the latter. Click **OK.** The Minitab counterpart to Figure 18.1 appears as shown here.

Fitting a Quadratic Trend Equation

Another possibility for fitting an equation to a time series is to use a quadratic model, as shown in Figure 18.1. This involves an equation estimating y as a function of x, but the method treats x and x^2 as two independent variables instead of just one. The quadratic trend equation relies on the multiple regression techniques of Chapter 16 and is described as follows:

The quadratic trend equation:

$$\hat{y} = b_0 + b_1x + b_2x^2 \qquad \text{where } \hat{y} = \text{the trend line estimate of } y, \quad x = \text{time period}$$

(x and x^2 are treated as separate variables.)

The quadratic trend equation includes a constant component (b_0), a linear component (b_1x), and a nonlinear component (b_2x^2). As such, it is especially appropriate when an upward trend in the time series is followed by a downward trend, or vice versa. This type of equation can also be applied whenever the time series either increases or decreases at an increasing rate.

Computer Solutions 18.1 shows the procedures and results when Excel and Minitab are used in fitting a quadratic trend equation to the hybrid vehicle sales data for 2000 (coded as 1) through 2005 (coded as 6). Using the number of decimal places shown in Figure 18.1, the quadratic trend equation is

$$\hat{y} = 54.068 - 45.418x + 11.377x^2$$

As with the linear trend equation, the estimate for a given year is made using the x value corresponding to the code for that year. For 2013, coded as $x = 14$, the estimated number of hybrid vehicles sold in the United States during that year would be $\hat{y} = 54.068 - 45.418(14) + 11.377(14)^2 = 1648.11$ thousand vehicles.

Note that the quadratic trend line seems to fit the data much more closely than a linear trend line would have. In Section 18.6, we will examine two methods for determining which type of trend equation best fits a set of time series data.

EXERCISES

18.1 What are the four components of a time series, and to what kind of variation in the series does each refer?

18.2 For each of the following, indicate which time series component is represented. Explain your reasoning.

a. The proprietor of a movie theater finds Saturday is the most popular day of the week for movie attendance.
b. Political advertising in the United States peaks every 4 years.
c. Sales of a nonaspirin pain reliever decline after the product receives negative publicity in a medical journal article.

18.3 The estimated trend value of community water consumption is 400,000 gallons for a given month. Assuming that the cyclical component is 120% of normal, with the seasonal and irregular components 70% and 110% of normal, respectively, use the multiplicative time series model in determining the quantity of water consumed during the month.

18.4 The trend equation $\hat{y} = 1200 + 35x$ has been fitted to a time series for industry worker days lost due to job-related injuries. If $x = 1$ for 1995, estimate the number of worker days lost during 2012.

18.5 Fit a linear trend equation to the following data describing average hourly earnings of U.S. production workers. What is the trend estimate for 2013? Source: *World Almanac and Book of Facts 2009*, p. 149.

Year:	2004	2005	2006	2007
Average Hourly Earnings:	$15.69	$16.13	$16.76	$17.42

18.6 Fit a linear trend equation to these data showing the value of shipments of food products in billions of dollars. What is the trend estimate for 2015? Source: Bureau of the Census, *Statistical Abstract of the United States 2009*, p. 624.

Year:	2004	2005	2006	2007
Shipments:	$512.3	$532.4	$537.8	$573.5

18.7 U.S. cellular phone subscribership has been reported as shown here for the 1996–2007 period. Source: *World Almanac and Book of Facts 2009*, p. 406.

Year	Subscribers (Millions)	Year	Subscribers (Millions)
1996	44.0	2002	140.8
1997	55.3	2003	158.7
1998	69.2	2004	182.1
1999	86.0	2005	207.9
2000	109.5	2006	233.0
2001	128.4	2007	255.4

a. Fit a linear trend equation to the time series. Using this equation, determine the trend estimate for 2013.
b. Fit a quadratic equation to the time series, then use the equation to determine the trend estimate for 2013.
c. Construct a graph of the time series along with the equations fitted in parts (a) and (b). Which equation appears to better fit the data?

18.8 From 2000 through 2007, the value of U.S. exports to Mexico was as shown below. Data are in billions of dollars. Source: *World Almanac and Book of Facts 2009*, p. 117.

Year:	2000	2001	2002	2003
Exports:	$111.3	$101.3	$97.5	$97.4

Year:	2004	2005	2006	2007
Exports:	$110.8	$120.4	$134.0	$136.1

a. Fit a linear trend equation to the time series. Using this equation, determine the trend estimate for 2015.
b. Fit a quadratic equation to the time series, then use the equation to determine the trend estimate for 2015.
c. Construct a graph of the time series along with the equations fitted in parts (a) and (b). Which equation appears to better fit the data?

SMOOTHING TECHNIQUES 18.3

While the preceding techniques fit an actual equation to the time series, the methods covered here smooth out short-term fluctuations in the data. These techniques function much like the shock absorbers on an automobile, dampening the sudden upward and downward "jolts" that occur over the series.

The Moving Average

The *moving average* replaces the original time series with another series, each point of which is the center of and the average of N points from the original series. For this reason, this technique is also known as the **centered moving average.** The purpose of the moving average is to take away the short-term seasonal and irregular variation, leaving behind a combined trend and cyclical movement. The technique will be explained through analysis of a time series for a state's electricity purchases.

We will examine the state's electricity purchases during 1995–2009, and there will be a lot of fluctuations from one year to the next. To smooth these fluctuations, we'll use a three-year moving average. Using several of the earlier data points as an example:

Year	Electricity Purchases (thousand kwh)	Three-Year Moving Total	Centered Three-Year Moving Average
1995	815	—	—
1996	826	2386	795.3 (or 2386/3)
1997	745	2347	782.3
1998	776	2359	786.3
1999	838		

The first point in the centered moving average is (815 + 826 + 745)/3, or 795.3. It corresponds to 1996 and is the average of the three years for which 1996 is at the center. The basis for the average now "moves," dropping the 1995 value and picking up the 1998 value. This second point includes purchases for 1996 through 1998, is centered at 1997, and has a value of 2347/3, or 782.3. The moving average can be viewed as the average weight of the occupants of a lifeboat that can hold only three people at a time.

Computer Solutions 18.2 shows the raw data, the procedures, and the results when Excel and Minitab are used in generating a three-year centered moving average of the electricity purchase data. Note that the Excel procedure is applicable only when the centered moving average is based on an odd number of periods; Computer Solutions 18.3 describes how to use Excel when the basis is an even number of periods. The same Minitab procedure is used regardless of the number of periods on which the moving average is based.

COMPUTER 18.2 SOLUTIONS

Centered Moving Average for Smoothing a Time Series

This Excel procedure applies when the centered moving average is based on an odd number of time periods. Computer Solutions 18.3 is applicable for Excel when the centered moving average is based on an even number of time periods. The Minitab procedure is applicable regardless of the number of time periods on which the centered moving average is based.

EXCEL

	A	B	C
1	Year	Electric	3-yrMA
2	1995	815	
3	1996	826	795.3
4	1997	745	782.3
5	1998	776	786.3
6	1999	838	817.0
7	2000	837	835.3
8	2001	831	842.0
9	2002	858	861.7
10	2003	896	893.3
11	2004	926	922.7
12	2005	946	939.7
13	2006	947	955.3
14	2007	973	965.7
15	2008	977	986.0
16	2009	1008	
17			
18			
19			

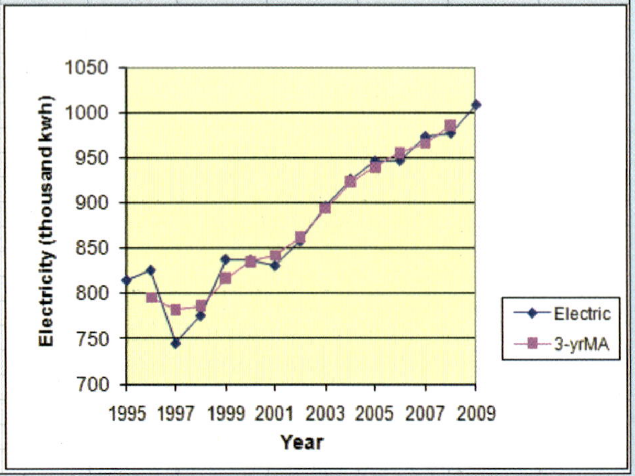

Excel Three-Period Centered Moving Average

1. The data in Table 18.1 are also in Excel file CX18EPWR, with years in A2:A16, electricity purchase data in B2:B16, and labels in A1 and B1. From the **Data** ribbon, click **Data Analysis.** Select **Moving Average.** Click **OK.**
2. Enter **B1:B16** into the **Input Range** box. Enter **3** into the **Interval** box. Select **Labels in First Row.** Enter **C1** into the **Output Range** box. (Do not select "Chart Output.") Click **OK.**
3. In column C, select the first two cells with "#NA" in them. Right-click and clear the contents of these cells. The first numerical value in column C should be C3, the centered moving average corresponding to 1996. Enter the label **3-yrMA** into **C1.**
4. Click on **B1** and drag to **C16** to select cells **B1:C16.** From the **Insert** ribbon, click **Line** from the **Charts** menu. From the **2-D Line** menu, click the fourth option (**Line with Markers**). Make formatting adjustments to the chart.

[*Excel tip:* Excel's default horizontal axis for this chart will show time periods 1, 2, and so on. *If you want to display the years, as shown here:* Right-click within the chart and click **Select Data.** In the **Select Data Source** menu, click the **Edit** button beneath **Horizontal (Category) Axis Labels.** When the **Axis Labels** box appears, click and drag to select cells **A2:A16.** As you click and drag, the years will appear within the box. Click **OK.** Click **OK.** *If you want to show every other year and have the years centered on the tick marks:* Right-click on the horizontal axis and click **Format Axis.** In the **Axis Options** menu, enter **2** into the **Interval between tick marks** box, enter **2** into the **Specify interval unit** box. In the **Position Axis** section, select **On Tick Marks.** Click **Close.**]

MINITAB

Minitab Three-Period Centered Moving Average

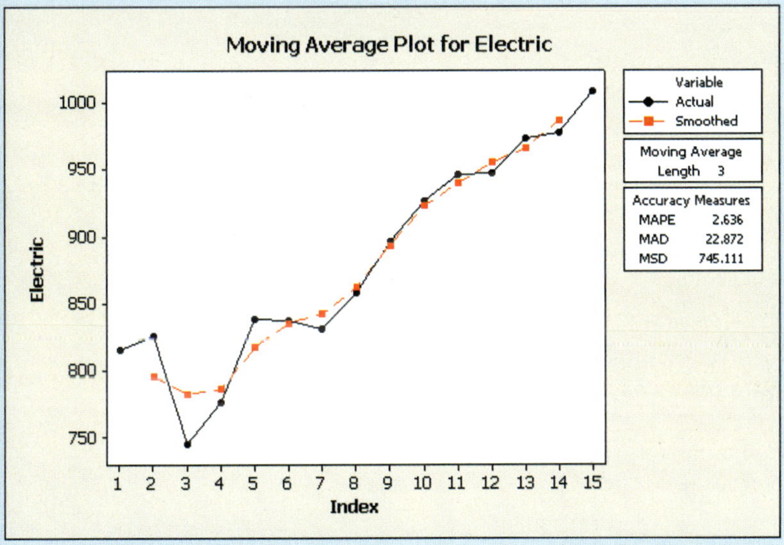

1. Using Minitab file CX18EPWR, with years in C1 and electricity purchase data in C2: Click **Stat.** Select **Time Series.** Click **Moving Average.**
2. Enter **C2** into the **Variable** box. Enter **3** into the **MA length** box. Select **Center the moving averages.**
3. Click **Graphs.** Select **Plot smoothed vs. actual.** Click **OK.** Click **OK.** The plot appears as shown here.

Constructing a Centered Moving Average with an Even Number of Periods

In the preceding example, the centered moving average had an *odd* number of periods as the base. With an *even* number of periods as the base, the method changes slightly so that each moving average point will correspond to an actual time period from the original series. The revised procedure is demonstrated in

COMPUTER 18.3 SOLUTIONS

Excel Centered Moving Average Based on Even Number of Periods

This Excel procedure is for centered moving averages based on an even number of periods.

EXCEL

	A	B	C	D
1	Year	Electric		4-yrMA
2	1995	815		
3	1996	826		
4	1997	745	790.50	793.38
5	1998	776	796.25	797.63
6	1999	838	799.00	809.75
7	2000	837	820.50	830.75
8	2001	831	841.00	848.25
9	2002	858	855.50	866.63
10	2003	896	877.75	892.13
11	2004	926	906.50	917.63
12	2005	946	928.75	938.38
13	2006	947	948.00	954.38
14	2007	973	960.75	968.50
15	2008	977	976.25	
16	2009	1008		
17				
18				
19				

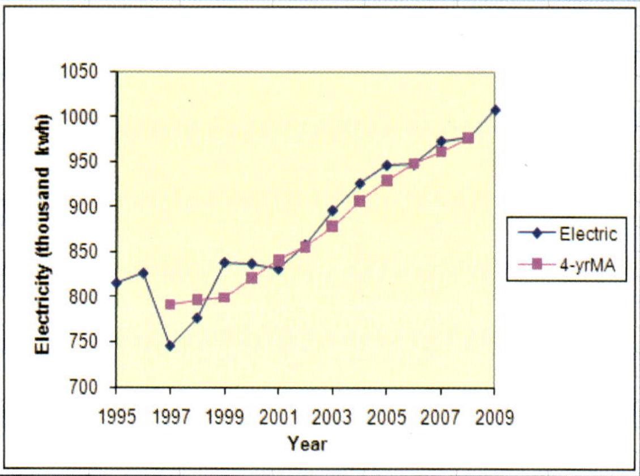

Excel Four-Period Centered Moving Average

1. This procedure corresponds to the hand calculations in Table 18.1. Use Excel file **CX18EPWR**, with years in A2:A16, electricity purchase data in B2:B16, and labels in A1 and B1. From the **Data** ribbon, click **Data Analysis.** Select **Moving Average.** Click **OK.**
2. Enter **B1:B16** into the **Input Range** box. Enter **4** into the **Interval** box. Select **Labels in First Row.** Enter **C1** into the **Output Range** box. (Do not select "Chart Output.") Click **OK.**
3. In column C, select the cells with "#NA" in them. Right-click and clear the contents of these cells. Cells C4:C15 will contain smoothed values, but they are not centered.
4. Next we do a two-period moving average to center the smoothed values: From the **Data** ribbon, click **Data Analysis.** Select **Moving Average.** Click **OK.**
5. Enter **C4:C15** into the **Input Range** box. Enter **2** into the **Interval** box. (Do not select "Labels in First Row" or "Chart Output.") Enter **D1** into the **Output Range** box. Click **OK.**
6. Enter the label **4-yrMA** into **D1.** Select cells D2:D12 and move them downward so the first value corresponds to 1997. The results, shown here, correspond to the calculations demonstrated in Table 18.1.
7. For the plot shown here, temporarily interchange columns C and D, so the variables Year, Electric, and 4-yrMA will be in adjacent columns.
8. Click on **B1** and drag to **C16** to select cells **B1:C16.** From the **Insert** ribbon, click **Line** from the **Charts** menu. From the **2-D Line** menu, click the fourth option **(Line with Markers).** Make formatting adjustments to the chart. If you prefer years instead of the default period-number horizontal axis, follow the procedure in the note accompanying Computer Solutions 18.2.

Table 18.1. In proceeding from column (1), each of the four-year moving totals in column (2) falls between two of the actual time periods. In turn, each average in column (3) falls between two figures from column (2). The result of these consecutive "in-between" placements is a moving average in column (4) that corresponds exactly to a time period of the original series.

Computer Solutions 18.3 shows the yearly data, the procedure and the results when we use a four-period centered moving average to smooth the electricity

TABLE 18.1

When a centered moving average is based on an even number of periods, the procedure is as shown for the example years shown here. Otherwise, the moving averages will not correspond exactly to actual time periods from the original series.

Year	(1) Electricity Purchases	(2) Four-Year Moving Total	(3) Average of the Four-Year Moving Totals	(4) Four-Year Centered Moving Average
1995	815			
1996	826			
		3162		
1997	745		3173.5 ÷ 4	→ 793.38
		3185		
1998	776		3190.5	797.63
		3196		
1999	838		3239.0	809.75
		3282		
2000	837		3323.0	830.75
		3364		
2001	831		3393.0	848.25
⋮	⋮	3422	⋮	⋮
		⋮		

purchases data. As noted previously, Excel requires a different method when we use an even number of periods in the moving average, while the Minitab procedure does not change.

Selecting the Number of Periods for a Centered Moving Average

As the cumbersome operations in Table 18.1 suggest, if you're not using a computer, it's desirable to select an odd number of periods unless there is some compelling reason to do otherwise. Such a reason could arise in the case of quarterly ($N = 4$) or monthly ($N = 12$) data where seasonal effects are strong. For example, with a 12-month base, each point in the moving average will include one January, one February, one March, and so on. In the case of retailers who experience the annual Christmas shopping frenzy, this seasonal "blip" will be dampened so that the overall trend can be better observed. In general, a moving average with a base equal to the duration of a strong cyclical or seasonal pattern will be highly useful in dampening fluctuations caused by the pattern.

Selecting the number of periods (N) in the base of a centered moving average is a subjective matter. A larger base provides greater dampening, but is less responsive to directional changes. In addition, an extremely large base will make less efficient use of the data because a greater number of the original values at the beginning and end of the series will not be represented by moving average points. In the Excel portion of Computer Solutions 18.2, there are 15 original data values, but the three-year moving average has just 13 points. You may wish to verify that a nine-year centered moving average would include just 7 points.

Exponential Smoothing

The fluctuations in a time series can also be dampened by a popular and relatively simple technique known as **exponential smoothing.** In this approach, each "smoothed" value is a weighted average of current and past values observed in the series. It can be summarized as follows:

The exponential smoothing of a time series:

$E_t = \alpha \cdot y_t + (1 - \alpha) \cdot E_{t-1}$ where E_t = the exponentially smoothed value for time period t

E_{t-1} = the exponentially smoothed value for the previous time period

y_t = the actual y value in the time series for time period t

α = the smoothing constant, and $0 \leq \alpha \leq 1$

Each exponentially smoothed (E) value is a weighted average of (1) the actual y value for that time period and (2) the exponentially smoothed value calculated for the previous time period. The larger the smoothing constant (α), the more importance is given to the actual y value for the time period.

To demonstrate how time series fluctuations can be dampened by exponential smoothing, we'll apply the technique to the electricity purchases that were the topic of the preceding section. The procedure is as follows:

1. Select the value of the smoothing constant, α. For this example, we'll arbitrarily select $\alpha = 0.3$.
2. Identify an initial value for the smoothed curve. Since each smoothed value will be (0.3 · the actual value) + ([1 − 0.3] · [the previous smoothed value]), we're stuck at the beginning unless we have a smoothed value with which to start the process. It's common practice to "prime the pump" by assuming that $E_1 = y_1$, so this will be our approach. As shown below, the smoothed value for time period 1 (1995) is 815 thousand kilowatt-hours. This initial value will become less important as its effect is "diluted" in subsequent calculations.
3. The calculations through 1998 are shown below.

Year	Electricity Purchases	For $\alpha = 0.3$, Exponentially Smoothed Purchases
1995	815	815
1996	826	818.3 = 0.3(826) + 0.7(815.0)
1997	745	796.3 = 0.3(745) + 0.7(818.3)
1998	776	790.2 = 0.3(776) + 0.7(796.3)

For 1995, E is set as $E = y = 815$ so the smoothed curve will have a starting point.

The fitted value for 1998 is 0.3(the actual purchases for 1998) + 0.7(the smoothed purchases for 1997).

The results of the remaining calculations are shown in the Excel portion of Computer Solutions 18.4, along with a set of exponentially smoothed values based on a larger smoothing constant, $\alpha = 0.6$. Notice that the curve with the smaller smoothing constant ($\alpha = 0.3$) more strongly dampens the fluctuations in the original series. However, it does so at the cost of lagging further behind when the series increases, decreases, or changes direction. In the Excel plots of Computer Solutions 18.4, both exponentially smoothed curves lag behind actual sales during the growth years from 2001 to 2009, much more so than either of the moving averages shown in Computer Solutions 18.2.

COMPUTER 18.4 SOLUTIONS

Exponentially Smoothing a Time Series

EXCEL

	A	B	C	D
1	Year	Electric	Alpha0.3	Alpha0.6
2	1995	815	815.0	815.0
3	1996	826	818.3	821.6
4	1997	745	796.3	775.6
5	1998	776	790.2	775.9
6	1999	838	804.6	813.1
7	2000	837	814.3	827.5
8	2001	831	819.3	829.6
9	2002	858	830.9	846.6
10	2003	896	850.4	876.3
11	2004	926	873.1	906.1
12	2005	946	895.0	930.0
13	2006	947	910.6	940.2
14	2007	973	929.3	959.9
15	2008	977	943.6	970.2
16	2009	1008	962.9	992.9
17				
18				
19				

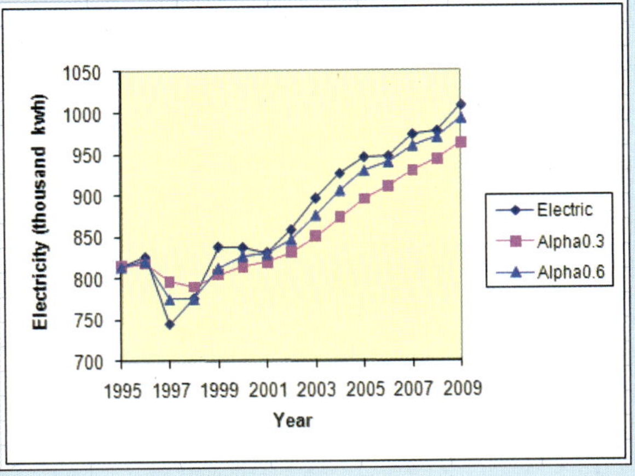

Excel Exponential Smoothing with Smoothing Constant, $\alpha = 0.3$

1. Using Excel file **CX18EPWR**, with years in A2:A16, electricity purchase data in B2:B16, and labels in A1 and B1: From the **Data** ribbon, click **Data Analysis.** Select **Exponential Smoothing.** Click **OK.**
2. Enter **B1:B16** into the **Input Range** box. Enter **0.7** into the **Damping factor** box. (*Note:* Excel's "damping factor" corresponds to $1 - \alpha$.) Select **Labels.** Enter **C1** into the **Output Range** box. (Do not select "Chart Output.") Click **OK.**
3. C1 will have "#NA" in it. Right-click and clear the contents of this cell. The first numerical value in column C should be C2. Enter the **Alpha0.3** label into **C1.** Select **C15** and, clicking at the lower right corner to produce the AutoFill "+" sign, **AutoFill** down to **C16.**
4. (Optional) We can also generate exponentially smoothed values corresponding to $\alpha = 0.6$: In step 2, enter **0.4** (i.e., $1 - \alpha$) into the **Damping factor** box, and enter **D1** into the **Output Range** box. In step 3, select and clear the "#NA" from D1, then enter the **Alpha0.6** label into **D1.** Select **D15** and, clicking at the lower right corner to produce the AutoFill "+" sign, **AutoFill** down to **D16.**
5. Click on **B1** and drag to **D16** to select cells **B1:D16.** From the **Insert** ribbon, click **Line** from the **Charts** menu. From the **2-D Line** menu, click the fourth option **(Line with Markers).** Make formatting adjustments to the chart. If you prefer years instead of the default period-number horizontal axis, follow the procedure in the note accompanying Computer Solutions 18.2.

MINITAB

Minitab Exponential Smoothing with Smoothing Constant, $\alpha = 0.3$

1. Using Minitab file **CX18EPWR**, with years in C1 and electricity purchase data in C2: Click **Stat.** Select **Time Series.** Click **Single Exp Smoothing.**
2. Enter **C2** into the **Variable** box. Select **Use:** and enter the smoothing constant **0.3** into the box. Click **Graphs.** Select **Plot smoothed versus actual.** Click **OK.** Select **Options** and enter **1** into the **Set initial smoothed value: Use average of first K observations** box. Click **OK.** Click **OK.** The plot appears as shown here.

(continued)

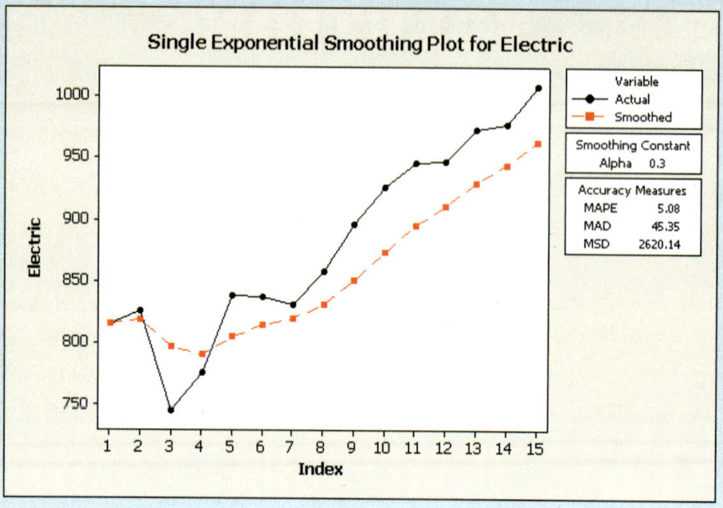

In the Excel portion of Computer Solutions 18.4, we must specify a "damping factor," calculated as $(1 - \alpha)$, then AutoFill to get the final smoothed value in the series—in this case, for 2009. Although Excel offers a plot along with the smoothed series, it assumes you are forecasting instead of just smoothing, and its plot automatically shifts the smoothed series one time period forward, leaving out the final smoothed value. The application of exponential smoothing to forecasting is covered in Section 18.5.

Minitab users can proceed relatively swiftly through the procedure described in Computer Solutions 18.4. However, to match the calculations in this section, be sure to specify that the first period be used in determining the initial smoothed value.

Selecting the value for the smoothing constant (α) is subjective and depends on the extent to which we wish to reduce the fluctuations in the data. At the extremes, a value of $\alpha = 0$ yields a horizontal line with no useful information, while $\alpha = 1$ results in an exponentially smoothed curve identical to the original time series. Naturally, neither extreme would serve the purpose of satisfactorily smoothing the series.

EXERCISES

18.9 The following data show residential and commercial natural gas consumption (quadrillion Btu) from 1992 through 2007. Source: Bureau of the Census, *Statistical Abstract of the United States 2009*, p. 567.

Year	Consumption	Year	Consumption
1992	29.6	2000	37.7
1993	30.6	2001	37.3
1994	30.8	2002	38.2
1995	33.3	2003	38.6
1996	33.0	2004	38.8
1997	33.0	2005	39.6
1998	32.8	2006	38.6
1999	36.0	2007	40.2

a. Construct a graph of the time series, then superimpose a three-year centered moving average over the original series.

b. Superimpose a five-year centered moving average over the original series. Why is the moving average "smoother" when $N = 5$?

18.10 For the data in Exercise 18.9, determine the centered moving average for $N = 2$; for $N = 4$.

18.11 According to the U.S. Energy Information Administration, a heating degree-day is a relative measurement of outdoor temperature that indicates how far the average daily temperature is below 65° Fahrenheit. The higher the number of heating degree-days, the greater the national demand for heating energy from sources like natural gas and electricity. The following monthly data show the number of heating degree-days from 2005 through 2008. Source: U.S. Energy Information Administration, *Annual Energy Review 2008*, p. 17.

	J	F	M	A	M	J
2005	859	676	648	305	186	25
2006	687	731	600	264	137	23
2007	841	853	502	372	111	24
2008	892	741	617	319	183	26

	J	A	S	O	N	D
2005	3	6	39	236	466	866
2006	2	9	82	304	467	690
2007	5	7	44	175	521	800
2008	5	13	52	281	534	831

a. Construct a graph of the time series. Does the overall trend appear to be upward or downward?
b. Determine the three-month and five-month centered moving average curves and superimpose each on the original time series.

18.12 For the data of Exercise 18.11:
a. Obtain the centered moving average for $N = 4$; for $N = 12$.
b. When $N = 12$ months, the centered moving average is relatively smooth compared to when $N = 4$. Aside from the fact that it includes a greater number of months at a time, is there any other reason why the moving average with $N = 12$ should be so smooth?

18.13 The following data describe the monthly volume for an urban courier service. Data are in thousands of pieces handled.

	J	F	M	A	M	J
2007	4.5	4.7	4.7	5.0	5.0	4.4
2008	5.4	4.9	5.3	5.6	5.4	5.4
2009	5.8	5.8	6.6	6.5	6.2	6.7

	J	A	S	O	N	D
2007	4.7	4.9	4.6	5.4	4.8	5.2
2008	5.5	5.4	5.7	6.3	5.3	6.3
2009	6.7	6.7	6.9	7.4	6.6	7.4

a. Construct a graph of the time series. Does the overall trend appear to be upward or downward?
b. Determine the three-month and five-month centered moving average curves and superimpose each on the original time series.

18.14 For the data of Exercise 18.13:
a. Obtain the centered moving average for $N = 4$; for $N = 12$.
b. When $N = 12$ months, the centered moving average is relatively smooth compared to $N = 4$. Aside from the fact that it includes a greater number of months at a time, is there any other reason why the moving average with $N = 12$ should be so smooth?

18.15 The following data describe U.S. manufacturers' shipments of general aviation jet aircraft from 1989 through 2008. Source: *2008 General Aviation Statistical Databook & Industry Outlook*, p. 20.

Year	Shipments	Year	Shipments
1989	157	1999	517
1990	168	2000	588
1991	186	2001	600
1992	171	2002	524
1993	198	2003	384
1994	222	2004	403
1995	246	2005	522
1996	233	2006	604
1997	342	2007	815
1998	413	2008	955

a. Construct a graph of the time series. Does the overall trend appear to be upward or downward?
b. Determine the three-year and five-year centered moving average curves and superimpose each on the original time series.

18.16 For the data of Exercise 18.11, fit an exponentially smoothed curve with smoothing constant $\alpha = 0.4$; with smoothing constant $\alpha = 0.7$. Which constant provides more "smoothing" and why? Which curve is quicker to "catch up" whenever the original series changes direction and why?

18.17 For the data of Exercise 18.13, fit an exponentially smoothed curve with smoothing constant $\alpha = 0.4$; with smoothing constant $\alpha = 0.7$. Which constant provides more "smoothing" and why? Which curve is quicker to "catch up" whenever the original series changes direction and why?

18.18 For the data of Exercise 18.15, determine the centered moving average for $N = 2$; for $N = 4$.

18.19 For the data of Exercise 18.15:
a. Fit an exponentially smoothed curve with smoothing constant $\alpha = 0.6$.
b. For these data, describe the appearance of the exponentially smoothed curve when the smoothing constant is $\alpha = 0.0$; when $\alpha = 1.0$. Why is it not useful to assign α either of these extreme values?

18.4 SEASONAL INDEXES

Depending on the nature of the time series variable, strong seasonal patterns may be present for certain months or quarters. The **seasonal index** is a factor that inflates or deflates a trend value to compensate for the effect of a pattern that repeats over a period of one year or less. Besides identifying the relative influence of various months or quarters, seasonal indexes are useful in the forecasting techniques of the next section.

To demonstrate the calculation of seasonal indexes, we will use a time series that has some very strong seasonal patterns. Shown in Figure 18.3, this is the quarterly sales of natural gas to customers in a northern city from 2005 through 2009. As the time series suggests, home heating and other natural gas demands are especially heavy during the first quarter of each year and tend to be relatively light during the third quarter.

The Ratio to Moving Average Method

The effect of seasonal fluctuations is quantified by using a technique called the **ratio to moving average** method. This method is explained in the context of the underlying data for Figure 18.3, with the following steps:

1. **Construct a centered moving average of the time series.** This is shown in columns (1)–(4) of Table 18.2; note that the procedure is similar to the moving average calculated earlier, in Table 18.1. Since data are quarterly, the moving average is based on four time periods. Had the data consisted of months, the moving average would have been based on 12 periods.
2. **Divide the original time series values by the corresponding centered moving average.** This yields a series of percentages of the moving average. The result of this operation is column (5) of Table 18.2. For example, the actual sales in

FIGURE 18.3

Time series plot of utility companies' sales of natural gas to customers in a northern city from 2005 through 2009.

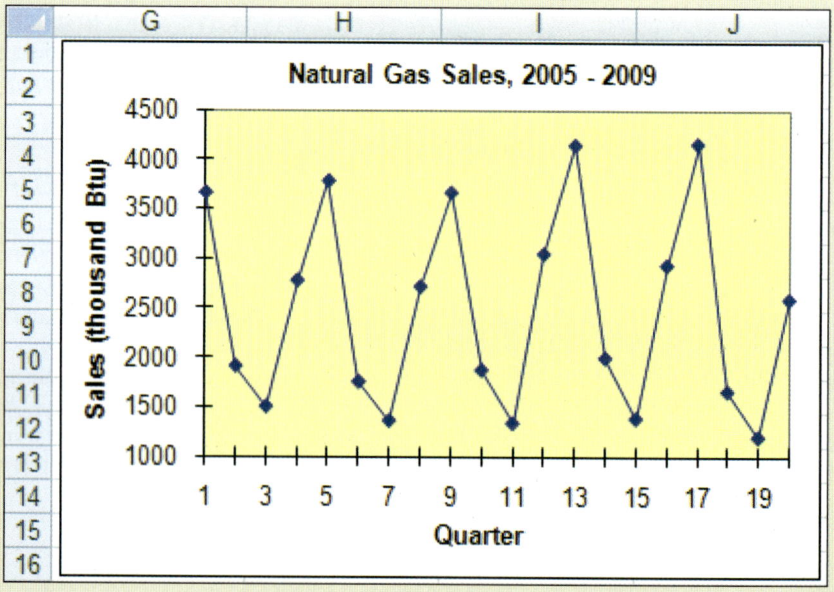

TABLE 18.2

The first stage of calculations in determining the seasonal indexes for the quarterly data of Figure 18.3. The remaining calculations and seasonal indexes are shown in Table 18.3.

Year	Quarter	(1) Sales	(2) Four-Quarter Moving Total	(3) Average of Moving Totals	(4) Centered Four-Quarter Moving Average	(5) Sales as a Percentage of Moving Average
2005	I	3661				
	II	1907				
			9,843			
	III	1500		9,903.0	2475.750	60.588%
			9,963			
	IV	2775		9,884.0	2471.000	112.303
			9,805			
2006	I	3781		9,734.5	2433.625	155.365
			9,664			
	II	1749		9,633.0	2408.250	72.625
			9,602			
	III	1359		9,541.5	2385.375	56.972
			9,481			
	IV	2713		9,541.0	2385.250	113.741
			9,601			
2007	I	3660		9,589.0	2397.250	152.675
			9,577			
	II	1869		9,741.5	2435.375	76.744
			9,906			
	III	1335		10,146.5	2536.625	52.629
			10,387			
	IV	3042		10,448.0	2612.000	116.462
			10,509			
2008	I	4141		10,532.5	2633.125	157.266
			10,556			
	II	1991		10,498.0	2624.500	75.862
			10,440			
	III	1382		10,446.5	2611.625	52.917
			10,453			
	IV	2926		10,284.0	2571.000	113.808
			10,115			
2009	I	4154		10,020.5	2505.125	165.820
			9,926			
	II	1653		9,755.0	2438.750	67.781
			9,584			
	III	1193				
	IV	2584				

the first quarter of 2006 were 3781 thousand Btu, which is 155.365% of the centered moving average (2433.625) for that quarter.

3. **For each quarter (or month), identify the mean percentage from those obtained in step 2.** This is done to reduce the effect of the irregular component of the time series, and the results are shown in the first box of Table 18.3. Other possibilities include using either the median or a trimmed mean that is calculated after the highest and lowest values are dropped.
4. **If the average value of the means is not 100, multiply each one by the same adjustment factor so that the average of 100 is obtained.** In other words, for quarterly data, the means should total 400. If this is not the case, inflate or

TABLE 18.3

The seasonal indexes for quarterly sales of natural gas to a northern city, based on the preliminary calculations of Table 18.2.

Sales as a % of the moving average:

	Quarter I	Quarter II	Quarter III	Quarter IV
2005			60.588	112.303
2006	155.365	72.625	56.972	113.741
2007	152.675	76.744	52.629	116.462
2008	157.266	75.862	52.917	113.808
2009	165.820	67.781		
Mean	157.782	73.253	55.777	114.079

↓

The sum of the means is 400.891

Multiplying each mean by the adjustment factor, $\frac{400}{400.891}$, the seasonal indexes are

I:	157.43
II:	73.09
III:	55.65
IV:	113.83

deflate the means accordingly. In Table 18.3, this is accomplished by multiplying each of the means by 400/400.891.

5. **Following the adjustment (if necessary) in step 4, the resulting values are the seasonal indexes.** For our example, these are shown in Table 18.3. The larger the seasonal index, the more influence that quarter has in terms of natural gas sales. As expected, the first quarter, and its heating requirements, has resulted in the highest seasonal index, 157.43. This means that 157.43% of the average quarterly gas purchases occur during the first quarter. Likewise, the third quarter accounts for only 55.65% of the average quarterly purchases.

Deseasonalizing the Time Series

Having determined the seasonal indexes, we can go back and *deseasonalize* the original time series. In this process, we remove the seasonal influences and generate a time series that is said to be *seasonally adjusted.*

From the standpoint of the classical time series model, the original and the *deseasonalized time series* can be summarized as follows:

1. **Original time series** Trend (T), cyclical (C), seasonal (S), irregular (I) components, or $T \cdot C \cdot S \cdot I$.
2. **Deseasonalized time series** Only trend, cyclical, and irregular components, or $T \cdot C \cdot I$.

For any time period, the deseasonalized value can also be expressed as $(T \cdot C \cdot S \cdot I)/S$, and dividing by S removes the seasonal influence from the series. In the denominator of the following formula, this seasonal influence (S) is represented as a proportion, or the seasonal index divided by 100:

Deseasonalized time series:

For each time period, the deseasonalized value is

$$D = \frac{\text{Value in the original time series}}{\text{(Seasonal index for that period)/100}}$$

The deseasonalized time series for the natural gas data is determined by the preceding formula and shown in Table 18.4. Throughout the series, each first quarter's sales are divided by 1.5743, each second quarter's sales by 0.7309, and so on. The deseasonalized sales in column (3) are the actual sales in column (1) divided by the (seasonal index)/100 in column (2). For example, the deseasonalized sales for the first quarter of 2009 are

$$D = \frac{\text{Actual sales}}{\text{(Seasonal index)/100}} = \frac{4154}{157.43/100} = 2638.6$$

The Excel portion of Computer Solutions 18.5 shows this and the other deseasonalized quarterly sales superimposed on the original time series of Figure 18.3. The end result of deseasonalizing the time series is a line that gives us a clearer view of the trend and cyclical components of the series.

Excel does not have a module dedicated to seasonal indexes, but we have provided Excel worksheet templates for determining seasonal indexes and deseasonalized values for quarterly and monthly series, and the Data Analysis Plus add-in can also be applied. Minitab does have a dedicated module but, as with Data Analysis Plus, the methodology differs from the technique described in this section and leads to seasonal index values that are slightly different from those obtained here.

TABLE 18.4

The deseasonalized time series for the natural gas sales of Table 18.2.

Year	Quarter	(1) Actual Sales	(2) (Seasonal Index)/100	(3) Deseasonalized Sales
2005	I	3661	1.5743	2325.5
	II	1907	0.7309	2609.1
	III	1500	0.5565	2695.4
	IV	2775	1.1383	2437.8
2006	I	3781	1.5743	2401.7
	II	1749	0.7309	2392.9
	III	1359	0.5565	2442.0
	IV	2713	1.1383	2383.4
2007	I	3660	1.5743	2324.8
	II	1869	0.7309	2557.1
	III	1335	0.5565	2398.9
	IV	3042	1.1383	2672.4
2008	I	4141	1.5743	2630.4
	II	1991	0.7309	2724.0
	III	1382	0.5565	2483.4
	IV	2926	1.1383	2570.5
2009	I	4154	1.5743	2638.6
	II	1653	0.7309	2261.6
	III	1193	0.5565	2143.8
	IV	2584	1.1383	2270.1

COMPUTER 18.5 SOLUTIONS

Determining Seasonal Indexes

EXCEL, USING WORKSHEET TEMPLATE

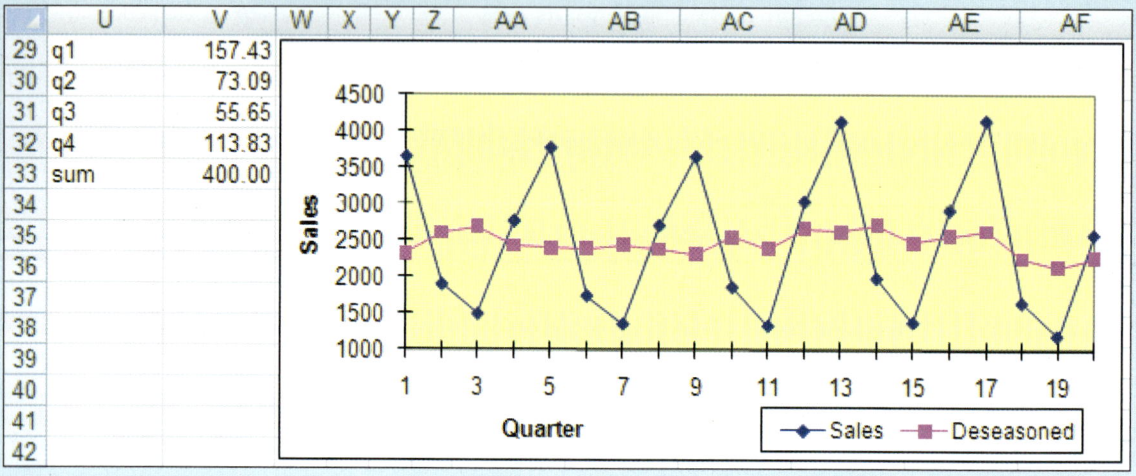

Seasonal Indexes for Quarters

1. Open Excel file **CX18GAS**, with quarters in B2:B21, gas sales data in C2:C21, and labels in B1 and C1. Simultaneously open Excel worksheet template **TMQUARTS**.
2. Working in **TMQUARTS**: Click on the row button for row 14, drag downward to select rows 14 through 37, then right-click and delete rows 14 through 37. There are now 20 rows in which to paste the 20 quarters of gas sales data.
3. Working in **CX18GAS**, click on C2 and drag to C21 to select cells C2:C21. Right-click and select **Copy.**
4. Working in **TMQUARTS**, click to select C6. Right-click and select **Paste.** The sales data are now in cells C6:C25. Click on cell D8 and, using the AutoFill "+" at the lower right corner, AutoFill downward over all existing values in column D. In column B, ensure that the quarters are numbered in sequence, from 1 to 20.
5. Within **TMQUARTS**, the seasonal indexes are shown in column V. Scroll downward in column R to view plots for the original time series and the deseasonalized time series. With slight formatting modifications, the deseasonalized plot appears as shown here. It has been moved next to the seasonal indexes for purposes of this Computer Solutions description.

Notes

1. In general, the specific rows for the locations of the indexes and plots will vary, depending on how many years' worth of quarterly data are being examined.
2. For monthly data, use Excel worksheet template **TMMONTHS**. The procedure will be similar to the one described here.

EXCEL, USING DATA ANALYSIS PLUS

Seasonal Indexes for Quarters

1. Open Excel file **CX18GAS**. The quarterly data are in C2:C21. Enter the label **Quarter** into cell **D1.** Enter the sequence of digits **1** through **4** into **D2:D5.** Continue repeating sequences of 1 through 4 into column D until you've put the sequence of digits 1 through 4 into cells D18:D21. Note that the repeating sequence of quarter identifications should be in the column immediately to the *right* of the column containing the data. If there is no blank column to the right, you will need to insert one.
2. From the **Add-Ins** ribbon, click **Data Analysis Plus.** Select **Seasonal Indexes.** Click **OK.**

3. Enter **C1:D21** into the **Input Range** box and select **Labels.** Click **OK.** The seasonal indexes are shown in the printout.

	A	B
1	Seasonal Indexes	
2		
3	Season	Index
4	1	1.5633
5	2	0.7411
6	3	0.5492
7	4	1.1463

Notes

1. The Data Analysis Plus methodology differs slightly from the one described in the text and used in the Excel worksheet template method described previously, and it will result in slightly different seasonal index values.
2. For monthly data, use the same procedure shown here, but repeat the sequence of digits 1 through 12 in the column immediately to the right of the column containing the data.

MINITAB

Minitab Seasonal Indexes, Quarters

1. Using Minitab file **CX18GAS**, with quarters in C2 and gas sales data in C3: Click **Stat.** Select **Time Series.** Click **Decomposition.**
2. Enter **C3** into the **Variable** box. Enter **4** into the **Seasonal length** box. Under **Model Type,** select **Multiplicative.** Under **Model Components,** select **Trend plus seasonal.** Click **Options** and enter **1** into the **First obs is in seasonal period:** box. Click **OK.** Click **OK.** The summary printout shown here is among the standard printout items, and a variety of tables and plots can be obtained as well.

```
Time Series Decomposition for Sales

Multiplicative Model

Data       Sales
Length     20
NMissing   0

Fitted Trend Equation
Yt = 2513.8 - 4.50850*t

Seasonal Indices
Period     Index
     1  1.56598
     2  0.74378
     3  0.55044
     4  1.13980
```

Notes

1. The Minitab methodology differs from the one described in the text and used in the Excel method above, and will result in slightly different seasonal index values.
2. For monthly data, use the same procedure, but in step 2 enter **12** into the **Seasonal length** box.

EXERCISES

18.20 In determining seasonal indexes for the months of the year, a statistician finds the total of the "unadjusted" seasonal indexes is 1180 instead of 12 · 100, or 1200. What correction must be made to arrive at the adjusted seasonal indexes?

18.21 The seasonal indexes for a convention center's bookings are 83, 120, 112, and 85 for quarters 1–4, respectively. What percentage of the center's annual bookings tend to occur during the second quarter?

18.22 The state of Texas has reported the following quarterly gross sales for the in-state accommodation and food services industry for the years 2004–2008.
Source: Texas Comptroller of Public Accounts site, ecpa.cpa.tx.us, August 22, 2009.

Gross Sales ($ Billions)

	I	II	III	IV
2004	$7.014	7.330	7.260	7.521
2005	7.441	7.904	7.716	8.228
2006	8.166	8.390	8.321	8.751
2007	8.642	9.322	9.136	9.599
2008	9.412	9.879	9.578	9.995

a. Construct the four-quarter centered moving average for these data and determine the percentages of the moving average for the quarters.
b. Determine the seasonal indexes for the quarters.
c. Use the seasonal indexes to deseasonalize the original time series.

18.23 A major amusement park has the following number of visitors each quarter from 2005 through 2009:

Number of Visitors (Thousands)

	I	II	III	IV
2005	155	231	270	105
2006	182	255	315	294
2007	160	250	280	297
2008	210	310	365	335
2009	225	325	384	386

a. Construct the four-quarter centered moving average for these data and determine the percentages of the moving average for the quarters.
b. Determine the seasonal indexes for the quarters.
c. Use the seasonal indexes to deseasonalize the original time series.

18.24 Given the degree-day data of Exercise 18.11, (a) determine the monthly seasonal indexes for January through December, then (b) use these indexes to deseasonalize the original time series.

18.25 Given the data of Exercise 18.13, (a) determine the monthly seasonal indexes for January through December, then (b) use these indexes to deseasonalize the original time series.

18.26 A mail-order firm has found the following seasonal indexes for the number of calls received each month. Below the seasonal indexes is the actual number of calls (thousands) per month during 2009.

a. Using the seasonal indexes, deseasonalize the data for 2009.
b. The president of the firm is quite pleased that the number of calls went up each month during the May through December portion of 2009. Based on the results obtained in part (a), is the firm's performance really on the increase?

Month:	J	F	M	A	M	J	J	A	S	O	N	D
Seasonal index:	110	105	90	60	70	75	85	95	100	105	140	165
Calls in 2009:	7.2	5.9	5.5	3.2	4.0	4.1	5.3	5.4	5.7	6.1	7.9	8.2

18.5 FORECASTING

In forecasting, we use data from the past in predicting the future value of the variable of interest, such as appliance sales, natural gas consumption, or the thickness of the ozone layer. The time series and seasonal index methods discussed earlier are often employed for this purpose. Using past data to predict future data is very much like facing backward in a car, then trying to figure out where you are going by looking at where you've been. There actually was such a car in the 1950s, and the analogy between it and time series-based forecasting is discussed in the opening vignette to this chapter.

Forecasts Using the Trend Equation

Once we've fitted a trend equation to the data, a forecast can be made by simply substituting a *future* value for the time period variable. The procedure is the same whether the fitted trend equation is linear or nonlinear.

For example, in Figure 18.1, we fitted a quadratic trend equation to hybrid vehicle sales in the United States from 2000 through 2005. Using this equation to make a forecast for 2013:

The fitted linear trend equation is

$$\hat{y} = 54.068 - 45.418x + 11.377x^2$$

where $\hat{y}$ = the trend line estimate of hybrid vehicle sales (thousands)
x = year code, with 2000 coded as 1.

↓

The forecast of hybrid vehicle sales for 2013 would be

$$\hat{y} = 54.068 - 45.418(14) + 11.377(14)^2, \quad \text{or 1648.11 thousand vehicles}$$

Forecasting with Exponential Smoothing

Besides smoothing a time series, exponential smoothing can also be used in making a forecast for one time period into the future. The approach is similar to the smoothing method of Section 18.3, except that F (for forecast value) replaces E (exponentially smoothed value) for each time period, and the time period (t) "shifts" one unit into the future:

Forecasting with exponential smoothing:

$$F_{t+1} = \alpha \cdot y_t + (1 - \alpha) \cdot F_t$$

where F_{t+1} = forecast for time period $t + 1$
y_t = actual y value for time period t
F_t = the value that had been forecast for time period t
α = the smoothing constant, and $0 \leq \alpha \leq 1$

As in our previous application of exponential smoothing, there is a smoothing constant, $0 \leq \alpha \leq 1$. If our forecast involves annual sales, the forecast for next year is a weighted average of the *actual* sales this year and the sales level that had been *forecast* for this year. If we choose $\alpha = 1$, we are essentially saying "next year will be the same as this year."

To illustrate the application of exponential smoothing to forecasting, we will use the electric power purchases data. These data included purchases from 1995 through 2009, and exponential smoothing will give us a forecast for 2010. For our example, we will use $\alpha = 0.6$ as the smoothing constant.

As shown below, exponential smoothing requires that we assume an initial forecast value in order to get things started. For this reason, the forecast for 1995 is assumed to have been 815, the same as the actual purchases for that year. Proceeding toward 2010, the influence of this choice will decrease, since its effect is diluted as more recent forecast values are introduced. For each year, the forecast is a weighted average of the actual and the forecasted level of purchases for the preceding year, and the weighting is according to the smoothing constant, $\alpha = 0.6$:

Year	y = Actual purchases	F = Purchases forecast for the year, with smoothing constant $\alpha = 0.6$
1995	815	815.0
1996	826	815.0 = 0.6(815) + 0.4(815.0)
1997	745	821.6 = 0.6(826) + 0.4(815.0)
1998	776	775.6 = 0.6(745) + 0.4(821.6)
⋮	⋮	⋮
2008	977	959.9 = 0.6(973) + 0.4(940.2)
2009	1008	970.2 = 0.6(977) + 0.4(959.9)
2010	—	992.9 = 0.6(1008) + 0.4(970.2)

To get the process started, the forecast for 1995 is assumed to have been 815.

The forecast for 2010 is a weighted average of the actual and the forecasted values for 2009.

The Excel portion of Computer Solutions 18.6 lists these and the other forecasted values for electricity purchases. In addition to the forecast and plot shown in Computer Solutions 18.6, Minitab can also generate a complete

COMPUTER 18.6 SOLUTIONS

Forecasting with Exponential Smoothing

EXCEL

	A	B	C
1	Year	Electric	Forecast
2	1995	815	815.0
3	1996	826	815.0
4	1997	745	821.6
5	1998	776	775.6
6	1999	838	775.9
7	2000	837	813.1
8	2001	831	827.5
9	2002	858	829.6
10	2003	896	846.6
11	2004	926	876.3
12	2005	946	906.1
13	2006	947	930.0
14	2007	973	940.2
15	2008	977	959.9
16	2009	1008	970.2
17	2010		992.9

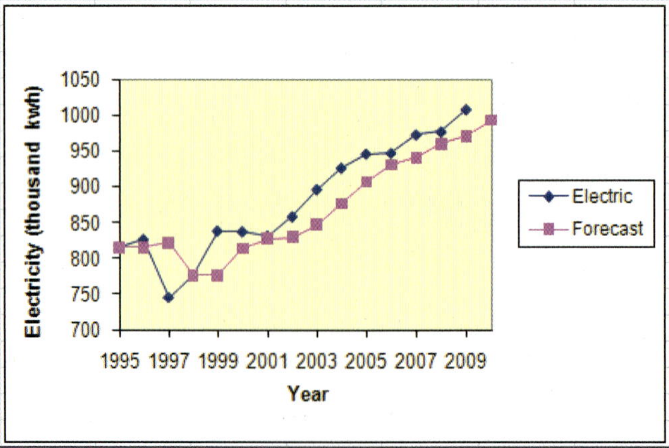

Excel Forecasting with Exponential Smoothing, $\alpha = 0.6$

1. Using Excel file **CX18EPWR**, with years in A2:A16, electricity purchase data in B2:B16, and labels in A1 and B1: From the **Data** ribbon, click **Data Analysis.** Select **Exponential Smoothing.** Click **OK.**

2. Enter **B1:B16** into the **Input Range** box. Enter **0.4** into the **Damping factor** box. (*Note:* Excel's "damping factor" corresponds to 1 − α.) Select **Labels.** Enter **C2** into the **Output Range** box. (Do not select "Chart Output.") Click **OK.**
3. Select **C2** and enter **815** (this replaces the "#NA" and provides a forecast for the first time period). Select **C16** and, clicking at the lower right corner to produce the AutoFill "+" sign, **AutoFill** down to **C17.** Enter **2010** into **A17.** Enter the **Forecast** label into **C1.**
4. Click on **B1** and drag to **C17** to select cells **B1:C17.** From the **Insert** ribbon, click **Line** from the **Charts** menu. From the **2-D Line** menu, click the fourth option (**Line with Markers**). Make formatting adjustments to the chart. If you prefer years instead of the default period-number horizontal axis, follow the procedure in the note accompanying Computer Solutions 18.2.

MINITAB

Minitab Exponential Smoothing with Smoothing Constant, α = 0.6

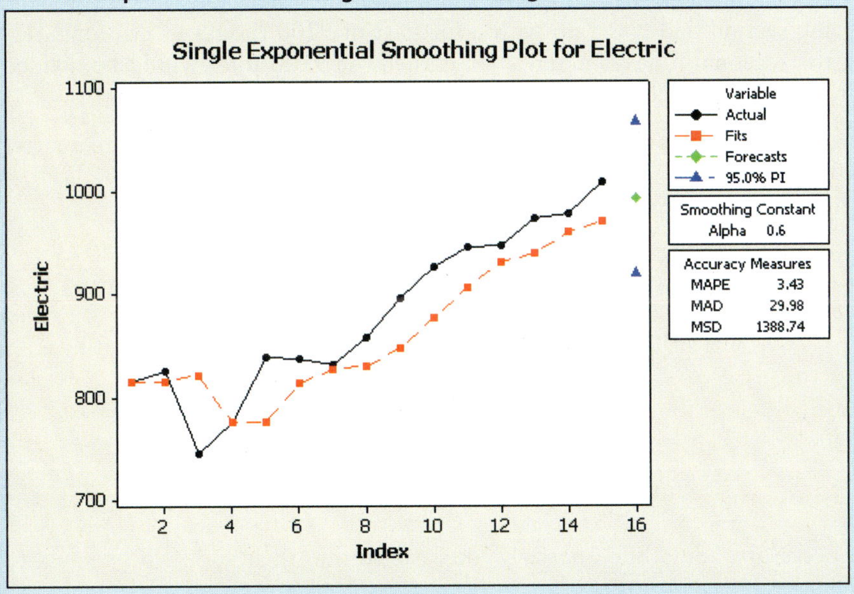

1. Using Minitab file **CX18EPWR**, with years in C1 and electricity purchase data in C2: Click **Stat.** Select **Time Series.** Click **Single Exp Smoothing.**
2. Enter **C2** into the **Variable** box. Select **Use:** and enter the smoothing constant **0.6** into the box. Select **Generate forecasts** and enter **1** into the **Number of forecasts** box. (Just leave the "Starting from origin" box blank and Minitab will make a forecast for the one period immediately following the last data value.)
3. Click **Graphs.** Select **Plot predicted vs. actual.** Click **OK.** Click **Options** and enter **1** into the **Set initial smoothed value: Use average of first K observations** box. Click **OK.** Click **OK.** The plot appears as shown here, with the forecast displayed as a single point beyond the data series.

table that includes time series values, smoothed values, forecast values, and residuals.

When exponential smoothing is used for forecasting, the forecast will tend to be too low for time series that are increasing (positive trend) and too high for time series that are decreasing (negative trend). This is because the forecast for a given period is always the weighted average of two values associated with the immediately preceding period. To correct this problem, the upward or downward trend of the time series can be introduced as an additional variable. Further details of this method, called the Holt-Winters forecasting model, can be found in more advanced texts on forecasting.

Seasonal Indexes in Forecasting

Once we have arrived at a forecast for a given year, we can use seasonal indexes to come up with a forecast for each month or quarter of the year. In doing so, we divide up the total forecast according to the relative sizes of the seasonal indexes. The procedure is as follows:

Dividing an annual forecast into seasonal components:

$$\text{Forecast for a season} = \frac{\text{Forecast for the year}}{\text{Number of seasons}} \cdot [(\text{Seasonal index})/100]$$

For the natural gas data, suppose that one of the forecasting methods of the preceding section had come up with a forecast of 9200 thousand Btu for 2015. Using the seasonal indexes determined earlier, the 9200 total would be divided up into the four quarters as follows:

1. If the forecast for the entire year = 9200 thousand Btu, the average quarter will be 9200/4, or 2300.
2. "Inflate" or "deflate" quarters according to their seasonal indexes:

Quarter:	I	II	III	IV
(Seasonal index)/100:	1.5743	0.7309	0.5565	1.1383

Quarter	
I	2300(1.5743) = 3620.89 thousand Btu
II	2300(0.7309) = 1681.07
III	2300(0.5565) = 1279.95
IV	2300(1.1383) = 2618.09
	9200.00

3. If necessary, adjust the quarterly forecasts so that their sum will be equal to the total forecast for the year.

The sum of these quarterly forecasts (9200.00) is equal to the 9200.00 forecasted for the entire year, so no further adjustment is necessary. (As an example, if the quarterly forecasts had totaled 9198.00, we would have had to multiply each of the quarterly estimates in step 2 by 9200/9198.00 so that the total for the year would be 9200.00.)

EXERCISES

18.27 The linear trend equation $\hat{y} = 120 + 4.8x$ has been developed, where $\hat{y}$ = estimated sales (thousands of dollars) and $x = 1$ for the year 1999. What level of sales would be forecast for 2013; for 2015?

18.28 An appliance repair shop owner has fitted the quadratic trend equation $\hat{y} = 90 + 0.9x + 3x^2$ to a time series of annual repair orders, with y = the number of repair orders and $x = 1$ for 2005. Forecast the number of repair orders for 2013; for 2015.

18.29 Based on annual truck rentals over the years, a rental firm has developed the quadratic trend equation $\hat{y} = 450 + 20x + 7.2x^2$, with $x = 1$ for 2005. Forecast y for 2013; for 2015.

18.30 When exponential smoothing is used in fitting a curve to a time series, the approach is slightly different from its application to forecasting. Compare the appropriate formulas and point out how they differ.

18.31 The U.S. trade deficit with Canada (billions of dollars) from 2000 through 2007 is reported as shown in the table. Using exponential smoothing and the smoothing constant $\alpha = 0.7$, what deficit would have been forecast for 2008? Source: *World Almanac and Book of Facts 2009*, p. 117.

Year:	2000	2001	2002	2003	2004	2005	2006	2007
Deficit:	$51.9	52.8	48.2	51.7	66.5	78.5	71.8	68.2

18.32 For the data of Exercise 18.7, use the weighting constant $\alpha = 0.5$ and exponential smoothing to determine the forecast for 2008.

18.33 For the data of Exercise 18.9, use the weighting constant $\alpha = 0.6$ and exponential smoothing to determine the forecast for 2008.

18.34 If a time series is such that sales are consistently increasing from one year to the next, will exponential smoothing tend to produce forecasts that are (a) over or (b) under the sales that are actually realized for the forecast period? Why?

18.35 A firm predicts it will use 560,000 gallons of water next year. If the seasonal index of water consumption is 135 for the second quarter of the year, what is the forecast for water consumption during that quarter?

18.36 The quarterly seasonal indexes for a firm's electricity consumption are 115, 92, 81, and 112 for quarters I–IV. It has been forecast that electricity consumption will be 850,000 kilowatt-hours during 2014. Forecast electricity consumption for each quarter of 2014.

EVALUATING ALTERNATIVE MODELS: *MAD* AND *MSE* (18.6)

When two or more models are fitted to the same time series, it's useful to have one or more criteria on which to compare them. In this section, we'll discuss two such criteria.

According to the **mean absolute deviation (*MAD*)** criterion, the best-fit model is the one having the lowest mean value for $|y_t - \hat{y}_t|$. The direction of an error, or deviation, is not considered, only its absolute value. The *MAD* criterion is as follows:

The mean absolute deviation (*MAD*) criterion:

From a given set of models or estimation equations fit to the same time series data, the model or equation that best fits the time series is the one with the lowest value of

$$MAD = \frac{\sum |y_t - \hat{y}_t|}{n}$$

where y_t = an observed value of y
$\hat{y}_t$ = the value of y that is predicted using the estimation equation or model
n = the number of time periods

Another criterion is **mean squared error (*MSE*),** in which the best-fit model or equation is the one having the lowest mean value for $(y_t - \hat{y}_t)^2$. This is analogous to the least-squares criterion used in determining the regression equations of Chapters 15–17. The MSE approach is described below:

The mean squared error (*MSE*) criterion:

From a given set of models or estimation equations fit to the same time series data, the model or equation that best fits the time series is the one with the lowest value of

$$MSE = \frac{\sum (y_t - \hat{y}_t)^2}{n}$$

where y_t = an observed value of y
$\hat{y}_t$ = the value of y that is predicted using the estimation equation or model
n = the number of time periods

Compared to the *MAD* criterion, the *MSE* approach places a greater penalty on estimates for which there is a large error. This is because *MSE* is the mean of the *squared* errors instead of simply the mean of the absolute values. Because of this, the *MSE* criterion is to be preferred whenever the cost of an error in estimation or forecasting increases in more than a direct proportion to the amount of the error.

Table 18.5 shows the application of the *MAD* and *MSE* criteria to the linear and quadratic estimation equations that were developed for the hybrid vehicle sales data in Section 18.2 of this chapter.

TABLE 18.5

MAD and *MSE* measures of fit to the hybrid vehicle sales data of Section 18.2.

A. Fit for the Linear Estimation Equation, $\hat{y} = -52.114 + 34.219x$

Year	Year Code, x_t	Actual Sales, y_t	Estimated Sales, $\hat{y}_t$	Error, $(y_t - \hat{y}_t)$	Absolute Error, $\lvert y_t - \hat{y}_t \rvert$	Squared Error, $(y_t - \hat{y}_t)^2$
2000	1	9.350	−17.895	27.245	27.245	742.290
2001	2	20.287	16.324	3.963	3.963	15.705
2002	3	35.000	50.543	−15.543	15.543	241.585
2003	4	47.530	84.762	−37.232	37.232	1386.222
2004	5	88.000	118.981	−30.981	30.981	959.822
2005	6	205.750	153.200	52.550	52.550	2761.503
					167.514	6107.127

$$MAD = \frac{\sum|y_t - \hat{y}_t|}{n} = \frac{167.514}{6} = 27.919 \qquad MSE = \frac{\sum(y_t - \hat{y}_t)^2}{n} = \frac{6107.127}{6} = 1017.854$$

B. Fit for the Quadratic Estimation Equation, $\hat{y} = 54.068 - 45.418x + 11.377x^2$

Year	Year Code, x_t	Actual Sales, y_t	Estimated Sales, $\hat{y}_t$	Error, $(y_t - \hat{y}_t)$	Absolute Error, $\lvert y_t - \hat{y}_t \rvert$	Squared Error, $(y_t - \hat{y}_t)^2$
2000	1	9.350	20.027	−10.677	10.677	113.998
2001	2	20.287	8.740	11.547	11.547	133.333
2002	3	35.000	20.207	14.793	14.793	218.833
2003	4	47.530	54.428	−6.898	6.898	47.582
2004	5	88.000	111.403	−23.403	23.403	547.700
2005	6	205.750	191.132	14.618	14.618	213.686
					81.936	1275.133

$$MAD = \frac{\sum|y_t - \hat{y}_t|}{n} = \frac{81.936}{6} = 13.656 \qquad MSE = \frac{\sum(y_t - \hat{y}_t)^2}{n} = \frac{1275.133}{6} = 212.522$$

Regardless of the criterion, the quadratic equation is a better fit to these data. This is not surprising, since the quadratic equation has the advantage of including both a linear and a nonlinear term.

EXERCISES

18.37 How do the *MAD* and *MSE* criteria differ in their approach to evaluating the fit of an estimation equation to a time series?

18.38 The following data are the wellhead prices for domestically produced natural gas, in dollars per thousand cubic feet, from 2001 through 2008. Given these data and the trend equations shown here, use the *MAD* criterion to determine which equation is the better fit. Repeat the evaluation, using the *MSE* criterion. Source: U.S. Energy Information Administration, *Annual Energy Review 2008*, p.199.

Year	$x =$ Year Code	$y =$ Price
2001	1	$4.00
2002	2	2.95
2003	3	4.88
2004	4	5.46
2005	5	7.33
2006	6	6.39
2007	7	6.37
2008	8	8.07

Linear:
$\hat{y} = 2.896 + 0.6189x$

Quadratic:
$\hat{y} = 2.543 + 0.8305x - 0.02351x^2$

18.39 The following data are the number of Harley-Davidson motorcycles sold worldwide for 2003 through 2006 (thousands). Given these data and the trend equations, use the *MAD* criterion to determine which equation is the better fit. Repeat the evaluation, using the *MSE* criterion. Source: Harley-Davidson, Inc., *2008 Annual Report*, p. 20.

Year	$x =$ Year Code	$y =$ Sales
2003	1	291.1
2004	2	317.3
2005	3	329.0
2006	4	349.2

Linear:
$\hat{y} = 275.15 + 18.60x$

Quadratic:
$\hat{y} = 267.65 + 26.10x - 1.50x^2$

18.40 Using the *MAD* criterion, determine which one of the equations developed in Exercise 18.7 is the better fit to the data in that exercise. Repeat the evaluation, using the *MSE* criterion.

18.41 Using the *MAD* criterion, determine which one of the equations developed in Exercise 18.8 is the better fit to the data in that exercise. Repeat the evaluation, using the *MSE* criterion.

AUTOCORRELATION, THE DURBIN-WATSON TEST, AND AUTOREGRESSIVE FORECASTING (18.7)

Autocorrelation and the Durbin-Watson Test

As we've discussed in earlier chapters, one of the assumptions required by the regression model is that the residuals (errors) be independent of each other. When this assumption is not met, the condition known as *autocorrelation* (or *serial correlation*) is present. Regression models based on time series data are especially susceptible to this condition because: (1) key independent (x) variables that have not been included in the model are likely to be related to time, and (2) the omission of one or more of these independent variables will tend to yield residuals (errors) that are also related to time.

Autocorrelation will tend to cause the calculated standard error of estimate (s_e) to underestimate the true value (σ_e). The partial regression coefficients (b_1, b_2, and so on) will also have standard errors that are understated, and the t values in their respective tests of significance will tend to be exaggerated. As a result, confidence and prediction intervals will be too narrow, and the respective independent variables will be exaggerated in their significance.

A common method of testing for autocorrelation is the *Durbin-Watson* test. Our discussion of this test will be in terms of its ability to detect the presence of *first-order* autocorrelation. This is the correlation between adjacent residuals or errors—that is, between e_t and e_{t-1}. (By extension, *second-order* autocorrelation would be the correlation between e_t and e_{t-2}. When we refer to autocorrelation in the remainder of this discussion, we will be referring to first-order autocorrelation.) The Durbin-Watson test treats the population of residuals as shown below.

Durbin-Watson test, description of residual (error) for time period t:

$$\varepsilon_t = \rho\varepsilon_{t-1} + z_t$$

where ε_t = residual associated with time period t
ε_{t-1} = residual associated with time period $t - 1$
ρ = the population correlation coefficient between values of ε_t and ε_{t-1}, with $-1 \le \rho \le +1$
z_t = a normally distributed random variable with $\mu = 0$

If ρ is zero, there is no autocorrelation and $\varepsilon_t = z_t$—that is, ε_t is nothing more than random variation with a mean of zero. Likewise, a ρ that is greater than 0 reflects positive autocorrelation and a ρ that is less than zero reflects negative autocorrelation. Examples of positive and negative autocorrelation are shown in Figure 18.4. The Durbin-Watson test is carried out as shown below.

Durbin-Watson test for autocorrelation:

- **Null and alternative hypotheses**

The test is

for positive autocorrelation	for negative autocorrelation	a two-tail test
H_0: $\rho \le 0$ H_1: $\rho > 0$	H_0: $\rho \ge 0$ H_1: $\rho < 0$	H_0: $\rho = 0$ H_1: $\rho \ne 0$

- **Test statistic**

$$d = \frac{\sum_{t=2}^{t=n} (e_t - e_{t-1})^2}{\sum_{t=1}^{t=n} e_t^2}$$

where e_t = observed residual (error) associated with time period t
e_{t-1} = observed residual (error) associated with time period $t - 1$
n = number of observations, with $n \ge 15$

- **Critical values and decision zones** Look up the appropriate d_L and d_U values in the Durbin-Watson tables in Appendix A. The actual critical values and decision zones are based on (1) the type of test (positive, negative, or two-tail), (2) α = the level of significance for the test, and (3) k = the number of independent variables in the regression model. The decision zones and appropriate conclusions are shown in Figure 18.5.

FIGURE 18.4

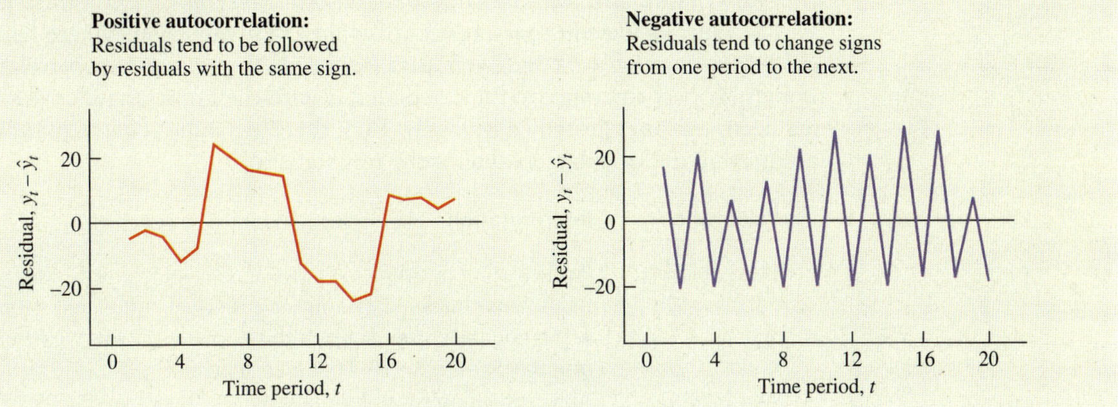

Examples of positive and negative autocorrelation of residuals

FIGURE 18.5

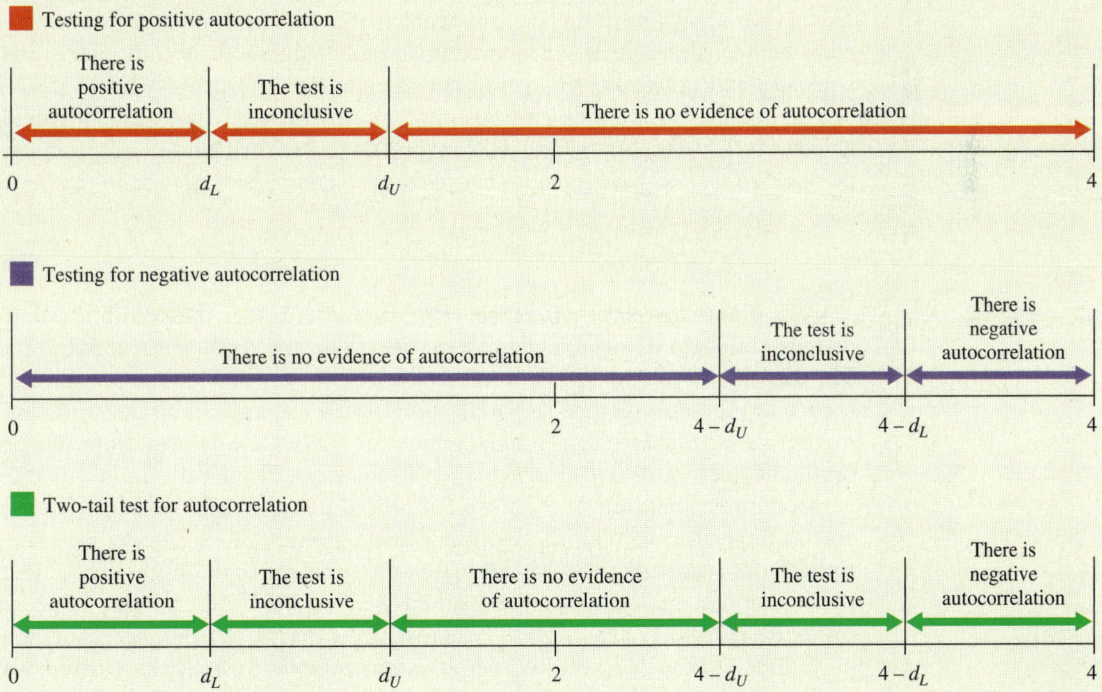

When proceeding with a Durbin-Watson test for autocorrelation, the calculated d statistic is compared to one of three sets of critical values and decision zones shown.

Appendix A provides three tables for the Durbin-Watson test. They correspond to the 0.05, 0.025, and 0.01 levels of significance for a one-tail (positive or negative) test, and to the 0.10, 0.05, and 0.02 levels for a two-tail test.

The Durbin-Watson test statistic (d) will have a value somewhere between 0 and 4, and a value of 2 indicates that the residuals are not autocorrelated. Although it's best to compare the calculated d with the appropriate critical values and decision zones described in Figure 18.5, there are some general guidelines for interpreting the calculated value of the test statistic, d:

Value of d	Interpretation
Close to 0.0	Positive autocorrelation: A positive residual in one period tends to be followed by a positive residual in the next period, and a negative residual tends to be followed by a negative residual. See Figure 18.4 for an example.
Approx. 2.0	Autocorrelation is weak or absent.
Close to 4.0	Negative autocorrelation: Residuals tend to change signs from one period to the next, as shown in Figure 18.4.

EXAMPLEEXAMPLE

EXAMPLE

Durbin-Watson Test

Table 18.6 shows how the Durbin-Watson test could be applied to a time series consisting of deseasonalized quarterly shipment data. Most computer statistical packages can spare you the trouble of calculating the Durbin-Watson d statistic, but the details are in Table 18.6 just to demonstrate how it's done. The calculated value ($d = 1.90$) is comfortably within the "there is no evidence of autocorrelation" decision zone, and we conclude that autocorrelation is not present in this series.

When autocorrelation is present, we should consider the possibility that one or more important variables may have been omitted from the regression model. The identification and inclusion of such variables can reduce the presence of autocorrelation and enhance the quality of confidence and prediction interval estimates and significance tests for the regression model. Transformations performed on the dependent or the independent variables can also be useful in this regard.

Computer Solutions 18.7 shows the procedures and results when we use Excel and Minitab in obtaining the Durbin-Watson statistic. In each case, we have performed the regression analysis and generated the residuals. With Excel, we must then specify the cells in which the residuals are located. With Minitab, we need only select the Durbin-Watson option when we are carrying out the regression analysis procedure. Although Excel and Minitab can provide us with the Durbin-Watson statistic, we must still refer to the printed tables in reaching a conclusion.

The Autoregressive Forecasting Model

When a time series exhibits autocorrelation, not only are the residuals correlated with each other, but the time series values themselves are correlated with each

TABLE 18.6

In this application of the Durbin-Watson test the calculated d (1.90) falls into the "there is no evidence of autocorrelation" decision zone of the two-tail test. The calculated d is available from most computer statistical packages, and Computer Solutions 18.7 shows how we can obtain it with Excel and Minitab.

For 16 quarters that have been deseasonalized.
The least-squares estimation equation is $\hat{y} = 34.0 + 0.7t$.

↓

H_0: $\rho = 0$ (there is no autocorrelation)
H_1: $\rho \neq 0$ (there is an autocorrelation)

↓

For a two-tail test at the 0.05 level. With $n = 16$ observations and $k = 1$ independent variable in the regression equation, reference to the Durbin-Watson tables in Appendix A gives $d_L = 0.98$ and $d_U = 1.24$.

↓

t	y_t	$\hat{y}_t$	e_t	e_t^2	e_{t-1}	$e_t - e_{t-1}$	$(e_t - e_{t-1})^2$
1	33.7	34.7	−1.0	1.00	—	—	—
2	32.7	35.4	−2.7	7.29	−1.0	−1.7	2.89
3	40.2	36.1	4.1	16.81	−2.7	6.8	46.24
4	42.5	36.8	5.7	32.49	4.1	1.6	2.56
5	36.6	37.5	−0.9	0.81	5.7	−6.6	43.56
6	36.4	38.2	−1.8	3.24	−0.9	−0.9	0.81
7	38.9	38.9	0.0	0.00	−1.8	1.8	3.24
8	37.6	39.6	−2.0	4.00	0.0	−2.0	4.00
9	36.3	40.3	−4.0	16.00	−2.0	−2.0	4.00
10	38.2	41.0	−2.8	7.84	−4.0	1.2	1.44
11	48.2	41.7	6.5	42.25	−2.8	9.3	86.49
12	41.0	42.4	−1.4	1.96	6.5	−7.9	62.41
13	40.9	43.1	−2.2	4.84	−1.4	−0.8	0.64
14	44.3	43.8	0.5	0.25	−2.2	2.7	7.29
15	43.8	44.5	−0.7	0.49	0.5	−1.2	1.44
16	48.0	45.2	2.8	7.84	−0.7	3.5	12.25
				147.11			279.26

The calculated Durbin-Watson statistic is $d = 279.26/147.11 - 1.90$.

↓

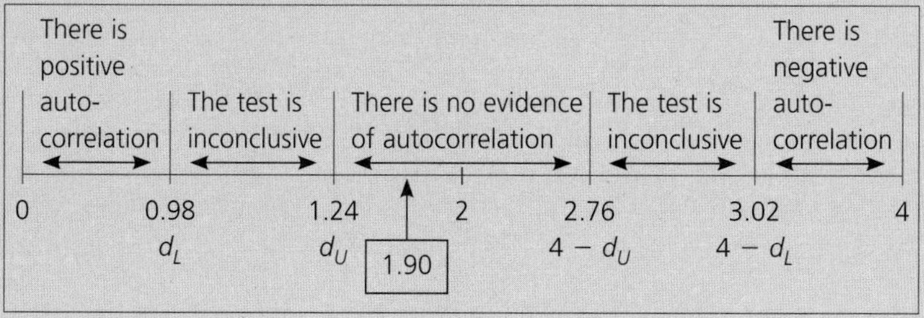

other. Although this can undermine the independent-residuals assumption of the conventional regression approach to time series, it provides an excellent opportunity for what is known as the **autoregressive forecasting model.**

In autoregressive forecasting, the dependent variable (y) is treated the same way it is in other regression models within this chapter. However, each independent

COMPUTER 18.7 SOLUTIONS

Durbin-Watson Test for Autocorrelation

EXCEL

	A	B	C
1	Durbin-Watson Statistic		
2			
3	d = 1.8983		

Excel Durbin-Watson Statistic

1. The first two columns of Table 18.6 are listed in columns A and B of Excel file **CX18SHIP**, with t values in A2:A17, y values in B2:B17, and labels in A1 and B1. First, we run the regression analysis: From the **Data** ribbon, click **Data Analysis.** Select **Regression.** Click **OK.**
2. Enter **B1:B17** into the **Input Y Range** box. Enter **A1:A17** into the **Input X Range** box. Select **Labels.** Enter **C1** into the **Output Range** box. Check the **Residuals** box. Click **OK.**
3. The output lists the residuals, in cells E25:E40. From the **Add-Ins** ribbon, click **Data Analysis Plus.** Click **Durbin-Watson Statistic.** Click **OK.** Do not select "Labels." Enter **E25:E40** into the **Input Range** box. Click **OK.** On a separate sheet, the Durbin-Watson statistic is reported as $d = 1.8983$.

MINITAB

Minitab Durbin-Watson Statistic

```
Regression Analysis: y versus t

The regression equation is y = 34.0 + 0.700 t

Predictor    Coef   SE Coef      T      P
Constant   34.008     1.700  20.01  0.000
t          0.6999    0.1758   3.98  0.001

...
...
...

Durbin-Watson statistic = 1.89832
```

Using Minitab file **CX18SHIP**, with t values in C1 and y values in C2, we run the regression analysis and specify that the output include the Durbin-Watson statistic: Click **Stat.** Select **Regression.** Click **Regression.** Enter **C2** into the **Response** box. Enter **C1** into the **Predictors** box. Click **Options.** Select **Durbin-Watson statistic.** Click **OK.** Click **OK.** The partial printout shown here includes the Durbin-Watson statistic, reported as $d = 1.89832$.

variable represents *previous values of the dependent variable.* We will examine two levels of the autoregressive forecasting model:

1. **First-order autoregressive model** The forecast for a given time period is based on the value observed in the preceding time period:

$$\hat{y}_t = b_0 + b_1 y_{t-1} \qquad \text{with } y_{t-1} = \text{the value of } y \text{ in period } t - 1$$

2. **Second-order autoregressive model** The forecast for a given time period is based on the values observed in the two previous time periods:

$$\hat{y}_t = b_0 + b_1 y_{t-1} + b_2 y_{t-2} \qquad \text{with } y_{t-1} = \text{the value of } y \text{ in period } t-1$$
$$y_{t-2} = \text{the value of } y \text{ in period } t-2$$

Table 18.7 shows annual billings (millions of dollars) for new U.S. manufactured general aviation aircraft shipped from 1986 (coded as $t = 1$) through 2005

TABLE 18.7

Annual billings (millions of dollars) for new U.S. manufactured general aviation aircraft shipped from 1986 through 2005. In these autoregressive forecasting models, the estimated value for a given year is based either on the value observed in the preceding year (first-order model) or on the values in the two preceding years (second-order model). The underlying data are in file **CX18AIR**.

Year	Year Code (t)	Billings (y_t)	y_{t-1}	y_{t-2}
1986	1	$1262		
1987	2	1364	$1262	
1988	3	1918	1364	$1262
1989	4	1804	1918	1364
1990	5	2008	1804	1918
1991	6	1968	2008	1804
1992	7	1840	1968	2008
1993	8	2144	1840	1968
1994	9	2357	2144	1840
1995	10	2842	2357	2144
1996	11	3048	2842	2357
1997	12	4580	3048	2842
1998	13	5761	4580	3048
1999	14	7843	5761	4580
2000	15	8558	7843	5761
2001	16	8641	8558	7843
2002	17	7719	8641	8558
2003	18	6434	7719	8641
2004	19	6816	6434	7719
2005	20	8667	6816	6434

First-order autoregressive model:
$\hat{y}_t = 443.9 + 0.987y_{t-1}$

Second-order autoregressive model:
$\hat{y}_t = 414.7 + 1.50y_{t-1} - 0.54y_{t-2}$

Data source: 2005 *General Aviation Statistical Databook & Industry Outlook*, p. 13.

(coded as $t = 20$). If we were to fit linear and quadratic trend equations to the time series, the results would be as follows:

Linear: The equation is $\hat{y}_t = -178.7 + 434.04t$ and the Durbin-Watson statistic = 0.47 (strong positive autocorrelation).

Quadratic: The equation is $\hat{y}_t = 557.3 + 233.3t + 9.558t^2$ and the Durbin-Watson statistic = 0.50 (strong positive autocorrelation).

In applying either of these estimation equations, we find very significant positive autocorrelation between consecutive values of the aircraft shipment billing data. A year in which billings exceed the model's estimate tends to be followed by *another* year in which billings are higher than estimated. Similarly, if the billings for a year are lower than the model has estimated, the billings for the next year also tend to be lower than estimated.

For each of the years in the time series, Table 18.7 shows the year code (t), the billing amount (y_t) for the year, the billing amount for the preceding year (y_{t-1}), and the billing amount for the period 2 years earlier (y_{t-2}). Notice how the billings in the fourth column are shifted downward one year from those in the third column, and how those in the fifth column are shifted downward two years from those in the third column. We simply apply either simple or multiple regression in obtaining the first-order and second-order autoregressive equations in the lower portion of the table.

The two equations at the bottom of Table 18.7 are the results when we apply the first-order and second-order autoregressive forecasting models to the aircraft billing data in the table. Using the mean absolute deviation criterion presented in Section 18.6, we find that MAD = \$603.1 million for the first-order model and \$575.0 million for the second-order model, indicating that the second-order model is a slightly better fit to the series. Given the available data and the autoregressive forecasting equations, the billings forecast for 2006 could be made as shown below:

Forecast for 2006, using the first-order autoregressive model:

$$\hat{y}_t = 443.9 + 0.987y_{t-1}$$

and

$$\hat{y}_{2006} = 443.9 + 0.987y_{2005} = 443.9 + 0.987(8667) = \$8998.2 \text{ million}$$

Forecast for 2006, using the second-order autoregressive model:

$$\hat{y}_t = 414.7 + 1.50y_{t-1} - 0.54y_{t-2}$$

and

$$\hat{y}_{2006} = 414.7 + 1.50y_{2005} - 0.54y_{2004} = 414.7 + 1.50(8667) - 0.54(6816)$$
$$= \$9734.6 \text{ million}$$

Because its *MAD* value is the lesser of the two, the forecast with the second-order autoregressive model would seem to be the preferred one. However, this does not mean we should automatically use a second-order autoregressive model instead of a first-order model. In fitting the second-order model, we must discard the initial *two* years in the series instead of just one. Referring to Table 18.7, note that years 1986 and 1987 would be considered as "missing data" in this model because no billing values were provided for years prior to 1986. Discarding the first two periods in a time series might not be acceptable if the series already happens to contain only a small number of periods.

Computer Solutions 18.8 describes the Excel and Minitab procedures for obtaining the first-order and second-order autoregressive equations shown at the bottom of Table 18.7. The Minitab procedure involves a special "Lag" module that makes it convenient to shift the data and create the y_{t-1} and y_{t-2} series that are lagged by one and two years, respectively.

COMPUTER 18.8 SOLUTIONS

Autoregressive Forecasting

EXCEL

Excel Autoregressive Forecasting [equations at the bottom of Table 18.7]

1. In Excel file CX18AIR, the billings time series in Table 18.7 is in C2:C21, with the label in C1. Select cells C2:C20 and copy them to cells D3:D21—this will be the series of (t − 1) values. Select cells C2:C19 and copy them to E4:E21—this will be the series of (t − 2) values.
2. The first-order autoregressive equation is obtained by performing a simple regression in which the Y input range is **C3:C21** and the X input range is **D3:D21.** When performing the regression, do not select "Labels." To obtain the residuals and predicted values, select **Residuals.**
3. The second-order autoregressive equation is obtained by performing a regression analysis in which the Y input range is **C4:C21** and the X input range is **D4:E21.** Do not select "Labels." To get the residuals and predicted values, select **Residuals.**

MINITAB

Minitab Autoregressive Forecasting [equations at the bottom of Table 18.7].

1. Using Minitab file CX18AIR, with y values in C3: Click **Stat.** Select **Time Series.** Click **Lag.** Enter **C3** into the **Series** box. Enter **C4** into the **Store lags in** box. Enter **1** into the **Lag** box. Click **OK.** The (t − 1) series will be stored in C4.
2. Repeat step 1, but enter **C5** into the **Store lags in** box and enter **2** into the **Lag** box. The (t − 2) series will be stored in C5.
3. To obtain the first-order autoregressive equation: Click **Stat.** Select **Regression.** Click **Regression.** Enter **C3** into the **Response** box. Enter **C4** into the **Predictors** box. Click **OK.**
4. To obtain the second order autoregressive equation, carry out step 3, but enter **C4 C5** into the **Predictors** box.

EXERCISES

18.42 When autocorrelation of the residuals is present, what effect can this have on interval estimation and significance tests regarding the regression model involved?

18.43 What is the Durbin-Watson test for autocorrelation, and how can it be useful in evaluating the relevance of a given regression model that has been fitted to a set of time series data?

18.44 The least-squares regression equation $\hat{y} = 98.1 + 5.0t$ has been fitted to the shipment data given here. Calculate the Durbin-Watson d statistic and test for positive autocorrelation of the residuals at the 0.05 level of significance. Comment on the appropriateness of using this model for this set of data.

t	y_t	t	y_t	t	y_t
1	97.0	6	125.5	11	150.6
2	109.7	7	132.9	12	157.8
3	110.3	8	139.7	13	158.3
4	121.5	9	145.6	14	166.9
5	130.4	10	149.4	15	175.8

18.45 The least-squares regression equation $\hat{y} = 15.2 + 10.7t$ has been fitted to the shipment data in the table. Calculate the Durbin-Watson d statistic and test for positive autocorrelation of the residuals at the 0.05 level of significance. Based on your conclusion, comment on the appropriateness of using this model for this set of data.

t	y_t	t	y_t	t	y_t
1	65.6	7	68.3	13	141.3
2	56.7	8	79.1	14	149.6
3	79.5	9	104.2	15	177.1
4	53.4	10	104.2	16	193.6
5	59.7	11	115.4	17	238.9
6	71.1	12	86.6	18	259.2

18.46 A regression model has 3 independent variables and 20 observations, and the calculated Durbin-Watson d statistic is 0.91. What, if any, conclusion will be reached in testing for autocorrelation of the residuals with a:

a. two-tail test at the 0.02 level of significance?
b. one-tail test for positive autocorrelation at the 0.05 level?

18.47 A regression model has 2 independent variables and 16 observations, and the calculated Durbin-Watson d statistic is 2.78. What, if any, conclusion will be reached in testing for autocorrelation of the residuals with a:

a. two-tail test at the 0.05 level of significance?
b. test for negative autocorrelation at the 0.01 level?

18.48 Analysis of a time series consisting of weekly visits (in thousands) to a corporate website has led to the following autoregressive forecasting equation: $\hat{y}_t = 34.50 + 0.58y_{t-1} - 0.72y_{t-2}$. If there were 260 thousand visitors this week and 245 thousand last week, how many visitors would the forecast predict for next week?

18.49 For the shipment data listed in Exercise 18.45, construct a first-order autoregressive forecasting equation and make a forecast for period 19.

18.50 According to the Investment Company Institute, the number of U.S. mutual shareholder accounts (millions) were as shown in the table for years 1990 (coded as $t = 1$) through 2007. Construct a second-order autoregressive equation and determine what the forecast would have been for 2008. Source: Investment Company Institute, *2008 Investment Company Fact Book,* p. 110.

t	y_t	t	y_t	t	y_t
1	61.9	7	149.9	13	251.1
2	68.3	8	170.3	14	260.7
3	79.9	9	194.0	15	269.5
4	94.0	10	226.2	16	275.5
5	114.4	11	244.7	17	288.6
6	131.2	12	248.7	18	299.0

18.51 Over the past 20 years, inventory carrying costs for a large tire manufacturing facility have been as shown. Data are in thousands of dollars. Construct a first-order autoregressive forecasting equation for these data, calculate the mean absolute deviation (MAD) for the fit of the forecast values to the actual values, then use the equation to forecast inventory carrying costs for time period 21.

t	y_t	t	y_t
1	4710.0	11	2828.5
2	4187.7	12	2734.8
3	4275.6	13	2743.9
4	4076.9	14	2789.6
5	3731.4	15	2939.4
6	3484.3	16	2770.6
7	3437.2	17	2815.0
8	3203.2	18	2808.4
9	3206.2	19	2938.1
10	2963.1	20	2918.6

18.52 Repeat Exercise 18.51, but for a second-order autoregressive forecasting equation. Compare the MAD values for forecasts generated by the two equations and indicate which one is the better fit to these data.

18.8 INDEX NUMBERS

In Section 18.4, we discussed seasonal indexes, measures intended to reflect the extent to which various parts of the year experience higher or lower levels of production, demand, or other kinds of economic activity. This section introduces another concept that includes the word "index," but this is where the similarity ends. Before introducing the notion of index numbers, let's consider a variety of settings in which they could be useful:

- A family of four has a household income of $80,000, up 15% from three years ago, but they still have a lot of trouble making ends meet. Is the family's purchasing power stronger or weaker than it was three years ago?

- Wishing to compare the "livability" of various metropolitan areas, a researcher collects data on crime, suicide, alcoholism, and unemployment rates in each.
- Critics point out that the \$8000 per semester tuition at a state university has escalated the cost of a college education beyond the reach of the typical family. University administrators, pointing to inflation, respond that a college education is more accessible than ever before.

For applications like these, it is useful to convert information to the form of an **index number.** An index number is a percentage that expresses a measurement in a given period in terms of the corresponding measurement for a selected base period. The measurement in the base period is defined as 100, and the measurement in a given period will be greater than or less than 100, depending on whether the measurement was greater than or less than that of the base period. For example:

In 2000 a company's inventory
costs were \$3.6 million. $\rightarrow I_{2000} = 100$ for the 2000 base period

In 2009 the company's inventory
costs were \$5.9 million. $\rightarrow I_{2009} = 163.9 \quad \left(\text{i.e., } \frac{5.9}{3.6} \cdot 100\right)$

In other words, if inventory costs in 2000 are defined as 100, 2009 costs can be expressed as 163.9, or 63.9% higher. Like other index numbers, this one is unitless. Note that both the numerator and the denominator are described in millions of dollars, and these cancel out during the division.

The base period for an index number should be a relatively *typical* period during which extremely unusual phenomena have not occurred. For example, it wouldn't be desirable to construct an index of oil prices using a base year during which there was either a severe oil glut or a shortage.

Index numbers are used in measuring changes over time for such diverse entities as stock values, prices, wages, production, and output. Some of them even combine many different measurements into a single value. If you happen to live in one of the colder regions of the nation, you're familiar with "wind chill," the translation of outside temperature and wind velocity into a single number representing what would be an equally uncomfortable temperature if the wind were calm. This particular indicator, originally based on Antarctic experiments of the 1930s, is not derived by the methods through which index numbers are usually determined, but it's yet another example of providing information by condensing two or more numbers into just one.

Many popular statistical indexes are in use, and you may already be familiar with such indexes as the Dow Jones Industrial Average, the Standard & Poor's Index of 500 Stocks, and the Consumer Price Index. The latter can be used in determining such things as whether the household earning \$80,000 has more purchasing power than three years ago.

The Consumer Price Index

The **Consumer Price Index (CPI)** describes the change in prices from one time period to another for a fixed "market basket" of goods and services. It is generated monthly by the U.S. Bureau of Labor Statistics and is probably the index with which both businesspersons and the general public are most familiar.

Derivation of the CPI

The CPI is based on periodic consumer expenditure surveys. At this writing, the base prices are for the 1982–1984 period. The change to the 1982–1984 base period was made on the basis of data from the 1982–1984 Consumer Expenditure Survey. From 1971 until 1987, the base period for the index had been 1967.

The "market basket" for the CPI includes items that reflect the price of bread, physicians' and dentists' fees, alcoholic and nonalcoholic beverages, clothing, transportation, shelter, and other goods and services purchased for day-to-day living. Every month, prices are obtained from 57,000 housing units and 19,000 establishments representing 85 areas across the United States.[1] The resulting data are used in generating price indexes for specific regions and cities as well as for various categories of products and services.

Applications

Economic Indicator. By reflecting changes in prices, the Consumer Price Index is the primary measurement of inflation in the United States and of the perceived success or failure of governmental policies to control inflation. Political candidates are eager to point out increases in the CPI when they wish to gain office and equally quick to call attention to any slowing in the growth of the CPI whenever they have been in office and are defending past policies.

Escalator. A common application of the Consumer Price Index is in the automatic adjustment of contracted wages or pension benefits through what are known as *escalator clauses* or *cost-of-living adjustments (COLAs)*. As prices go up, the wages or benefits are increased, either (1) in proportion to the change in the CPI, or (2) according to a formula that has been included in the collective bargaining agreement or set forth in the relevant legislation.

Following a 1987 work stoppage, 4600 workers at Rohr Industries' aircraft parts plants in California got a new three-year contract negotiated by the Machinists Union. Among other items, the contract called for hourly wage increases based on the Consumer Price Index. The U.S. Department of Labor summed up the contract details:

Three-year agreement negotiated 2/87 also provided: lump sum payments equal to 10 percent of pay 1st year and 6 percent in the 2nd and 3rd years; current COLA rolled into base rates and cost-of-living adjustment formula continued to provide 1 cent for each 0.3–point movement in the BLS-CPI; $24 (was $20) monthly pension benefit rate; 100 percent (was 95) employer payment for preferred provider option-health care; $1 million (was $250,000) lifetime major medical maximum; $15,000 (was $10,000) life insurance. . . .[2]

Deflator. Using the Consumer Price Index, current prices and wages can be deflated to *real prices* or *real wages*. Compared to their "current" counterparts, these let us more readily determine whether we have more earning power than in some previous period. If a person's real income has not kept pace with real prices,

[1] *The New York Times Almanac 2009,* p. 338.
[2] Bureau of Labor Statistics, U.S. Department of Labor, *Current Wage Developments,* April 1987, p. 10.

his or her economic condition and standard of living will have decreased. Current (nominal) income can be converted to real (constant) income as follows:

Converting current income to real income:

$$\text{Real income} = \frac{\text{Current income}}{\text{Consumer Price Index}} \cdot 100$$

In 2007 the median U.S. family income was $61,355, compared to the 1970 figure of $9867. With 1982–1984 = 100 as the base, the 2007 Consumer Price Index was 207.3, compared to 38.8 in 1970.[3] Had real income increased or decreased from 1970 to 2007? To get the answer, we need only compare the real income levels for the two years:

- 2007

  Median family income = $61,355 and CPI (1982–1984 base) = 207.3

  $$\text{Real income (in 1982–1984 dollars)} = \frac{\$61{,}355}{207.3} \cdot 100 = \$29{,}597$$

- 1970

  Median family income = $9867 and CPI (1982–1984 base) = 38.8

  $$\text{Real income (in 1982–1984 dollars)} = \frac{\$9867}{38.8} \cdot 100 = \$25{,}430$$

In terms of current dollars, the 2007 median family income was 6.22 times as great as in 1970. The 2007 CPI, however, was just 5.34 times as large. Because the current income growth exceeded the CPI growth, the median family had a higher real income in 2007 than in 1970. Comparing real income for the two years, we find the $29,597 median family real income of 2007 was 16.4% greater than the $25,430 of 1970. Statistics In Action 18.1 provides a glimpse of some prices from years ago and takes a "time-traveler" approach to the CPI.

Other Indexes

The federal government and other institutions employ a great many special-purpose indexes for specific applications. The following are typical examples:

- **The Producer Price Index (PPI)** Dating from 1980, this is the oldest continuous statistical series published by the U.S. Bureau of Labor Statistics. It reflects the prices of commodities produced or processed in the United States.
- **Index of Industrial Production** Provided by the Board of Governors of the Federal Reserve System since the 1920s, this index measures changes in the physical volume or quantity of output of manufacturing, mining, and electric and gas utilities.
- **Employment Cost Index (ECI)** Compiled by the U.S. Bureau of Labor Statistics, the ECI is a quarterly measure of the total cost of employing labor. It includes wages and salaries, plus employers' costs for employee benefits such as paid leave, severance pay, health and life insurance, pension and saving plans, Social Security, and unemployment insurance.
- **Standard & Poor's Composite Index of Stock Prices** From Standard & Poor's Corporation, the index reflects the composite price of 500 common stocks.
- **New York Stock Exchange (NYSE) Composite Index** Compiled by the New York Stock Exchange, this index measures changes in the aggregate value of all stocks listed on the NYSE.

[3] *The New York Times Almanac 2009*, pp. 336–337.

STATISTICS IN ACTION 18.1

The CPI Time Machine

How would it feel to be able to buy first-class postage for 3 cents, a quart of milk for 15 cents, a new car for $1649, and a new house for $12,638? Seems like a pretty good deal, at least until you consider that your salary is a mere $2359—per year.

It was 1946. Timex watches and Tide detergent were brand new, less than 1% of households had a TV, and the hit song of the year was "Zip-A-Dee-Doo-Dah." The milk, car, house, and salary figures above were the averages for that year. The minimum wage was 40 cents an hour, but would go up to 65 cents later in the year.

Technology hasn't yet made it possible for us to time-travel, but—using the Consumer Price Index (CPI) discussed in this section—we can translate those long-ago prices and salaries into their modern-day equivalents. For example, that $1649 car would be about $17,530 in terms of 2007 dollars, and the 15-cent quart of milk would have been the equivalent of $1.59 in 2007. The multiplication factor we're using is the ratio of the Consumer Price Indexes for 2007 and 1946: 207.3 and 19.5, respectively.

Source: "A Snapshot of What Life Was Like in '46," *USA Today*, February 12, 1996, p. 6D; *New York Times Almanac 2009*, p. 338.

Shifting the Base of an Index

Changing the base of an index can be desirable for two purposes:

1. **To better reflect the mix of public tastes or economic activities from which the index is compiled** In this application, the change to a more recent base year involves a reweighting of the components comprising the index.
2. **To convert the index to the same base period as another series so that the two may be more easily compared** For example, Table 18.8 lists 1990 through 2007 values for two indexes: (a) the Producer Price Index for commercial natural gas and (b) the Producer Price Index for heating equipment. The former has 1990 as its base; the latter, 1982.

The two indexes will be more comparable if they have a common base year, and the 1990 base year of the commercial natural gas price index has been arbitrarily selected for this purpose. As shown in column (3) of Table 18.8, the shift is made by simply dividing each index number by the index for the new base year, then multiplying the quotient by 100. Following the conversion, it is apparent that the natural gas prices increased 135.4% from 1990 to 2007, while the price of heating equipment went up by 48.4% during the same period.

TABLE 18.8

As reported, these indexes have base years of 1990 and 1982, respectively. By shifting the base year for heating equipment as shown here, the two series will be more comparable, with each having 1990 as its base year.

	(1) Producer Price Index, Commercial Natural Gas	(2) Producer Price Index, Heating Equipment	(3)
Year	(1990 = 100)	(1982 = 100)	(1990 = 100)
1990	100.0	131.6	100.0
2000	134.7	155.6	118.2
2005	232.5	179.9	136.7
2007	235.4	195.3	148.4

$$148.4 = \frac{195.3}{131.6} \cdot 100$$

Data source: Bureau of the Census, U.S. Department of Commerce, *Statistical Abstract of the United States 2009*, pp. 475, 588.

EXERCISES

18.53 What is an index number, and what is its value for the base period?

18.54 What is the Consumer Price Index and what does it reflect?

18.55 The Consumer Price Index and average annual wages in selected industries were as shown in the table. Source: Bureau of the Census, *Statistical Abstract of the United States 2009*, pp. 409, 465.

	Average Wages ($/Hour)			
	Manufacturing	Construction	Leisure and Hospitality	CPI (1982–1984 = 100)
1990	$10.78	$13.42	$6.02	130.7
2000	14.32	17.48	8.32	172.2
2007	17.26	20.95	10.41	207.3

From 1990 to 2007, the average wages went up for workers in all three industries. However, which wage-earning groups fared best and worst in terms of their percentage change in spending power during this period?

18.56 An executive's salary increased from $95,000 to $130,000 between 2000 and 2007, but the Consumer Price Index went from 172.2 to 207.3 during this time span. In terms of real income, what was the percentage increase or decrease in the executive's real earnings?

18.57 Use the Consumer Price Index to convert the 2000 salary of the executive in Exercise 18.56 to its 2007 equivalent.

18.58 The Producer Price Index values for plumbing fixtures and brass fittings from 2002 through 2007 are shown here. Convert the index numbers so that the base period will be 2002 instead of 1982. Source: Bureau of the Census, *Statistical Abstract of the United States 2009*, p. 588.

2002	181.9	2005	197.6
2003	183.4	2006	207.2
2004	188.3	2007	220.8

SUMMARY 18.9

Time series model and its components

The analysis of time series data is useful in identifying past patterns and tendencies as well as helping to forecast future values. In the classical time series model, the time series is made up of trend (T), cyclical (C), seasonal (S), and irregular (I) components. Although other possibilities exist, a popular approach to combining these components is the multiplicative approach, or $y = T \cdot C \cdot S \cdot I$. The model assumes that any observed value of y is the result of influences exerted by the four components.

Trend equations and smoothing techniques

Trend equations can be fitted to a time series, and two approaches are examined: (1) the linear trend equation, $\hat{y} = b_0 + b_1x$; and (2) the quadratic trend equation, $\hat{y} = b_0 + b_1x + b_2x^2$. The time series can also be smoothed by using a centered moving average or by exponential smoothing. When moving averages are used, the procedure will differ slightly, depending on whether the number of periods in the basis of the average is odd or even.

Seasonal indexes

Strong seasonal patterns may be present for certain months or quarters, and seasonal indexes can be calculated in order to inflate or deflate a trend value to compensate for seasonal effects. Seasonal indexes can be determined through the ratio to moving average method, and a time series can be deseasonalized by using the seasonal indexes for the respective months or quarters of the year.

Seasonal indexes and forecasting

In forecasting based on time series information, data from the past are used in predicting the future value of the variable of interest. Two approaches to time series forecasting are the extrapolation of a trend equation fitted to the series and the use of exponential smoothing. Once a forecast has been made for a given year, seasonal indexes can be used in determining the forecast for each month or quarter of the year.

- **Testing for fit and autocorrelation**

When two or more models are fitted to the same time series, it's useful to have one or more criteria on which to compare the closeness with which they fit the data. Two such criteria are the mean absolute deviation (*MAD*) and mean squared error (*MSE*). The Durbin-Watson test can be used to determine whether the residuals in a regression model are autocorrelated.

- **Autoregressive forecasting**

When values in a time series are highly correlated with each other, autoregressive forecasting models can be useful. In autoregressive forecasting, each independent variable represents the value of the dependent variable for a previous time period. First-order and second-order autoregressive forecasting models are discussed.

- **Index numbers**

An index number is a percentage that compares a measurement to its value for a selected base period. The Consumer Price Index describes the change in prices from one period to the next for a fixed "market basket" of goods and services. The CPI is used as a measurement of inflation, as a "cost of living" adjustment in labor contracts, and as a means of deflating current prices or wages to "real" prices or wages.

EQUATIONS

Classical Time Series Model

$$y = T \cdot C \cdot S \cdot I$$

where y = observed value of the time series variable
T = trend component
C = cyclical component
S = seasonal component
I = irregular component

Linear Trend Equation

$$\hat{y} = b_0 + b_1x$$

where $\hat{y}$ = the trend line estimate of y
x = time period

Quadratic Trend Equation

$$\hat{y} = b_0 + b_1x + b_2x^2$$

where $\hat{y}$ = the trend line estimate of y
x = time period

(x and x^2 are treated as separate variables.)

Exponential Smoothing of a Time Series

$$E_t = \alpha \cdot y_t + (1 - \alpha) \cdot E_{t-1}$$

where E_t = the exponentially smoothed value for time period t
E_{t-1} = the exponentially smoothed value for the previous time period
y_t = the actual y value in the time series for time period t
α = the smoothing constant, and $0 \le \alpha \le 1$

Deseasonalized Time Series

For each time period, the deseasonalized value is

$$D = \frac{\text{Value in the original time series}}{(\text{Seasonal index for that period})/100}$$

Forecasting with Exponential Smoothing

$$F_{t+1} = \alpha \cdot y_t + (1 - \alpha) \cdot F_t$$

where F_{t+1} = forecast for time period $t + 1$
y_t = actual y value for time period t
F_t = the value that had been forecast for time period t
α = the smoothing constant, and $0 \leq \alpha \leq 1$

Dividing an Annual Forecast into Seasonal Components

$$\text{Forecast for a season} = \frac{\text{Forecast for the year}}{\text{Number of seasons}} \cdot [(\text{Seasonal index})/100]$$

Mean Absolute Deviation (*MAD*) Criterion for Evaluating Fit

$$MAD = \frac{\sum |y_t - \hat{y}_t|}{n}$$

where y_t = an observed value of y
$\hat{y}_t$ = the value of y that is predicted using the estimation equation or model
n = the number of time periods

Mean Squared Error (*MSE*) Criterion for Evaluating Fit

$$MSE = \frac{\sum (y_t - \hat{y}_t)^2}{n}$$

where y_t = an observed value of y
$\hat{y}_t$ = the value of y that is predicted using the estimation equation or model
n = the number of time periods

Durbin-Watson Test, Description of Residual (Error) for Time Period *t*

$$\varepsilon_t = \rho\varepsilon_{t-1} + z_t$$

where ε_t = residual associated with time period t
ε_{t-1} = residual associated with time period $t - 1$
ρ = the population correlation coefficient between values of ε_t and ε_{t-1}, with $-1 \leq \rho \leq +1$
z_t = a normally distributed random variable with $\mu = 0$

Durbin-Watson Test for Autocorrelation

- Null and alternative hypotheses

The test is

for positive autocorrelation	for negative autocorrelation	a two-tail test
H_0: $\rho \leq 0$ H_1: $\rho > 0$	H_0: $\rho \geq 0$ H_1: $\rho < 0$	H_0: $\rho = 0$ H_1: $\rho \neq 0$

- Test statistic

$$d = \frac{\sum_{t=2}^{t=n} (e_t - e_{t-1})^2}{\sum_{t=1}^{t=n} e_t^2}$$

where e_t = observed residual (error) associated with time period t
e_{t-1} = observed residual (error) associated with time period $t - 1$
n = number of observations, with $n \geq 15$

- **Critical values and decision zones** Look up the appropriate d_L and d_U values in the Durbin-Watson tables in Appendix A. The actual critical values and decision zones are based on (1) the type of test (positive, negative, or two-tail), (2) α = the level of significance for the test, and (3) k = the number of independent variables in the regression model. The decision zones and appropriate conclusions are shown in Figure 18.5. Appendix A provides three tables for the Durbin-Watson test. They correspond to the 0.05, 0.025, and 0.01 levels of significance for a one-tail (positive or negative) test, and to the 0.10, 0.05, and 0.02 levels for a two-tail test.

Autoregressive Forecasting Model

- **First-order autoregressive model** The forecast for a given time period is based on the value observed in the preceding time period:

$$\hat{y}_t = b_0 + b_1 y_{t-1} \qquad \text{with } y_{t-1} = \text{the value of } y \text{ in period } t - 1$$

- **Second-order autoregressive model** The forecast for a given time period is based on the values observed in the two previous time periods:

$$\hat{y}_t = b_0 + b_1 y_{t-1} + b_2 y_{t-2} \qquad \text{with } y_{t-1} = \text{the value of } y \text{ in period } t - 1$$
$$y_{t-2} = \text{the value of } y \text{ in period } t - 2$$

Converting Current Income to Real Income

$$\text{Real Income} = \frac{\text{Current income}}{\text{Consumer Price Index}} \cdot 100$$

CHAPTER EXERCISES

18.59 Fit a linear trend equation to the following data describing the number of active U.S. Army personnel from 2001 through 2007. What is the trend estimate for 2015? Source: Bureau of the Census, *Statistical Abstract of the United States 2009*, p. 328.

Year	Army Personnel (thousands)
2001	481
2002	487
2003	499
2004	500
2005	493
2006	505
2007	522

18.60 Given the time series in Exercise 18.59, and using exponential smoothing with $\alpha = 0.7$, what would have been the forecast for 2008?

18.61 SUBWAY has grown to include more than 31,000 restaurants worldwide since they opened their first one in 1965. The accompanying data show the number of SUBWAY restaurants worldwide from 1985 through 2002. Fit a quadratic equation to these data and estimate the number of SUBWAY restaurants worldwide in 2014. Source: SUBWAY.com, July 3, 2003.

Year	SUBWAY Restaurants	Year	SUBWAY Restaurants
1985	596	1994	9893
1986	991	1995	11,420
1987	1810	1996	12,516
1988	2888	1997	13,066
1989	4071	1998	13,600
1990	5144	1999	14,162
1991	6187	2000	14,662
1992	7327	2001	15,762
1993	8450	2002	17,482

18.62 Given the time series in Exercise 18.61, and using exponential smoothing with $\alpha = 0.8$, what would have been the forecast for 2003?

18.63 From 2001 through 2007, average monthly cable TV bills were as shown here. Source: Bureau of the Census, *Statistical Abstract of the United States 2009*, p. 703.

Year	Average Bill
2001	$32.87
2002	34.71
2003	36.59
2004	38.14
2005	39.63
2006	41.17
2007	42.72

a. Fit a linear trend equation to the time series. Using this equation, determine the trend estimate for 2015.
b. Fit a quadratic equation to the time series, then use the equation to determine the trend estimate for 2015.
c. Which of the preceding equations is the better fit to the time series if the *MAD* criterion is used? If the *MSE* criterion is used?

18.64 The annual number of international visitors to the United States from 2000 through 2007 is shown here. Source: *World Almanac and Book of Facts 2009*, p. 127.

Year	Visitors (millions)
2000	50.9
2001	44.9
2002	41.9
2003	41.2
2004	46.1
2005	49.4
2006	51.0
2007	56.0

a. Fit a linear trend equation to the time series. Using this equation, determine the trend estimate for 2015.
b. Fit a quadratic equation to the time series, then use the equation to determine the trend estimate for 2015.
c. Which of the preceding equations is the better fit to the time series if the *MAD* criterion is used? If the *MSE* criterion is used?

18.65 The following data show U.S. retail sales of canoes from 1991 through 2008, with data in thousands of boats. Source: National Marine Manufacturers Association, *2008 Recreational Boating Statistical Abstract*, p. 72.

Year	Sales (thousands)	Year	Sales (thousands)
1991	72.3	2000	111.8
1992	78.0	2001	105.8
1993	89.7	2002	100.0
1994	99.8	2003	86.7
1995	97.8	2004	93.9
1996	92.9	2005	77.2
1997	103.6	2006	99.9
1998	107.8	2007	99.6
1999	121.0	2008	73.7

a. Construct a graph of the time series. Does the overall trend appear to be upward or downward?
b. Construct a three-year centered moving average for this series.
c. Using the constant $\alpha = 0.3$, fit an exponentially smoothed curve to the original time series.
d. Repeat part (c) using $\alpha = 0.7$. How has the new value for the constant affected the smoothness of the fitted curve?

18.66 Given the time series in Exercise 18.65, and using exponential smoothing with $\alpha = 0.5$, what would have been the forecast for 2009?

18.67 The following data are quarterly sales for the J. C. Penney Company for 2000 through 2002. Determine the quarterly seasonal indexes, then use these indexes to deseasonalize the original data. Source: J.C. Penney, Inc., *2001 Annual Report*, p. 31, *2002 Annual Report*, p. 39.

Gross Sales (Billions)

	I	II	III	IV
2000	$7.53	7.21	7.54	9.57
2001	7.52	7.21	7.73	9.54
2002	7.73	7.20	7.87	9.55

18.68 Construct three-period and four-period centered moving averages for the quarterly J. C. Penney sales in Exercise 18.67. Are there any other retailing companies for whom a four-period moving average might have such a great effect on smoothing quarterly sales data? If so, give a real or hypothetical example of such a company.

18.69 The following data describe quarterly residential natural gas sales from 2005 through 2008. Determine the quarterly seasonal indexes, then use these indexes to deseasonalize the original data. Source: U.S. Energy Information Administration, eia.doe.gov, August 22, 2009.

Residential Gas Sales (Trillions of Cubic Feet)

	I	II	III	IV
2005	2329	784	353	1360
2006	2042	700	349	1277
2007	2317	761	346	1294
2008	2351	762	345	1408

18.70 Construct four-period and five-period centered moving averages for the quarterly residential gas sales in Exercise 18.69. Are there any other energy industries for whom a four-period moving average might have such a great effect on smoothing quarterly sales data? If so, give a real or hypothetical example of such an industry.

18.71 The Walt Disney Company has reported the following quarterly revenues (in billions of dollars) for its Parks & Resorts operating segment for fiscal years 1998 through 2002. Determine the quarterly seasonal indexes, then use these indexes to deseasonalize the original data. Source: The Walt Disney Company, *Fact Book*, editions 1998–2002.

Quarter

	I	II	III	IV
1998	$1.260	1.244	1.504	1.524
1999	1.434	1.408	1.711	1.553
2000	1.577	1.571	1.940	1.715
2001	1.729	1.650	1.946	1.684
2002	1.433	1.525	1.847	1.660

18.72 Construct four-period and five-period centered moving averages for the quarterly revenues in Exercise 18.71.

Are there any other entertainment or recreation companies for whom a four-period moving average might have such a great effect on smoothing quarterly revenue data? If so, give a real or hypothetical example of such a company.

18.73 The actual number of help-wanted ads in a local newspaper was 1682, 1963, 2451, and 3205 for quarters I through IV, respectively, of the preceding year. The corresponding deseasonalized values are 2320, 2362, 2205, and 2414, respectively. Based on this information, what seasonal index is associated with each quarter?

18.74 The trend equation $\hat{y} = 2050 + 12x$ has been developed for the number of vehicles using the ferry service at a national seashore each year. (The years have been coded so that 1997 corresponds to $x = 1$.) If the seasonal indexes for quarters I through IV are 55.0, 130.0, 145.0, and 70.0, respectively, estimate the usage volume for each quarter of 2014.

18.75 The trend equation $\hat{y} = 150 + 12x + 0.01x^2$ has been developed for annual unit sales of a filtering system component used by municipal water authorities. The years have been coded so that 2001 corresponds to $x = 1$. If the seasonal indexes for quarters I through IV are 85.0, 110.0, 125.0, and 80.0, respectively, estimate the unit sales for each quarter of 2015.

18.76 The U.S. Energy Information Administration reports the following prices for regular unleaded gasoline in Germany from 2003 through 2008. Source: U.S. Energy Information Administration, *Annual Energy Review 2008*, p. 321.

	2003	2004	2005	2006	2007	2008
Price per Gallon	\$4.59	\$5.24	\$5.66	\$6.03	\$6.88	\$7.75

For the price data given here:

a. Fit a linear trend equation to the time series. Using this equation, determine the trend estimate for 2014.
b. Fit a quadratic equation to the time series, then use the equation to determine the trend estimate for 2014.
c. Which of the preceding equations is the better fit to the time series if the *MAD* criterion is used? If the *MSE* criterion is used?

18.77 Ford Motor Company reports the average labor cost per hour (earnings plus benefits) for 1993–2002. Fit a quadratic trend equation to the data, fit a linear trend equation to the data, and then use the *MSE* criterion to determine which equation is the better fit. Source: Ford Motor Company, *2002 Annual Report*, p. 91.

Year	Labor Cost	Year	Labor Cost
1993	\$39.06/hour	1998	45.72
1994	40.94	1999	47.37
1995	40.45	2000	48.44
1996	41.77	2001	47.73
1997	43.55	2002	52.65

18.78 A Caribbean cruise line booked the number of passengers (in thousands) described below for each quarter from 2005 through 2009.

	I	II	III	IV
2005	27	23	21	24
2006	32	26	18	29
2007	28	25	16	30
2008	37	31	21	34
2009	38	33	23	39

a. Combining the quarterly bookings for each year, fit a linear trend equation and forecast the number of bookings for 2013.
b. Determine the seasonal indexes and break the 2013 forecast in part (a) into its quarterly components.
c. Combining the quarterly bookings for each year, use exponential smoothing, with $\alpha = 0.3$, to forecast the number of bookings for 2010.
d. Use the seasonal indexes to break the 2010 forecast in part (c) into its quarterly components.

18.79 Use the seasonal indexes obtained in Exercise 18.78 to generate a deseasonalized series of quarterly bookings for those data.

18.80 Monthly loan applications for a local bank from 2006 through 2009 were as shown in the table.

	J	F	M	A	M	J
2006	30	28	32	33	32	32
2007	22	26	30	36	35	40
2008	19	22	30	31	31	36
2009	29	28	42	47	49	53

	J	A	S	O	N	D
2006	39	44	48	47	33	28
2007	41	36	32	34	23	19
2008	39	37	36	36	32	29
2009	51	55	50	46	36	33

a. Combining the monthly applications for each year, fit a linear trend equation and forecast the number of applications for 2012.
b. Determine the seasonal indexes and break the 2012 forecast in part (a) into its monthly components.
c. Combining monthly applications for each year, use exponential smoothing and $\alpha = 0.4$ to obtain a forecast for 2010.
d. Use the seasonal indexes to break the 2010 forecast in part (c) into its monthly components.

18.81 Use the seasonal indexes obtained in Exercise 18.80 to generate a deseasonalized series of monthly loan applications for the bank.

18.82 Apply the Durbin-Watson test and the 0.05 level of significance in examining the data series in Exercise 18.65 for positive autocorrelation. Based on your conclusion, comment on the appropriateness of the linear regression model for these data.

18.83 Apply the Durbin-Watson test and the 0.05 level of significance in examining the data series in Exercise 18.61 for positive autocorrelation. Based on your conclusion, comment on the appropriateness of the linear regression model for these data.

18.84 The Producer Price Index for finished consumer goods from 1980 through 2007 is in file **XR18084**, with 1982 = 100 and 1980 coded as $t = 1$. Construct a first-order autoregressive forecasting equation for these data, calculate the mean absolute deviation (MAD) for the fit of the forecast values to the actual values, then use the equation to forecast the PPI for 2008. Source: Bureau of the Census, *Statistical Abstract of the United States 2009,* p. 474.

18.85 The Consumer Price Index from 1961 through 2007 is in file **XR18085**, with 1982–1984 = 100 and 1961 coded as $t = 1$. Construct a first-order autoregressive forecasting equation for these data, calculate the mean absolute deviation (MAD) for the fit of the forecast values to the actual values, then use the equation to forecast the CPI for 2008. Source: *The New York Times Almanac 2009,* p. 338; and Bureau of the Census, *Statistical Abstract of the United States 1999,* p. 495.

18.86 Repeat Exercise 18.85, but for a second-order autoregressive forecasting equation. Compare the MAD values for forecasts generated by the two equations and indicate which one is the better fit to these data.

18.87 From 1996 to 2007, the median weekly earnings of workers represented by unions went from \$610 to \$857, and the Consumer Price Index (1982–1984 = 100) went from 156.9 to 207.3. What was the percentage increase or decrease in real wages for these workers? Source: *World Almanac and Book of Facts 2009,* pp. 85, 150.

18.88 Convert each of the following indexes so that the new base period will be 2004.

	A	B
2001	101.3	100.0
2002	100.0	102.9
2003	103.4	104.7
2004	105.6	103.2
2005	107.9	107.5
2006	110.3	126.4
2007	116.2	122.1
2008	125.3	129.0
	(2002 = 100)	(2001 = 100)

INTEGRATED CASE

Thorndike Sports Equipment (Video Unit Five)

Negotiations are taking place for the next labor agreement between Thorndike Sports Equipment and its manufacturing employees. The company and the union have agreed to consider linking hourly wages to the predicted Consumer Price Index (CPI) for each of the four years of the new contract. The latest proposal would specify an hourly wage that is 10% of the estimated CPI (e.g., if the estimated CPI for a given year were 230.5, the hourly wage for that contract year would be \$23.05).

Table 18.9 shows the CPI for 1961 through 2007, along with the year and the year code. The data are also in file **THORN18**. The base period for the CPI figures is 1982–1984, for which the average is 100. Assuming that the new contract will be in force from 2011 through 2014, use a computer statistical package to help determine the hourly wage that would be paid in each year of the contract if the estimated CPI is an extrapolation based on the following:

1. A simple linear estimation equation ($\hat{y} = b_0 + b_1x$) fitted to the 1961–2007 CPI data.
2. A quadratic estimation equation ($\hat{y} = b_0 + b_1x + b_2x^2$) fitted to the 1961–2007 CPI data.

Several additional questions regarding the estimation equations and the contract negotiations:

3. Which equation would each negotiating party tend to prefer as the basis of the hourly wages to be paid during the contract years?
4. Using the mean absolute deviation (MAD) criterion, which equation is the better fit to the CPI time series data?
5. Assuming (a) the hourly wages are linked to the equation identified in the preceding question, (b) the estimated CPI for 2014 actually occurs, and (c) the average Thorndike worker made \$12.40 per hour in 1987, will his or her real earnings increase or decrease from 1987 to 2014? Explain.

TABLE 18.9

Values of the Consumer Price Index (CPI) for 1961 through 2007. The base period is 1982–1984 = 100.

Year	Year Code	CPI	Year	Year Code	CPI
1961	1	29.9	1985	25	107.6
1962	2	30.2	1986	26	109.6
1963	3	30.6	1987	27	113.6
1964	4	31.0	1988	28	118.3
1965	5	31.5	1989	29	124.0
1966	6	32.4	1990	30	130.7
1967	7	33.4	1991	31	136.2
1968	8	34.8	1992	32	140.3
1969	9	36.7	1993	33	144.5
1970	10	38.8	1994	34	148.2
1971	11	40.5	1995	35	152.4
1972	12	41.8	1996	36	156.9
1973	13	44.4	1997	37	160.5
1974	14	49.3	1998	38	163.0
1975	15	53.8	1999	39	166.6
1976	16	56.9	2000	40	172.2
1977	17	60.6	2001	41	177.1
1978	18	65.2	2002	42	179.9
1979	19	72.6	2003	43	184.0
1980	20	82.4	2004	44	188.9
1981	21	90.9	2005	45	195.3
1982	22	96.5	2006	46	201.6
1983	23	99.6	2007	47	207.3
1984	24	103.9			

Sources: Bureau of the Census, U.S. Department of Commerce, *Statistical Abstract of the United States 1999,* p. 495; and *New York Times Almanac 2009*, p.338.

CHAPTER 19

Decision Theory

LAWSUITS—TAKE THE CASH OR TRY FOR THE CURTAIN?

After Jill became pregnant, her company decided it no longer needed her services. Upon being released, Jill accused the company of gender bias and filed a wrongful-termination suit against it. The company is now trying to settle out of court and has offered Jill $160,000.

Should Jill take the settlement offer, or might it be more attractive to go to court and take her chances with the trial? Jill's attorney believes she has a 60% chance of winning her case if it proceeds to the trial stage. He also informs her that the median jury award in employment-practices liability cases has been estimated as $250,000.

Essentially, Jill's alternatives can be viewed as (a) the sure $160,000 versus (b) the 60% chance at $250,000 (which includes a 40% chance at $0). Based on the expected payoff decision criterion discussed in this chapter, Jill should take the money and run. After reading that section, you can verify our advice to Jill for yourself. By the way, Jill should be willing to pay up to $54,000 for a perfect forecast as to what the jury would decide if she were to take the case to trial. That's known as the *expected value of perfect information*, and it's in this chapter, too.

Source: The median award value is from "Report: EPL Jury Awards Up 18%," *Insurance Journal*, insurancejournal.com, August 29, 2009.

"When you come to a fork in the road, take it."
—Yogi Berra

LEARNING OBJECTIVES

After reading this chapter, you should be able to:

- Express a decision situation in terms of decision alternatives, states of nature, and payoffs.
- Differentiate between non-Bayesian and Bayesian decision criteria.
- Determine the expected payoff for a decision alternative.
- Calculate and interpret the expected value of perfect information.
- Express and analyze the decision situation in terms of opportunity loss and expected opportunity loss.
- Apply incremental analysis to inventory-level decisions.

(19.1) INTRODUCTION

While it's been said that nothing is certain except death and taxes, there is one more thing that's certain: uncertainty. A fish market owner isn't sure how many customers will decide to buy a flounder today, a pharmaceutical company isn't completely certain that a new drug will be safe for every one of 20 million users, and a computer firm doesn't have complete certainty that a new computer in the final design stages will still be competitive when it is finally manufactured. In each of these cases, a decision must be made. For the fish market owner, it's how many flounder to stock; for the pharmaceutical and computer firms, it's whether to proceed with the new product.

Both decision making and uncertainty are also discussed in other chapters. In hypothesis testing, we had to decide whether to reject the null hypothesis, and uncertainty was reflected in the Type I and Type II errors. In the next chapter, we will examine how control charts can help us decide whether a process is in control, and we will briefly discuss how acceptance sampling can assist in deciding whether an entire shipment of products or components should be accepted.

The biggest difference between these discussions and the topics of this chapter is that we will be concentrating here on decisions made *without* the benefit of knowledge gained from sampling. While the possibility of gathering further information will be considered, this will be presented in terms of, "What is the most we should consider paying for additional information to reduce our uncertainty in making the decision?"

We will begin by structuring the decision situation and expressing it as either a payoff table or a decision tree diagram. The discussion then turns to two general approaches by which the actual decision might be reached. The chapter concludes with a discussion of incremental analysis, a technique especially useful for inventory decisions.

(19.2) STRUCTURING THE DECISION SITUATION

Before analyzing a decision situation, we must have a structure to which its various components can be fitted. These components, and the structure itself, are described in this section.

Alternatives

Decision alternatives are the choices available to the decision maker. For the computer firm, decision alternatives could be (1) plan to introduce the new computer, and (2) drop plans to introduce the new computer. Only one alternative can be selected from among those available.

STATISTICS IN ACTION 19.1

New Technologies Take Film out of the Picture

In June 2009, Kodak announced the end of the line for Kodachrome film, an American icon in family and professional photography for 74 years. As it neared the end of its product life, Kodachrome was less than 1% of the company's still-photography film sales.

The major culprit in the demise of this legend was the state of nature known as *technology*. Not only had digital photography become the dominant mode of capturing and storing images, other types of film required considerably less complexity in both their manufacture and their processing. At this writing, there is only one photographic laboratory in the entire United States that still processes Kodachrome film.

Kodak itself went through a troubled technological period during the early 2000s, as its very survival depended on a successful transition from the film age to the digital age. Today, digital photography sales account for more than 70% of Kodak's revenue, and the company has become a leading competitor in the digital photography movement that many believe had threatened its corporate existence.

Source: "Kodak Kills Kodachrome Film After 74 Years," Reuters.com, June 22, 2009.

States of Nature

After a decision alternative has been selected, the outcome is subject to chance and no longer under the control of the decision maker. For example, the new computer could be either competitive or noncompetitive in technology and features versus other brands on the market at the time of its introduction. Each of these possibilities is a **state of nature,** a future environment that has a chance of occurring. Statistics in Action 19.1 describes how the occurrence of a given state of nature spelled doom for a part of Americana that had been around for 74 years. The states of nature must be both mutually exclusive (only one will occur) and exhaustive (one of them must occur).

Having introduced the concept of states of nature, we can now describe three different levels of doubt regarding the decision situation itself:

RISK Both the possible states of nature and their exact probabilities of occurrence are known. This is typical of many gambling situations, such as playing roulette. This is an extreme rarity in business decisions, since the exact probabilities are generally not known.

UNCERTAINTY The possible states of nature are known, but their exact probabilities are unknown. Most business decision situations fall into this category, and the probabilities must be estimated based on past experience or executive judgment.

IGNORANCE Neither the possible states of nature nor their exact probabilities are known. This is another rarity, as only a fugitive from Mars would have absolutely *no* idea what states of nature might occur after a decision has been made.

The Payoff Table

The **payoff table** is a table in which the rows are decision alternatives, the columns are states of nature, and the entry at each intersection of a row and column is a numerical **payoff** such as a profit or loss. The table may also include estimated probabilities for the states of nature.

EXAMPLEEXAMPLEEXAMPLEEXAMPLEEXAMPLEEXA

EXAMPLE

Payoff Table

A German automobile manufacturer currently produces its best-selling U.S. model in Germany, but the relative strength of the Euro versus the U.S. dollar has been making the car very expensive for the market in which it competes. To ensure a lower, more stable price, the company is considering the possibility of manufacturing cars for U.S. consumption at one of its South American plants.

SOLUTION

Table 19.1 is a payoff table that summarizes this decision situation. It includes the following components:

- **Alternatives** (1) Continue producing cars in Germany for the U.S. market; or (2) manufacture cars for U.S. consumption in the South American plant.
- **States of nature** Over the next two model years, the relative strength of the dollar versus the Euro will be (1) weak, (2) moderate, or (3) strong.
- **Probabilities** Corporate economists have estimated the probabilities for the relative strength of the dollar as 0.3 (weak), 0.5 (moderate), and 0.2 (strong).
- **Payoffs** The estimated profit for this model depends on which alternative is selected and which state of nature occurs. For example, if manufacturing is switched to the South American plant and the dollar is strong, the firm's profit is estimated as $16 million over the two-year period.

The same decision situation could also be described by a **decision tree,** a diagram in which alternatives, states of nature and their probabilities, and payoffs are displayed in graphical form. Figure 19.1 is the decision tree counterpart to the payoff table of Table 19.1. It provides the same information, but in a graphical form. The first two branches are the decision alternatives. Depending on which alternative is selected, either node 1 or node 2 will be encountered next, and the probabilities and the states of nature are the same for each, just as in the payoff table.

In the decision tree, each square node shows the making of a decision, and each circular node shows the occurrence of a state of nature. The payoff table will be our primary method for describing decision situations, but the decision tree approach will be revisited for some later applications.

TABLE 19.1

The payoff table includes the decision alternatives, the possible states of nature, and the payoff for each combination of the alternatives and states of nature. Estimated probabilities for the states of nature may also be included, as shown here. Profits are in millions of dollars.

	State of Nature Compared to the Euro, the Strength of the U.S. Dollar Will Be		
Alternative	**Weak (Prob. = 0.3)**	**Moderate (Prob. = 0.5)**	**Strong (Prob. = 0.2)**
German Production	$10	$15	$25
South American Production	$20	$18	$16

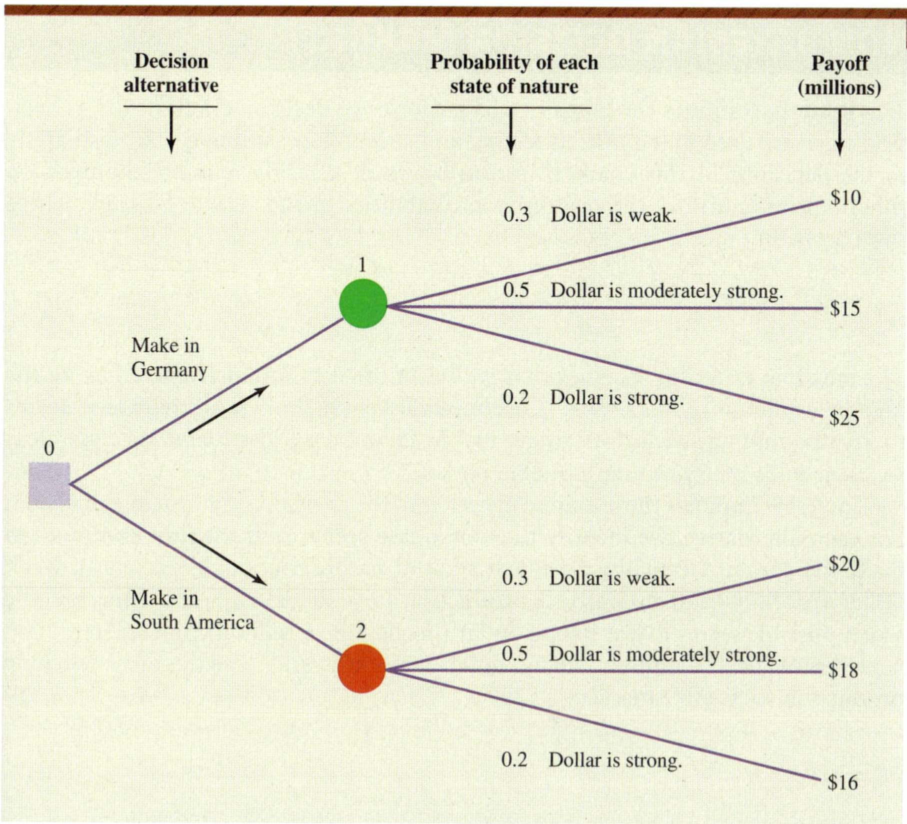

FIGURE 19.1

The decision tree counterpart to the payoff table of Table 19.1.

EXERCISES

19.1 What are the differences among risk, uncertainty, and ignorance? Into which category of decision situation do most business decisions fall?

19.2 What is meant when we say that the states of nature in the payoff table must be mutually exclusive and exhaustive?

19.3 The manager of a small local airport must extend a runway if business jets are to be accommodated. However, the airport is practically surrounded by a suburban housing development, and residents are already complaining to the city council about the volume and noise of existing air traffic at the facility. Identify the decision alternatives and the possible states of nature that might occur after the decision is made.

19.4 A management consultant, unable to locate receipts or records for several business trips during the past year, makes some rough estimates of these expenditures and considers claiming them as business expenses on his income tax form. If he makes the claims and is not audited, he will owe $12,000 in taxes. If he makes the claims and the IRS audits his return, he will have to pay an extra $8,000 in taxes and penalties. If he ignores the trips, he will owe $14,000 in taxes. Construct a payoff table and a decision tree diagram for this decision situation.

19.5 A trainee in a realty firm is being paid a straight salary, but has the opportunity to (a) continue with straight salary, (b) change to straight commission, or (c) change to a combination that includes a lower base salary and a commission on each sale. What states of nature might be useful in constructing a payoff table for this situation?

19.6 Having been injured while trying to use his rotary lawn mower as a snowblower, a consumer has sued the company and is offered the chance to settle out of court. If the case goes to trial, the jury could award him either nothing or a large amount in damages. Construct a payoff table for this consumer, including probabilities and payoff amounts that you feel might be realistic.

19.7 A Florida orange grower is considering the installation of a heating system to protect his crop in the event of frost during the coming winter. What would be the rows and columns of a payoff table applicable to this decision?

19.3 NON-BAYESIAN DECISION MAKING

This section examines the first of two categories of decision criteria. Just as decision rules are used in hypothesis testing and acceptance sampling, they also apply to the decisions in this chapter. In *non-Bayesian decision making*, the decision rule does not consider the respective probabilities of the states of nature. Three such decision criteria are discussed here.

Maximin

The **maximin** criterion specifies that we select the decision alternative having the highest minimum payoff. This is a conservative, pessimistic strategy that seems to assume that nature is "out to get us." In fact, the word *maximin* is equivalent to "*max*imize the *min*imum possible payoff."

Table 19.2 applies the maximin criterion to the decision situation in Table 19.1. For each alternative, we identify the worst-case scenario. If the firm continues to make the cars in Germany, the worst state of nature that could occur is a weak dollar (profit = \$10 million). If production is switched to South America, the worst possible scenario is a strong dollar (profit = \$16 million). Since \$16 million is preferable to \$10 million, the maximin criterion would have the firm switching production to South America.

Maximax

The **maximax** criterion specifies that we select the decision alternative having the highest maximum payoff. This is an optimistic strategy, making the assumption that nature will be generous to us. The word *maximax* is equivalent to "*max*imize the *max*imum possible payoff."

Table 19.3 applies the maximax criterion to the decision situation in Table 19.1. For each alternative, the best-case scenario is identified. If the firm continues German production for the United States, the best state of nature that could occur is a strong dollar (profit = \$25 million). If production is switched to South America, the best possible scenario is a weak dollar (profit = \$20 million). Because \$25 million is better than \$20 million, the maximax criterion would specify that German production continue.

TABLE 19.2

Using the maximin criterion, production of vehicles for the U.S. market would be switched to the South American plant. The maximin criterion pessimistically assumes that the worst possible state of nature will occur after the decision has been made.

	State of Nature: Compared to the Euro, the Strength of the U.S. Dollar Will Be			
Alternative	Weak (Prob. = 0.3)	Moderate (Prob. = 0.5)	Strong (Prob. = 0.2)	Minimum payoff
German Production	\$10	\$15	\$25	\$10 million
→ South American Production	\$20	\$18	\$16	\$16 million ←

\$16 million is the highest minimum payoff.

Alternative	State of Nature: Compared to the Euro, the Strength of the U.S. Dollar Will Be — Weak (Prob. = 0.3)	Moderate (Prob. = 0.5)	Strong (Prob. = 0.2)	Maximum payoff
→ German Production	$10	$15	$25	$25 million ←
South American Production	$20	$18	$16	$20 million

$25 million is the highest maximum payoff.

TABLE 19.3
Using the maximax criterion, production of vehicles for the U.S. market would remain in Germany. This criterion is optimistic, since it assumes that the best possible state of nature will occur after the decision has been made.

Minimax Regret

The **minimax regret** criterion is the equivalent of being paranoid about "Monday-morning quarterbacks" who, with the benefit of hindsight, come along later and tell you what you *should* have done. In using minimax regret, we must first construct a *regret table*. This is a table in which, for each state of nature, the entries are **regrets,** the differences between each payoff and the best possible payoff if that state of nature were to occur. The regret table is the same as the opportunity loss table, to be discussed in Section 19.5.

Table 19.4 is the regret table for the manufacturing decision described in Table 19.1. If the firm knew for sure that the dollar was going to be moderately

Alternative	State of Nature: Compared to the Euro, the Strength of the U.S. Dollar Will Be — Weak (Prob. = 0.3)	Moderate (Prob. = 0.5)	Strong (Prob. = 0.2)	Maximum regret
German Production	$10	$3	$0	$10 million
→ South American Production	$0	$0	$9	$ 9 million ←

If the dollar is moderately strong, profit from German production ($15 million) will be $3 million less than what could have been made with South American production ($18 million).

If the firm makes the cars in Germany and this state of nature occurs, the amount of the regret will be $3 million.

$9 million is the lowest maximum regret.

TABLE 19.4
The regret table for the manufacturing plant decision of Table 19.1. Using the minimax regret decision criterion, production will be switched to South America.

strong, it would switch production to the South American plant and realize a profit of $18 million. However, if this state of nature were to occur and the firm had continued German production, profits would only be $15 million. The entry in the first row, second column of the regret table is thus ($18 million − $15 million), or $3 million. Using the same logic, there will be zero regret if the dollar is moderately strong and production is switched to the South American plant—for this state of nature, switching production out of Germany would be the better choice.

The minimax regret criterion specifies selection of the alternative with the lowest maximum regret. In Table 19.4, the regret could be as high as $10 million if production stays in Germany, while the maximum regret would be just $9 million if the cars are made in South America. If the minimax regret criterion is used, the firm will move production to the South American plant.

EXERCISES

19.8 For Exercise 19.4, which decision alternative will be selected if the consultant uses the maximin criterion? The maximax criterion? The minimax regret criterion?

19.9 Dave's local college football team is playing for the national championship in next month's bowl game, and he's thinking about having "We're number one" T-shirts imprinted in honor of the team. It's too late to get the shirts before game day, but if he orders today, he can get them within a few days after the game. The shirts must be ordered in multiples of 1000, and Dave has estimated the amount of profit he will make for possible outcomes of the game. Given the following payoff table, what will Dave do if he uses the maximin criterion? The maximax criterion? The minimax regret criterion?

	The Team		
	Wins	**Loses**	**Ties**
Order no shirts	$0	$0	$0
Order 1000 shirts	$800	−$200	$200
Order 2000 shirts	$1600	−$400	$300

19.10 Three different designs are being considered for a new electronics product, and profits will depend on the combination of product design and market condition. The following payoff table summarizes the decision situation, with amounts in millions of dollars. What decision will be made using the maximin criterion? The maximax criterion? The minimax regret criterion?

	Market Condition		
	I	**II**	**III**
Design A	$25	$14	$6
Design B	$15	$30	$18
Design C	$5	$10	$40

19.11 For each of the following, indicate whether the behavior represents the maximin criterion or the maximax criterion. Explain your reasoning.

a. A trucker is behind schedule and knows he is approaching an area where police radar is often encountered, but he continues driving at 15 mph over the speed limit.
b. A struggling retailer, in danger of defaulting on next month's payroll, hires a public relations firm to carry out an intensive promotional campaign for his store. If the campaign is not extremely successful, the retailer won't be able to pay the PR firm, either.
c. The owner of a roller-skating rink buys liability insurance.

19.12 When Fred takes the direct route to work, he gets there in 30 minutes if no train is at the railroad crossing. When a train is using the crossing, his travel time is 15 minutes longer. His alternative is a route that takes 40 minutes to travel, but has no traffic signals or other delays. Assuming that shorter transit times are more desirable, which route will Fred select if he uses the maximin criterion? The maximax criterion? The minimax regret criterion?

19.13 An antique collector is shipping $50,000 worth of items from Europe to America and must decide whether to insure the shipment for its full value. An insurance firm will provide this protection for a fee of $1000. What will be the collector's decision if she uses the maximin criterion? The maximax criterion? The minimax regret criterion?

19.14 An auto manufacturer has preliminary plans to introduce a new diesel engine capable of extremely high fuel economy. The success of the engine depends on whether the U.S. Environmental Protection Agency goes ahead with stringent particulate emission limits that are scheduled for the coming model year. The payoff table, with amounts in millions of dollars for the corporation's profit, is shown here. What decision will be made using the maximin criterion? The maximax criterion? The minimax regret criterion?

	The Strict Emission Limits Are	
	Enacted	Not Enacted
Continue engine development	−\$10	\$30
Halt engine development	\$15	\$15

BAYESIAN DECISION MAKING (19.4)

Whereas the decision criteria just discussed ignore the probabilities for the respective states of nature, **Bayesian decision making** takes them into account. Specifically, the alternative chosen is the one with the best expected payoff.

Expected Payoff

The **expected payoff** of a decision alternative is the sum of the products of the payoffs and the state of nature probabilities. Each decision alternative is treated as a discrete probability distribution with an expected value that can be calculated using the method introduced in Chapter 6:

Expected payoff for a decision alternative:

Expected payoff $= \Sigma p_i \cdot m_i$ **where** $p_i =$ **the probability that state of nature** i **will occur**

$m_i =$ **the payoff if this alternative is selected and state of nature** i **occurs**

For the automobile production decision we've been following, the expected payoffs for the alternatives can be calculated as follows:

German production:

$$\text{Expected payoff} = \Sigma p_i \cdot m_i = 0.3(10) + 0.5(15) + 0.2(25) = \$15.5 \text{ million}$$

South American production:

$$\text{Expected payoff} = \Sigma p_i \cdot m_i = 0.3(20) + 0.5(18) + 0.2(16) = \$18.2 \text{ million}$$

As Table 19.5 shows, the alternative with the higher expected payoff is South American production, with an expected profit of \$18.2 million. Although \$18.2 million is not one of the entries in the payoff table, this would be the result if this decision situation were to be replayed a great many times—on the average, the profit would be \$18.2 million. Expected payoff is sometimes referred to as *expected monetary value* (*EMV*).

Figure 19.2 shows the decision tree diagram corresponding to Table 19.5. In using the decision tree, calculations proceed from right to left (i.e., from future to present), and the expected payoff for each alternative is obtained as in the calculations just presented. The "X" through the "German production" branch indicates that this alternative has an inferior expected payoff and would not be selected.

TABLE 19.5

Using the expected payoff criterion, production will be switched to South America and the expected profit will be $18.2 million.

Alternative	State of Nature: Compared to the Euro, the Strength of the U.S. Dollar Will Be			Expected payoff
	Weak (Prob. = 0.3)	Moderate (Prob. = 0.5)	Strong (Prob. = 0.2)	
German Production	$10	$15	$25	$15.5 million
→ South American Production	$20	$18	$16	$18.2 million

If production is switched to South America, the expected payoff is

$$\Sigma p_i \cdot m_i = 0.3(20) + 0.5(18) + 0.2(16) = \$18.2 \text{ million}$$

Using the expected payoff criterion, this alternative would be selected.

FIGURE 19.2

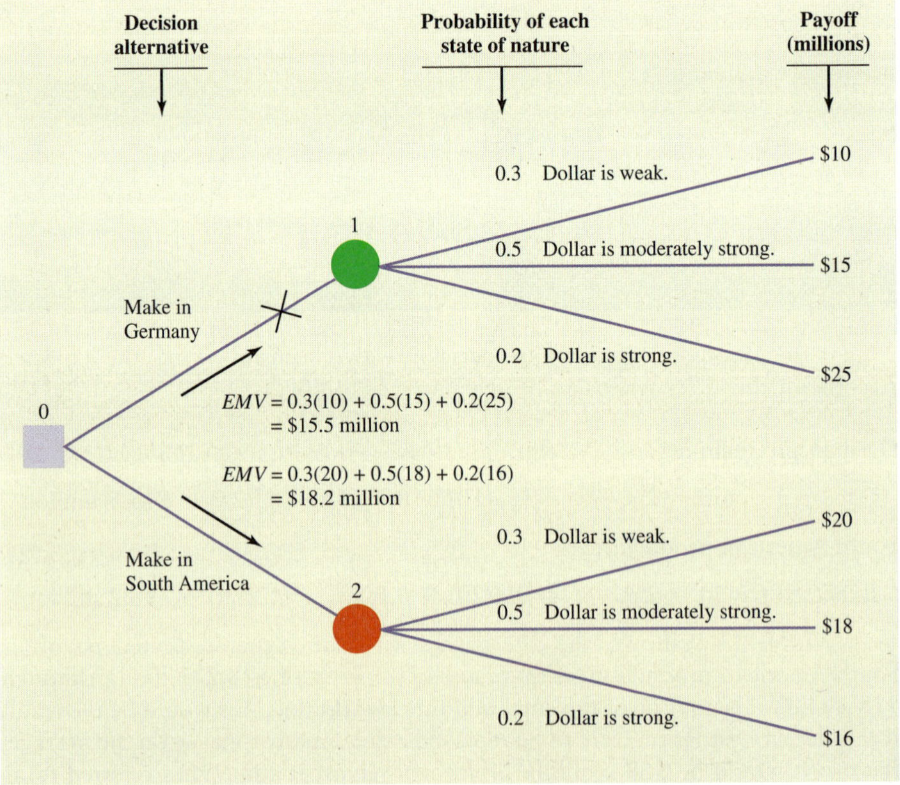

The decision tree summary for the table and expected payoff calculations of Table 19.5. The "X" through the German production alternative reflects that its expected payoff makes it inferior to the other alternative.

The Expected Value of Perfect Information

In decision making under uncertainty, it's often possible to invest in research that will give us a better idea as to which one of the states of nature will actually occur. For example, before deciding whether to continue making cars for the U.S.

market in Germany, the automobile company might like to consult with outside economic experts and forecasters regarding the predicted strength of the U.S. dollar versus the Euro.

When we are considering whether to collect additional information before making the decision, an important question arises: How much should we pay for such information? To help answer this question, we can determine the **expected value of perfect information (*EVPI*)**. This is the difference between (1) the expected payoff if we are guaranteed perfect knowledge about the future and (2) the expected payoff under uncertainty:

The expected value of perfect information, *EVPI*:

$$EVPI = \text{Expected payoff with perfect information} - \text{Expected payoff with present information}$$

For the auto company, the expected payoff with present information has already been calculated as \$18.2 million. As Table 19.5 showed, this is the expected payoff for the alternative with the greatest expected payoff—that is, switching production of the car to South America.

If perfect information were available on the future relative strength of the U.S. dollar versus the Euro, the company could make its decision under *certainty*. It would know which state of nature was going to occur and would make the best choice for this state of nature. For example, if the firm had 100% confidence that the "weak dollar" state were going to occur, it would switch production to South America and have a profit of \$20 million.

In calculating the expected value *of* perfect information, we must first determine the expected payoff *with* perfect information. The perfect information has not yet been received, so we operate under the assumption that it might predict any one of the three states of nature. The probabilities of these states of nature have been estimated as 0.3, 0.5, and 0.2:

- **Probability = 0.3 that perfect information will say "weak dollar."** In this case, the firm switches production to South America and makes a profit of \$20 million.
- **Probability = 0.5 that perfect information will say "moderately strong dollar."** In this case, the firm switches production to South America and makes a profit of \$18 million.
- **Probability = 0.2 that perfect information will say "strong dollar."** In this case, the firm continues German production and makes a profit of \$25 million.

Then the expected payoff with perfect information is

$$0.3(20) + 0.5(18) + 0.2(25) = \$20.0 \text{ million}$$

The expected payoff with present information has already been calculated as \$18.2 million. This is the expected payoff for the alternative with the highest expected payoff under uncertainty (i.e., switch production to South America). Now we can calculate the expected value of perfect information:

$$\begin{aligned} EVPI &= \text{Expected payoff with perfect information} - \text{Expected payoff with present information} \\ &= \$20.0 \text{ million} - \$18.2 \text{ million} = \$1.8 \text{ million} \end{aligned}$$

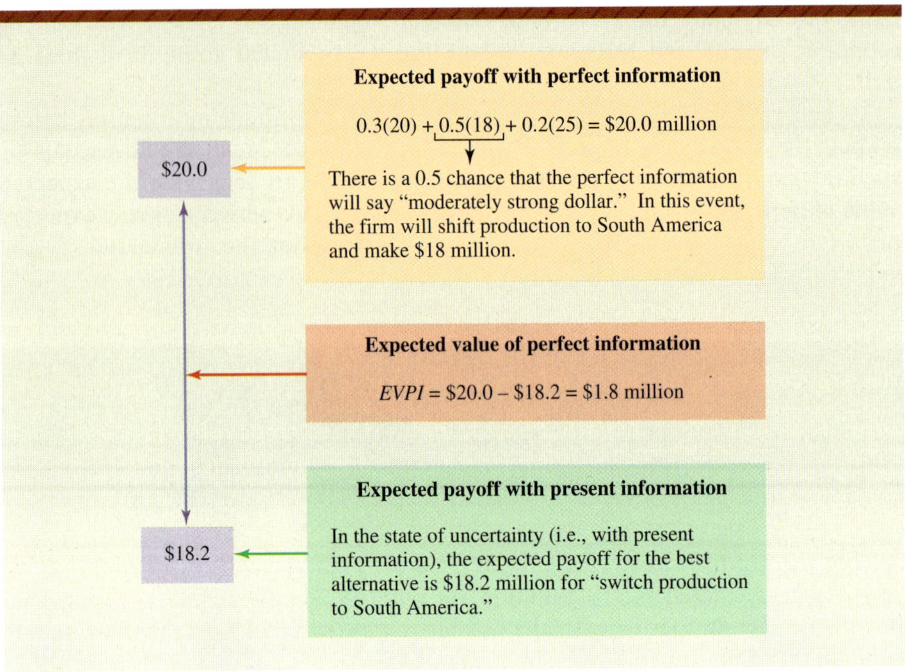

FIGURE 19.3

For the decision situation of Table 19.5, the expected value of perfect information is $1.8 million. This is the upper limit the firm should consider spending for any information to reduce uncertainty.

As calculated here and shown in Figure 19.3, the expected value of perfect information is $1.8 million. This is the upper limit the firm should consider spending for the purpose of reducing uncertainty.

EXERCISES

19.15 What is the most important difference between Bayesian and non-Bayesian decision criteria?

19.16 A financial institution is evaluating the loan application of an entrepreneur who has come up with a revolutionary product for which he needs developmental funding. The entrepreneur has practically no assets for collateral, but offers the institution a portion of the profits if the product is successful. The loan officer has constructed the following payoff table, with institution profits in thousands of dollars and the probabilities estimated for three states of nature.

	The Product's Level of Success Is		
	Low (0.1)	Moderate (0.3)	High (0.6)
Loan the full amount requested	−$25	$20	$50
Loan half the amount requested	−$10	$10	$20
Do not grant the loan	$0	$0	$0

Using the expected payoff criterion, which alternative will be selected and what is its expected monetary value? Determine the expected value of perfect information and interpret its meaning.

19.17 In Exercise 19.4, if the probability that his return will be audited is 0.1, which alternative will the consultant choose if he maximizes expected monetary value? What is the most he should be willing to pay for more information regarding his likelihood of being audited?

19.18 In Exercise 19.9, assume that the team has a 0.6 probability of winning, a 0.3 probability of losing, and a 0.1 probability for a tie game. Which alternative should Dave select if he wishes to maximize his expected payoff, and what is the expected value of perfect information?

19.19 In Exercise 19.10, the probabilities for the three market conditions are estimated as 0.1, 0.6, and 0.3, respectively. Which design should be selected in order to maximize the firm's expected profit? What is the most the firm should be willing to pay for a research study designed to reduce its uncertainty about market conditions?

19.20 For the following payoff table, determine the expected value of perfect information. Does the result seem unusual? If so, why did this occur?

	State of Nature	
	I (0.6)	II (0.4)
Alternative A	\$80	\$30
Alternative B	\$12	\$20

19.21 In Exercise 19.12, there is a 0.1 chance that a train will be using the railroad crossing. Given this information, which route should Fred select, and what is the expected number of minutes his trip will require? What is the expected number of minutes Fred would save if he could predict with certainty whether a train will be at the crossing?

19.22 In Exercise 19.13, if x is the probability that the shipment is lost, for what value of x will the alternatives have the same EMV? Assuming this value for x, what is the expected value of perfect information?

19.23 In business research, it's extremely rare that any study will provide a perfect assessment of the future. Why then should we bother determining the expected value of perfect information when such information is practically never available?

THE OPPORTUNITY LOSS APPROACH (19.5)

Another way to use probabilities in identifying the best alternative and finding the expected value of perfect information is to structure the decision in terms of opportunity loss. **Opportunity loss** is another term for regret, discussed earlier in the chapter, and expected opportunity loss is calculated in a manner similar to that of expected payoff:

Expected opportunity loss (*EOL*) for a decision alternative:

$EOL = \sum p_i \cdot l_i$ where p_i = the probability that state of nature i will occur

l_i = the opportunity loss if this alternative is selected and state of nature i occurs

The results of the opportunity loss approach will be the same as those obtained in the expected payoff method of the preceding section. In particular,

1. the alternative having the maximum expected payoff will also have the minimum expected opportunity loss (EOL), and
2. the expected opportunity loss for this alternative is the expected value of perfect information.

TABLE 19.6

The expected opportunity loss is the same as the expected value of perfect information. The expected payoff and expected opportunity loss approaches will always lead to the same alternative being selected.

Alternative	State of Nature: Compared to the Euro, the Strength of the U.S. Dollar Will Be			Expected opportunity loss
	Weak (Prob. = 0.3)	Moderate (Prob. = 0.5)	Strong (Prob. = 0.2)	
German Production	$10	$3	$0	$4.5 million
→ South American Production	$0	$0	$9	$1.8 million

If production is switched to South America, the expected opportunity loss is

$$\sum p_i \cdot l_i = 0.3(0) + 0.5(0) + 0.2(9) = \$1.8 \text{ million}$$

Table 19.6 is the opportunity loss approach to the auto maker's decision situation. Note that the table entries are the same as for the regret table of Table 19.4—this is because opportunity loss and regret are the same thing.

In Table 19.6, "South American production" has the minimum expected opportunity loss, $1.8 million. This indicates that (1) "South American production" is the better alternative and (2) the absence of perfect information can be expected to cost the company $1.8 million. Again, for any decision situation the expected payoff and expected opportunity loss approaches will always lead to the same alternative being selected. Just as opportunity loss and regret are the same thing, expected opportunity loss and the expected value of perfect information are equivalent.

EXERCISES

19.24 In what way are the minimax regret criterion and the expected opportunity loss (*EOL*) criterion similar? In what ways do they differ?

19.25 An investor intends to purchase stock in one of three companies. Company A is very sensitive to the overall national economy, company C is relatively insensitive to the economic environment, and company B is about average. The investor has constructed the following payoff table for his profit, in thousands of dollars.

	The State of the Economy Will Be		
	Weak (0.2)	**Moderate (0.4)**	**Strong (0.4)**
Invest in company A	−$40	$30	$60
Invest in company B	−$20	$20	$45
Invest in company C	$24	$28	$30

Using the opportunity loss approach, which alternative will be selected, and what is its expected opportunity loss? What is the most the investor should be willing to pay for perfect information about the future state of the economy?

19.26 In Exercise 19.10, with the probabilities for the three market conditions estimated as 0.1, 0.6, and 0.3, respectively, what is the expected opportunity loss associated with each of the three decision alternatives? In what way is the lowest of these expected opportunity losses related to the expected value of perfect information?

19.27 In Exercise 19.12, with a probability of 0.1 that a train will be using the crossing, compute the expected opportunity loss for each of Fred's alternatives. Explain the meaning of the *EOL* for each alternative.

INCREMENTAL ANALYSIS AND INVENTORY DECISIONS (19.6)

In some cases, the number of decision alternatives and states of nature may be extremely large. This can occur, for example, if a retailer can stock anywhere from 0 to 1000 units and demand might also be anywhere from 0 to 1000. A payoff table could end up having 1000 rows and 1000 columns, and calculations to find the best decision alternative would be extremely tedious. In such applications, we can use **incremental analysis,** where units are sequentially considered, with each unit in the sequence being stocked only if the expected payoff for stocking it exceeds the expected payoff for not stocking it. Incremental analysis involves the following key terms:

MARGINAL PROFIT The profit if one more unit is stocked and sold.

MARGINAL LOSS The cost if one more unit is stocked and *not* sold.

EXAMPLE

Incremental Analysis

A supermarket receives weekly milk shipments from its dairy products supplier. The weekly demand for milk can be approximated by a normal distribution having a mean of 2000 gallons and a standard deviation of 300 gallons. The supermarket pays $2.10 per gallon and makes a profit of $0.50 on each gallon sold. At the end of the week, the supplier repurchases all unsold milk for $0.60 per gallon.

SOLUTION

A payoff table for this situation could have over 1000 rows and columns, with each one representing a single gallon of milk. Using incremental analysis, we can determine the alternative with the maximum expected payoff (or minimum expected opportunity loss) much more easily.

EXAMPLEEXAMPLEEXAMPLEEXAMPLEEXAMI

TABLE 19.7

In incremental analysis, item i is stocked if the expected payoff from stocking it exceeds the expected payoff from not stocking it.

A. Should the 1st Gallon of Milk Be Stocked? **Yes**

	The Week's Demand Will Be		
	≥1 Gallon (Prob. = 1.00)	<1 Gallon (Prob. = 0.00)	Expected Payoff
Stock the 1st Gallon	\$0.50	−\$1.50	\$0.50
Don't Stock It	\$0.00	\$0.00	\$0.00

B. Should the 2nd Gallon of Milk Be Stocked? **Yes**

	The Week's Demand Will Be		
	≥2 Gallons (Prob. = 1.00)	<2 Gallons (Prob. = 0.00)	Expected Payoff
Stock the 2nd Gallon	\$0.50	−\$1.50	\$0.50
Don't Stock It	\$0.00	\$0.00	\$0.00

C. Should the *i*th Gallon of Milk Be Stocked?

Yes, if p = the probability that it will be sold is $\geq p_c$

$$\text{where } p_c = \frac{\text{Marginal loss}}{\text{Marginal profit} + \text{Marginal loss}} = \frac{\$1.50}{\$0.50 + \$1.50} = 0.75$$

	The Week's Demand Will Be		
	≥ i Gallons (Prob. = p)	< i Gallons (Prob. = $1 - p$)	Expected Payoff
Stock the ith Gallon	\$0.50	−\$1.50	$\$0.50p - \$1.50(1 - p)$
Don't Stock It	\$0.00	\$0.00	\$0.00

To show how it works, let's first consider whether the *first* gallon of milk should be stocked. If it's stocked and sold, the marginal profit is \$0.50. If it's stocked and *not* sold, the marginal loss is \$1.50. For a normal distribution with $\mu = 2000$ and $\sigma = 300$, it's virtually certain that at least one gallon of milk will be sold. Table 19.7 is a series of payoff tables for this decision setting, and part A shows that the expected payoff from stocking the first gallon easily exceeds the expected payoff if it is not stocked.

In part B of Table 19.7, we move on to consider the second gallon. As with the first gallon, the probability that demand will be at least two gallons is virtually certain. If the second gallon is stocked, the expected payoff will be \$0.50 higher than if it is not stocked.

Since it would be inefficient to continue evaluating one gallon at a time, let's introduce a time-saver. The key is the answer to the question, "What probability of selling one more gallon results in the expected payoff for stocking this gallon being at least as high as the expected payoff for not stocking it?" Given the marginal profit if the ith gallon is sold and the marginal loss if it is not, this probability can be determined as follows:

Critical probability for incremental analysis:

Stock the ith unit if Prob(demand $\geq i$) is at least p_c, with

$$p_c = \frac{ml}{mp + ml}$$

where p_c = the critical probability
mp = the marginal profit if the unit is stocked and sold
ml = the marginal loss if the unit is stocked and not sold

With marginal profit mp = \$0.50 and marginal loss ml = \$1.50, the critical probability for this decision situation will be

$$p_c = \frac{ml}{mp + ml} = \frac{\$1.50}{\$0.50 + \$1.50} = 0.75$$

Thus, as long as there is at least a 75% chance that the ith gallon will be sold, that gallon should be stocked. This can also be found by setting the expected payoff in the first row of Table 19.7, part C, to be ≥ 0, or

$$\$0.50p - \$1.50(1 - p) \geq \$0$$
$$2.00p - 1.50 \geq 0$$

and p must be ≥ 0.75, the value of p_c

To find which gallon of milk corresponds to a 0.75 probability of being sold, we use the standard normal distribution as shown in Figure 19.4. For a right-tail area of 0.75, the cumulative area to the left will be 0.25. This corresponds to $z = -0.67$ standard deviation units, and

$$z = \frac{x - \mu}{\sigma} \quad \text{or} \quad -0.67 = \frac{x \text{ gallons} - 2000 \text{ gallons}}{300 \text{ gallons}}$$

and

$$x = 1799 \text{ gallons}$$

The probability is 0.75 that the demand will be high enough to include the 1799th gallon, and this is the quantity the supermarket should stock. Stocking the 1800th gallon would not be advisable, since the probability of a demand at least this high would be slightly less than 0.75 and the expected payoff for stocking the 1800th gallon would be less than the expected payoff for not stocking it.

FIGURE 19.4

The probability is 0.75 that demand will include the 1799th gallon of milk. If the 1800th gallon is stocked, there is a less than 75% chance it will be sold. Using incremental analysis, expected profit is maximized if 1799 gallons are stocked.

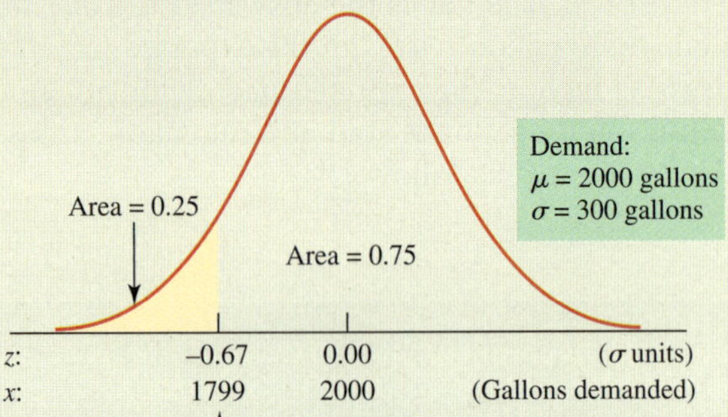

For a left-tail area of 0.25, z must be −0.67.
If the xth gallon has a 75% chance of being sold,
x is expressed by

$$z = \frac{x - \mu}{\sigma} \text{ or } -0.67 = \frac{x - 2000}{300} \text{ and } x = 1799 \text{ gallons}$$

EXERCISES

19.28 A newsstand manager must decide how many daily papers to stock. Past experience has shown that demand can be approximated by a normal distribution with a mean of 400 papers and a standard deviation of 80 papers. The manager pays 30 cents for each paper, sells it for 50 cents, and any papers not sold are picked up by the distributor, with the manager receiving 5 cents for each of them. If the manager wants to maximize his expected profit, how many newspapers should be stocked?

19.29 The manager of a fish market pays 80 cents each for cod and sells them for $1.50 each. Fish left over at the end of the day are discarded. The daily demand can be approximated by a normal distribution having a mean of 300 and a standard deviation of 50. To maximize his expected profit, how many cod should the manager buy before the market opens each day?

19.30 For Exercise 19.29, how many cod should be purchased if leftover fish can be sold to a fertilizer company for 30 cents each?

19.31 An athletic director must order football programs well in advance of game day. She pays $1.00 for each program, and the selling price to fans is $5.00. Programs not sold at the game can be sold to the college bookstore for $0.30 each. Demand at the game is normally distributed, with $\mu = 3000$ and $\sigma = 500$. To maximize expected payoff, how many programs should the athletic director order?

19.32 Repeat Exercise 19.31, under the assumption that the bookstore manager has refused to purchase unsold programs for future games.

19.7 SUMMARY

• Structuring the decision situation

The decision situation can be structured according to the decision alternatives and the possible states of nature. The states of nature are both exhaustive (one of them must occur) and mutually exclusive (only one can occur). The decision situation may involve risk, uncertainty, or ignorance, depending on how much is known about the states of nature and their exact probabilities. Most business decisions are made under uncertainty, where the states of nature are known, but their exact probabilities are not.

The payoff table rows represent the decision alternatives, the columns show the states of nature, and the entries are payoffs for the various intersections of alternatives and states of nature. The decision situation can also be described by a decision tree, in which alternatives and states of nature are displayed in graphical form. Both the payoff table and the decision tree can include estimated probabilities for the states of nature.

• Bayesian and non-Bayesian decision criteria

In non-Bayesian decision making, there is no attempt to consider the probabilities of the respective states of nature. Three such decision rules are maximin, maximax, and minimax regret. Bayesian decision making takes these probabilities into account, and the alternative with the highest expected payoff is selected. If it is possible to obtain information that will reduce uncertainty, the expected value of perfect information (*EVPI*) can be calculated. This is the amount by which the expected payoff could be improved if perfect information were available, and it is the most that should be paid for any information designed to reduce uncertainty.

• Opportunity loss approach

In the opportunity loss table, each entry is the difference between a payoff and the best possible payoff for that state of nature. The results of the opportunity loss approach are the same as for the expected payoff method, and the alternative with the minimum expected opportunity loss will have the maximum expected payoff. Also, the minimum expected opportunity loss is equivalent to the expected value of perfect information.

• Incremental stocking decisions

For some situations, such as inventory stocking decisions, the number of alternatives and states of nature may be extremely large. In such cases, incremental analysis can be applied. In this method, units are sequentially considered, with each unit in the sequence being stocked only if the expected payoff for stocking it exceeds the expected payoff for not stocking it.

EQUATIONS

Expected Payoff for a Decision Alternative

$$\text{Expected payoff} = \sum p_i \cdot m_i$$

where p_i = the probability that state of nature i will occur

m_i = the payoff if this alternative is selected and state of nature i occurs

Expected Value of Perfect Information, *EVPI*

$$EVPI = \begin{matrix}\text{Expected payoff with}\\ \text{perfect information}\end{matrix} - \begin{matrix}\text{Expected payoff with}\\ \text{present information}\end{matrix}$$

Expected Opportunity Loss (*EOL*) for a Decision Alternative

$$EOL = \sum p_i \cdot l_i$$

where p_i = the probability that state of nature i will occur

l_i = the opportunity loss if this alternative is selected and state of nature i occurs

Critical Probability for Incremental Analysis

Stock the ith unit if Prob(demand $\geq i$) is at least p_c, with

$$p_c = \frac{ml}{mp + ml}$$

where p_c = the critical probability
mp = the marginal profit if the unit is stocked and sold
ml = the marginal loss if the unit is stocked and not sold

CHAPTER EXERCISES

19.33 For a given perishable product, a retailer pays \$5 for each unit, then sells them for \$10 each. At the end of the day, units not sold at the store are disposed of, and the retailer receives just \$1 for each. Given the following probability distribution describing daily demand, how many units should be stocked?

x	Prob. (Demand = x)	x	Prob. (Demand = x)
0	0.01	7	0.18
1	0.02	8	0.14
2	0.04	9	0.09
3	0.07	10	0.05
4	0.09	11	0.02
5	0.12	12	0.01
6	0.15	13	0.01

19.34 Dennis Dunkwright is a former NBA player and free spirit whose face and unusual antics are well known to young Americans across the land. A marketer of athletic shoes is considering the possibility of retaining Dennis to appear in a series of high-profile television ads over the coming fiscal year. If Dennis is associated with their company and stays out of legal and substance abuse difficulties during the year, the company estimates its profits will be \$4.4 million for the year. However, if Dennis runs afoul of the law or does anything else that might make the company look bad, the estimated profit would be just \$0.5 million. If the company remains with its current spokesperson, who is lesser known but more stable, it estimates its annual profit for the year to be \$2.2 million. If the company believes there to be a 0.20 probability that Dennis will get into trouble during the year, what strategy should be selected if the company wishes to maximize its expected profit? What would be the expected value of a perfect forecast for the future behavior of their potential new spokesperson?

19.35 In Exercise 19.34, what strategy would the company select if it were using the maximin criterion? The maximax criterion? The minimax regret criterion?

19.36 An investor is trying to decide between two promising technology stocks to round out his portfolio. The remainder of his portfolio consists of relatively conservative companies and will not enter into his decision as to which technology company will be selected. DiskWorth is a well-established company engaged in the design and manufacture of data storage disks, while ComTranDat is a start-up firm in the midst of developing what could be a wireless communication and data transmission system superior to those currently used or under development by the leaders in its industry. Depending on whether the technology sector weakens, remains the same, or strengthens over the next 10 years, the investor estimates that his profits with DiskWorth would be \$20,000, \$30,000, and \$40,000, for the respective scenarios. The corresponding profits for ComTranDat are estimated to be −\$20,000, \$10,000, and \$90,000, respectively. If the technology sector has a 20% probability of weakening over the next 10 years, a 50% probability of remaining the same, and a 30% probability of strengthening, which stock purchase should the investor make if he wishes to maximize his expected profits over the 10-year period? What is the most he should be willing to pay for a perfect forecast for the future strength of the technology sector?

19.37 In Exercise 19.36, which company would the investor select if it were using the maximin criterion? The maximax criterion? The minimax regret criterion?

19.38 A ski resort operator must decide before the winter season whether he will lease a snow-making machine. If he has no machine, he will make \$20,000 if the winter is mild, \$30,000 if it is typical, and \$50,000 if the winter is severe. If he decides to lease the machine, his profits for these conditions will be \$30,000, \$35,000, and \$40,000, respectively. The probability of a mild winter is 0.3, with a 0.5 chance of a typical winter and a 0.2 chance of a severe winter. If the operator wants to maximize his expected profit, should he lease the machine? What is the most he should be willing to pay for a perfect forecast?

19.39 In Exercise 19.38, the operator says that he has no idea regarding probabilities for the various weather conditions for the coming winter. What are some other criteria by which he might reach a decision, and what decision would be reached using each of them?

19.40 A driver in the Indianapolis 500 auto race has the lead but is nearly out of fuel with eight laps remaining in the race. The second-place car is just 10 seconds behind and has plenty of fuel. The driver in the lead car elects to try for the finish line without making a fuel stop. Does this behavior seem to be a maximin choice or a maximax choice? Explain.

INTEGRATED CASE

Thorndike Sports Equipment (Video Unit Seven)

For several years, Thorndike Sports Equipment has been one of the suppliers of refreshments to the Titans of Tennis national tournament. Luke Thorndike doesn't really like to call the company's product a "refreshment," since it's been scientifically designed to replenish the fluids and vital body chemicals lost through perspiration and exercise. According to the elder Thorndike, Thornado is not "just another sugar-fluffy soft drink. It helps athletes perform at their best."

Naturally, each competing athlete is given a complimentary carton of Thornado. However, the Thorndikes must decide how many cans of the product to send to the tournament for fan consumption. The Thorndikes make a profit of $0.25 on each can of Thornado purchased at the tournament. Each can that is not purchased is distributed to local sports clubs in the tournament's host city. Luke Thorndike believes in helping those in need, but is quick to point out that it costs him a total of $0.80 to produce and distribute each of these unsold cans of Thornado.

Luke believes in charity, but he is also rather frugal in making it available. He's asked Ted to figure out how many cans of Thornado to send to the Titans of Tennis tournament in order to make as much profit as possible. Ted Thorndike, having analyzed past sales patterns for Thornado at this tournament, has found annual tournament sales of the product to be closely approximated by a normal distribution with $\mu = 4500$ cans and $\sigma = 300$ cans. In order to satisfy his grandfather's dictate that profit be maximized, how many cans of Thornado should Ted plan on shipping to the Titans of Tennis?

APPENDIX A

STATISTICAL TABLES

TABLE A.1

Binomial Distribution, Individual Probabilities for x = number of successes in n trials, prob($x = k$)

$n = 2$

π:	0.1	0.2	0.3	0.4	0.5	0.6	0.7	0.8	0.9
k: 0	0.8100	0.6400	0.4900	0.3600	0.2500	0.1600	0.0900	0.0400	0.0100
1	0.1800	0.3200	0.4200	0.4800	0.5000	0.4800	0.4200	0.3200	0.1800
2	0.0100	0.0400	0.0900	0.1600	0.2500	0.3600	0.4900	0.6400	0.8100

$n = 3$

π:	0.1	0.2	0.3	0.4	0.5	0.6	0.7	0.8	0.9
k: 0	0.7290	0.5120	0.3430	0.2160	0.1250	0.0640	0.0270	0.0080	0.0010
1	0.2430	0.3840	0.4410	0.4320	0.3750	0.2880	0.1890	0.0960	0.0270
2	0.0270	0.0960	0.1890	0.2880	0.3750	0.4320	0.4410	0.3840	0.2430
3	0.0010	0.0080	0.0270	0.0640	0.1250	0.2160	0.3430	0.5120	0.7290

$n = 4$

π:	0.1	0.2	0.3	0.4	0.5	0.6	0.7	0.8	0.9
k: 0	0.6561	0.4096	0.2401	0.1296	0.0625	0.0256	0.0081	0.0016	0.0001
1	0.2916	0.4096	0.4116	0.3456	0.2500	0.1536	0.0756	0.0256	0.0036
2	0.0486	0.1536	0.2646	0.3456	0.3750	0.3456	0.2646	0.1536	0.0486
3	0.0036	0.0256	0.0756	0.1536	0.2500	0.3456	0.4116	0.4096	0.2916
4	0.0001	0.0016	0.0081	0.0256	0.0625	0.1296	0.2401	0.4096	0.6561

$n = 5$

π:	0.1	0.2	0.3	0.4	0.5	0.6	0.7	0.8	0.9
k: 0	0.5905	0.3277	0.1681	0.0778	0.0313	0.0102	0.0024	0.0003	0.0000
1	0.3281	0.4096	0.3601	0.2592	0.1562	0.0768	0.0284	0.0064	0.0005
2	0.0729	0.2048	0.3087	0.3456	0.3125	0.2304	0.1323	0.0512	0.0081
3	0.0081	0.0512	0.1323	0.2304	0.3125	0.3456	0.3087	0.2048	0.0729
4	0.0005	0.0064	0.0284	0.0768	0.1562	0.2592	0.3601	0.4096	0.3281
5	0.0000	0.0003	0.0024	0.0102	0.0313	0.0778	0.1681	0.3277	0.5905

Source: Probabilities generated by Minitab, then compiled as shown here.

TABLE A.1

(continued)

$n = 6$

π:	0.1	0.2	0.3	0.4	0.5	0.6	0.7	0.8	0.9
k: 0	0.5314	0.2621	0.1176	0.0467	0.0156	0.0041	0.0007	0.0001	0.0000
1	0.3543	0.3932	0.3025	0.1866	0.0937	0.0369	0.0102	0.0015	0.0001
2	0.0984	0.2458	0.3241	0.3110	0.2344	0.1382	0.0595	0.0154	0.0012
3	0.0146	0.0819	0.1852	0.2765	0.3125	0.2765	0.1852	0.0819	0.0146
4	0.0012	0.0154	0.0595	0.1382	0.2344	0.3110	0.3241	0.2458	0.0984
5	0.0001	0.0015	0.0102	0.0369	0.0937	0.1866	0.3025	0.3932	0.3543
6	0.0000	0.0001	0.0007	0.0041	0.0156	0.0467	0.1176	0.2621	0.5314

$n = 7$

π:	0.1	0.2	0.3	0.4	0.5	0.6	0.7	0.8	0.9
k: 0	0.4783	0.2097	0.0824	0.0280	0.0078	0.0016	0.0002	0.0000	
1	0.3720	0.3670	0.2471	0.1306	0.0547	0.0172	0.0036	0.0004	0.0000
2	0.1240	0.2753	0.3177	0.2613	0.1641	0.0774	0.0250	0.0043	0.0002
3	0.0230	0.1147	0.2269	0.2903	0.2734	0.1935	0.0972	0.0287	0.0026
4	0.0026	0.0287	0.0972	0.1935	0.2734	0.2903	0.2269	0.1147	0.0230
5	0.0002	0.0043	0.0250	0.0774	0.1641	0.2613	0.3177	0.2753	0.1240
6	0.0000	0.0004	0.0036	0.0172	0.0547	0.1306	0.2471	0.3670	0.3720
7		0.0000	0.0002	0.0016	0.0078	0.0280	0.0824	0.2097	0.4783

$n = 8$

π:	0.1	0.2	0.3	0.4	0.5	0.6	0.7	0.8	0.9
k: 0	0.4305	0.1678	0.0576	0.0168	0.0039	0.0007	0.0001	0.0000	
1	0.3826	0.3355	0.1977	0.0896	0.0313	0.0079	0.0012	0.0001	
2	0.1488	0.2936	0.2965	0.2090	0.1094	0.0413	0.0100	0.0011	0.0000
3	0.0331	0.1468	0.2541	0.2787	0.2187	0.1239	0.0467	0.0092	0.0004
4	0.0046	0.0459	0.1361	0.2322	0.2734	0.2322	0.1361	0.0459	0.0046
5	0.0004	0.0092	0.0467	0.1239	0.2187	0.2787	0.2541	0.1468	0.0331
6	0.0000	0.0011	0.0100	0.0413	0.1094	0.2090	0.2965	0.2936	0.1488
7		0.0001	0.0012	0.0079	0.0313	0.0896	0.1977	0.3355	0.3826
8		0.0000	0.0001	0.0007	0.0039	0.0168	0.0576	0.1678	0.4305

$n = 9$

π:	0.1	0.2	0.3	0.4	0.5	0.6	0.7	0.8	0.9
k: 0	0.3874	0.1342	0.0404	0.0101	0.0020	0.0003	0.0000		
1	0.3874	0.3020	0.1556	0.0605	0.0176	0.0035	0.0004	0.0000	
2	0.1722	0.3020	0.2668	0.1612	0.0703	0.0212	0.0039	0.0003	0.0000
3	0.0446	0.1762	0.2668	0.2508	0.1641	0.0743	0.0210	0.0028	0.0001
4	0.0074	0.0661	0.1715	0.2508	0.2461	0.1672	0.0735	0.0165	0.0008
5	0.0008	0.0165	0.0735	0.1672	0.2461	0.2508	0.1715	0.0661	0.0074
6	0.0001	0.0028	0.0210	0.0743	0.1641	0.2508	0.2668	0.1762	0.0446
7	0.0000	0.0003	0.0039	0.0212	0.0703	0.1612	0.2668	0.3020	0.1722
8		0.0000	0.0004	0.0035	0.0176	0.0605	0.1556	0.3020	0.3874
9			0.0000	0.0003	0.0020	0.0101	0.0404	0.1342	0.3874

TABLE A.1
(continued)

$n = 10$

π:	0.1	0.2	0.3	0.4	0.5	0.6	0.7	0.8	0.9
k: 0	0.3487	0.1074	0.0282	0.0060	0.0010	0.0001	0.0000		
1	0.3874	0.2684	0.1211	0.0403	0.0098	0.0016	0.0001	0.0000	
2	0.1937	0.3020	0.2335	0.1209	0.0439	0.0106	0.0014	0.0001	
3	0.0574	0.2013	0.2668	0.2150	0.1172	0.0425	0.0090	0.0008	0.0000
4	0.0112	0.0881	0.2001	0.2508	0.2051	0.1115	0.0368	0.0055	0.0001
5	0.0015	0.0264	0.1029	0.2007	0.2461	0.2007	0.1029	0.0264	0.0015
6	0.0001	0.0055	0.0368	0.1115	0.2051	0.2508	0.2001	0.0881	0.0112
7	0.0000	0.0008	0.0090	0.0425	0.1172	0.2150	0.2668	0.2013	0.0574
8		0.0001	0.0014	0.0106	0.0439	0.1209	0.2335	0.3020	0.1937
9		0.0000	0.0001	0.0016	0.0098	0.0403	0.1211	0.2684	0.3874
10			0.0000	0.0001	0.0010	0.0060	0.0282	0.1074	0.3487

$n = 11$

π:	0.1	0.2	0.3	0.4	0.5	0.6	0.7	0.8	0.9
k: 0	0.3138	0.0859	0.0198	0.0036	0.0005	0.0000			
1	0.3835	0.2362	0.0932	0.0266	0.0054	0.0007	0.0000		
2	0.2131	0.2953	0.1998	0.0887	0.0269	0.0052	0.0005	0.0000	
3	0.0710	0.2215	0.2568	0.1774	0.0806	0.0234	0.0037	0.0002	
4	0.0158	0.1107	0.2201	0.2365	0.1611	0.0701	0.0173	0.0017	0.0000
5	0.0025	0.0388	0.1321	0.2207	0.2256	0.1471	0.0566	0.0097	0.0003
6	0.0003	0.0097	0.0566	0.1471	0.2256	0.2207	0.1321	0.0388	0.0025
7	0.0000	0.0017	0.0173	0.0701	0.1611	0.2365	0.2201	0.1107	0.0158
8		0.0002	0.0037	0.0234	0.0806	0.1774	0.2568	0.2215	0.0710
9		0.0000	0.0005	0.0052	0.0269	0.0887	0.1998	0.2953	0.2131
10			0.0000	0.0007	0.0054	0.0266	0.0932	0.2362	0.3835
11				0.0000	0.0005	0.0036	0.0198	0.0859	0.3138

$n = 12$

π:	0.1	0.2	0.3	0.4	0.5	0.6	0.7	0.8	0.9
k: 0	0.2824	0.0687	0.0138	0.0022	0.0002	0.0000			
1	0.3766	0.2062	0.0712	0.0174	0.0029	0.0003	0.0000		
2	0.2301	0.2835	0.1678	0.0639	0.0161	0.0025	0.0002	0.0000	
3	0.0852	0.2362	0.2397	0.1419	0.0537	0.0125	0.0015	0.0001	
4	0.0213	0.1329	0.2311	0.2128	0.1208	0.0420	0.0078	0.0005	
5	0.0038	0.0532	0.1585	0.2270	0.1934	0.1009	0.0291	0.0033	0.0000
6	0.0005	0.0155	0.0792	0.1766	0.2256	0.1766	0.0792	0.0155	0.0005
7	0.0000	0.0033	0.0291	0.1009	0.1934	0.2270	0.1585	0.0532	0.0038
8		0.0005	0.0078	0.0420	0.1208	0.2128	0.2311	0.1329	0.0213
9		0.0001	0.0015	0.0125	0.0537	0.1419	0.2397	0.2362	0.0852
10		0.0000	0.0002	0.0025	0.0161	0.0639	0.1678	0.2835	0.2301
11			0.0000	0.0003	0.0029	0.0174	0.0712	0.2062	0.3766
12				0.0000	0.0002	0.0022	0.0138	0.0687	0.2824

TABLE A.1
(continued)

n = 13

π:	0.1	0.2	0.3	0.4	0.5	0.6	0.7	0.8	0.9
k: 0	0.2542	0.0550	0.0097	0.0013	0.0001	0.0000			
1	0.3672	0.1787	0.0540	0.0113	0.0016	0.0001	0.0000		
2	0.2448	0.2680	0.1388	0.0453	0.0095	0.0012	0.0001		
3	0.0997	0.2457	0.2181	0.1107	0.0349	0.0065	0.0006	0.0000	
4	0.0277	0.1535	0.2337	0.1845	0.0873	0.0243	0.0034	0.0001	
5	0.0055	0.0691	0.1803	0.2214	0.1571	0.0656	0.0142	0.0011	0.0000
6	0.0008	0.0230	0.1030	0.1968	0.2095	0.1312	0.0442	0.0058	0.0001
7	0.0001	0.0058	0.0442	0.1312	0.2095	0.1968	0.1030	0.0230	0.0008
8	0.0000	0.0011	0.0142	0.0656	0.1571	0.2214	0.1803	0.0691	0.0055
9		0.0001	0.0034	0.0243	0.0873	0.1845	0.2337	0.1535	0.0277
10		0.0000	0.0006	0.0065	0.0349	0.1107	0.2181	0.2457	0.0997
11			0.0001	0.0012	0.0095	0.0453	0.1388	0.2680	0.2448
12			0.0000	0.0001	0.0016	0.0113	0.0540	0.1787	0.3672
13				0.0000	0.0001	0.0013	0.0097	0.0550	0.2542

n = 14

π:	0.1	0.2	0.3	0.4	0.5	0.6	0.7	0.8	0.9
k: 0	0.2288	0.0440	0.0068	0.0008	0.0001	0.0000			
1	0.3559	0.1539	0.0407	0.0073	0.0009	0.0001			
2	0.2570	0.2501	0.1134	0.0317	0.0056	0.0005	0.0000		
3	0.1142	0.2501	0.1943	0.0845	0.0222	0.0033	0.0002		
4	0.0349	0.1720	0.2290	0.1549	0.0611	0.0136	0.0014	0.0000	
5	0.0078	0.0860	0.1963	0.2066	0.1222	0.0408	0.0066	0.0003	
6	0.0013	0.0322	0.1262	0.2066	0.1833	0.0918	0.0232	0.0020	0.0000
7	0.0002	0.0092	0.0618	0.1574	0.2095	0.1574	0.0618	0.0092	0.0002
8	0.0000	0.0020	0.0232	0.0918	0.1833	0.2066	0.1262	0.0322	0.0013
9		0.0003	0.0066	0.0408	0.1222	0.2066	0.1963	0.0860	0.0078
10		0.0000	0.0014	0.0136	0.0611	0.1549	0.2290	0.1720	0.0349
11			0.0002	0.0033	0.0222	0.0845	0.1943	0.2501	0.1142
12			0.0000	0.0005	0.0056	0.0317	0.1134	0.2501	0.2570
13				0.0001	0.0009	0.0073	0.0407	0.1539	0.3559
14				0.0000	0.0001	0.0008	0.0068	0.0440	0.2288

n = 15

π:	0.1	0.2	0.3	0.4	0.5	0.6	0.7	0.8	0.9
k: 0	0.2059	0.0352	0.0047	0.0005	0.0000				
1	0.3432	0.1319	0.0305	0.0047	0.0005	0.0000			
2	0.2669	0.2309	0.0916	0.0219	0.0032	0.0003	0.0000		
3	0.1285	0.2501	0.1700	0.0634	0.0139	0.0016	0.0001		
4	0.0428	0.1876	0.2186	0.1268	0.0417	0.0074	0.0006	0.0000	
5	0.0105	0.1032	0.2061	0.1859	0.0916	0.0245	0.0030	0.0001	
6	0.0019	0.0430	0.1472	0.2066	0.1527	0.0612	0.0116	0.0007	
7	0.0003	0.0138	0.0811	0.1771	0.1964	0.1181	0.0348	0.0035	0.0000
8	0.0000	0.0035	0.0348	0.1181	0.1964	0.1771	0.0811	0.0138	0.0003
9		0.0007	0.0116	0.0612	0.1527	0.2066	0.1472	0.0430	0.0019
10		0.0001	0.0030	0.0245	0.0916	0.1859	0.2061	0.1032	0.0105
11		0.0000	0.0006	0.0074	0.0417	0.1268	0.2186	0.1876	0.0428
12			0.0001	0.0016	0.0139	0.0634	0.1700	0.2501	0.1285
13			0.0000	0.0003	0.0032	0.0219	0.0916	0.2309	0.2669
14				0.0000	0.0005	0.0047	0.0305	0.1319	0.3432
15					0.0000	0.0005	0.0047	0.0352	0.2059

TABLE A.1
(continued)

n = 16

π:	0.1	0.2	0.3	0.4	0.5	0.6	0.7	0.8	0.9
k: 0	0.1853	0.0281	0.0033	0.0003	0.0000				
1	0.3294	0.1126	0.0228	0.0030	0.0002	0.0000			
2	0.2745	0.2111	0.0732	0.0150	0.0018	0.0001			
3	0.1423	0.2463	0.1465	0.0468	0.0085	0.0008	0.0000		
4	0.0514	0.2001	0.2040	0.1014	0.0278	0.0040	0.0002		
5	0.0137	0.1201	0.2099	0.1623	0.0667	0.0142	0.0013	0.0000	
6	0.0028	0.0550	0.1649	0.1983	0.1222	0.0392	0.0056	0.0002	
7	0.0004	0.0197	0.1010	0.1889	0.1746	0.0840	0.0185	0.0012	0.0000
8	0.0001	0.0055	0.0487	0.1417	0.1964	0.1417	0.0487	0.0055	0.0001
9	0.0000	0.0012	0.0185	0.0840	0.1746	0.1889	0.1010	0.0197	0.0004
10		0.0002	0.0056	0.0392	0.1222	0.1983	0.1649	0.0550	0.0028
11		0.0000	0.0013	0.0142	0.0667	0.1623	0.2099	0.1201	0.0137
12			0.0002	0.0040	0.0278	0.1014	0.2040	0.2001	0.0514
13			0.0000	0.0008	0.0085	0.0468	0.1465	0.2463	0.1423
14				0.0001	0.0018	0.0150	0.0732	0.2111	0.2745
15				0.0000	0.0002	0.0030	0.0228	0.1126	0.3294
16					0.0000	0.0003	0.0033	0.0281	0.1853

n = 17

π:	0.1	0.2	0.3	0.4	0.5	0.6	0.7	0.8	0.9
k: 0	0.1668	0.0225	0.0023	0.0002	0.0000				
1	0.3150	0.0957	0.0169	0.0019	0.0001	0.0000			
2	0.2800	0.1914	0.0581	0.0102	0.0010	0.0001			
3	0.1556	0.2393	0.1245	0.0341	0.0052	0.0004	0.0000		
4	0.0605	0.2093	0.1868	0.0796	0.0182	0.0021	0.0001		
5	0.0175	0.1361	0.2081	0.1379	0.0472	0.0081	0.0006	0.0000	
6	0.0039	0.0680	0.1784	0.1839	0.0944	0.0242	0.0026	0.0001	
7	0.0007	0.0267	0.1201	0.1927	0.1484	0.0571	0.0095	0.0004	
8	0.0001	0.0084	0.0644	0.1606	0.1855	0.1070	0.0276	0.0021	0.0000
9	0.0000	0.0021	0.0276	0.1070	0.1855	0.1606	0.0644	0.0084	0.0001
10		0.0004	0.0095	0.0571	0.1484	0.1927	0.1201	0.0267	0.0007
11		0.0001	0.0026	0.0242	0.0944	0.1839	0.1784	0.0680	0.0039
12		0.0000	0.0006	0.0081	0.0472	0.1379	0.2081	0.1361	0.0175
13			0.0001	0.0021	0.0182	0.0796	0.1868	0.2093	0.0605
14			0.0000	0.0004	0.0052	0.0341	0.1245	0.2393	0.1556
15				0.0001	0.0010	0.0102	0.0581	0.1914	0.2800
16				0.0000	0.0001	0.0019	0.0169	0.0957	0.3150
17					0.0000	0.0002	0.0023	0.0225	0.1668

TABLE A.1
(continued)

n = 18

π:	0.1	0.2	0.3	0.4	0.5	0.6	0.7	0.8	0.9
k: 0	0.1501	0.0180	0.0016	0.0001	0.0000				
1	0.3002	0.0811	0.0126	0.0012	0.0001				
2	0.2835	0.1723	0.0458	0.0069	0.0006	0.0000			
3	0.1680	0.2297	0.1046	0.0246	0.0031	0.0002			
4	0.0700	0.2153	0.1681	0.0614	0.0117	0.0011	0.0000		
5	0.0218	0.1507	0.2017	0.1146	0.0327	0.0045	0.0002		
6	0.0052	0.0816	0.1873	0.1655	0.0708	0.0145	0.0012	0.0000	
7	0.0010	0.0350	0.1376	0.1892	0.1214	0.0374	0.0046	0.0001	
8	0.0002	0.0120	0.0811	0.1734	0.1669	0.0771	0.0149	0.0008	
9	0.0000	0.0033	0.0386	0.1284	0.1855	0.1284	0.0386	0.0033	0.0000
10		0.0008	0.0149	0.0771	0.1669	0.1734	0.0811	0.0120	0.0002
11		0.0001	0.0046	0.0374	0.1214	0.1892	0.1376	0.0350	0.0010
12		0.0000	0.0012	0.0145	0.0708	0.1655	0.1873	0.0816	0.0052
13			0.0002	0.0045	0.0327	0.1146	0.2017	0.1507	0.0218
14			0.0000	0.0011	0.0117	0.0614	0.1681	0.2153	0.0700
15				0.0002	0.0031	0.0246	0.1046	0.2297	0.1680
16				0.0000	0.0006	0.0069	0.0458	0.1723	0.2835
17					0.0001	0.0012	0.0126	0.0811	0.3002
18					0.0000	0.0001	0.0016	0.0180	0.1501

n = 19

π:	0.1	0.2	0.3	0.4	0.5	0.6	0.7	0.8	0.9
k: 0	0.1351	0.0144	0.0011	0.0001					
1	0.2852	0.0685	0.0093	0.0008	0.0000				
2	0.2852	0.1540	0.0358	0.0046	0.0003	0.0000			
3	0.1796	0.2182	0.0869	0.0175	0.0018	0.0001			
4	0.0798	0.2182	0.1491	0.0467	0.0074	0.0005	0.0000		
5	0.0266	0.1636	0.1916	0.0933	0.0222	0.0024	0.0001		
6	0.0069	0.0955	0.1916	0.1451	0.0518	0.0085	0.0005		
7	0.0014	0.0443	0.1525	0.1797	0.0961	0.0237	0.0022	0.0000	
8	0.0002	0.0166	0.0981	0.1797	0.1442	0.0532	0.0077	0.0003	
9	0.0000	0.0051	0.0514	0.1464	0.1762	0.0976	0.0220	0.0013	
10		0.0013	0.0220	0.0976	0.1762	0.1464	0.0514	0.0051	0.0000
11		0.0003	0.0077	0.0532	0.1442	0.1797	0.0981	0.0166	0.0002
12		0.0000	0.0022	0.0237	0.0961	0.1797	0.1525	0.0443	0.0014
13			0.0005	0.0085	0.0518	0.1451	0.1916	0.0955	0.0069
14			0.0001	0.0024	0.0222	0.0933	0.1916	0.1636	0.0266
15			0.0000	0.0005	0.0074	0.0467	0.1491	0.2182	0.0798
16				0.0001	0.0018	0.0175	0.0869	0.2182	0.1796
17				0.0000	0.0003	0.0046	0.0358	0.1540	0.2852
18					0.0000	0.0008	0.0093	0.0685	0.2852
19						0.0001	0.0011	0.0144	0.1351

TABLE A.1
(continued)

$n = 20$

π:	0.1	0.2	0.3	0.4	0.5	0.6	0.7	0.8	0.9
k: 0	0.1216	0.0115	0.0008	0.0000					
1	0.2702	0.0576	0.0068	0.0005	0.0000				
2	0.2852	0.1369	0.0278	0.0031	0.0002				
3	0.1901	0.2054	0.0716	0.0123	0.0011	0.0000			
4	0.0898	0.2182	0.1304	0.0350	0.0046	0.0003			
5	0.0319	0.1746	0.1789	0.0746	0.0148	0.0013	0.0000		
6	0.0089	0.1091	0.1916	0.1244	0.0370	0.0049	0.0002		
7	0.0020	0.0545	0.1643	0.1659	0.0739	0.0146	0.0010	0.0000	
8	0.0004	0.0222	0.1144	0.1797	0.1201	0.0355	0.0039	0.0001	
9	0.0001	0.0074	0.0654	0.1597	0.1602	0.0710	0.0120	0.0005	
10	0.0000	0.0020	0.0308	0.1171	0.1762	0.1171	0.0308	0.0020	0.0000
11		0.0005	0.0120	0.0710	0.1602	0.1597	0.0654	0.0074	0.0001
12		0.0001	0.0039	0.0355	0.1201	0.1797	0.1144	0.0222	0.0004
13		0.0000	0.0010	0.0146	0.0739	0.1659	0.1643	0.0545	0.0020
14			0.0002	0.0049	0.0370	0.1244	0.1916	0.1091	0.0089
15			0.0000	0.0013	0.0148	0.0746	0.1789	0.1746	0.0319
16				0.0003	0.0046	0.0350	0.1304	0.2182	0.0898
17				0.0000	0.0011	0.0123	0.0716	0.2054	0.1901
18					0.0002	0.0031	0.0278	0.1369	0.2852
19					0.0000	0.0005	0.0068	0.0576	0.2702
20						0.0000	0.0008	0.0115	0.1216

$n = 25$

π:	0.1	0.2	0.3	0.4	0.5	0.6	0.7	0.8	0.9
k: 0	0.0718	0.0038	0.0001						
1	0.1994	0.0236	0.0014	0.0000					
2	0.2659	0.0708	0.0074	0.0004	0.0000				
3	0.2265	0.1358	0.0243	0.0019	0.0001				
4	0.1384	0.1867	0.0572	0.0071	0.0004				
5	0.0646	0.1960	0.1030	0.0199	0.0016	0.0000			
6	0.0239	0.1633	0.1472	0.0442	0.0053	0.0002			
7	0.0072	0.1108	0.1712	0.0800	0.0143	0.0009	0.0000		
8	0.0018	0.0623	0.1651	0.1200	0.0322	0.0031	0.0001		
9	0.0004	0.0294	0.1336	0.1511	0.0609	0.0088	0.0004		
10	0.0001	0.0118	0.0916	0.1612	0.0974	0.0212	0.0013	0.0000	
11	0.0000	0.0040	0.0536	0.1465	0.1328	0.0434	0.0042	0.0001	
12		0.0012	0.0268	0.1140	0.1550	0.0760	0.0115	0.0003	
13		0.0003	0.0115	0.0760	0.1550	0.1140	0.0268	0.0012	
14		0.0001	0.0042	0.0434	0.1328	0.1465	0.0536	0.0040	0.0000
15		0.0000	0.0013	0.0212	0.0974	0.1612	0.0916	0.0118	0.0001
16			0.0004	0.0088	0.0609	0.1511	0.1336	0.0294	0.0004
17			0.0001	0.0031	0.0322	0.1200	0.1651	0.0623	0.0018
18			0.0000	0.0009	0.0143	0.0800	0.1712	0.1108	0.0072
19				0.0002	0.0053	0.0442	0.1472	0.1633	0.0239
20				0.0000	0.0016	0.0199	0.1030	0.1960	0.0646
21					0.0004	0.0071	0.0572	0.1867	0.1384
22					0.0001	0.0019	0.0243	0.1358	0.2265
23					0.0000	0.0004	0.0074	0.0708	0.2659
24						0.0000	0.0014	0.0236	0.1994
25							0.0001	0.0038	0.0718

TABLE A.2

Binomial Distribution, Cumulative Probabilities for x = number of successes in n trials, prob($x \le k$)

$n = 2$

π:	0.1	0.2	0.3	0.4	0.5	0.6	0.7	0.8	0.9
k: 0	0.8100	0.6400	0.4900	0.3600	0.2500	0.1600	0.0900	0.0400	0.0100
1	0.9900	0.9600	0.9100	0.8400	0.7500	0.6400	0.5100	0.3600	0.1900
2	1.0000	1.0000	1.0000	1.0000	1.0000	1.0000	1.0000	1.0000	1.0000

$n = 3$

π:	0.1	0.2	0.3	0.4	0.5	0.6	0.7	0.8	0.9
k: 0	0.7290	0.5120	0.3430	0.2160	0.1250	0.0640	0.0270	0.0080	0.0010
1	0.9720	0.8960	0.7840	0.6480	0.5000	0.3520	0.2160	0.1040	0.0280
2	0.9990	0.9920	0.9730	0.9360	0.8750	0.7840	0.6570	0.4880	0.2710
3	1.0000	1.0000	1.0000	1.0000	1.0000	1.0000	1.0000	1.0000	1.0000

$n = 4$

π:	0.1	0.2	0.3	0.4	0.5	0.6	0.7	0.8	0.9
k: 0	0.6561	0.4096	0.2401	0.1296	0.0625	0.0256	0.0081	0.0016	0.0001
1	0.9477	0.8192	0.6517	0.4752	0.3125	0.1792	0.0837	0.0272	0.0037
2	0.9963	0.9728	0.9163	0.8208	0.6875	0.5248	0.3483	0.1808	0.0523
3	0.9999	0.9984	0.9919	0.9744	0.9375	0.8704	0.7599	0.5904	0.3439
4	1.0000	1.0000	1.0000	1.0000	1.0000	1.0000	1.0000	1.0000	1.0000

$n = 5$

π:	0.1	0.2	0.3	0.4	0.5	0.6	0.7	0.8	0.9
k: 0	0.5905	0.3277	0.1681	0.0778	0.0313	0.0102	0.0024	0.0003	0.0000
1	0.9185	0.7373	0.5282	0.3370	0.1875	0.0870	0.0308	0.0067	0.0005
2	0.9914	0.9421	0.8369	0.6826	0.5000	0.3174	0.1631	0.0579	0.0086
3	0.9995	0.9933	0.9692	0.9130	0.8125	0.6630	0.4718	0.2627	0.0815
4	1.0000	0.9997	0.9976	0.9898	0.9688	0.9222	0.8319	0.6723	0.4095
5		1.0000	1.0000	1.0000	1.0000	1.0000	1.0000	1.0000	1.0000

$n = 6$

π:	0.1	0.2	0.3	0.4	0.5	0.6	0.7	0.8	0.9
k: 0	0.5314	0.2621	0.1176	0.0467	0.0156	0.0041	0.0007	0.0001	0.0000
1	0.8857	0.6554	0.4202	0.2333	0.1094	0.0410	0.0109	0.0016	0.0001
2	0.9841	0.9011	0.7443	0.5443	0.3437	0.1792	0.0705	0.0170	0.0013
3	0.9987	0.9830	0.9295	0.8208	0.6563	0.4557	0.2557	0.0989	0.0159
4	0.9999	0.9984	0.9891	0.9590	0.8906	0.7667	0.5798	0.3446	0.1143
5	1.0000	0.9999	0.9993	0.9959	0.9844	0.9533	0.8824	0.7379	0.4686
6		1.0000	1.0000	1.0000	1.0000	1.0000	1.0000	1.0000	1.0000

Source: Probabilities generated by Minitab, then compiled as shown here.

TABLE A.2
(continued)

n = 7

π:	0.1	0.2	0.3	0.4	0.5	0.6	0.7	0.8	0.9
k: 0	0.4783	0.2097	0.0824	0.0280	0.0078	0.0016	0.0002	0.0000	
1	0.8503	0.5767	0.3294	0.1586	0.0625	0.0188	0.0038	0.0004	0.0000
2	0.9743	0.8520	0.6471	0.4199	0.2266	0.0963	0.0288	0.0047	0.0002
3	0.9973	0.9667	0.8740	0.7102	0.5000	0.2898	0.1260	0.0333	0.0027
4	0.9998	0.9953	0.9712	0.9037	0.7734	0.5801	0.3529	0.1480	0.0257
5	1.0000	0.9996	0.9962	0.9812	0.9375	0.8414	0.6706	0.4233	0.1497
6		1.0000	0.9998	0.9984	0.9922	0.9720	0.9176	0.7903	0.5217
7			1.0000	1.0000	1.0000	1.0000	1.0000	1.0000	1.0000

n = 8

π:	0.1	0.2	0.3	0.4	0.5	0.6	0.7	0.8	0.9
k: 0	0.4305	0.1678	0.0576	0.0168	0.0039	0.0007	0.0001	0.0000	
1	0.8131	0.5033	0.2553	0.1064	0.0352	0.0085	0.0013	0.0001	
2	0.9619	0.7969	0.5518	0.3154	0.1445	0.0498	0.0113	0.0012	0.0000
3	0.9950	0.9437	0.8059	0.5941	0.3633	0.1737	0.0580	0.0104	0.0004
4	0.9996	0.9896	0.9420	0.8263	0.6367	0.4059	0.1941	0.0563	0.0050
5	1.0000	0.9988	0.9887	0.9502	0.8555	0.6846	0.4482	0.2031	0.0381
6		0.9999	0.9987	0.9915	0.9648	0.8936	0.7447	0.4967	0.1869
7		1.0000	0.9999	0.9993	0.9961	0.9832	0.9424	0.8322	0.5695
8			1.0000	1.0000	1.0000	1.0000	1.0000	1.0000	1.0000

n = 9

π:	0.1	0.2	0.3	0.4	0.5	0.6	0.7	0.8	0.9
k: 0	0.3874	0.1342	0.0404	0.0101	0.0020	0.0003	0.0000		
1	0.7748	0.4362	0.1960	0.0705	0.0195	0.0038	0.0004	0.0000	
2	0.9470	0.7382	0.4628	0.2318	0.0898	0.0250	0.0043	0.0003	0.0000
3	0.9917	0.9144	0.7297	0.4826	0.2539	0.0994	0.0253	0.0031	0.0001
4	0.9991	0.9804	0.9012	0.7334	0.5000	0.2666	0.0988	0.0196	0.0009
5	0.9999	0.9969	0.9747	0.9006	0.7461	0.5174	0.2703	0.0856	0.0083
6	1.0000	0.9997	0.9957	0.9750	0.9102	0.7682	0.5372	0.2618	0.0530
7		1.0000	0.9996	0.9962	0.9805	0.9295	0.8040	0.5638	0.2252
8			1.0000	0.9997	0.9980	0.9899	0.9596	0.8658	0.6126
9				1.0000	1.0000	1.0000	1.0000	1.0000	1.0000

n = 10

π:	0.1	0.2	0.3	0.4	0.5	0.6	0.7	0.8	0.9
k: 0	0.3487	0.1074	0.0282	0.0060	0.0010	0.0001	0.0000		
1	0.7361	0.3758	0.1493	0.0464	0.0107	0.0017	0.0001	0.0000	
2	0.9298	0.6778	0.3828	0.1673	0.0547	0.0123	0.0016	0.0001	
3	0.9872	0.8791	0.6496	0.3823	0.1719	0.0548	0.0106	0.0009	0.0000
4	0.9984	0.9672	0.8497	0.6331	0.3770	0.1662	0.0473	0.0064	0.0001
5	0.9999	0.9936	0.9527	0.8338	0.6230	0.3669	0.1503	0.0328	0.0016
6	1.0000	0.9991	0.9894	0.9452	0.8281	0.6177	0.3504	0.1209	0.0128
7		0.9999	0.9984	0.9877	0.9453	0.8327	0.6172	0.3222	0.0702
8		1.0000	0.9999	0.9983	0.9893	0.9536	0.8507	0.6242	0.2639
9			1.0000	0.9999	0.9990	0.9940	0.9718	0.8926	0.6513
10				1.0000	1.0000	1.0000	1.0000	1.0000	1.0000

TABLE A.2
(continued)

n = 11

π:	0.1	0.2	0.3	0.4	0.5	0.6	0.7	0.8	0.9
k: 0	0.3138	0.0859	0.0198	0.0036	0.0005	0.0000			
1	0.6974	0.3221	0.1130	0.0302	0.0059	0.0007	0.0000		
2	0.9104	0.6174	0.3127	0.1189	0.0327	0.0059	0.0006	0.0000	
3	0.9815	0.8389	0.5696	0.2963	0.1133	0.0293	0.0043	0.0002	
4	0.9972	0.9496	0.7897	0.5328	0.2744	0.0994	0.0216	0.0020	0.0000
5	0.9997	0.9883	0.9218	0.7535	0.5000	0.2465	0.0782	0.0117	0.0003
6	1.0000	0.9980	0.9784	0.9006	0.7256	0.4672	0.2103	0.0504	0.0028
7		0.9998	0.9957	0.9707	0.8867	0.7037	0.4304	0.1611	0.0185
8		1.0000	0.9994	0.9941	0.9673	0.8811	0.6873	0.3826	0.0896
9			1.0000	0.9993	0.9941	0.9698	0.8870	0.6779	0.3026
10				1.0000	0.9995	0.9964	0.9802	0.9141	0.6862
11					1.0000	1.0000	1.0000	1.0000	1.0000

n = 12

π:	0.1	0.2	0.3	0.4	0.5	0.6	0.7	0.8	0.9
k: 0	0.2824	0.0687	0.0138	0.0022	0.0002	0.0000			
1	0.6590	0.2749	0.0850	0.0196	0.0032	0.0003	0.0000		
2	0.8891	0.5583	0.2528	0.0834	0.0193	0.0028	0.0002	0.0000	
3	0.9744	0.7946	0.4925	0.2253	0.0730	0.0153	0.0017	0.0001	
4	0.9957	0.9274	0.7237	0.4382	0.1938	0.0573	0.0095	0.0006	0.0000
5	0.9995	0.9806	0.8822	0.6652	0.3872	0.1582	0.0386	0.0039	0.0001
6	0.9999	0.9961	0.9614	0.8418	0.6128	0.3348	0.1178	0.0194	0.0005
7	1.0000	0.9994	0.9905	0.9427	0.8062	0.5618	0.2763	0.0726	0.0043
8		0.9999	0.9983	0.9847	0.9270	0.7747	0.5075	0.2054	0.0256
9		1.0000	0.9998	0.9972	0.9807	0.9166	0.7472	0.4417	0.1109
10			1.0000	0.9997	0.9968	0.9804	0.9150	0.7251	0.3410
11				1.0000	0.9998	0.9978	0.9862	0.9313	0.7176
12					1.0000	1.0000	1.0000	1.0000	1.0000

n = 13

π:	0.1	0.2	0.3	0.4	0.5	0.6	0.7	0.8	0.9
k: 0	0.2542	0.0550	0.0097	0.0013	0.0001	0.0000			
1	0.6213	0.2336	0.0637	0.0126	0.0017	0.0001	0.0000		
2	0.8661	0.5017	0.2025	0.0579	0.0112	0.0013	0.0001		
3	0.9658	0.7473	0.4206	0.1686	0.0461	0.0078	0.0007	0.0000	
4	0.9935	0.9009	0.6543	0.3530	0.1334	0.0321	0.0040	0.0002	
5	0.9991	0.9700	0.8346	0.5744	0.2905	0.0977	0.0182	0.0012	0.0000
6	0.9999	0.9930	0.9376	0.7712	0.5000	0.2288	0.0624	0.0070	0.0001
7	1.0000	0.9988	0.9818	0.9023	0.7095	0.4256	0.1654	0.0300	0.0009
8		0.9998	0.9960	0.9679	0.8666	0.6470	0.3457	0.0991	0.0065
9		1.0000	0.9993	0.9922	0.9539	0.8314	0.5794	0.2527	0.0342
10			0.9999	0.9987	0.9888	0.9421	0.7975	0.4983	0.1339
11			1.0000	0.9999	0.9983	0.9874	0.9363	0.7664	0.3787
12				1.0000	0.9999	0.9987	0.9903	0.9450	0.7458
13					1.0000	1.0000	1.0000	1.0000	1.0000

TABLE A.2
(continued)

n = 14

π:	0.1	0.2	0.3	0.4	0.5	0.6	0.7	0.8	0.9
k: 0	0.2288	0.0440	0.0068	0.0008	0.0001	0.0000			
1	0.5846	0.1979	0.0475	0.0081	0.0009	0.0001			
2	0.8416	0.4481	0.1608	0.0398	0.0065	0.0006	0.0000		
3	0.9559	0.6982	0.3552	0.1243	0.0287	0.0039	0.0002		
4	0.9908	0.8702	0.5842	0.2793	0.0898	0.0175	0.0017	0.0000	
5	0.9985	0.9561	0.7805	0.4859	0.2120	0.0583	0.0083	0.0004	
6	0.9998	0.9884	0.9067	0.6925	0.3953	0.1501	0.0315	0.0024	0.0000
7	1.0000	0.9976	0.9685	0.8499	0.6047	0.3075	0.0933	0.0116	0.0002
8		0.9996	0.9917	0.9417	0.7880	0.5141	0.2195	0.0439	0.0015
9		1.0000	0.9983	0.9825	0.9102	0.7207	0.4158	0.1298	0.0092
10			0.9998	0.9961	0.9713	0.8757	0.6448	0.3018	0.0441
11			1.0000	0.9994	0.9935	0.9602	0.8392	0.5519	0.1584
12				0.9999	0.9991	0.9919	0.9525	0.8021	0.4154
13				1.0000	0.9999	0.9992	0.9932	0.9560	0.7712
14					1.0000	1.0000	1.0000	1.0000	1.0000

n = 15

π:	0.1	0.2	0.3	0.4	0.5	0.6	0.7	0.8	0.9
k: 0	0.2059	0.0352	0.0047	0.0005	0.0000				
1	0.5490	0.1671	0.0353	0.0052	0.0005	0.0000			
2	0.8159	0.3980	0.1268	0.0271	0.0037	0.0003	0.0000		
3	0.9444	0.6482	0.2969	0.0905	0.0176	0.0019	0.0001		
4	0.9873	0.8358	0.5155	0.2173	0.0592	0.0093	0.0007	0.0000	
5	0.9978	0.9389	0.7216	0.4032	0.1509	0.0338	0.0037	0.0001	
6	0.9997	0.9819	0.8689	0.6098	0.3036	0.0950	0.0152	0.0008	
7	1.0000	0.9958	0.9500	0.7869	0.5000	0.2131	0.0500	0.0042	0.0000
8		0.9992	0.9848	0.9050	0.6964	0.3902	0.1311	0.0181	0.0003
9		0.9999	0.9963	0.9662	0.8491	0.5968	0.2784	0.0611	0.0022
10		1.0000	0.9993	0.9907	0.9408	0.7827	0.4845	0.1642	0.0127
11			0.9999	0.9981	0.9824	0.9095	0.7031	0.3518	0.0556
12			1.0000	0.9997	0.9963	0.9729	0.8732	0.6020	0.1841
13				1.0000	0.9995	0.9948	0.9647	0.8329	0.4510
14					1.0000	0.9995	0.9953	0.9648	0.7941
15						1.0000	1.0000	1.0000	1.0000

TABLE A.2

(continued)

n = 16

π:	0.1	0.2	0.3	0.4	0.5	0.6	0.7	0.8	0.9
k: 0	0.1853	0.0281	0.0033	0.0003	0.0000				
1	0.5147	0.1407	0.0261	0.0033	0.0003	0.0000			
2	0.7892	0.3518	0.0994	0.0183	0.0021	0.0001			
3	0.9316	0.5981	0.2459	0.0651	0.0106	0.0009	0.0000		
4	0.9830	0.7982	0.4499	0.1666	0.0384	0.0049	0.0003		
5	0.9967	0.9183	0.6598	0.3288	0.1051	0.0191	0.0016	0.0000	
6	0.9995	0.9733	0.8247	0.5272	0.2272	0.0583	0.0071	0.0002	
7	0.9999	0.9930	0.9256	0.7161	0.4018	0.1423	0.0257	0.0015	0.0000
8	1.0000	0.9985	0.9743	0.8577	0.5982	0.2839	0.0744	0.0070	0.0001
9		0.9998	0.9929	0.9417	0.7728	0.4728	0.1753	0.0267	0.0006
10		1.0000	0.9984	0.9809	0.8949	0.6712	0.3402	0.0817	0.0033
11			0.9997	0.9951	0.9616	0.8334	0.5501	0.2018	0.0170
12			1.0000	0.9991	0.9894	0.9349	0.7541	0.4019	0.0684
13				0.9999	0.9979	0.9817	0.9006	0.6482	0.2108
14				1.0000	0.9997	0.9967	0.9739	0.8593	0.4853
15					1.0000	0.9997	0.9967	0.9719	0.8147
16						1.0000	1.0000	1.0000	1.0000

n = 17

π:	0.1	0.2	0.3	0.4	0.5	0.6	0.7	0.8	0.9
k: 0	0.1668	0.0225	0.0023	0.0002	0.0000				
1	0.4818	0.1182	0.0193	0.0021	0.0001	0.0000			
2	0.7618	0.3096	0.0774	0.0123	0.0012	0.0001			
3	0.9174	0.5489	0.2019	0.0464	0.0064	0.0005	0.0000		
4	0.9779	0.7582	0.3887	0.1260	0.0245	0.0025	0.0001		
5	0.9953	0.8943	0.5968	0.2639	0.0717	0.0106	0.0007	0.0000	
6	0.9992	0.9623	0.7752	0.4478	0.1662	0.0348	0.0032	0.0001	
7	0.9999	0.9891	0.8954	0.6405	0.3145	0.0919	0.0127	0.0005	
8	1.0000	0.9974	0.9597	0.8011	0.5000	0.1989	0.0403	0.0026	0.0000
9		0.9995	0.9873	0.9081	0.6855	0.3595	0.1046	0.0109	0.0001
10		0.9999	0.9968	0.9652	0.8338	0.5522	0.2248	0.0377	0.0008
11		1.0000	0.9993	0.9894	0.9283	0.7361	0.4032	0.1057	0.0047
12			0.9999	0.9975	0.9755	0.8740	0.6113	0.2418	0.0221
13			1.0000	0.9995	0.9936	0.9536	0.7981	0.4511	0.0826
14				0.9999	0.9988	0.9877	0.9226	0.6904	0.2382
15				1.0000	0.9999	0.9979	0.9807	0.8818	0.5182
16					1.0000	0.9998	0.9977	0.9775	0.8332
17						1.0000	1.0000	1.0000	1.0000

TABLE A.2
(continued)

n = 18

π:	0.1	0.2	0.3	0.4	0.5	0.6	0.7	0.8	0.9
k: 0	0.1501	0.0180	0.0016	0.0001	0.0000				
1	0.4503	0.0991	0.0142	0.0013	0.0001				
2	0.7338	0.2713	0.0600	0.0082	0.0007	0.0000			
3	0.9018	0.5010	0.1646	0.0328	0.0038	0.0002			
4	0.9718	0.7164	0.3327	0.0942	0.0154	0.0013	0.0000		
5	0.9936	0.8671	0.5344	0.2088	0.0481	0.0058	0.0003		
6	0.9988	0.9487	0.7217	0.3743	0.1189	0.0203	0.0014	0.0000	
7	0.9998	0.9837	0.8593	0.5634	0.2403	0.0576	0.0061	0.0002	
8	1.0000	0.9957	0.9404	0.7368	0.4073	0.1347	0.0210	0.0009	
9		0.9991	0.9790	0.8653	0.5927	0.2632	0.0596	0.0043	0.0000
10		0.9998	0.9939	0.9424	0.7597	0.4366	0.1407	0.0163	0.0002
11		1.0000	0.9986	0.9797	0.8811	0.6257	0.2783	0.0513	0.0012
12			0.9997	0.9942	0.9519	0.7912	0.4656	0.1329	0.0064
13			1.0000	0.9987	0.9846	0.9058	0.6673	0.2836	0.0282
14				0.9998	0.9962	0.9672	0.8354	0.4990	0.0982
15				1.0000	0.9993	0.9918	0.9400	0.7287	0.2662
16					0.9999	0.9987	0.9858	0.9009	0.5497
17					1.0000	0.9999	0.9984	0.9820	0.8499
18						1.0000	1.0000	1.0000	1.0000

n = 19

π:	0.1	0.2	0.3	0.4	0.5	0.6	0.7	0.8	0.9
k: 0	0.1351	0.0144	0.0011	0.0001					
1	0.4203	0.0829	0.0104	0.0008	0.0000				
2	0.7054	0.2369	0.0462	0.0055	0.0004	0.0000			
3	0.8850	0.4551	0.1332	0.0230	0.0022	0.0001			
4	0.9648	0.6733	0.2822	0.0696	0.0096	0.0006	0.0000		
5	0.9914	0.8369	0.4739	0.1629	0.0318	0.0031	0.0001		
6	0.9983	0.9324	0.6655	0.3081	0.0835	0.0116	0.0006		
7	0.9997	0.9767	0.8180	0.4878	0.1796	0.0352	0.0028	0.0000	
8	1.0000	0.9933	0.9161	0.6675	0.3238	0.0885	0.0105	0.0003	
9		0.9984	0.9674	0.8139	0.5000	0.1861	0.0326	0.0016	
10		0.9997	0.9895	0.9115	0.6762	0.3325	0.0839	0.0067	0.0000
11		1.0000	0.9972	0.9648	0.8204	0.5122	0.1820	0.0233	0.0003
12			0.9994	0.9884	0.9165	0.6919	0.3345	0.0676	0.0017
13			0.9999	0.9969	0.9682	0.8371	0.5261	0.1631	0.0086
14			1.0000	0.9994	0.9904	0.9304	0.7178	0.3267	0.0352
15				0.9999	0.9978	0.9770	0.8668	0.5449	0.1150
16				1.0000	0.9996	0.9945	0.9538	0.7631	0.2946
17					1.0000	0.9992	0.9896	0.9171	0.5797
18						0.9999	0.9989	0.9856	0.8649
19						1.0000	1.0000	1.0000	1.0000

TABLE A.2

(continued)

n = 20

π:	0.1	0.2	0.3	0.4	0.5	0.6	0.7	0.8	0.9
k: 0	0.1216	0.0115	0.0008	0.0000					
1	0.3917	0.0692	0.0076	0.0005	0.0000				
2	0.6769	0.2061	0.0355	0.0036	0.0002				
3	0.8670	0.4114	0.1071	0.0160	0.0013	0.0000			
4	0.9568	0.6296	0.2375	0.0510	0.0059	0.0003			
5	0.9887	0.8042	0.4164	0.1256	0.0207	0.0016	0.0000		
6	0.9976	0.9133	0.6080	0.2500	0.0577	0.0065	0.0003		
7	0.9996	0.9679	0.7723	0.4159	0.1316	0.0210	0.0013	0.0000	
8	0.9999	0.9900	0.8867	0.5956	0.2517	0.0565	0.0051	0.0001	
9	1.0000	0.9974	0.9520	0.7553	0.4119	0.1275	0.0171	0.0006	
10		0.9994	0.9829	0.8725	0.5881	0.2447	0.0480	0.0026	0.0000
11		0.9999	0.9949	0.9435	0.7483	0.4044	0.1133	0.0100	0.0001
12		1.0000	0.9987	0.9790	0.8684	0.5841	0.2277	0.0321	0.0004
13			0.9997	0.9935	0.9423	0.7500	0.3920	0.0867	0.0024
14			1.0000	0.9984	0.9793	0.8744	0.5836	0.1958	0.0113
15				0.9997	0.9941	0.9490	0.7625	0.3704	0.0432
16				1.0000	0.9987	0.9840	0.8929	0.5886	0.1330
17					0.9998	0.9964	0.9645	0.7939	0.3231
18					1.0000	0.9995	0.9924	0.9308	0.6083
19						1.0000	0.9992	0.9885	0.8784
20							1.0000	1.0000	1.0000

n = 25

π:	0.1	0.2	0.3	0.4	0.5	0.6	0.7	0.8	0.9
k: 0	0.0718	0.0038	0.0001	0.0000					
1	0.2712	0.0274	0.0016	0.0001					
2	0.5371	0.0982	0.0090	0.0004	0.0000				
3	0.7636	0.2340	0.0332	0.0024	0.0001				
4	0.9020	0.4207	0.0905	0.0095	0.0005	0.0000			
5	0.9666	0.6167	0.1935	0.0294	0.0020	0.0001			
6	0.9905	0.7800	0.3407	0.0736	0.0073	0.0003			
7	0.9977	0.8909	0.5118	0.1536	0.0216	0.0012	0.0000		
8	0.9995	0.9532	0.6769	0.2735	0.0539	0.0043	0.0001		
9	0.9999	0.9827	0.8106	0.4246	0.1148	0.0132	0.0005		
10	1.0000	0.9944	0.9022	0.5858	0.2122	0.0344	0.0018	0.0000	
11		0.9985	0.9558	0.7323	0.3450	0.0778	0.0060	0.0001	
12		0.9996	0.9825	0.8462	0.5000	0.1538	0.0175	0.0004	
13		0.9999	0.9940	0.9222	0.6550	0.2677	0.0442	0.0015	
14		1.0000	0.9982	0.9656	0.7878	0.4142	0.0978	0.0056	0.0000
15			0.9995	0.9868	0.8852	0.5754	0.1894	0.0173	0.0001
16			0.9999	0.9957	0.9461	0.7265	0.3231	0.0468	0.0005
17			1.0000	0.9988	0.9784	0.8464	0.4882	0.1091	0.0023
18				0.9997	0.9927	0.9264	0.6593	0.2200	0.0095
19				0.9999	0.9980	0.9706	0.8065	0.3833	0.0334
20				1.0000	0.9995	0.9905	0.9095	0.5793	0.0980
21					0.9999	0.9976	0.9668	0.7660	0.2364
22					1.0000	0.9996	0.9910	0.9018	0.4629
23						0.9999	0.9984	0.9726	0.7288
24						1.0000	0.9999	0.9962	0.9282
25							1.0000	1.0000	1.0000

TABLE A.3

Poisson Distribution, Individual Probabilities for x = number of occurrences, prob($x = k$)

λ:	0.1	0.2	0.3	0.4	0.5	0.6	0.7	0.8	0.9	1.0
k: 0	0.9048	0.8187	0.7408	0.6703	0.6065	0.5488	0.4966	0.4493	0.4066	0.3679
1	0.0905	0.1637	0.2222	0.2681	0.3033	0.3293	0.3476	0.3595	0.3659	0.3679
2	0.0045	0.0164	0.0333	0.0536	0.0758	0.0988	0.1217	0.1438	0.1647	0.1839
3	0.0002	0.0011	0.0033	0.0072	0.0126	0.0198	0.0284	0.0383	0.0494	0.0613
4	0.0000	0.0001	0.0003	0.0007	0.0016	0.0030	0.0050	0.0077	0.0111	0.0153
5		0.0000	0.0000	0.0001	0.0002	0.0004	0.0007	0.0012	0.0020	0.0031
6				0.0000	0.0000	0.0000	0.0001	0.0002	0.0003	0.0005
7							0.0000	0.0000	0.0000	0.0001
8										0.0000

λ:	1.1	1.2	1.3	1.4	1.5	1.6	1.7	1.8	1.9	2.0
k: 0	0.3329	0.3012	0.2725	0.2466	0.2231	0.2019	0.1827	0.1653	0.1496	0.1353
1	0.3662	0.3614	0.3543	0.3452	0.3347	0.3230	0.3106	0.2975	0.2842	0.2707
2	0.2014	0.2169	0.2303	0.2417	0.2510	0.2584	0.2640	0.2678	0.2700	0.2707
3	0.0738	0.0867	0.0998	0.1128	0.1255	0.1378	0.1496	0.1607	0.1710	0.1804
4	0.0203	0.0260	0.0324	0.0395	0.0471	0.0551	0.0636	0.0723	0.0812	0.0902
5	0.0045	0.0062	0.0084	0.0111	0.0141	0.0176	0.0216	0.0260	0.0309	0.0361
6	0.0008	0.0012	0.0018	0.0026	0.0035	0.0047	0.0061	0.0078	0.0098	0.0120
7	0.0001	0.0002	0.0003	0.0005	0.0008	0.0011	0.0015	0.0020	0.0027	0.0034
8	0.0000	0.0000	0.0001	0.0001	0.0001	0.0002	0.0003	0.0005	0.0006	0.0009
9			0.0000	0.0000	0.0000	0.0000	0.0001	0.0001	0.0001	0.0002
10							0.0000	0.0000	0.0000	0.0000

λ:	2.1	2.2	2.3	2.4	2.5	2.6	2.7	2.8	2.9	3.0
k: 0	0.1225	0.1108	0.1003	0.0907	0.0821	0.0743	0.0672	0.0608	0.0550	0.0498
1	0.2572	0.2438	0.2306	0.2177	0.2052	0.1931	0.1815	0.1703	0.1596	0.1494
2	0.2700	0.2681	0.2652	0.2613	0.2565	0.2510	0.2450	0.2384	0.2314	0.2240
3	0.1890	0.1966	0.2033	0.2090	0.2138	0.2176	0.2205	0.2225	0.2237	0.2240
4	0.0992	0.1082	0.1169	0.1254	0.1336	0.1414	0.1488	0.1557	0.1622	0.1680
5	0.0417	0.0476	0.0538	0.0602	0.0668	0.0735	0.0804	0.0872	0.0940	0.1008
6	0.0146	0.0174	0.0206	0.0241	0.0278	0.0319	0.0362	0.0407	0.0455	0.0504
7	0.0044	0.0055	0.0068	0.0083	0.0099	0.0118	0.0139	0.0163	0.0188	0.0216
8	0.0011	0.0015	0.0019	0.0025	0.0031	0.0038	0.0047	0.0057	0.0068	0.0081
9	0.0003	0.0004	0.0005	0.0007	0.0009	0.0011	0.0014	0.0018	0.0022	0.0027
10	0.0001	0.0001	0.0001	0.0002	0.0002	0.0003	0.0004	0.0005	0.0006	0.0008
11	0.0000	0.0000	0.0000	0.0000	0.0000	0.0001	0.0001	0.0001	0.0002	0.0002
12						0.0000	0.0000	0.0000	0.0000	0.0001
13										0.0000

Source: Probabilities generated by Minitab, then compiled as shown here.

TABLE A.3

(continued)

λ:	3.1	3.2	3.3	3.4	3.5	3.6	3.7	3.8	3.9	4.0
k: 0	0.0450	0.0408	0.0369	0.0334	0.0302	0.0273	0.0247	0.0224	0.0202	0.0183
1	0.1397	0.1304	0.1217	0.1135	0.1057	0.0984	0.0915	0.0850	0.0789	0.0733
2	0.2165	0.2087	0.2008	0.1929	0.1850	0.1771	0.1692	0.1615	0.1539	0.1465
3	0.2237	0.2226	0.2209	0.2186	0.2158	0.2125	0.2087	0.2046	0.2001	0.1954
4	0.1733	0.1781	0.1823	0.1858	0.1888	0.1912	0.1931	0.1944	0.1951	0.1954
5	0.1075	0.1140	0.1203	0.1264	0.1322	0.1377	0.1429	0.1477	0.1522	0.1563
6	0.0555	0.0608	0.0662	0.0716	0.0771	0.0826	0.0881	0.0936	0.0989	0.1042
7	0.0246	0.0278	0.0312	0.0348	0.0385	0.0425	0.0466	0.0508	0.0551	0.0595
8	0.0095	0.0111	0.0129	0.0148	0.0169	0.0191	0.0215	0.0241	0.0269	0.0298
9	0.0033	0.0040	0.0047	0.0056	0.0066	0.0076	0.0089	0.0102	0.0116	0.0132
10	0.0010	0.0013	0.0016	0.0019	0.0023	0.0028	0.0033	0.0039	0.0045	0.0053
11	0.0003	0.0004	0.0005	0.0006	0.0007	0.0009	0.0011	0.0013	0.0016	0.0019
12	0.0001	0.0001	0.0001	0.0002	0.0002	0.0003	0.0003	0.0004	0.0005	0.0006
13	0.0000	0.0000	0.0000	0.0000	0.0001	0.0001	0.0001	0.0001	0.0002	0.0002
14					0.0000	0.0000	0.0000	0.0000	0.0000	0.0001
15										0.0000

λ:	4.1	4.2	4.3	4.4	4.5	4.6	4.7	4.8	4.9	5.0
k: 0	0.0166	0.0150	0.0136	0.0123	0.0111	0.0101	0.0091	0.0082	0.0074	0.0067
1	0.0679	0.0630	0.0583	0.0540	0.0500	0.0462	0.0427	0.0395	0.0365	0.0337
2	0.1393	0.1323	0.1254	0.1188	0.1125	0.1063	0.1005	0.0948	0.0894	0.0842
3	0.1904	0.1852	0.1798	0.1743	0.1687	0.1631	0.1574	0.1517	0.1460	0.1404
4	0.1951	0.1944	0.1933	0.1917	0.1898	0.1875	0.1849	0.1820	0.1789	0.1755
5	0.1600	0.1633	0.1662	0.1687	0.1708	0.1725	0.1738	0.1747	0.1753	0.1755
6	0.1093	0.1143	0.1191	0.1237	0.1281	0.1323	0.1362	0.1398	0.1432	0.1462
7	0.0640	0.0686	0.0732	0.0778	0.0824	0.0869	0.0914	0.0959	0.1002	0.1044
8	0.0328	0.0360	0.0393	0.0428	0.0463	0.0500	0.0537	0.0575	0.0614	0.0653
9	0.0150	0.0168	0.0188	0.0209	0.0232	0.0255	0.0281	0.0307	0.0334	0.0363
10	0.0061	0.0071	0.0081	0.0092	0.0104	0.0118	0.0132	0.0147	0.0164	0.0181
11	0.0023	0.0027	0.0032	0.0037	0.0043	0.0049	0.0056	0.0064	0.0073	0.0082
12	0.0008	0.0009	0.0011	0.0013	0.0016	0.0019	0.0022	0.0026	0.0030	0.0034
13	0.0002	0.0003	0.0004	0.0005	0.0006	0.0007	0.0008	0.0009	0.0011	0.0013
14	0.0001	0.0001	0.0001	0.0001	0.0002	0.0002	0.0003	0.0003	0.0004	0.0005
15	0.0000	0.0000	0.0000	0.0000	0.0001	0.0001	0.0001	0.0001	0.0001	0.0002
16					0.0000	0.0000	0.0000	0.0000	0.0000	0.0000

λ:	5.1	5.2	5.3	5.4	5.5	5.6	5.7	5.8	5.9	6.0
k: 0	0.0061	0.0055	0.0050	0.0045	0.0041	0.0037	0.0033	0.0030	0.0027	0.0025
1	0.0311	0.0287	0.0265	0.0244	0.0225	0.0207	0.0191	0.0176	0.0162	0.0149
2	0.0793	0.0746	0.0701	0.0659	0.0618	0.0580	0.0544	0.0509	0.0477	0.0446
3	0.1348	0.1293	0.1239	0.1185	0.1133	0.1082	0.1033	0.0985	0.0938	0.0892
4	0.1719	0.1681	0.1641	0.1600	0.1558	0.1515	0.1472	0.1428	0.1383	0.1339
5	0.1753	0.1748	0.1740	0.1728	0.1714	0.1697	0.1678	0.1656	0.1632	0.1606
6	0.1490	0.1515	0.1537	0.1555	0.1571	0.1584	0.1594	0.1601	0.1605	0.1606
7	0.1086	0.1125	0.1163	0.1200	0.1234	0.1267	0.1298	0.1326	0.1353	0.1377
8	0.0692	0.0731	0.0771	0.0810	0.0849	0.0887	0.0925	0.0962	0.0998	0.1033
9	0.0392	0.0423	0.0454	0.0486	0.0519	0.0552	0.0586	0.0620	0.0654	0.0688
10	0.0200	0.0220	0.0241	0.0262	0.0285	0.0309	0.0334	0.0359	0.0386	0.0413
11	0.0093	0.0104	0.0116	0.0129	0.0143	0.0157	0.0173	0.0190	0.0207	0.0225
12	0.0039	0.0045	0.0051	0.0058	0.0065	0.0073	0.0082	0.0092	0.0102	0.0113
13	0.0015	0.0018	0.0021	0.0024	0.0028	0.0032	0.0036	0.0041	0.0046	0.0052
14	0.0006	0.0007	0.0008	0.0009	0.0011	0.0013	0.0015	0.0017	0.0019	0.0022
15	0.0002	0.0002	0.0003	0.0003	0.0004	0.0005	0.0006	0.0007	0.0008	0.0009
16	0.0001	0.0001	0.0001	0.0001	0.0001	0.0002	0.0002	0.0002	0.0003	0.0003
17	0.0000	0.0000	0.0000	0.0000	0.0000	0.0001	0.0001	0.0001	0.0001	0.0001
18						0.0000	0.0000	0.0000	0.0000	0.0000

TABLE A.3
(continued)

λ:	6.1	6.2	6.3	6.4	6.5	6.6	6.7	6.8	6.9	7.0
k: 0	0.0022	0.0020	0.0018	0.0017	0.0015	0.0014	0.0012	0.0011	0.0010	0.0009
1	0.0137	0.0126	0.0116	0.0106	0.0098	0.0090	0.0082	0.0076	0.0070	0.0064
2	0.0417	0.0390	0.0364	0.0340	0.0318	0.0296	0.0276	0.0258	0.0240	0.0223
3	0.0848	0.0806	0.0765	0.0726	0.0688	0.0652	0.0617	0.0584	0.0552	0.0521
4	0.1294	0.1249	0.1205	0.1162	0.1118	0.1076	0.1034	0.0992	0.0952	0.0912
5	0.1579	0.1549	0.1519	0.1487	0.1454	0.1420	0.1385	0.1349	0.1314	0.1277
6	0.1605	0.1601	0.1595	0.1586	0.1575	0.1562	0.1546	0.1529	0.1511	0.1490
7	0.1399	0.1418	0.1435	0.1450	0.1462	0.1472	0.1480	0.1486	0.1489	0.1490
8	0.1066	0.1099	0.1130	0.1160	0.1188	0.1215	0.1240	0.1263	0.1284	0.1304
9	0.0723	0.0757	0.0791	0.0825	0.0858	0.0891	0.0923	0.0954	0.0985	0.1014
10	0.0441	0.0469	0.0498	0.0528	0.0558	0.0588	0.0618	0.0649	0.0679	0.0710
11	0.0244	0.0265	0.0285	0.0307	0.0330	0.0353	0.0377	0.0401	0.0426	0.0452
12	0.0124	0.0137	0.0150	0.0164	0.0179	0.0194	0.0210	0.0227	0.0245	0.0263
13	0.0058	0.0065	0.0073	0.0081	0.0089	0.0099	0.0108	0.0119	0.0130	0.0142
14	0.0025	0.0029	0.0033	0.0037	0.0041	0.0046	0.0052	0.0058	0.0064	0.0071
15	0.0010	0.0012	0.0014	0.0016	0.0018	0.0020	0.0023	0.0026	0.0029	0.0033
16	0.0004	0.0005	0.0005	0.0006	0.0007	0.0008	0.0010	0.0011	0.0013	0.0014
17	0.0001	0.0002	0.0002	0.0002	0.0003	0.0003	0.0004	0.0004	0.0005	0.0006
18	0.0000	0.0001	0.0001	0.0001	0.0001	0.0001	0.0001	0.0002	0.0002	0.0002
19		0.0000	0.0000	0.0000	0.0000	0.0000	0.0001	0.0001	0.0001	0.0001
20							0.0000	0.0000	0.0000	0.0000

λ:	7.1	7.2	7.3	7.4	7.5	7.6	7.7	7.8	7.9	8.0
k: 0	0.0008	0.0007	0.0007	0.0006	0.0006	0.0005	0.0005	0.0004	0.0004	0.0003
1	0.0059	0.0054	0.0049	0.0045	0.0041	0.0038	0.0035	0.0032	0.0029	0.0027
2	0.0208	0.0194	0.0180	0.0167	0.0156	0.0145	0.0134	0.0125	0.0116	0.0107
3	0.0492	0.0464	0.0438	0.0413	0.0389	0.0366	0.0345	0.0324	0.0305	0.0286
4	0.0874	0.0836	0.0799	0.0764	0.0729	0.0696	0.0663	0.0632	0.0602	0.0573
5	0.1241	0.1204	0.1167	0.1130	0.1094	0.1057	0.1021	0.0986	0.0951	0.0916
6	0.1468	0.1445	0.1420	0.1394	0.1367	0.1339	0.1311	0.1282	0.1252	0.1221
7	0.1489	0.1486	0.1481	0.1474	0.1465	0.1454	0.1442	0.1428	0.1413	0.1396
8	0.1321	0.1337	0.1351	0.1363	0.1373	0.1381	0.1388	0.1392	0.1395	0.1396
9	0.1042	0.1070	0.1096	0.1121	0.1144	0.1167	0.1187	0.1207	0.1224	0.1241
10	0.0740	0.0770	0.0800	0.0829	0.0858	0.0887	0.0914	0.0941	0.0967	0.0993
11	0.0478	0.0504	0.0531	0.0558	0.0585	0.0613	0.0640	0.0667	0.0695	0.0722
12	0.0283	0.0303	0.0323	0.0344	0.0366	0.0388	0.0411	0.0434	0.0457	0.0481
13	0.0154	0.0168	0.0181	0.0196	0.0211	0.0227	0.0243	0.0260	0.0278	0.0296
14	0.0078	0.0086	0.0095	0.0104	0.0113	0.0123	0.0134	0.0145	0.0157	0.0169
15	0.0037	0.0041	0.0046	0.0051	0.0057	0.0062	0.0069	0.0075	0.0083	0.0090
16	0.0016	0.0019	0.0021	0.0024	0.0026	0.0030	0.0033	0.0037	0.0041	0.0045
17	0.0007	0.0008	0.0009	0.0010	0.0012	0.0013	0.0015	0.0017	0.0019	0.0021
18	0.0003	0.0003	0.0004	0.0004	0.0005	0.0006	0.0006	0.0007	0.0008	0.0009
19	0.0001	0.0001	0.0001	0.0002	0.0002	0.0002	0.0003	0.0003	0.0003	0.0004
20	0.0000	0.0000	0.0001	0.0001	0.0001	0.0001	0.0001	0.0001	0.0001	0.0002
21			0.0000	0.0000	0.0000	0.0000	0.0000	0.0000	0.0001	0.0001
22									0.0000	0.0000

TABLE A.3
(continued)

λ:	8.1	8.2	8.3	8.4	8.5	8.6	8.7	8.8	8.9	9.0
k: 0	0.0003	0.0003	0.0002	0.0002	0.0002	0.0002	0.0002	0.0002	0.0001	0.0001
1	0.0025	0.0023	0.0021	0.0019	0.0017	0.0016	0.0014	0.0013	0.0012	0.0011
2	0.0100	0.0092	0.0086	0.0079	0.0074	0.0068	0.0063	0.0058	0.0054	0.0050
3	0.0269	0.0252	0.0237	0.0222	0.0208	0.0195	0.0183	0.0171	0.0160	0.0150
4	0.0544	0.0517	0.0491	0.0466	0.0443	0.0420	0.0398	0.0377	0.0357	0.0337
5	0.0882	0.0849	0.0816	0.0784	0.0752	0.0722	0.0692	0.0663	0.0635	0.0607
6	0.1191	0.1160	0.1128	0.1097	0.1066	0.1034	0.1003	0.0972	0.0941	0.0911
7	0.1378	0.1358	0.1338	0.1317	0.1294	0.1271	0.1247	0.1222	0.1197	0.1171
8	0.1395	0.1392	0.1388	0.1382	0.1375	0.1366	0.1356	0.1344	0.1332	0.1318
9	0.1256	0.1269	0.1280	0.1290	0.1299	0.1306	0.1311	0.1315	0.1317	0.1318
10	0.1017	0.1040	0.1063	0.1084	0.1104	0.1123	0.1140	0.1157	0.1172	0.1186
11	0.0749	0.0776	0.0802	0.0828	0.0853	0.0878	0.0902	0.0925	0.0948	0.0970
12	0.0505	0.0530	0.0555	0.0579	0.0604	0.0629	0.0654	0.0679	0.0703	0.0728
13	0.0315	0.0334	0.0354	0.0374	0.0395	0.0416	0.0438	0.0459	0.0481	0.0504
14	0.0182	0.0196	0.0210	0.0225	0.0240	0.0256	0.0272	0.0289	0.0306	0.0324
15	0.0098	0.0107	0.0116	0.0126	0.0136	0.0147	0.0158	0.0169	0.0182	0.0194
16	0.0050	0.0055	0.0060	0.0066	0.0072	0.0079	0.0086	0.0093	0.0101	0.0109
17	0.0024	0.0026	0.0029	0.0033	0.0036	0.0040	0.0044	0.0048	0.0053	0.0058
18	0.0011	0.0012	0.0014	0.0015	0.0017	0.0019	0.0021	0.0024	0.0026	0.0029
19	0.0005	0.0005	0.0006	0.0007	0.0008	0.0009	0.0010	0.0011	0.0012	0.0014
20	0.0002	0.0002	0.0002	0.0003	0.0003	0.0004	0.0004	0.0005	0.0005	0.0006
21	0.0001	0.0001	0.0001	0.0001	0.0001	0.0002	0.0002	0.0002	0.0002	0.0003
22	0.0000	0.0000	0.0000	0.0000	0.0001	0.0001	0.0001	0.0001	0.0001	0.0001
23					0.0000	0.0000	0.0000	0.0000	0.0000	0.0000

λ:	9.1	9.2	9.3	9.4	9.5	9.6	9.7	9.8	9.9	10.0
k: 0	0.0001	0.0001	0.0001	0.0001	0.0001	0.0001	0.0001	0.0001	0.0001	0.0000
1	0.0010	0.0009	0.0009	0.0008	0.0007	0.0007	0.0006	0.0005	0.0005	0.0005
2	0.0046	0.0043	0.0040	0.0037	0.0034	0.0031	0.0029	0.0027	0.0025	0.0023
3	0.0140	0.0131	0.0123	0.0115	0.0107	0.0100	0.0093	0.0087	0.0081	0.0076
4	0.0319	0.0302	0.0285	0.0269	0.0254	0.0240	0.0226	0.0213	0.0201	0.0189
5	0.0581	0.0555	0.0530	0.0506	0.0483	0.0460	0.0439	0.0418	0.0398	0.0378
6	0.0881	0.0851	0.0822	0.0793	0.0764	0.0736	0.0709	0.0682	0.0656	0.0631
7	0.1145	0.1118	0.1091	0.1064	0.1037	0.1010	0.0982	0.0955	0.0928	0.0901
8	0.1302	0.1286	0.1269	0.1251	0.1232	0.1212	0.1191	0.1170	0.1148	0.1126
9	0.1317	0.1315	0.1311	0.1306	0.1300	0.1293	0.1284	0.1274	0.1263	0.1251
10	0.1198	0.1210	0.1219	0.1228	0.1235	0.1241	0.1245	0.1249	0.1250	0.1251
11	0.0991	0.1012	0.1031	0.1049	0.1067	0.1083	0.1098	0.1112	0.1125	0.1137
12	0.0752	0.0776	0.0799	0.0822	0.0844	0.0866	0.0888	0.0908	0.0928	0.0948
13	0.0526	0.0549	0.0572	0.0594	0.0617	0.0640	0.0662	0.0685	0.0707	0.0729
14	0.0342	0.0361	0.0380	0.0399	0.0419	0.0439	0.0459	0.0479	0.0500	0.0521
15	0.0208	0.0221	0.0235	0.0250	0.0265	0.0281	0.0297	0.0313	0.0330	0.0347
16	0.0118	0.0127	0.0137	0.0147	0.0157	0.0168	0.0180	0.0192	0.0204	0.0217
17	0.0063	0.0069	0.0075	0.0081	0.0088	0.0095	0.0103	0.0111	0.0119	0.0128
18	0.0032	0.0035	0.0039	0.0042	0.0046	0.0051	0.0055	0.0060	0.0065	0.0071
19	0.0015	0.0017	0.0019	0.0021	0.0023	0.0026	0.0028	0.0031	0.0034	0.0037
20	0.0007	0.0008	0.0009	0.0010	0.0011	0.0012	0.0014	0.0015	0.0017	0.0019
21	0.0003	0.0003	0.0004	0.0004	0.0005	0.0006	0.0006	0.0007	0.0008	0.0009
22	0.0001	0.0001	0.0002	0.0002	0.0002	0.0002	0.0003	0.0003	0.0004	0.0004
23	0.0000	0.0001	0.0001	0.0001	0.0001	0.0001	0.0001	0.0001	0.0002	0.0002
24		0.0000	0.0000	0.0000	0.0000	0.0000	0.0000	0.0001	0.0001	0.0001
25								0.0000	0.0000	0.0000

TABLE A.4

Poisson Distribution, Cumulative Probabilities for x = number of occurrences, prob($x \le k$)

λ:	0.1	0.2	0.3	0.4	0.5	0.6	0.7	0.8	0.9	1.0
k: 0	0.9048	0.8187	0.7408	0.6703	0.6065	0.5488	0.4966	0.4493	0.4066	0.3679
1	0.9953	0.9825	0.9631	0.9384	0.9098	0.8781	0.8442	0.8088	0.7725	0.7358
2	0.9998	0.9989	0.9964	0.9921	0.9856	0.9769	0.9659	0.9526	0.9371	0.9197
3	1.0000	0.9999	0.9997	0.9992	0.9982	0.9966	0.9942	0.9909	0.9865	0.9810
4		1.0000	1.0000	0.9999	0.9998	0.9996	0.9992	0.9986	0.9977	0.9963
5				1.0000	1.0000	1.0000	0.9999	0.9998	0.9997	0.9994
6							1.0000	1.0000	1.0000	0.9999
7										1.0000

λ:	1.1	1.2	1.3	1.4	1.5	1.6	1.7	1.8	1.9	2.0
k: 0	0.3329	0.3012	0.2725	0.2466	0.2231	0.2019	0.1827	0.1653	0.1496	0.1353
1	0.6990	0.6626	0.6268	0.5918	0.5578	0.5249	0.4932	0.4628	0.4337	0.4060
2	0.9004	0.8795	0.8571	0.8335	0.8088	0.7834	0.7572	0.7306	0.7037	0.6767
3	0.9743	0.9662	0.9569	0.9463	0.9344	0.9212	0.9068	0.8913	0.8747	0.8571
4	0.9946	0.9923	0.9893	0.9857	0.9814	0.9763	0.9704	0.9636	0.9559	0.9473
5	0.9990	0.9985	0.9978	0.9968	0.9955	0.9940	0.9920	0.9896	0.9868	0.9834
6	0.9999	0.9997	0.9996	0.9994	0.9991	0.9987	0.9981	0.9974	0.9966	0.9955
7	1.0000	1.0000	0.9999	0.9999	0.9998	0.9997	0.9996	0.9994	0.9992	0.9989
8			1.0000	1.0000	1.0000	1.0000	0.9999	0.9999	0.9998	0.9998
9							1.0000	1.0000	1.0000	1.0000

λ:	2.1	2.2	2.3	2.4	2.5	2.6	2.7	2.8	2.9	3.0
k: 0	0.1225	0.1108	0.1003	0.0907	0.0821	0.0743	0.0672	0.0608	0.0550	0.0498
1	0.3796	0.3546	0.3309	0.3084	0.2873	0.2674	0.2487	0.2311	0.2146	0.1991
2	0.6496	0.6227	0.5960	0.5697	0.5438	0.5184	0.4936	0.4695	0.4460	0.4232
3	0.8386	0.8194	0.7993	0.7787	0.7576	0.7360	0.7141	0.6919	0.6696	0.6472
4	0.9379	0.9275	0.9162	0.9041	0.8912	0.8774	0.8629	0.8477	0.8318	0.8153
5	0.9796	0.9751	0.9700	0.9643	0.9580	0.9510	0.9433	0.9349	0.9258	0.9161
6	0.9941	0.9925	0.9906	0.9884	0.9858	0.9828	0.9794	0.9756	0.9713	0.9665
7	0.9985	0.9980	0.9974	0.9967	0.9958	0.9947	0.9934	0.9919	0.9901	0.9881
8	0.9997	0.9995	0.9994	0.9991	0.9989	0.9985	0.9981	0.9976	0.9969	0.9962
9	0.9999	0.9999	0.9999	0.9998	0.9997	0.9996	0.9995	0.9993	0.9991	0.9989
10	1.0000	1.0000	1.0000	1.0000	0.9999	0.9999	0.9999	0.9998	0.9998	0.9997
11					1.0000	1.0000	1.0000	1.0000	0.9999	0.9999
12									1.0000	1.0000

λ:	3.1	3.2	3.3	3.4	3.5	3.6	3.7	3.8	3.9	4.0
k: 0	0.0450	0.0408	0.0369	0.0334	0.0302	0.0273	0.0247	0.0224	0.0202	0.0183
1	0.1847	0.1712	0.1586	0.1468	0.1359	0.1257	0.1162	0.1074	0.0992	0.0916
2	0.4012	0.3799	0.3594	0.3397	0.3208	0.3027	0.2854	0.2689	0.2531	0.2381
3	0.6248	0.6025	0.5803	0.5584	0.5366	0.5152	0.4942	0.4735	0.4532	0.4335
4	0.7982	0.7806	0.7626	0.7442	0.7254	0.7064	0.6872	0.6678	0.6484	0.6288
5	0.9057	0.8946	0.8829	0.8705	0.8576	0.8441	0.8301	0.8156	0.8006	0.7851
6	0.9612	0.9554	0.9490	0.9421	0.9347	0.9267	0.9182	0.9091	0.8995	0.8893
7	0.9858	0.9832	0.9802	0.9769	0.9733	0.9692	0.9648	0.9599	0.9546	0.9489
8	0.9953	0.9943	0.9931	0.9917	0.9901	0.9883	0.9863	0.9840	0.9815	0.9786
9	0.9986	0.9982	0.9978	0.9973	0.9967	0.9960	0.9952	0.9942	0.9931	0.9919
10	0.9996	0.9995	0.9994	0.9992	0.9990	0.9987	0.9984	0.9981	0.9977	0.9972
11	0.9999	0.9999	0.9998	0.9998	0.9997	0.9996	0.9995	0.9994	0.9993	0.9991
12	1.0000	1.0000	1.0000	0.9999	0.9999	0.9999	0.9999	0.9998	0.9998	0.9997
13				1.0000	1.0000	1.0000	1.0000	1.0000	0.9999	0.9999
14									1.0000	1.0000

Source: Probabilities generated by Minitab, then compiled as shown here.

TABLE A.4
(continued)

λ:	4.1	4.2	4.3	4.4	4.5	4.6	4.7	4.8	4.9	5.0
k: 0	0.0166	0.0150	0.0136	0.0123	0.0111	0.0101	0.0091	0.0082	0.0074	0.0067
1	0.0845	0.0780	0.0719	0.0663	0.0611	0.0563	0.0518	0.0477	0.0439	0.0404
2	0.2238	0.2102	0.1974	0.1851	0.1736	0.1626	0.1523	0.1425	0.1333	0.1247
3	0.4142	0.3954	0.3772	0.3594	0.3423	0.3257	0.3097	0.2942	0.2793	0.2650
4	0.6093	0.5898	0.5704	0.5512	0.5321	0.5132	0.4946	0.4763	0.4582	0.4405
5	0.7693	0.7531	0.7367	0.7199	0.7029	0.6858	0.6684	0.6510	0.6335	0.6160
6	0.8786	0.8675	0.8558	0.8436	0.8311	0.8180	0.8046	0.7908	0.7767	0.7622
7	0.9427	0.9361	0.9290	0.9214	0.9134	0.9049	0.8960	0.8867	0.8769	0.8666
8	0.9755	0.9721	0.9683	0.9642	0.9597	0.9549	0.9497	0.9442	0.9382	0.9319
9	0.9905	0.9889	0.9871	0.9851	0.9829	0.9805	0.9778	0.9749	0.9717	0.9682
10	0.9966	0.9959	0.9952	0.9943	0.9933	0.9922	0.9910	0.9896	0.9880	0.9863
11	0.9989	0.9986	0.9983	0.9980	0.9976	0.9971	0.9966	0.9960	0.9953	0.9945
12	0.9997	0.9996	0.9995	0.9993	0.9992	0.9990	0.9988	0.9986	0.9983	0.9980
13	0.9999	0.9999	0.9998	0.9998	0.9997	0.9997	0.9996	0.9995	0.9994	0.9993
14	1.0000	1.0000	1.0000	0.9999	0.9999	0.9999	0.9999	0.9999	0.9998	0.9998
15				1.0000	1.0000	1.0000	1.0000	1.0000	0.9999	0.9999
16									1.0000	1.0000

λ:	5.1	5.2	5.3	5.4	5.5	5.6	5.7	5.8	5.9	6.0
k: 0	0.0061	0.0055	0.0050	0.0045	0.0041	0.0037	0.0033	0.0030	0.0027	0.0025
1	0.0372	0.0342	0.0314	0.0289	0.0266	0.0244	0.0224	0.0206	0.0189	0.0174
2	0.1165	0.1088	0.1016	0.0948	0.0884	0.0824	0.0768	0.0715	0.0666	0.0620
3	0.2513	0.2381	0.2254	0.2133	0.2017	0.1906	0.1800	0.1700	0.1604	0.1512
4	0.4231	0.4061	0.3895	0.3733	0.3575	0.3422	0.3272	0.3127	0.2987	0.2851
5	0.5984	0.5809	0.5635	0.5461	0.5289	0.5119	0.4950	0.4783	0.4619	0.4457
6	0.7474	0.7324	0.7171	0.7017	0.6860	0.6703	0.6544	0.6384	0.6224	0.6063
7	0.8560	0.8449	0.8335	0.8217	0.8095	0.7970	0.7841	0.7710	0.7576	0.7440
8	0.9252	0.9181	0.9106	0.9027	0.8944	0.8857	0.8766	0.8672	0.8574	0.8472
9	0.9644	0.9603	0.9559	0.9512	0.9462	0.9409	0.9352	0.9292	0.9228	0.9161
10	0.9844	0.9823	0.9800	0.9775	0.9747	0.9718	0.9686	0.9651	0.9614	0.9574
11	0.9937	0.9927	0.9916	0.9904	0.9890	0.9875	0.9859	0.9841	0.9821	0.9799
12	0.9976	0.9972	0.9967	0.9962	0.9955	0.9949	0.9941	0.9932	0.9922	0.9912
13	0.9992	0.9990	0.9988	0.9986	0.9983	0.9980	0.9977	0.9973	0.9969	0.9964
14	0.9997	0.9997	0.9996	0.9995	0.9994	0.9993	0.9991	0.9990	0.9988	0.9986
15	0.9999	0.9999	0.9999	0.9998	0.9998	0.9998	0.9997	0.9996	0.9996	0.9995
16	1.0000	1.0000	1.0000	0.9999	0.9999	0.9999	0.9999	0.9999	0.9999	0.9998
17				1.0000	1.0000	1.0000	1.0000	1.0000	1.0000	0.9999
18										1.0000

TABLE A.4
(continued)

λ:	6.1	6.2	6.3	6.4	6.5	6.6	6.7	6.8	6.9	7.0
k: 0	0.0022	0.0020	0.0018	0.0017	0.0015	0.0014	0.0012	0.0011	0.0010	0.0009
1	0.0159	0.0146	0.0134	0.0123	0.0113	0.0103	0.0095	0.0087	0.0080	0.0073
2	0.0577	0.0536	0.0498	0.0463	0.0430	0.0400	0.0371	0.0344	0.0320	0.0296
3	0.1425	0.1342	0.1264	0.1189	0.1118	0.1052	0.0988	0.0928	0.0871	0.0818
4	0.2719	0.2592	0.2469	0.2351	0.2237	0.2127	0.2022	0.1920	0.1823	0.1730
5	0.4298	0.4141	0.3988	0.3837	0.3690	0.3547	0.3406	0.3270	0.3137	0.3007
6	0.5902	0.5742	0.5582	0.5423	0.5265	0.5108	0.4953	0.4799	0.4647	0.4497
7	0.7301	0.7160	0.7017	0.6873	0.6728	0.6581	0.6433	0.6285	0.6136	0.5987
8	0.8367	0.8259	0.8148	0.8033	0.7916	0.7796	0.7673	0.7548	0.7420	0.7291
9	0.9090	0.9016	0.8939	0.8858	0.8774	0.8686	0.8596	0.8502	0.8405	0.8305
10	0.9531	0.9486	0.9437	0.9386	0.9332	0.9274	0.9214	0.9151	0.9084	0.9015
11	0.9776	0.9750	0.9723	0.9693	0.9661	0.9627	0.9591	0.9552	0.9510	0.9467
12	0.9900	0.9887	0.9873	0.9857	0.9840	0.9821	0.9801	0.9779	0.9755	0.9730
13	0.9958	0.9952	0.9945	0.9937	0.9929	0.9920	0.9909	0.9898	0.9885	0.9872
14	0.9984	0.9981	0.9978	0.9974	0.9970	0.9966	0.9961	0.9956	0.9950	0.9943
15	0.9994	0.9993	0.9992	0.9990	0.9988	0.9986	0.9984	0.9982	0.9979	0.9976
16	0.9998	0.9997	0.9997	0.9996	0.9996	0.9995	0.9994	0.9993	0.9992	0.9990
17	0.9999	0.9999	0.9999	0.9999	0.9998	0.9998	0.9998	0.9997	0.9997	0.9996
18	1.0000	1.0000	1.0000	1.0000	0.9999	0.9999	0.9999	0.9999	0.9999	0.9999
19					1.0000	1.0000	1.0000	1.0000	1.0000	1.0000

λ:	7.1	7.2	7.3	7.4	7.5	7.6	7.7	7.8	7.9	8.0
k: 0	0.0008	0.0007	0.0007	0.0006	0.0006	0.0005	0.0005	0.0004	0.0004	0.0003
1	0.0067	0.0061	0.0056	0.0051	0.0047	0.0043	0.0039	0.0036	0.0033	0.0030
2	0.0275	0.0255	0.0236	0.0219	0.0203	0.0188	0.0174	0.0161	0.0149	0.0138
3	0.0767	0.0719	0.0674	0.0632	0.0591	0.0554	0.0518	0.0485	0.0453	0.0424
4	0.1641	0.1555	0.1473	0.1395	0.1321	0.1249	0.1181	0.1117	0.1055	0.0996
5	0.2881	0.2759	0.2640	0.2526	0.2414	0.2307	0.2203	0.2103	0.2006	0.1912
6	0.4349	0.4204	0.4060	0.3920	0.3782	0.3646	0.3514	0.3384	0.3257	0.3134
7	0.5838	0.5689	0.5541	0.5393	0.5246	0.5100	0.4956	0.4812	0.4670	0.4530
8	0.7160	0.7027	0.6892	0.6757	0.6620	0.6482	0.6343	0.6204	0.6065	0.5925
9	0.8202	0.8096	0.7988	0.7877	0.7764	0.7649	0.7531	0.7411	0.7290	0.7166
10	0.8942	0.8867	0.8788	0.8707	0.8622	0.8535	0.8445	0.8352	0.8257	0.8159
11	0.9420	0.9371	0.9319	0.9265	0.9208	0.9148	0.9085	0.9020	0.8952	0.8881
12	0.9703	0.9673	0.9642	0.9609	0.9573	0.9536	0.9496	0.9454	0.9409	0.9362
13	0.9857	0.9841	0.9824	0.9805	0.9784	0.9762	0.9739	0.9714	0.9687	0.9658
14	0.9935	0.9927	0.9918	0.9908	0.9897	0.9886	0.9873	0.9859	0.9844	0.9827
15	0.9972	0.9969	0.9964	0.9959	0.9954	0.9948	0.9941	0.9934	0.9926	0.9918
16	0.9989	0.9987	0.9985	0.9983	0.9980	0.9978	0.9974	0.9971	0.9967	0.9963
17	0.9996	0.9995	0.9994	0.9993	0.9992	0.9991	0.9989	0.9988	0.9986	0.9984
18	0.9998	0.9998	0.9998	0.9997	0.9997	0.9996	0.9996	0.9995	0.9994	0.9993
19	0.9999	0.9999	0.9999	0.9999	0.9999	0.9999	0.9998	0.9998	0.9998	0.9997
20	1.0000	1.0000	1.0000	1.0000	1.0000	1.0000	0.9999	0.9999	0.9999	0.9999
21							1.0000	1.0000	1.0000	1.0000

TABLE A.4
(continued)

λ:	8.1	8.2	8.3	8.4	8.5	8.6	8.7	8.8	8.9	9.0
k: 0	0.0003	0.0003	0.0002	0.0002	0.0002	0.0002	0.0002	0.0002	0.0001	0.0001
1	0.0028	0.0025	0.0023	0.0021	0.0019	0.0018	0.0016	0.0015	0.0014	0.0012
2	0.0127	0.0118	0.0109	0.0100	0.0093	0.0086	0.0079	0.0073	0.0068	0.0062
3	0.0396	0.0370	0.0346	0.0323	0.0301	0.0281	0.0262	0.0244	0.0228	0.0212
4	0.0940	0.0887	0.0837	0.0789	0.0744	0.0701	0.0660	0.0621	0.0584	0.0550
5	0.1822	0.1736	0.1653	0.1573	0.1496	0.1422	0.1352	0.1284	0.1219	0.1157
6	0.3013	0.2896	0.2781	0.2670	0.2562	0.2457	0.2355	0.2256	0.2160	0.2068
7	0.4391	0.4254	0.4119	0.3987	0.3856	0.3728	0.3602	0.3478	0.3357	0.3239
8	0.5786	0.5647	0.5507	0.5369	0.5231	0.5094	0.4958	0.4823	0.4689	0.4557
9	0.7041	0.6915	0.6788	0.6659	0.6530	0.6400	0.6269	0.6137	0.6006	0.5874
10	0.8058	0.7955	0.7850	0.7743	0.7634	0.7522	0.7409	0.7294	0.7178	0.7060
11	0.8807	0.8731	0.8652	0.8571	0.8487	0.8400	0.8311	0.8220	0.8126	0.8030
12	0.9313	0.9261	0.9207	0.9150	0.9091	0.9029	0.8965	0.8898	0.8829	0.8758
13	0.9628	0.9595	0.9561	0.9524	0.9486	0.9445	0.9403	0.9358	0.9311	0.9261
14	0.9810	0.9791	0.9771	0.9749	0.9726	0.9701	0.9675	0.9647	0.9617	0.9585
15	0.9908	0.9898	0.9887	0.9875	0.9862	0.9848	0.9832	0.9816	0.9798	0.9780
16	0.9958	0.9953	0.9947	0.9941	0.9934	0.9926	0.9918	0.9909	0.9899	0.9889
17	0.9982	0.9979	0.9977	0.9973	0.9970	0.9966	0.9962	0.9957	0.9952	0.9947
18	0.9992	0.9991	0.9990	0.9989	0.9987	0.9985	0.9983	0.9981	0.9978	0.9976
19	0.9997	0.9997	0.9996	0.9995	0.9995	0.9994	0.9993	0.9992	0.9991	0.9989
20	0.9999	0.9999	0.9998	0.9998	0.9998	0.9998	0.9997	0.9997	0.9996	0.9996
21	1.0000	1.0000	0.9999	0.9999	0.9999	0.9999	0.9999	0.9999	0.9998	0.9998
22			1.0000	1.0000	1.0000	1.0000	1.0000	1.0000	0.9999	0.9999
23									1.0000	1.0000

λ:	9.1	9.2	9.3	9.4	9.5	9.6	9.7	9.8	9.9	10.0
k: 0	0.0001	0.0001	0.0001	0.0001	0.0001	0.0001	0.0001	0.0001	0.0001	0.0000
1	0.0011	0.0010	0.0009	0.0009	0.0008	0.0007	0.0007	0.0006	0.0005	0.0005
2	0.0058	0.0053	0.0049	0.0045	0.0042	0.0038	0.0035	0.0033	0.0030	0.0028
3	0.0198	0.0184	0.0172	0.0160	0.0149	0.0138	0.0129	0.0120	0.0111	0.0103
4	0.0517	0.0486	0.0456	0.0429	0.0403	0.0378	0.0355	0.0333	0.0312	0.0293
5	0.1098	0.1041	0.0986	0.0935	0.0885	0.0838	0.0793	0.0750	0.0710	0.0671
6	0.1978	0.1892	0.1808	0.1727	0.1649	0.1574	0.1502	0.1433	0.1366	0.1301
7	0.3123	0.3010	0.2900	0.2792	0.2687	0.2584	0.2485	0.2388	0.2294	0.2202
8	0.4426	0.4296	0.4168	0.4042	0.3918	0.3796	0.3676	0.3558	0.3442	0.3328
9	0.5742	0.5611	0.5479	0.5349	0.5218	0.5089	0.4960	0.4832	0.4705	0.4579
10	0.6941	0.6820	0.6699	0.6576	0.6453	0.6329	0.6205	0.6080	0.5955	0.5830
11	0.7932	0.7832	0.7730	0.7626	0.7520	0.7412	0.7303	0.7193	0.7081	0.6968
12	0.8684	0.8607	0.8529	0.8448	0.8364	0.8279	0.8191	0.8101	0.8009	0.7916
13	0.9210	0.9156	0.9100	0.9042	0.8981	0.8919	0.8853	0.8786	0.8716	0.8645
14	0.9552	0.9517	0.9480	0.9441	0.9400	0.9357	0.9312	0.9265	0.9216	0.9165
15	0.9760	0.9738	0.9715	0.9691	0.9665	0.9638	0.9609	0.9579	0.9546	0.9513
16	0.9878	0.9865	0.9852	0.9838	0.9823	0.9806	0.9789	0.9770	0.9751	0.9730
17	0.9941	0.9934	0.9927	0.9919	0.9911	0.9902	0.9892	0.9881	0.9870	0.9857
18	0.9973	0.9969	0.9966	0.9962	0.9957	0.9952	0.9947	0.9941	0.9935	0.9928
19	0.9988	0.9986	0.9985	0.9983	0.9980	0.9978	0.9975	0.9972	0.9969	0.9965
20	0.9995	0.9994	0.9993	0.9992	0.9991	0.9990	0.9989	0.9987	0.9986	0.9984
21	0.9998	0.9998	0.9997	0.9997	0.9996	0.9996	0.9995	0.9995	0.9994	0.9993
22	0.9999	0.9999	0.9999	0.9999	0.9999	0.9998	0.9998	0.9998	0.9997	0.9997
23	1.0000	1.0000	1.0000	1.0000	0.9999	0.9999	0.9999	0.9999	0.9999	0.9999
24					1.0000	1.0000	1.0000	1.0000	1.0000	1.0000

TABLE A.5

Table of Random Digits

Row	1–10		11–20		21–30		31–40		41–50	
1	37220	84861	59998	77311	87754	04899	75699	22423	01200	79606
2	31618	06840	45167	13747	74382	39250	37038	42810	81183	27226
3	53778	71813	50306	47415	21682	74984	69038	80269	18955	47908
4	82652	58162	90352	10751	49968	06611	23573	29523	78542	49685
5	40372	53109	76995	24681	42711	06899	14208	63897	60834	46618
6	24952	30711	22535	51397	37277	04343	84920	11970	28098	41367
7	94953	96367	87143	71467	31194	39897	56605	21810	72064	11296
8	98972	12203	90759	56157	40644	16283	80816	07379	87171	89016
9	61759	32900	05299	56687	92355	11684	23278	82492	00366	81502
10	14338	44401	22630	06918	86157	02229	68995	88455	35210	84775
11	03990	64484	50620	04134	04132	35323	39948	53446	13150	01522
12	04003	21357	21560	26820	97543	15607	75220	27896	29722	45389
13	75942	36578	85584	55496	88476	01255	75094	26800	78580	86252
14	29945	17580	17770	56883	90165	54800	47262	14471	66496	66618
15	05675	07093	47456	58942	13685	58422	50987	46420	60965	07193
16	38770	60956	63383	54619	28131	43657	25686	30869	44495	68087
17	93578	84503	84911	86482	45267	23662	59118	35926	11854	12602
18	54526	35249	68501	02206	77793	34752	67528	51706	94351	65383
19	66019	99264	83410	16479	55742	75660	24750	16698	15630	33348
20	30995	81658	22563	34561	29004	64725	80137	97704	63540	29097
21	54873	00592	16042	46776	31730	87258	15861	03115	95167	17410
22	69350	23570	89681	00936	22521	88911	10908	04029	21081	42224
23	43775	21148	77549	86081	02143	29803	17758	81909	09509	64255
24	94238	08655	67935	26234	17011	44426	18663	93938	12767	38048
25	90329	65940	30771	99446	08749	86843	72979	03701	72777	17442
26	49606	51896	13569	58172	37655	22835	26546	08220	22221	62889
27	85876	62869	92898	06662	89382	20703	70312	34032	40788	07235
28	99021	17433	31031	92399	28930	46651	81288	32707	50799	59045
29	43672	46199	83184	90585	32998	82557	82994	70297	79910	86498
30	77915	14650	48304	10282	16356	50667	23469	12116	42141	20876
31	62815	20732	36276	19416	09552	52494	85690	61579	99879	80023
32	35431	28351	10559	52307	49924	97564	68685	49850	62185	36563
33	26212	06143	84840	76247	08208	68191	73511	51271	28635	33390
34	03772	70603	28430	72720	70275	87396	54825	60139	48572	90762
35	22831	69776	07506	32013	51906	49757	41867	15688	53729	44526
36	91527	15810	88468	47664	28231	75104	24370	63076	53040	01302
37	98889	36471	23606	86428	36786	79805	17891	92903	52891	03626
38	80596	42027	72309	71805	39628	59368	15410	41685	05732	87997
39	06160	85702	50617	52953	85886	99013	52135	35314	13276	64765
40	07145	80861	37672	03747	76227	52880	65131	92937	32659	44959
41	25756	94616	75611	70478	82600	89252	99473	75367	36786	08165
42	31698	84269	82241	05714	68314	46971	27440	12794	51833	88879
43	65292	54687	65808	38799	44635	83610	83257	79134	29222	11974
44	55947	27034	40437	52112	94440	36443	89401	58176	74072	57605
45	21916	48584	21519	80149	38375	27493	94284	94171	65668	45875
46	38532	95321	76039	97791	57165	33428	37024	73345	32198	48764
47	33462	61546	54421	41760	68011	47564	77265	91710	89205	04352
48	13526	46548	09115	99063	45898	59162	70038	88042	85648	47177
49	34828	14870	70575	24312	79631	75697	72404	31103	00016	42082
50	89895	82390	89421	56029	61854	05548	21777	51553	17064	33859

Source: Random digits generated by Minitab.

TABLE A.5

(continued)

Row	1–10		11–20		21–30		31–40		41–50	
51	96018	27032	35948	81017	87632	68779	98468	82882	58835	61583
52	08657	49477	20792	55367	27957	21669	25807	01687	06853	87214
53	05183	80239	20082	32806	84896	36574	99899	02874	58987	07423
54	30558	18243	96761	22184	05203	62794	49547	62406	82189	38127
55	85574	37307	36048	72513	49070	82728	86719	96008	85102	66717
56	30554	02449	74975	54343	32643	06124	18521	63560	51187	23974
57	08408	15911	13377	74630	53089	11912	93262	10961	33576	60466
58	62944	23357	49036	41265	88109	02667	75931	89523	53776	14953
59	93160	32676	91167	63054	46459	76406	02739	44198	16693	85504
60	99581	93196	04941	28933	30480	81685	14525	34978	68211	18591
61	71317	85120	37427	57745	78399	49835	67713	10736	70576	81235
62	88100	72776	19450	63109	23254	45376	89233	40445	09325	73602
63	70782	68998	78042	28590	94522	49559	77357	16808	45458	35337
64	35340	66670	34011	47582	58480	32166	22558	69670	72272	26486
65	96757	46833	55325	73157	57438	18733	95369	55382	97468	19294
66	86797	63659	61773	64818	35218	19597	21190	18369	84201	85388
67	69259	17468	69838	56975	02929	21112	16636	61211	15809	48807
68	83837	17354	59419	24965	90989	83896	39712	78988	95621	81415
69	43638	54811	63054	08267	02125	17892	50132	01293	90524	30664
70	67436	14892	65654	88989	63574	39532	26906	72282	03567	54279
71	08336	51472	04056	15208	42517	97140	87426	55647	76220	70525
72	72678	31240	47192	49329	64702	25685	72651	25217	65463	65810
73	13063	46035	35899	22682	86499	28475	89917	11856	11202	45920
74	70022	46043	37033	04781	53014	68215	46567	83996	51586	00305
75	72803	61399	52929	75693	38989	14261	39257	38112	24227	15175
76	11566	26680	54965	42491	37382	11800	49989	96604	71890	62676
77	63085	07137	63898	54353	44912	00010	04212	51192	86710	38272
78	43029	84646	21389	71859	85939	95973	79164	23010	21267	39741
79	25172	64085	98417	45034	55105	15375	50465	75316	92495	93023
80	22565	88294	35872	45035	69125	46818	73424	10302	89830	66504
81	08342	34507	14599	76497	42542	52684	56830	80974	79301	25622
82	32972	02526	37798	66645	79124	60382	84256	55650	85417	83653
83	67256	76680	09063	90514	16663	29278	75623	96264	96627	21865
84	32925	24366	96567	21296	53412	46323	35922	56796	82847	43730
85	61270	37743	04870	55927	32722	51403	88391	35180	18657	14786
86	49847	35977	16510	95436	68933	06504	13832	58070	25616	00710
87	71292	25605	48086	25880	62403	90032	37255	49115	91855	85141
88	00829	65292	74516	37134	04001	09070	74628	70384	14075	85153
89	59329	18382	27521	73483	37372	94615	89458	56168	99200	58918
90	33251	02545	18701	39495	48746	23063	44545	99331	37707	11114
91	12488	88006	10476	66065	31729	75517	76418	49074	46531	71802
92	87379	02837	55836	46426	04337	20093	73444	93438	98875	92489
93	87793	28895	36704	80744	60137	62342	61414	31190	37768	17153
94	20540	32139	16496	72337	38768	82413	02398	89143	54376	16104
95	34086	44938	99218	17223	92393	39053	93918	56496	28382	77369
96	10403	83656	78289	65316	56677	99757	51479	03639	05957	62608
97	37124	75750	63234	00172	17453	87288	99592	98549	30527	03473
98	51218	08866	01031	39402	97757	19459	19458	52282	34979	82757
99	95060	16896	75433	40816	50161	87246	69040	64414	14555	27653
100	68291	35003	94288	95139	60876	61419	00615	49678	68123	55444

TABLE A.6

Critical Values of *F*

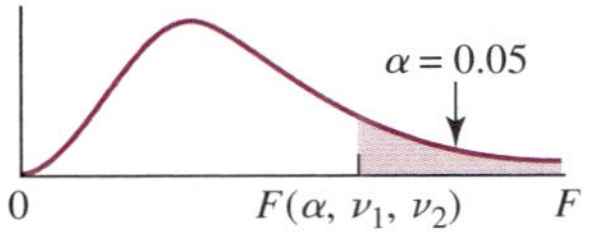

$v_1 = df$, numerator

$v_2 = df$, denominator	1	2	3	4	5	6	7	8	9	10	12	15	20	24	30	40	60	120	∞
1	161.4	199.5	215.7	224.6	230.2	234.0	236.8	238.9	240.5	241.9	243.9	245.9	248.0	249.1	250.1	251.1	252.2	253.3	254.3
2	18.51	19.00	19.16	19.25	19.30	19.33	19.35	19.37	19.38	19.40	19.41	19.43	19.45	19.45	19.46	19.47	19.48	19.49	19.50
3	10.13	9.55	9.28	9.12	9.01	8.94	8.89	8.85	8.81	8.79	8.74	8.70	8.66	8.64	8.62	8.59	8.57	8.55	8.53
4	7.71	6.94	6.59	6.39	6.26	6.16	6.09	6.04	6.00	5.96	5.91	5.86	5.80	5.77	5.75	5.72	5.69	5.66	5.63
5	6.61	5.79	5.41	5.19	5.05	4.95	4.88	4.82	4.77	4.74	4.68	4.62	4.56	4.53	4.50	4.46	4.43	4.40	4.36
6	5.99	5.14	4.76	4.53	4.39	4.28	4.21	4.15	4.10	4.06	4.00	3.94	3.87	3.84	3.81	3.77	3.74	3.70	3.67
7	5.59	4.74	4.35	4.12	3.97	3.87	3.79	3.73	3.68	3.64	3.57	3.51	3.44	3.41	3.38	3.34	3.30	3.27	3.23
8	5.32	4.46	4.07	3.84	3.69	3.58	3.50	3.44	3.39	3.35	3.28	3.22	3.15	3.12	3.08	3.04	3.01	2.97	2.93
9	5.12	4.26	3.86	3.63	3.48	3.37	3.29	3.23	3.18	3.14	3.07	3.01	2.94	2.90	2.86	2.83	2.79	2.75	2.71
10	4.96	4.10	3.71	3.48	3.33	3.22	3.14	3.07	3.02	2.98	2.91	2.85	2.77	2.74	2.70	2.66	2.62	2.58	2.54
11	4.84	3.98	3.59	3.36	3.20	3.09	3.01	2.95	2.90	2.85	2.79	2.72	2.65	2.61	2.57	2.53	2.49	2.45	2.40
12	4.75	3.89	3.49	3.26	3.11	3.00	2.91	2.85	2.80	2.75	2.69	2.62	2.54	2.51	2.47	2.43	2.38	2.34	2.30
13	4.67	3.81	3.41	3.18	3.03	2.92	2.83	2.77	2.71	2.67	2.60	2.53	2.46	2.42	2.38	2.34	2.30	2.25	2.21
14	4.60	3.74	3.34	3.11	2.96	2.85	2.76	2.70	2.65	2.60	2.53	2.46	2.39	2.35	2.31	2.27	2.22	2.18	2.13
15	4.54	3.68	3.29	3.06	2.90	2.79	2.71	2.64	2.59	2.54	2.48	2.40	2.33	2.29	2.25	2.20	2.16	2.11	2.07
16	4.49	3.63	3.24	3.01	2.85	2.74	2.66	2.59	2.54	2.49	2.42	2.35	2.28	2.24	2.19	2.15	2.11	2.06	2.01
17	4.45	3.59	3.20	2.96	2.81	2.70	2.61	2.55	2.49	2.45	2.38	2.31	2.23	2.19	2.15	2.10	2.06	2.01	1.96
18	4.41	3.55	3.16	2.93	2.77	2.66	2.58	2.51	2.46	2.41	2.34	2.27	2.19	2.15	2.11	2.06	2.02	1.97	1.92
19	4.38	3.52	3.13	2.90	2.74	2.63	2.54	2.48	2.42	2.38	2.31	2.23	2.16	2.11	2.07	2.03	1.98	1.93	1.88
20	4.35	3.49	3.10	2.87	2.71	2.60	2.51	2.45	2.39	2.35	2.28	2.20	2.12	2.08	2.04	1.99	1.95	1.90	1.84
21	4.32	3.47	3.07	2.84	2.68	2.57	2.49	2.42	2.37	2.32	2.25	2.18	2.10	2.05	2.01	1.96	1.92	1.87	1.81
22	4.30	3.44	3.05	2.82	2.66	2.55	2.46	2.40	2.34	2.30	2.23	2.15	2.07	2.03	1.98	1.94	1.89	1.84	1.78
23	4.28	3.42	3.03	2.80	2.64	2.53	2.44	2.37	2.32	2.27	2.20	2.13	2.05	2.01	1.96	1.91	1.86	1.81	1.76
24	4.26	3.40	3.01	2.78	2.62	2.51	2.42	2.36	2.30	2.25	2.18	2.11	2.03	1.98	1.94	1.89	1.84	1.79	1.73
25	4.24	3.39	2.99	2.76	2.60	2.49	2.40	2.34	2.28	2.24	2.16	2.09	2.01	1.96	1.92	1.87	1.82	1.77	1.71
26	4.23	3.37	2.98	2.74	2.59	2.47	2.39	2.32	2.27	2.22	2.15	2.07	1.99	1.95	1.90	1.85	1.80	1.75	1.69
27	4.21	3.35	2.96	2.73	2.57	2.46	2.37	2.31	2.25	2.20	2.13	2.06	1.97	1.93	1.88	1.84	1.79	1.73	1.67
28	4.20	3.34	2.95	2.71	2.56	2.45	2.36	2.29	2.24	2.19	2.12	2.04	1.96	1.91	1.87	1.82	1.77	1.71	1.65
29	4.18	3.33	2.93	2.70	2.55	2.43	2.35	2.28	2.22	2.18	2.10	2.03	1.94	1.90	1.85	1.81	1.75	1.70	1.64
30	4.17	3.32	2.92	2.69	2.53	2.42	2.33	2.27	2.21	2.16	2.09	2.01	1.93	1.89	1.84	1.79	1.74	1.68	1.62
40	4.08	3.23	2.84	2.61	2.45	2.34	2.25	2.18	2.12	2.08	2.00	1.92	1.84	1.79	1.74	1.69	1.64	1.58	1.51
60	4.00	3.15	2.76	2.53	2.37	2.25	2.17	2.10	2.04	1.99	1.92	1.84	1.75	1.70	1.65	1.59	1.53	1.47	1.39
120	3.92	3.07	2.68	2.45	2.29	2.17	2.09	2.02	1.96	1.91	1.83	1.75	1.66	1.61	1.55	1.50	1.43	1.35	1.25
∞	3.84	3.00	2.60	2.37	2.21	2.10	2.01	1.94	1.88	1.83	1.75	1.67	1.57	1.52	1.46	1.39	1.32	1.22	1.00

TABLE A.6

(continued)

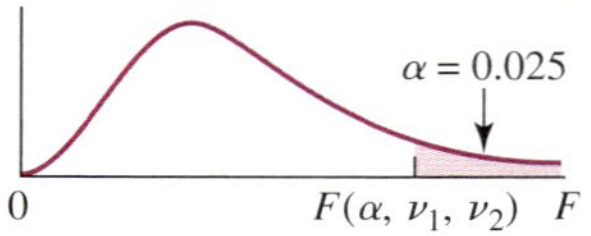

$v_2 = df$, denominator	$v_1 = df$, numerator: 1	2	3	4	5	6	7	8	9	10	12	15	20	24	30	40	60	120	∞
1	647.8	799.5	864.2	899.6	921.8	937.1	948.2	956.7	963.3	968.6	976.7	984.9	993.1	997.2	1001	1006	1010	1014	1018
2	38.51	39.00	39.17	39.25	39.30	39.33	39.36	39.37	39.39	39.40	39.41	39.43	39.45	39.46	39.46	39.47	39.48	39.49	39.50
3	17.44	16.04	15.44	15.10	14.88	14.73	14.62	14.54	14.47	14.42	14.34	14.25	14.17	14.12	14.08	14.04	13.99	13.95	13.90
4	12.22	10.65	9.98	9.60	9.36	9.20	9.07	8.98	8.90	8.84	8.75	8.66	8.56	8.51	8.46	8.41	8.36	8.31	8.26
5	10.01	8.43	7.76	7.39	7.15	6.98	6.85	6.76	6.68	6.62	6.52	6.43	6.33	6.28	6.23	6.18	6.12	6.07	6.02
6	8.81	7.26	6.60	6.23	5.99	5.82	5.70	5.60	5.52	5.46	5.37	5.27	5.17	5.12	5.07	5.01	4.96	4.90	4.85
7	8.07	6.54	5.89	5.52	5.29	5.12	4.99	4.90	4.82	4.76	4.67	4.57	4.47	4.42	4.36	4.31	4.25	4.20	4.14
8	7.57	6.06	5.42	5.05	4.82	4.65	4.53	4.43	4.36	4.30	4.20	4.10	4.00	3.95	3.89	3.84	3.78	3.73	3.67
9	7.21	5.71	5.08	4.72	4.48	4.32	4.20	4.10	4.03	3.96	3.87	3.77	3.67	3.61	3.56	3.51	3.45	3.39	3.33
10	6.94	5.46	4.83	4.47	4.24	4.07	3.95	3.85	3.78	3.72	3.62	3.52	3.42	3.37	3.31	3.26	3.20	3.14	3.08
11	6.72	5.26	4.63	4.28	4.04	3.88	3.76	3.66	3.59	3.53	3.43	3.33	3.23	3.17	3.12	3.06	3.00	2.94	2.88
12	6.55	5.10	4.47	4.12	3.89	3.73	3.61	3.51	3.44	3.37	3.28	3.18	3.07	3.02	2.96	2.91	2.85	2.79	2.72
13	6.41	4.97	4.35	4.00	3.77	3.60	3.48	3.39	3.31	3.25	3.15	3.05	2.95	2.89	2.84	2.78	2.72	2.66	2.60
14	6.30	4.86	4.24	3.89	3.66	3.50	3.38	3.29	3.21	3.15	3.05	2.95	2.84	2.79	2.73	2.67	2.61	2.55	2.49
15	6.20	4.77	4.15	3.80	3.58	3.41	3.29	3.20	3.12	3.06	2.96	2.86	2.76	2.70	2.64	2.59	2.52	2.46	2.40
16	6.12	4.69	4.08	3.73	3.50	3.34	3.22	3.12	3.05	2.99	2.89	2.79	2.68	2.63	2.57	2.51	2.45	2.38	2.32
17	6.04	4.62	4.01	3.66	3.44	3.28	3.16	3.06	2.98	2.92	2.82	2.72	2.62	2.56	2.50	2.44	2.38	2.32	2.25
18	5.98	4.56	3.95	3.61	3.38	3.22	3.10	3.01	2.93	2.87	2.77	2.67	2.56	2.50	2.44	2.38	2.32	2.26	2.19
19	5.92	4.51	3.90	3.56	3.33	3.17	3.05	2.96	2.88	2.82	2.72	2.62	2.51	2.45	2.39	2.33	2.27	2.20	2.13
20	5.87	4.46	3.86	3.51	3.29	3.13	3.01	2.91	2.84	2.77	2.68	2.57	2.46	2.41	2.35	2.29	2.22	2.16	2.09
21	5.83	4.42	3.82	3.48	3.25	3.09	2.97	2.87	2.80	2.73	2.64	2.53	2.42	2.37	2.31	2.25	2.18	2.11	2.04
22	5.79	4.38	3.78	3.44	3.22	3.05	2.93	2.84	2.76	2.70	2.60	2.50	2.39	2.33	2.27	2.21	2.14	2.08	2.00
23	5.75	4.35	3.75	3.41	3.18	3.02	2.90	2.81	2.73	2.67	2.57	2.47	2.36	2.30	2.24	2.18	2.11	2.04	1.97
24	5.72	4.32	3.72	3.38	3.15	2.99	2.87	2.78	2.70	2.64	2.54	2.44	2.33	2.27	2.21	2.15	2.08	2.01	1.94
25	5.69	4.29	3.69	3.35	3.13	2.97	2.85	2.75	2.68	2.61	2.51	2.41	2.30	2.24	2.18	2.12	2.05	1.98	1.91
26	5.66	4.27	3.67	3.33	3.10	2.94	2.82	2.73	2.65	2.59	2.49	2.39	2.28	2.22	2.16	2.09	2.03	1.95	1.88
27	5.63	4.24	3.65	3.31	3.08	2.92	2.80	2.71	2.63	2.57	2.47	2.36	2.25	2.19	2.13	2.07	2.00	1.93	1.85
28	5.61	4.22	3.63	3.29	3.06	2.90	2.78	2.69	2.61	2.55	2.45	2.34	2.23	2.17	2.11	2.05	1.98	1.91	1.83
29	5.59	4.20	3.61	3.27	3.04	2.88	2.76	2.67	2.59	2.53	2.43	2.32	2.21	2.15	2.09	2.03	1.96	1.89	1.81
30	5.57	4.18	3.59	3.25	3.03	2.87	2.75	2.65	2.57	2.51	2.41	2.31	2.20	2.14	2.07	2.01	1.94	1.87	1.79
40	5.42	4.05	3.46	3.13	2.90	2.74	2.62	2.53	2.45	2.39	2.29	2.18	2.07	2.01	1.94	1.88	1.80	1.72	1.64
60	5.29	3.93	3.34	3.01	2.79	2.63	2.51	2.41	2.33	2.27	2.17	2.06	1.94	1.88	1.82	1.74	1.67	1.58	1.48
120	5.15	3.80	3.23	2.89	2.67	2.52	2.39	2.30	2.22	2.16	2.05	1.94	1.82	1.76	1.69	1.61	1.53	1.43	1.31
∞	5.02	3.69	3.12	2.79	2.57	2.41	2.29	2.19	2.11	2.05	1.94	1.83	1.71	1.64	1.57	1.48	1.39	1.27	1.00

(continued)

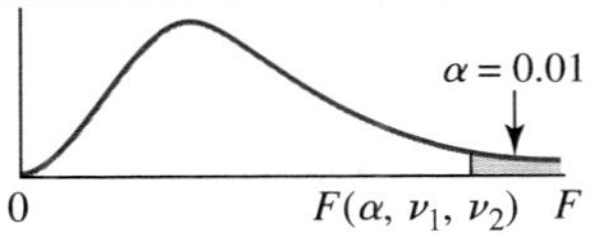

$v_1 = df$, numerator

$v_2 = df$, denominator	1	2	3	4	5	6	7	8	9	10	12	15	20	24	30	40	60	120	∞
1	4052	4999.5	5403	5625	5764	5859	5928	5982	6022	6056	6106	6157	6209	6235	6261	6287	6313	6339	6366
2	98.50	99.00	99.17	99.25	99.30	99.33	99.36	99.37	99.39	99.40	99.42	99.43	99.45	99.46	99.47	99.47	99.48	99.49	99.50
3	34.12	30.82	29.46	28.71	28.24	27.91	27.67	27.49	27.35	27.23	27.05	26.87	26.69	26.60	26.50	26.41	26.32	26.22	26.13
4	21.20	18.00	16.69	15.98	15.52	15.21	14.98	14.80	14.66	14.55	14.37	14.20	14.02	13.93	13.84	13.75	13.65	13.56	13.46
5	16.26	13.27	12.06	11.39	10.97	10.67	10.46	10.29	10.16	10.05	9.89	9.72	9.55	9.47	9.38	9.29	9.20	9.11	9.02
6	13.75	10.92	9.78	9.15	8.75	8.47	8.26	8.10	7.98	7.87	7.72	7.56	7.40	7.31	7.23	7.14	7.06	6.97	6.88
7	12.25	9.55	8.45	7.85	7.46	7.19	6.99	6.84	6.72	6.62	6.47	6.31	6.16	6.07	5.99	5.91	5.82	5.74	5.65
8	11.26	8.65	7.59	7.01	6.63	6.37	6.18	6.03	5.91	5.81	5.67	5.52	5.36	5.28	5.20	5.12	5.03	4.95	4.86
9	10.56	8.02	6.99	6.42	6.06	5.80	5.61	5.47	5.35	5.26	5.11	4.96	4.81	4.73	4.65	4.57	4.48	4.40	4.31
10	10.04	7.56	6.55	5.99	5.64	5.39	5.20	5.06	4.94	4.85	4.71	4.56	4.41	4.33	4.25	4.17	4.08	4.00	3.91
11	9.65	7.21	6.22	5.67	5.32	5.07	4.89	4.74	4.63	4.54	4.40	4.25	4.10	4.02	3.94	3.86	3.78	3.69	3.60
12	9.33	6.93	5.95	5.41	5.06	4.82	4.64	4.50	4.39	4.30	4.16	4.01	3.86	3.78	3.70	3.62	3.54	3.45	3.36
13	9.07	6.70	5.74	5.21	4.86	4.62	4.44	4.30	4.19	4.10	3.96	3.82	3.66	3.59	3.51	3.43	3.34	3.25	3.17
14	8.86	6.51	5.56	5.04	4.69	4.46	4.28	4.14	4.03	3.94	3.80	3.66	3.51	3.43	3.35	3.27	3.18	3.09	3.00
15	8.68	6.36	5.42	4.89	4.56	4.32	4.14	4.00	3.89	3.80	3.67	3.52	3.37	3.29	3.21	3.13	3.05	2.96	2.87
16	8.53	6.23	5.29	4.77	4.44	4.20	4.03	3.89	3.78	3.69	3.55	3.41	3.26	3.18	3.10	3.02	2.93	2.84	2.75
17	8.40	6.11	5.18	4.67	4.34	4.10	3.93	3.79	3.68	3.59	3.46	3.31	3.16	3.08	3.00	2.92	2.83	2.75	2.65
18	8.29	6.01	5.09	4.58	4.25	4.01	3.84	3.71	3.60	3.51	3.37	3.23	3.08	3.00	2.92	2.84	2.75	2.66	2.57
19	8.18	5.93	5.01	4.50	4.17	3.94	3.77	3.63	3.52	3.43	3.30	3.15	3.00	2.92	2.84	2.76	2.67	2.58	2.49
20	8.10	5.85	4.94	4.43	4.10	3.87	3.70	3.56	3.46	3.37	3.23	3.09	2.94	2.86	2.78	2.69	2.61	2.52	2.42
21	8.02	5.78	4.87	4.37	4.04	3.81	3.64	3.51	3.40	3.31	3.17	3.03	2.88	2.80	2.72	2.64	2.55	2.46	2.36
22	7.95	5.72	4.82	4.31	3.99	3.76	3.59	3.45	3.35	3.26	3.12	2.98	2.83	2.75	2.67	2.58	2.50	2.40	2.31
23	7.88	5.66	4.76	4.26	3.94	3.71	3.54	3.41	3.30	3.21	3.07	2.93	2.78	2.70	2.62	2.54	2.45	2.35	2.26
24	7.82	5.61	4.72	4.22	3.90	3.67	3.50	3.36	3.26	3.17	3.03	2.89	2.74	2.66	2.58	2.49	2.40	2.31	2.21
25	7.77	5.57	4.68	4.18	3.85	3.63	3.46	3.32	3.22	3.13	2.99	2.85	2.70	2.62	2.54	2.45	2.36	2.27	2.17
26	7.72	5.53	4.64	4.14	3.82	3.59	3.42	3.29	3.18	3.09	2.96	2.81	2.66	2.58	2.50	2.42	2.33	2.23	2.13
27	7.68	5.49	4.60	4.11	3.78	3.56	3.39	3.26	3.15	3.06	2.93	2.78	2.63	2.55	2.47	2.38	2.29	2.20	2.10
28	7.64	5.45	4.57	4.07	3.75	3.53	3.36	3.23	3.12	3.03	2.90	2.75	2.60	2.52	2.44	2.35	2.26	2.17	2.06
29	7.60	5.42	4.54	4.04	3.73	3.50	3.33	3.20	3.09	3.00	2.87	2.73	2.57	2.49	2.41	2.33	2.23	2.14	2.03
30	7.56	5.39	4.51	4.02	3.70	3.47	3.30	3.17	3.07	2.98	2.84	2.70	2.55	2.47	2.39	2.30	2.21	2.11	2.01
40	7.31	5.18	4.31	3.83	3.51	3.29	3.12	2.99	2.89	2.80	2.66	2.52	2.37	2.29	2.20	2.11	2.02	1.92	1.80
60	7.08	4.98	4.13	3.65	3.34	3.12	2.95	2.82	2.72	2.63	2.50	2.35	2.20	2.12	2.03	1.94	1.84	1.73	1.60
120	6.85	4.79	3.95	3.48	3.17	2.96	2.79	2.66	2.56	2.47	2.34	2.19	2.03	1.95	1.86	1.76	1.66	1.53	1.38
∞	6.63	4.61	3.78	3.32	3.02	2.80	2.64	2.51	2.41	2.32	2.18	2.04	1.88	1.79	1.70	1.59	1.47	1.32	1.00

Source: Standard Mathematical Tables, 26th ed., William H. Beyer (ed.), CRC Press, Inc., Boca Raton, FL, 1983.

TABLE A.7

The Chi-Square Distribution

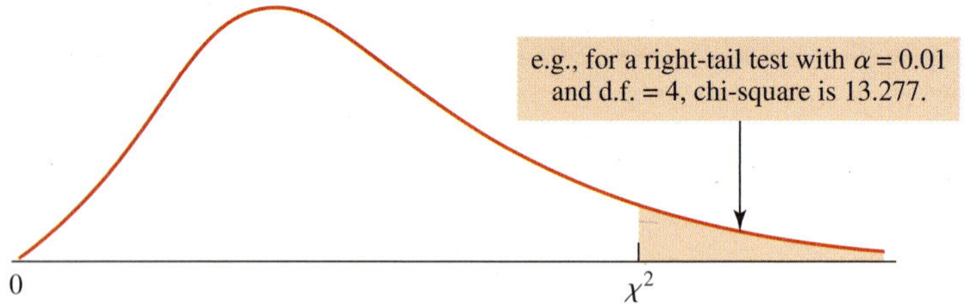

For α = Right-Tail Area of

d.f.	0.99	0.975	0.95	0.90	0.10	0.05	0.025	0.01
1	0.00016	0.00098	0.00039	0.0158	2.706	3.841	5.024	6.635
2	0.0201	0.0506	0.103	0.211	4.605	5.991	7.378	9.210
3	0.115	0.216	0.352	0.584	6.251	7.815	9.348	11.345
4	0.297	0.484	0.711	1.064	7.779	9.488	11.143	13.277
5	0.554	0.831	1.145	1.610	9.236	11.070	12.833	15.086
6	0.872	1.237	1.635	2.204	10.645	12.592	14.449	16.812
7	1.239	1.690	2.167	2.833	12.017	14.067	16.013	18.475
8	1.646	2.180	2.733	3.490	13.362	15.507	17.535	20.090
9	2.088	2.700	3.325	4.168	14.684	16.919	19.023	21.666
10	2.558	3.247	3.940	4.865	15.987	18.307	20.483	23.209
11	3.053	3.816	4.575	5.578	17.275	19.675	21.920	24.725
12	3.571	4.404	5.226	6.304	18.549	21.026	23.337	26.217
13	4.107	5.009	5.892	7.042	19.812	22.362	24.736	27.688
14	4.660	5.629	6.571	7.790	21.064	23.685	26.119	29.141
15	5.229	6.262	7.261	8.547	22.307	24.996	27.488	30.578
16	5.812	6.908	7.962	9.312	23.542	26.296	28.845	32.000
17	6.408	7.564	8.672	10.085	24.769	27.587	30.191	33.409
18	7.015	8.231	9.390	10.865	25.989	28.869	31.526	34.805
19	7.633	8.907	10.117	11.651	27.204	30.144	32.852	36.191
20	8.260	9.591	10.851	12.443	28.412	31.410	34.170	37.566
21	8.897	10.283	11.591	13.240	29.615	32.671	35.479	38.932
22	9.542	10.982	12.338	14.042	30.813	33.924	36.781	40.290
23	10.916	11.689	13.091	14.848	32.007	35.172	38.076	41.638
24	10.856	12.401	13.848	15.659	33.196	36.415	39.364	42.980
25	11.524	13.120	14.611	16.473	34.382	37.652	40.647	44.314
26	12.198	13.844	15.379	17.292	35.563	38.885	41.923	45.642
27	12.879	14.573	16.151	18.114	36.741	40.113	43.195	46.963
28	13.565	15.308	16.928	18.939	37.916	41.337	44.461	48.278
29	14.256	16.047	17.708	19.768	39.087	42.557	45.722	49.588
30	14.953	16.791	18.493	20.599	40.256	43.773	46.979	50.892
40	22.164	24.433	26.509	29.051	51.805	55.759	59.342	63.691
50	29.707	32.357	34.764	37.689	63.167	67.505	71.420	76.154
60	37.485	40.482	43.188	46.459	74.397	79.082	83.298	88.381
70	45.442	48.758	51.739	55.329	85.527	90.531	95.023	100.42
80	53.540	57.153	60.391	64.278	96.578	101.88	106.63	112.33
90	61.754	65.647	69.126	73.291	107.57	113.15	118.14	124.12
100	70.065	74.222	77.930	82.358	118.50	124.34	129.56	135.81

Source: Chi-square values generated by Minitab, then rounded as shown.

TABLE A.8

Wilcoxon Signed Rank Test, Lower and Upper Critical Values

Two-Tail Test: One-Tail Test:	$\alpha = 0.20$ $\alpha = 0.10$	$\alpha = 0.10$ $\alpha = 0.05$	$\alpha = 0.05$ $\alpha = 0.025$	$\alpha = 0.02$ $\alpha = 0.01$	$\alpha = 0.01$ $\alpha = 0.005$
$n = 4$	1, 9	0, 10	0, 10	0, 10	0, 10
5	3, 12	1, 14	0, 15	0, 15	0, 15
6	4, 17	3, 18	1, 20	0, 21	0, 21
7	6, 22	4, 24	3, 25	1, 27	0, 28
8	9, 27	6, 30	4, 32	2, 34	1, 35
9	11, 34	9, 36	6, 39	4, 41	2, 43
10	15, 40	11, 44	9, 46	6, 49	4, 51
11	18, 48	14, 52	11, 55	8, 58	6, 60
12	22, 56	18, 60	14, 64	10, 68	8, 70
13	27, 64	22, 69	18, 73	13, 78	10, 81
14	32, 73	26, 79	22, 83	16, 89	13, 92
15	37, 83	31, 89	26, 94	20, 100	16, 104
16	43, 93	36, 100	30, 106	24, 112	20, 116
17	49, 104	42, 111	35, 118	28, 125	24, 129
18	56, 115	48, 123	41, 130	33, 138	28, 143
19	63, 127	54, 136	47, 143	38, 152	33, 157
20	70, 140	61, 149	53, 157	44, 166	38, 172

Source: Adapted from Roger C. Pfaffenberger and James H. Patterson, *Statistical Methods for Business and Economics* (Homewood, Ill.: Richard D. Irwin, Inc., 1987), p. 110, and R. L. McCornack, "Extended Tables of the Wilcoxon Matched Pairs Signed Rank Statistics," *Journal of the American Statistical Association* 60 (1965), 864–871.

TABLE A.9

Wilcoxon Rank Sum Test, Lower and Upper Critical Values

$\alpha = 0.025$ (one-tail) or $\alpha = 0.05$ (two-tail)

n_1:	3	4	5	6	7	8	9	10
n_2: 3	5, 16	6, 18	6, 21	7, 23	7, 26	8, 28	8, 31	9, 33
4	6, 18	11, 25	12, 28	12, 32	13, 35	14, 38	15, 41	16, 44
5	6, 21	12, 28	18, 37	19, 41	20, 45	21, 49	22, 53	24, 56
6	7, 23	12, 32	19, 41	26, 52	28, 56	29, 61	31, 65	32, 70
7	7, 26	13, 35	20, 45	28, 56	37, 68	39, 73	41, 78	43, 83
8	8, 28	14, 38	21, 49	29, 61	39, 73	49, 87	51, 93	54, 98
9	8, 31	15, 41	22, 53	31, 65	41, 78	51, 93	63, 108	66, 114
10	9, 33	16, 44	24, 56	32, 70	43, 83	54, 98	66, 114	79, 131

(Note: n_1 is the smaller of the two samples—i.e., $n_1 \leq n_2$.)

$\alpha = 0.05$ (one-tail) or $\alpha = 0.10$ (two-tail)

n_1:	3	4	5	6	7	8	9	10
n_2: 3	6, 15	7, 17	7, 20	8, 22	9, 24	9, 27	10, 29	11, 31
4	7, 17	12, 24	13, 27	14, 30	15, 33	16, 36	17, 39	18, 42
5	7, 20	13, 27	19, 36	20, 40	22, 43	24, 46	25, 50	26, 54
6	8, 22	14, 30	20, 40	28, 50	30, 54	32, 58	33, 63	35, 67
7	9, 24	15, 33	22, 43	30, 54	39, 66	41, 71	43, 76	46, 80
8	9, 27	16, 36	24, 46	32, 58	41, 71	52, 84	54, 90	57, 95
9	10, 29	17, 39	25, 50	33, 63	43, 76	54, 90	66, 105	69, 111
10	11, 31	18, 42	26, 54	35, 67	46, 80	57, 95	69, 111	83, 127

(Note: n_1 is the smaller of the two samples—i.e., $n_1 \leq n_2$.)

Source: F. Wilcoxon and R. A. Wilcox, *Some Approximate Statistical Procedures* (New York: American Cyanamid Company, 1964), pp. 20–23.

TABLE A.10

Critical Values of *D* for the Kolmogorov–Smirnov Test of Normality

Sample Size *n*	Significance Level α 0.20	0.15	0.10	0.05	0.01
4	.300	.319	.352	.381	.417
5	.285	.299	.315	.337	.405
6	.265	.277	.294	.319	.364
7	.247	.258	.276	.300	.348
8	.233	.244	.261	.285	.331
9	.223	.233	.249	.271	.311
10	.215	.224	.239	.258	.294
11	.206	.217	.230	.249	.284
12	.199	.212	.223	.242	.275
13	.190	.202	.214	.234	.268
14	.183	.194	.207	.227	.261
15	.177	.187	.201	.220	.257
16	.173	.182	.195	.213	.250
17	.169	.177	.189	.206	.245
18	.166	.173	.184	.200	.239
19	.163	.169	.179	.195	.235
20	.160	.166	.174	.190	.231
25	.142	.147	.158	.173	.200
30	.131	.136	.144	.161	.187
Over 30	$\frac{.736}{\sqrt{n}}$	$\frac{.768}{\sqrt{n}}$	$\frac{.805}{\sqrt{n}}$	$\frac{.886}{\sqrt{n}}$	$\frac{1.031}{\sqrt{n}}$

Source: From H. W. Lilliefors, "On the Kolmogorov–Smirnov Test for Normality with Mean and Variance Unknown," *Journal of the American Statistical Association*, 62 (1967), pp. 399–402. As adapted by Conover, *Practical Nonparametric Statistics* (New York: John Wiley, 1971), p. 398.

TABLE A.11

Critical Values of Spearman's Rank Correlation Coefficient, r_s, One-Tail Test. (For a two-tail test, the listed values correspond to the 2α level of significance.)

n	$\alpha = 0.05$	$\alpha = 0.025$	$\alpha = 0.01$	$\alpha = 0.005$
5	0.900	—	—	—
6	0.829	0.886	0.943	—
7	0.714	0.786	0.893	—
8	0.643	0.738	0.833	0.881
9	0.600	0.683	0.783	0.833
10	0.564	0.648	0.745	0.794
11	0.523	0.623	0.736	0.818
12	0.497	0.591	0.703	0.780
13	0.475	0.566	0.673	0.745
14	0.457	0.545	0.646	0.716
15	0.441	0.525	0.623	0.689
16	0.425	0.507	0.601	0.666
17	0.412	0.490	0.582	0.645
18	0.399	0.476	0.564	0.625
19	0.388	0.462	0.549	0.608
20	0.377	0.450	0.534	0.591
21	0.368	0.438	0.521	0.576
22	0.359	0.428	0.508	0.562
23	0.351	0.418	0.496	0.549
24	0.343	0.409	0.485	0.537
25	0.336	0.400	0.475	0.526
26	0.329	0.392	0.465	0.515
27	0.323	0.385	0.456	0.505
28	0.317	0.377	0.448	0.496
29	0.311	0.370	0.440	0.487
30	0.305	0.364	0.432	0.478

Source: E. G. Olds, "Distribution of Sums of Squares of Rank Differences for Small Samples," *Annals of Mathematical Statistics* 9 (1938).

TABLE A.12

Values of d_L and d_U for the Durbin–Watson Test for $\alpha = 0.05$

n = number of observations
k = number of independent variables

	$k = 1$		$k = 2$		$k = 3$		$k = 4$		$k = 5$	
n	d_L	d_U	d_L	d_U	d_L	d_U	d_L	d_U	d_L	d_U
15	1.08	1.36	0.95	1.54	0.82	1.75	0.69	1.97	0.56	2.21
16	1.10	1.37	0.98	1.54	0.86	1.73	0.74	1.93	0.62	2.15
17	1.13	1.38	1.02	1.54	0.90	1.71	0.78	1.90	0.67	2.10
18	1.16	1.39	1.05	1.53	0.93	1.69	0.82	1.87	0.71	2.06
19	1.18	1.40	1.08	1.53	0.97	1.68	0.86	1.85	0.75	2.02
20	1.20	1.41	1.10	1.54	1.00	1.68	0.90	1.83	0.79	1.99
21	1.22	1.42	1.13	1.54	1.03	1.67	0.93	1.81	0.83	1.96
22	1.24	1.43	1.15	1.54	1.05	1.66	0.96	1.80	0.86	1.94
23	1.26	1.44	1.17	1.54	1.08	1.66	0.99	1.79	0.90	1.92
24	1.27	1.45	1.19	1.55	1.10	1.66	1.01	1.78	0.93	1.90
25	1.29	1.45	1.21	1.55	1.12	1.66	1.04	1.77	0.95	1.89
26	1.30	1.46	1.22	1.55	1.14	1.65	1.06	1.76	0.98	1.88
27	1.32	1.47	1.24	1.56	1.16	1.65	1.08	1.76	1.01	1.86
28	1.33	1.48	1.26	1.56	1.18	1.65	1.10	1.75	1.03	1.85
29	1.34	1.48	1.27	1.56	1.20	1.65	1.12	1.74	1.05	1.84
30	1.35	1.49	1.28	1.57	1.21	1.65	1.14	1.74	1.07	1.83
31	1.36	1.50	1.30	1.57	1.23	1.65	1.16	1.74	1.09	1.83
32	1.37	1.50	1.31	1.57	1.24	1.65	1.18	1.73	1.11	1.82
33	1.38	1.51	1.32	1.58	1.26	1.65	1.19	1.73	1.13	1.81
34	1.39	1.51	1.33	1.58	1.27	1.65	1.21	1.73	1.15	1.81
35	1.40	1.52	1.34	1.58	1.28	1.65	1.22	1.73	1.16	1.80
36	1.41	1.52	1.35	1.59	1.29	1.65	1.24	1.73	1.18	1.80
37	1.42	1.53	1.36	1.59	1.31	1.66	1.25	1.72	1.19	1.80
38	1.43	1.54	1.37	1.59	1.32	1.66	1.26	1.72	1.21	1.79
39	1.43	1.54	1.38	1.60	1.33	1.66	1.27	1.72	1.22	1.79
40	1.44	1.54	1.39	1.60	1.34	1.66	1.29	1.72	1.23	1.79
45	1.48	1.57	1.43	1.62	1.38	1.67	1.34	1.72	1.29	1.78
50	1.50	1.59	1.46	1.63	1.42	1.67	1.38	1.72	1.34	1.77
55	1.53	1.60	1.49	1.64	1.45	1.68	1.41	1.72	1.38	1.77
60	1.55	1.62	1.51	1.65	1.48	1.69	1.44	1.73	1.41	1.77
65	1.57	1.63	1.54	1.66	1.50	1.70	1.47	1.73	1.44	1.77
70	1.58	1.64	1.55	1.67	1.52	1.70	1.49	1.74	1.46	1.77
75	1.60	1.65	1.57	1.68	1.54	1.71	1.51	1.74	1.49	1.77
80	1.61	1.66	1.59	1.69	1.56	1.72	1.53	1.74	1.51	1.77
85	1.62	1.67	1.60	1.70	1.57	1.72	1.55	1.75	1.52	1.77
90	1.63	1.68	1.61	1.70	1.59	1.73	1.57	1.75	1.54	1.78
95	1.64	1.69	1.62	1.71	1.60	1.73	1.58	1.75	1.56	1.78
100	1.65	1.69	1.63	1.72	1.61	1.74	1.59	1.76	1.57	1.78

Source: From J. Durbin and G.S. Watson, "Testing for Serial Correlation in Least Squares Regression," *Biometrika*, 38 June, 1951.

TABLE A.12

(continued)

Values of d_L and d_U for the Durbin–Watson Test for $\alpha = 0.025$

n = number of observations
k = number of independent variables

	$k = 1$		$k = 2$		$k = 3$		$k = 4$		$k = 5$	
n	d_L	d_U	d_L	d_U	d_L	d_U	d_L	d_U	d_L	d_U
15	0.95	1.23	0.83	1.40	0.71	1.61	0.59	1.84	0.48	2.09
16	0.98	1.24	0.86	1.40	0.75	1.59	0.64	1.80	0.53	2.03
17	1.01	1.25	0.90	1.40	0.79	1.58	0.68	1.77	0.57	1.98
18	1.03	1.26	0.93	1.40	0.82	1.56	0.72	1.74	0.62	1.93
19	1.06	1.28	0.96	1.41	0.86	1.55	0.76	1.72	0.66	1.90
20	1.08	1.28	0.99	1.41	0.89	1.55	0.79	1.70	0.70	1.87
21	1.10	1.30	1.01	1.41	0.92	1.54	0.83	1.69	0.73	1.84
22	1.12	1.31	1.04	1.42	0.95	1.54	0.86	1.68	0.77	1.82
23	1.14	1.32	1.06	1.42	0.97	1.54	0.89	1.67	0.80	1.80
24	1.16	1.33	1.08	1.43	1.00	1.54	0.91	1.66	0.83	1.79
25	1.18	1.34	1.10	1.43	1.02	1.54	0.94	1.65	0.86	1.77
26	1.19	1.35	1.12	1.44	1.04	1.54	0.96	1.65	0.88	1.76
27	1.21	1.36	1.13	1.44	1.06	1.54	0.99	1.64	0.91	1.75
28	1.22	1.37	1.15	1.45	1.08	1.54	1.01	1.64	0.93	1.74
29	1.24	1.38	1.17	1.45	1.10	1.54	1.03	1.63	0.96	1.73
30	1.25	1.38	1.18	1.46	1.12	1.54	1.05	1.63	0.98	1.73
31	1.26	1.39	1.20	1.47	1.13	1.55	1.07	1.63	1.00	1.72
32	1.27	1.40	1.21	1.47	1.15	1.55	1.08	1.63	1.02	1.71
33	1.28	1.41	1.22	1.48	1.16	1.55	1.10	1.63	1.04	1.71
34	1.29	1.41	1.24	1.48	1.17	1.55	1.12	1.63	1.06	1.70
35	1.30	1.42	1.25	1.48	1.19	1.55	1.13	1.63	1.07	1.70
36	1.31	1.43	1.26	1.49	1.20	1.56	1.15	1.63	1.09	1.70
37	1.32	1.43	1.27	1.49	1.21	1.56	1.16	1.62	1.10	1.70
38	1.33	1.44	1.28	1.50	1.23	1.56	1.17	1.62	1.12	1.70
39	1.34	1.44	1.29	1.50	1.24	1.56	1.19	1.63	1.13	1.69
40	1.35	1.45	1.30	1.51	1.25	1.57	1.20	1.63	1.15	1.69
45	1.39	1.48	1.34	1.53	1.30	1.58	1.25	1.63	1.21	1.69
50	1.42	1.50	1.38	1.54	1.34	1.59	1.30	1.64	1.26	1.69
55	1.45	1.52	1.41	1.56	1.37	1.60	1.33	1.64	1.30	1.69
60	1.47	1.54	1.44	1.57	1.40	1.61	1.37	1.65	1.33	1.69
65	1.49	1.55	1.46	1.59	1.43	1.62	1.40	1.66	1.36	1.69
70	1.51	1.57	1.48	1.60	1.45	1.63	1.42	1.66	1.39	1.70
75	1.53	1.58	1.50	1.61	1.47	1.64	1.45	1.67	1.42	1.70
80	1.54	1.59	1.52	1.62	1.49	1.65	1.47	1.67	1.44	1.70
85	1.56	1.60	1.53	1.63	1.51	1.65	1.49	1.68	1.46	1.71
90	1.57	1.61	1.55	1.64	1.53	1.66	1.50	1.69	1.48	1.71
95	1.58	1.62	1.56	1.65	1.54	1.67	1.52	1.69	1.50	1.71
100	1.59	1.63	1.57	1.65	1.55	1.67	1.53	1.70	1.51	1.72

TABLE A.12

(continued)

Values of d_L and d_U for the Durbin–Watson Test for $\alpha = 0.01$

n = number of observations
k = number of independent variables

	$k = 1$		$k = 2$		$k = 3$		$k = 4$		$k = 5$	
n	d_L	d_U	d_L	d_U	d_L	d_U	d_L	d_U	d_L	d_U
15	0.81	1.07	0.70	1.25	0.59	1.46	0.49	1.70	0.39	1.96
16	0.84	1.09	0.74	1.25	0.63	1.44	0.53	1.66	0.44	1.90
17	0.87	1.10	0.77	1.25	0.67	1.43	0.57	1.63	0.48	1.85
18	0.90	1.12	0.80	1.26	0.71	1.42	0.61	1.60	0.52	1.80
19	0.93	1.13	0.83	1.26	0.74	1.41	0.65	1.58	0.56	1.77
20	0.95	1.15	0.86	1.27	0.77	1.41	0.68	1.57	0.60	1.74
21	0.97	1.16	0.89	1.27	0.80	1.41	0.72	1.55	0.63	1.71
22	1.00	1.17	0.91	1.28	0.83	1.40	0.75	1.54	0.66	1.69
23	1.02	1.19	0.94	1.29	0.86	1.40	0.77	1.53	0.70	1.67
24	1.04	1.20	0.96	1.30	0.88	1.41	0.80	1.53	0.72	1.66
25	1.05	1.21	0.98	1.30	0.90	1.41	0.83	1.52	0.75	1.65
26	1.07	1.22	1.00	1.31	0.93	1.41	0.85	1.52	0.78	1.64
27	1.09	1.23	1.02	1.32	0.95	1.41	0.88	1.51	0.81	1.63
28	1.10	1.24	1.04	1.32	0.97	1.41	0.90	1.51	0.83	1.62
29	1.12	1.25	1.05	1.33	0.99	1.42	0.92	1.51	0.85	1.61
30	1.13	1.26	1.07	1.34	1.01	1.42	0.94	1.51	0.88	1.61
31	1.15	1.27	1.08	1.34	1.02	1.42	0.96	1.51	0.90	1.60
32	1.16	1.28	1.10	1.35	1.04	1.43	0.98	1.51	0.92	1.60
33	1.17	1.29	1.11	1.36	1.05	1.43	1.00	1.51	0.94	1.59
34	1.18	1.30	1.13	1.36	1.07	1.43	1.01	1.51	0.95	1.59
35	1.19	1.31	1.14	1.37	1.08	1.44	1.03	1.51	0.97	1.59
36	1.21	1.32	1.15	1.38	1.10	1.44	1.04	1.51	0.99	1.59
37	1.22	1.32	1.16	1.38	1.11	1.45	1.06	1.51	1.00	1.59
38	1.23	1.33	1.18	1.39	1.12	1.45	1.07	1.52	1.02	1.58
39	1.24	1.34	1.19	1.39	1.14	1.45	1.09	1.52	1.03	1.58
40	1.25	1.34	1.20	1.40	1.15	1.46	1.10	1.52	1.05	1.58
45	1.29	1.38	1.24	1.42	1.20	1.48	1.16	1.53	1.11	1.58
50	1.32	1.40	1.28	1.45	1.24	1.49	1.20	1.54	1.16	1.59
55	1.36	1.43	1.32	1.47	1.28	1.51	1.25	1.55	1.21	1.59
60	1.38	1.45	1.35	1.48	1.32	1.52	1.28	1.56	1.25	1.60
65	1.41	1.47	1.38	1.50	1.35	1.53	1.31	1.57	1.28	1.61
70	1.43	1.49	1.40	1.52	1.37	1.55	1.34	1.58	1.31	1.61
75	1.45	1.50	1.42	1.53	1.39	1.56	1.37	1.59	1.34	1.62
80	1.47	1.52	1.44	1.54	1.42	1.57	1.39	1.60	1.36	1.62
85	1.48	1.53	1.46	1.55	1.43	1.58	1.41	1.60	1.39	1.63
90	1.50	1.54	1.47	1.56	1.45	1.59	1.43	1.61	1.41	1.64
95	1.51	1.55	1.49	1.57	1.47	1.60	1.45	1.62	1.42	1.64
100	1.52	1.56	1.50	1.58	1.48	1.60	1.46	1.63	1.44	1.65

TABLE A.13

Factors for Determining 3-Sigma Control Limits, Mean and Range Control Charts

Number of Observations in Each Sample	Factor for Determining Control Limits, Control Chart for the Mean	Factors for Determining Control Limits, Control Chart for the Range	
n	A_2	D_3	D_4
2	1.880	0	3.267
3	1.023	0	2.575
4	0.729	0	2.282
5	0.577	0	2.115
6	0.483	0	2.004
7	0.419	0.076	1.924
8	0.373	0.136	1.864
9	0.337	0.184	1.816
10	0.308	0.223	1.777
11	0.285	0.256	1.744
12	0.266	0.284	1.716
13	0.249	0.308	1.692
14	0.235	0.329	1.671
15	0.223	0.348	1.652

Source: From E. S. Pearson, "The Percentage Limits for the Distribution of Range in Samples from a Normal Population," *Biometrika* 24 (1932): 416.

APPENDIX B

SELECTED ANSWERS

Answers to Selected Odd-Numbered Exercises

Chapter 2

2.3 a. 45.59 million **b.** lower limit is 35, upper limit is under 45 **c.** 10 years **d.** 40 years **2.5 a.** 183.53 thousand **b.** lower limit is 45, upper limit is under 55 **c.** 10 years **d.** 50 years **2.45 a.** $y = 323.53 + 0.8112x$ **b.** yes, direct **2.55 a.** 1342 cities **b.** 2815 cities **c.** 253 cities; 8.77% **d.** 175,000 **2.67 a.** yes **b.** lawjudge = 0.788 + 0.840*acad **2.69 a.** yes **b.** Canada = 1.089 + 1.754*U.S. **2.71 c.** highest: mechanical 2 and electrical 1; lowest: mechanical 1 and electrical 3

Chapter 3

3.1 $\bar{x}$ = \$20.449, median = \$20.495 **3.3** $\bar{x}$ = 57.05 visitors, median = 57.50, mode = 63 **3.5** $\bar{x}$ = 6.93, median = 7.095 **3.7** $\bar{x}$ = 38.1, median = 36.5 **3.9** 83.2 **3.11 a.** mean **b.** median **3.13** $\bar{x}$ = 398.86, median = 396.75 **3.15** females: $\bar{x}$ = 40.62, median = 39.00; males: $\bar{x}$ = 41.08, median = 41.50 **3.17** range = 39, MAD = 10.35 visitors, s = 12.40, s^2 = 153.84 **3.19 a.** μ = 29.11 million, median = 22.4 million, range = 44.5 million, midrange = 39.45 million **b.** MAD = 12.99 million **c.** σ = 15.44 million, σ^2 = 238.396 **3.21 a.** $\bar{x}$ = 27.2 mpg, median = 28 mpg, range = 30 mpg, midrange = 25 mpg **b.** MAD = 5.6 mpg **c.** s = 8.052, s^2 = 64.84 **3.23** $Q_1 = 7$, $Q_2 = 18$, $Q_3 = 30$, interquartile range = 23, quartile deviation = 11.5 **3.25 a.** $\bar{x}$ = 23.35, median = 22.86, range = 26.13, midrange = 26.465 **b.** MAD = 4.316 **c.** s = 5.49, s^2 = 30.14 **3.27 a.** $\bar{x}$ = 90.771, median = 91.4, range = 40.4, midrange = 92.0 **b.** MAD = 6.70 **c.** s = 8.36, s^2 = 69.90 **3.29 a.** 84% **b.** 88.89% **c.** 96% **3.31** 90%, yes **3.33 a.** 68% **b.** 2.5% **c.** 84% **d.** 13.5% **3.35** Barnsboro **3.39 a.** approximately 7.00 and 1.504 **3.41** approximately 21.75 and 15.65 **3.43** $r = -0.8$ **3.45** lawjudge = 0.7880 + 0.8404*acad, r^2 = 0.9344, r = 0.967 **3.47** generic = −3.5238 + 0.377*brand, r^2 = 0.7447, r = 0.863 **3.49** \$4.69 **3.51** $\bar{x}$ = 117.75, median = 117.5, no **3.53** no, positively skewed **3.55 a.** $\bar{x}$ becomes 3.1 lbs., s remains 0.5 lbs. **b.** 4.1 lbs. **3.57 a.** $\bar{x}$ = 2.099, median = 2.115, range = 0.46, midrange = 2.08 **b.** MAD = 0.137 **c.** s = 0.156, s^2 = 0.02446 **3.59** 120, 116, 124, 20, symmetrical **3.61** greater variation for data in exercise 3.60

Chapter 4

4.3 a. secondary **b.** secondary **4.11** response error **4.13** telephone **4.39** systematic **4.43 a.** sample **b.** sample **c.** sample **d.** census **4.69 a.** 54 **b.** 146

Chapter 5

5.1 subjective **5.7** decrease for men, increase for women **5.9** 0.36 **5.13 a.** 0 **b.** 196 **c.** 147 **d.** 372 **5.19** 0.945, 0.035 **5.21** 0.35, 0.65 **5.23** 0.797 **5.25 b.** 0.13 **c.** 0.30 **d.** 0.89 **5.27** 0.91 **5.29 a.** 0.947 **b.** 0.710 **c.** 0.351 **d.** 0.992 **5.31** no, no **5.33 a.** 0.16 **b.** 0.31 **c.** 0.12 **d.** 0.41 **5.35** 0.851 **5.37 a.** 0.025 **b.** 0.709 **c.** 0.266 **5.39** 0.175 **5.43** 0.154 **5.45 a.** 0.51 **b.** 0.85 **c.** 0.77 **5.47 a.** 0.5 **b.** 0.455 **5.49 a.** 0.1 **b.** 0.233 **5.51** 256 **5.53** 36 **5.57** 20 **5.59** $7.9019*10^{27}$ **5.61 a.** 0.000000005 **b.** 0.0625 **c.** 0.062500005 **d.** 0 **5.63 a.** 0.001 **b.** classical **c.** yes **5.65 a.** 3 **b.** 3 **c.** 1 **5.67 a.** 0.005 **b.** 0.855 **c.** 0.14 **d.** 0.995 **5.69 a.** 0.85 **b.** 0.278 **c.** 0.046 **5.71 a.** 0.7 **b.** 0.9 **c.** 0.667 **5.73 a.** 0.973 **b.** $4.741*10^{-12}$ **c.** 10 **5.75** 22 **5.77** 0.296, 0.963 **5.79 a.** 0.9 **b.** 0.429 **5.81** 120 **5.83** 1320 **5.85** 10,000

Chapter 6

6.3 a. discrete **b.** continuous **c.** discrete **d.** continuous **6.5** $\mu = 7.2$, $\sigma = 2.18$, $\sigma^2 = 4.76$ **6.7** $\mu = 1.9$, $\sigma = 1.14$, $\sigma^2 = 1.29$ **6.9** \$10 **6.11** $E(x) = 3.1$ **6.13** \$0.50 **6.15** \$37.9 million, \$41.9 million, \$40.3 million, minor **6.19 a.** 3.6 **b.** 1.59 **c.** 0.2397 **d.** 0.9133 **e.** 0.5075 **6.21 a.** 0.0102 **b.** 0.2304 **c.** 0.3456 **d.** 0.0778 **6.23 a.** 0.9703 **b.** 0.0294 **c.** 0.0003 **d.** 0.0000 **6.25 a.** 0.9375 **b.** 0.6875 **6.27** 0.1250, 0.1250, 0.3750 **6.29** 0.3439 **6.31 a.** 0.0014 **b.** 0.0139 **c.** yes **6.35 a.** 0.1000 **b.** 0.6000 **c.** 0.3000 **6.37** 0.9000 **6.39 a.** 0.1353 **b.** 0.2707 **c.** 0.8571 **d.** 0.5940 **6.41 a.** 4.9 **b.** 0.1460 **c.** 0.1753 **d.** 0.9382 **e.** 0.8547 **6.43 a.** 1.8 **b.** 0.2975 **c.** 0.0723 **d.** 0.9974 **e.** 0.5268 **6.45** 0.0047, should consider slight decrease **6.47 a.** 10.0 **b.** 0.0901 **c.** 0.1251 **d.** 0.7916 **e.** 0.7311 **6.49** 0.2275, \$45,000 **6.51** not merely a coincidence **6.55** 0.9098 **6.57** no **6.61** 87.95% **6.63** 0.3684 **6.65** 0.1837 **6.67** 0.6065 **6.69** 0.3125 **6.71** 0.1904 **6.73** 0.0839 **6.75** 0.9872 **6.77** 0.7748 **6.79** 0.0012, not believable

Chapter 7

7.9 a. 0.5 **b.** approx 0.683 **c.** approx 0.8415 **d.** 0 **e.** approx 0.4775 **f.** approx 0.9775 **7.11 a.** 0.5 **b.** approx 0.955 **c.** approx 0.683 **d.** approx 0.9985 **7.13 a.** approx 0.1585 **b.** approx 0.683 **c.** approx 0.0225 **d.** approx 0.819 **7.15 a.** approx 0.0015 **b.** approx 0.1585 **7.17 a.** −0.67, 0.67 **b.** −1.28, 1.28 **c.** −0.74, 0.74 **7.19 a.** −2.00 **b.** −0.80 **c.** 0.00 **d.** 3.40 **e.** 4.60 **7.21 a.** 0.3643 **b.** 0.1357 **c.** 0.9115 **7.23 a.** 0.8730 **b.** 0.2272 **c.** 0.1091 **7.25 a.** 0.52 **b.** −1.28 **c.** 0.10 **d.** 0.52 **7.27 a.** 0.0874 **b.** 0.8790

c. 0.8413 **7.29 a.** 0.4207 **b.** 0.3050 **c.** 0.3446 **7.31** \$409,350 **7.33 a.** 0.4207 **b.** \$9545 **7.35** 3.22 minutes **7.37** 70,500 **7.41 a.** 10.0, 2.739 **b.** 0.1098, 0.2823, 0.3900, 0.1003 **7.43 a.** 12.0, 1.549 **b.** 0.2501 **c.** 0.2510 **d.** 0.9463 **7.45** 0.4350 **7.47** 0.4191 **7.53 a.** 0.5488 **b.** 0.4493 **c.** 0.3679 **d.** 0.3012 **7.55** 0.2865 **7.57** 0.619, 0.383, 8779.7 hours **7.61** 11.51% **7.63** 0.3911 **7.65** no **7.67** 0.0668, 0.0228 **7.69** 0.1587 **7.71** 0.8023 **7.73** not credible **7.75** 0.6604 **7.77** 0.4584, 0.2101 **7.79** 0.4307 **7.81** 0.0918, 0.0446 **7.83** 20.615 **7.85 a.** 0.4628 **b.** 0.7772 **c.** 12,900 **7.87** 0.4724

Chapter 8

8.3 0.18, 0.15, 1000 **8.7 a.** 25 **b.** 10 **c.** 5 **d.** 3.162 **8.9 a.** 0.8413 **b.** 0.6853 **c.** 0.9938 **8.11** 0.8849, 0.8823 **8.13** concern is justified **8.15** 0.1056 **8.17** 0.1949 **8.19** 0.9976 **8.21** $\pi = 0.46, \sigma_p \leq 0.0895$ **8.23 a.** 0.9616 **b.** 0.5222 **c.** 0.9616 **8.25 a.** 0.40 **b.** 0.35 **c.** 0.035 **d.** 0.9236 **8.27** 0.9938 **8.29** 0.0000; district's claim is much more credible **8.35** 0.1170 **8.37** 0.8023 **8.43 a.** 0.0183 **b.** not credible **8.45 b.** 0.0446 **8.47 a.** 2.40 **b.** 0.0082 **c.** no **8.51 a.** 0.10 **b.** 0.0062 **c.** no **8.53** 0.0643 **8.55** 0.0000 **8.57** 0.0228

Chapter 9

9.7 a. 2.625 **b.** 4.554 **9.11 a.** 0.45 **b.** (0.419, 0.481) **c.** 95%, 0.95 **9.15** 90%: (236.997, 243.003); 95%: (236.422, 243.578) **9.17** 90%: (82.92, 87.08); 95%: (82.52, 87.48) **9.19** (149.006, 150.994) **9.23** (99.897, 100.027); yes **9.27** 1.313 **9.29 a.** 1.292 **b.** −1.988 **c.** 2.371 **9.31** 95%: (46.338, 54.362); 99%: (44.865, 55.835) **9.33 a.** (22.16, 27.84) **b.** yes **9.35** (28.556, 29.707) **9.37** (1520.96, 1549.04) **9.39** (17.43, 21.97); no **9.41** (14.646, 15.354); yes **9.43** (0.429, 0.491) **9.45** (0.153, 0.247); not credible **9.47** (0.449, 0.511) **9.49** (0.579, 0.625) **9.51 a.** (0.527, 0.613) **9.53** (0.649, 0.711) **9.55** (0.566, 0.634); no; may not succeed **9.57** (0.311, 0.489) **9.61** 92 **9.63** 1068 **9.65** 863 **9.67** 601 **9.71** 95%: (0.522, 0.578); 99%: (0.513, 0.587) **9.73** 458 **9.75** 92 **9.77** 462 **9.79** 1226 **9.81** 0.01 **9.83** (135.211, 138.789) **9.85** 246 **9.87** 90%: (44.91, 49.96); 95%: (44.39, 50.47) **9.89** 411 **9.91** \$1200 **9.93** 95%: (0.347, 0.433); 99%: (0.334, 0.446) **9.95** 4161 **9.97** 722 **9.99** 1068 **9.101** 1110 **9.103** 3736 **9.105** 90%: (0.375, 0.425); 95%: (0.370, 0.430) **9.107** (0.018, 0.062) **9.109** (\$95.56, \$98.45) **9.111** (64.719, 68.301); funds are not endangered

Chapter 10

10.3 a. no **b.** yes **c.** yes **d.** no **e.** yes **f.** no **10.5** type I **10.7** type II **10.13** type I **10.17 a.** numerically high **b.** numerically low **10.23** no; do not reject H_0 **10.25 a.** 0.0618 **b.** 0.1515 **c.** 0.0672 **10.27** reject H_0; p-value = 0.021 **10.29** do not reject H_0; p-value = 0.035 **10.31** no; do not reject H_0; p-value = 0.052 **10.33** p-value = 0.316; do not reject H_0 **10.35 a.** do not reject H_0 **b.** reject H_0 **c.** do not reject H_0 **d.** reject H_0 **10.37** (1.995, 2.055); do not reject H_0; same **10.41** do not reject H_0 **10.43** no; do not reject H_0 **10.45** no; reject H_0 **10.47** yes; reject H_0 **10.49** do not reject H_0 **10.51** yes; reject H_0 **10.53** (86.29, 92.71); no; yes **10.55** (35.657, 37.943); no; yes **10.57** 0.03; reject H_0 **10.61** reject H_0 **10.63** reject H_0 **10.65** no; reject H_0; 0.003 **10.67** yes; reject H_0; p-value = 0.005 **10.69** reject H_0; p-value = 0.014 **10.71** yes; p-value = 0.220 **10.73** no; do not reject H_0; p-value = 0.079 **10.75** (0.735, 0.805); yes; yes **10.77** (0.402, 0.518); yes; yes **10.81** alpha unchanged, beta decreases **10.83** 0.9871 **10.87 a.** 2.33 **b.** 0.036 **10.93** reject H_0, has increased **10.95 a.** 0.05 level: reject H_0 **b.** 95% CI: (236,279; 255,321) **10.97** yes **10.99** reject H_0 **10.101** reject H_0; statement not credible **10.103** do not reject H_0; claim is credible **10.105 a.** 0.9505 **b.** 0.8212 **c.** 0.5753 **d.** 0.2946 **e.** 0.1020 **10.107 b.** e.g., 0.005 **c.** e.g., 0.02 **d.** 0.024 **10.109** yes; reject H_0; p-value = 0.015 **10.111** no; do not reject H_0; p-value = 0.059 **10.113** do not reject H_0; p-value = 0.282

Chapter 11

11.3 do not reject H_0, > 0.20 **11.5** yes; reject H_0 **11.7** do not reject H_0; between 0.05 and 0.10; (−8.872, 0.472) **11.9** reject H_0; between 0.10 and 0.05; 90% CI: (−1.318, −0.082) **11.11** claim could be valid; reject H_0; between 0.01 and 0.025 **11.13** yes; p-value = 0.031 **11.15** (−24.23, 1.03); yes; yes **11.17** (−3.33, 31.20); yes; yes **11.19** do not reject H_0 **11.21** yes, do not reject H_0; > 0.20; 95% CI: (−23.43, 6.43) **11.23** do not reject H_0; 0.1400; (−117.18, 17.18); yes; yes **11.25** do not reject H_0; (−2.03, −1.37); yes; yes **11.27** reject H_0; 0.000 **11.29** reject H_0; 0.006 **11.33** reject H_0 **11.35** yes, do not reject H_0; 0.2538; 95% CI: (−23.10, 6.10) **11.37** do not reject H_0; 0.1967; (−7.77, 37.77) **11.39** reject H_0; 0.0122 **11.41** reject H_0; 0.001 **11.43** do not reject H_0; 0.12 **11.45** dependent **11.47** reject H_0 **11.49** reject H_0; between 0.005 and 0.01 **11.51** do not reject H_0; 0.170; (−4.663, 1.774) **11.53** reject H_0 **11.55** no; do not reject H_0 **11.57** do not reject H_0 **11.59** yes; reject H_0; 0.0607; (−0.150, −0.010) **11.61** reject H_0; 0.0583 **11.63** do not reject H_0; 0.057; (−0.031, 0.211) **11.65** do not reject H_0; 0.1866; (−0.055, 0.005) **11.67** do not reject H_0; no, no **11.69** yes; do not reject H_0; no **11.71** yes; reject H_0 **11.73** reject H_0 **11.75** suspicion confirmed; reject H_0; between 0.025 and 0.05 **11.77** do not reject H_0; (−505.32, 65.32) **11.79** reject H_0; < 0.005 **11.81** yes; reject H_0; 0.0141 **11.83** reject H_0; table: < 0.01; computer: 0.0075; (0.13, 6.67) **11.85** do not reject H_0; (−1.13, 0.13) **11.87** yes; reject H_0 **11.89** no; do not reject H_0 **11.91** yes; reject H_0; < 0.005 **11.93** yes; reject H_0; 0.001 **11.95** yes; do not reject H_0; no **11.97** do not reject H_0; 0.140; (−1.95, 12.96) **11.99** not supported; do not reject H_0; 0.135 **11.101** do not reject H_0; 0.414; (−0.1204, 0.0404)

Chapter 12

12.7 designed **12.21** do not reject H_0; > 0.05 **12.23** do not reject H_0; > 0.05 **12.25** do not reject H_0 **12.27 b.** reject H_0 **12.29 b.** do not reject H_0 **c.** μ_1: (48.914, 61.086); μ_2: (42.714, 54.886); μ_3: (39.514, 51.686) **12.35** μ_1: (16.271, 19.009); μ_2: (13.901, 16.899); μ_3: (15.690, 18.060) **12.47** reject H_0; between 0.025 and 0.05 **12.49** do not reject H_0; do not reject H_0 **12.51** do not reject H_0; do not reject H_0 **12.53** reject H_0 **12.55** reject H_0 **12.59** yes; reject H_0 **12.69** Factor A, do not reject H_0; Factor B, reject H_0; Interaction, reject H_0

12.71 Factor A, do not reject H_0; Factor B, reject H_0; Interaction, do not reject H_0 **12.73** Factor A, reject H_0; Factor B, reject H_0; Interaction, reject H_0 **12.75** Assembly, do not reject H_0; Music, reject H_0; Interaction, reject H_0; Method 1, (35.846, 41.654); Method 2, (34.096, 39.904); Classical, (29.096, 34.904); Rock, (40.846, 46.654) **12.77** Bag, do not reject H_0; Dress, reject H_0; Interaction, reject H_0; Carry, (24.798, 30.536); Don't Carry, (27.298, 33.036); Sloppy, (41.736, 48.764); Casual, (17.736, 24.764); Dressy, (16.736, 23.764) **12.79** Keyboard, reject H_0; Wordpack, do not reject H_0; Interaction, reject H_0 **12.83** randomized block **12.85** independent: faceplate design; dependent: time to complete task; designed **12.87** no, randomized block procedure should be used **12.89** reject H_0 **12.91** do not reject H_0 **12.93** reject H_0 **12.95** do not reject H_0 **12.97** reject H_0 **12.99** Style, do not reject H_0; Darkness, reject H_0; Interaction, do not reject H_0; Style 1, (25.693, 30.307); Style 2, (23.693, 28.307); Light, (27.425, 33.075); Medium, (22.175, 27.825); Dark, (22.925, 28.575) **12.101** Position, reject H_0; Display, do not reject H_0; Interaction, reject H_0; Position 1, (42.684, 47.982); Position 2, (47.129, 52.427); Display 1, (42.089, 48.577); Display 2, (43.923, 50.411); Display 3, (46.923, 53.411)

Chapter 13

13.7 a. 3.490 **b.** 13.362 **c.** 2.733 **d.** 15.507 **e.** 17.535 **f.** 2.180 **13.9 a.** A = 8.547, B = 22.307 **b.** A = 7.261, B = 24.996 **c.** A = 6.262, B = 27.488 **d.** A = 5.229, B = 30.578 **13.13** 4 **13.15 a.** 6 **b.** 12.592 **c.** do not reject H_0 **13.17** yes; reject H_0 **13.19** do not reject H_0 **13.21** do not reject H_0 **13.23** do not reject H_0 **13.25** reject H_0 **13.29 a.** 6 **b.** 2 **c.** 12 **d.** 8 **e.** 12 **f.** 4 **13.31 a.** 12.833 **b.** 15.507 **c.** 16.812 **d.** 7.779 **13.33** no, reject H_0; between 0.025 and 0.01 **13.35** no; reject H_0; between 0.05 and 0.025 **13.37** no; reject H_0; less than 0.01 **13.39** yes; do not reject H_0 **13.41** no; reject H_0 **13.43** do not reject H_0 **13.45** do not reject H_0 **13.47** reject H_0 **13.49** yes; do not reject H_0 **13.53** (15.096, 43.011) **13.55** (2.620, 9.664) **13.57** (0.1346, 0.2791) **13.59** do not reject H_0 **13.61** do not reject H_0; between 0.05 and 0.025 **13.63** (0.0329, 0.0775) **13.65** do not reject H_0 **13.67** reject H_0 **13.69** yes; do not reject H_0 **13.71** do not reject H_0; between 0.10 and 0.05 **13.73** no; do not reject H_0; between 0.10 and 0.05 **13.75** (0.02001, 0.05694) **13.77** reject H_0 **13.79** do not reject H_0 **13.81** reject H_0 **13.83** reject H_0 **13.85** do not reject H_0 **13.87** probably not; reject H_0

Chapter 14

14.7 reject H_0; < 0.005 **14.9** yes; reject H_0; 0.017 **14.11** do not reject H_0; 0.137 **14.13** reject H_0 **14.15** yes, reject H_0 **14.17** do not reject H_0; between 0.05 and 0.10 **14.19** no; do not reject H_0; 0.137 **14.23** do not reject H_0 **14.25** yes; do not reject H_0; 0.2443 **14.29** reject H_0 **14.31** no; reject H_0 **14.33** no; reject H_0; 0.058 **14.37** yes; do not reject H_0; between 0.10 and 0.90 **14.39** no; do not reject H_0; 0.393 **14.43** do not reject H_0 **14.45** reject H_0 **14.49** do not reject H_0 **14.51** do not reject H_0 **14.53** do not reject H_0 **14.55** 0.83, yes **14.57** 0.343, no **14.59** reject H_0; 0.0287 **14.61** no, 0.619 **14.63** do not reject H_0 **14.65** yes; reject H_0; 0.0056 **14.67** no; do not reject H_0; between 0.05 and 0.10 **14.69** yes; do not reject H_0; between 0.025 and 0.05 **14.71** yes; do not reject H_0; between 0.025 and 0.05 **14.75** claim not credible, reject H_0; between 0.05 and 0.025 **14.77** 0.868, yes **14.79** no; do not reject H_0; 0.343 **14.81** do not reject H_0; 0.210 **14.83** they can tell the difference; reject H_0; 0.016 **14.85** do not reject H_0; 0.486

Chapter 15

15.5 second **15.7** second **15.9 a.** Shares = 44.3 +38.756*Years **b.** 431.9 **15.11** totgross = 110.02 + 1.0485*2wks; $214.87 million **15.13** Acres = 349,550 − 7851*Rain; 208,223 acres **15.21 a.** $\hat{y} = 21.701 - 1.354x$ **b.** 3.617 **c.** (8.107, 16.339) **d.** (4.647, 14.383) **e.** interval d is wider **15.23 a.** Rating = 54.97 + 7.166*TD% **b.** 90.796 **c.** 1.386 **d.** (86.728, 94.863) **e.** (88.833, 92.758) **15.25** 91.48; (−33.9, 510.1) **15.27** Revenue $= -1.66*10^8 + 64{,}076{,}803*$Dealers; CI: $(5.88*10^9, 6.60*10^9)$; PI: $(4.92*10^9, 7.57*10^9)$ **15.29** CI: (89,209; 295,832); PI: (−155,578; 540,619) **15.33** 0.81 **15.37 a.** Coll = 24.48 + 0.955*Comp **b.** 0.860, 0.740 **c.** 110.44 **15.39 a.** %Over = −82.19 + 0.04236*Cals **b.** 0.931, 0.866 **c.** 44.89% **15.41** $r = 0.800$; $r^2 = 0.640$ **15.43** Forgross = −69.382 + 1.5266*Domgross; $r = 0.7356$; $r^2 = 0.5411$ **15.45** no **15.47** > 0.10 **15.49 a.** reject H_0 **b.** reject H_0 **c.** (0.556, 1.354) **15.51 a.** reject H_0 **b.** reject H_0 **c.** (0.392, 9.608) **15.53** 61.2, 58.8 **15.55** 774.80, 536.01, 238.79; 0.692; do not reject H_0 **15.57** Gallons = 5,921,560.92 + 10.44806*Hours; $r = 0.993$; $r^2 = 0.986$; no; (8.746, 12.150) **15.59** NetIncome = 59.8006 + 0.0382*Revenue; $r = 0.7492$; $r^2 = 0.5612$; no; (0.031, 0.045) **15.71** NetIncome = 0.211 + 0.0999*TotRev; $2.009 billion **15.73 a.** OneYr% = −2.01 +1.45*ThreeYr% **b.** 5.264% **c.** 8.172% **15.75** NetIncome = 2.301 + 0.116*OpRev; $5.789 billion **15.77 a.** Amount = −2366.0 + 623.93*Policies **b.** $r = 0.99$; $r^2 = 0.984$ **c.** $16352 million **15.79 a.** Fuel = 48.065 + 0.042244*Miles **b.** 0.992, 0.984 **c.** 179.02 billion gallons **15.81 a.** RearFull = 318 + 2.612*RearCorner **b.** 0.397, 0.157 **c.** $2408 **15.83 a.** Strength = 60.02 + 10.507*Temptime **b.** 53.0% **c.** 0.017 **d.** 0.017 **e.** (2.445, 18.569) **15.85 a.** Rolresis = 9.450 − 0.08113*psi **b.** 23.9% **c.** 0.029 **d.** 0.029 **e.** (−0.15290, −0.00936) **15.87** CI: (8.26, 28.87); PI: (−6.66, 43.78) **15.89** CI: (49.710, 54.647); PI: (46.776, 57.581) **15.91 a.** GPA = −0.6964 + 0.0033282*SAT; 2.965 **b.** 69.5% **c.** CI: (2.527, 3.402); PI: (1.651, 4.278) **15.93 a.** Pay% = 7.020 + 1.516*Rate%; 19.148% **b.** 98.0%; yes **c.** CI: (18.9874, 19.3084); PI: (18.6159, 19.6799) **15.95 a.** Estp/e = 52.56 − 0.0959*Revgrow%; 38.17 **b.** 0.2%; no **c.** CI: (−0.74, 77.07); PI: (−94.67, 171.01)

Chapter 16

16.9 a. 300, 7, 13 **b.** 399 **16.11 a.** $\hat{y} = 10.687 + 2.157x_1 + 0.0416x_2$ **c.** 24.59 **16.13 a.** $\hat{y} = -127.19 + 7.611x_1 + 0.3567x_2$ **c.** −17.79 **16.15 b.** 454.42 **16.17 a.** 130.0 **b.** 3.195 **c.** (98.493, 101.507) **d.** (93.259, 106.741)

16.19 a. CalcFin = −26.6 + 0.776*MathPro + 0.0820*SATQ; 90% CI: (64.01, 73.46) **b.** 90% PI: (59.59, 77.88) **16.21 a.** (81.588, 90.978) **b.** (77.565, 95.002) **16.27** 0.716 **16.29 a.** yes **b.** β_1, reject H_0; β_2, do not reject H_0 **d.** β_1, (0.54, 3.77); β_2, (−0.07, 0.15) **16.31** β_1, (0.0679, 0.4811); β_2, (0.1928, 0.5596); β_3, (0.1761, 0.4768) **16.33 a.** yes **b.** each is significant **16.35 a.** $\hat{y} = -40{,}855{,}482 + 44{,}281.6x_1 + 152{,}760.2x_2$ **b.** yes **c.** β_1, reject H_0; β_2, do not reject H_0 **d.** β_1, (41,229.4, 47,333.8); β_2, (−4,446,472, 4,751,992) **16.45 a.** $\hat{y} = -0.8271 + 0.007163x_1 + 0.01224x_2$ **b.** β_1, (0.00615, 0.00818); β_2, (0.00198, 0.02249); **c.** 0.959 **16.51** Speed = 67.6 − 3.21*Occupants − 6.63*Seatbelt **16.53 a.** $\hat{y} = 99.865 + 1.236x_1 + 0.822x_2$ **b.** 125.816 lbs. **c.** (124.194, 127.438) **d.** (124.439, 127.193) **e.** β_1, (0.92, 1.55); β_2, (0.09, 1.56) **16.55 a.** gpa = −1.984 + 0.00372*sat + 0.00658*rank **b.** 2.634 **c.** (1.594, 3.674) **d.** (2.365, 2.904) **e.** β_1, (0.000345, 0.007093); β_2, (−0.010745, 0.023915) **16.57** $137,289

Chapter 17

17.3 negative, negative **17.5** positive, negative, positive **17.7** $Avgrate = 336.094 − 8.239*%Occup + 0.07709*%Occup²; 49.1% **17.9** 0to60 = 26.8119 − 0.153866*hp + 0.0003083*hp²; 8.396 seconds; yes **17.11** Forgross = 860.8 − 4.152*Domgross + 0.007689*Domgross²; $430.4 million; yes **17.13** second-order with interaction **17.15** $percall = 61.2 + 25.63*yrs + 6.41*score −1.82*yrs² − 0.058*score² + 0.29*yrs*score; $R^2 = 0.949$; yes **17.17 a.** oprev = −231.2 + 0.129*employs + 0.00565*departs; yes **b.** oprev = 399.2 + 0.0745*employs + 0.00087*departs + 0.00000014*employs*departs; R^2 increases from 0.958 to 0.986 **17.19** 0to60 = 25.4 − 0.161*hp − 0.00030*curbwt + 0.000028*hp*curbwt; $R^2 = 0.734$; yes **17.23** two **17.25** 600 customers **17.27** price = −30.77 + 4.975*gb + 54.20*highrpm; $54.20 **17.29** productivity = 75.4 + 1.59*yrsexp − 7.36*metha + 9.73*methb; $R^2 = 0.741$ **17.31** $\hat{y} = 0.66(1.38)^x$ **17.33** $\hat{y} = 42.668371(1.0169419)^x$; $R^2 = 0.496$; $138.30 **17.35** log revenue = −0.1285 + 1.0040 log employs − 0.1121 log departs; revenue = 0.7439*employs$^{1.0040}$*departs$^{-0.1121}$; $546.2 million **17.41** yes **17.43** may be present **17.45** will not be a problem **17.49 a.** x_5, x_2, x_9 **b.** $\hat{y} = 106.85 - 0.35x_5 - 0.33x_2$ **c.** 0.05 level: x_5, x_2, x_9; 0.02 level: x_2 **17.59** Pages = −141.5 + 422.5x − 61.01x^2 + 2.749x^3; 3727 **17.61** $\hat{y} = 10.705 + 0.974x - 0.015x^2$; yes; 83.6% **17.63** Wchill = −11.296 + 1.320 Temp − 0.456 Wind; Wchill = −11.296 + 1.185 Temp − 0.456 Wind + 0.00542 TempxWind; from 0.995 to 0.997 **17.65** productivity = 19.09 + 0.211*backlog + 0.577*female; $R^2 = 0.676$; yes **17.67** Log_AllLoans = −1.0130 + 2.0567Log_CredCard + 0.0635Log_Resid; AllLoans = 0.09705*CredCard$^{2.0567}$*Resid$^{0.0635}$; 1.76%. **17.69** yes; OpCost/Hr = 525.8 +3.21*Gal/Hr − 0.68*Range; 90.68% **17.71 a.** yes, test 1 **b.** final = 14.79 + 0.885*test1; $R^2 = 0.8568$ **17.73** Rating = 2.225 + 4.1793*YdsperAtt − 4.1763*Int_pct +3.3231*TD_pct + 0.8305*Comp_pct; 100.00% (rounded)

Chapter 18

18.3 369,600 gallons **18.5** with $x = 1$ for 2004, Earnings = 15.045 + 0.582x; $20.865. **18.7 a.** Subs = 13.371 + 19.357x; 361.8 million **b.** Subs = 29.993 + 12.233x + 0.54798x^2; 427.7 million **c.** quadratic **18.17** the 0.4 curve; the 0.7 curve **18.21** 30% **18.23 b.** I , 74.72; II, 103.98; III, 123.76; IV, 97.54 **18.25 a.** J, 100.50; F, 94.33; M, 103.16; A, 103.60; M, 98.06; J, 100.55; J, 98.17; A, 98.38; S, 96.86; O, 108.84; N, 93.20; D, 104.35 **18.27** $192.0 thousand, $201.6 thousand **18.29** 1213.2; 1541.2 **18.31** $69.43 billion **18.33** 39.655 quadrillion Btu **18.35** 189,000 gallons **18.39** quadratic, quadratic **18.41** quadratic, quadratic **18.45** 0.58, positive autocorrelation **18.47 a.** inconclusive **b.** inconclusive **18.49** $y_t = -1.51 + 1.119y_{t-1}$; 288.5 **18.51** $y_t = 525.6 + 0.812y_{t-1}$; 2895.5; MAD = 100.5 **18.53** 100 **18.55** best: leisure and hospitality; worst: construction **18.57** $114,364. **18.59** Army = 476.29 + 5.4643x; 558.25 thousand **18.61** Restaurants = −1408.0 + 1206.50x − 9.92905x^2; 25,851 restaurants **18.63 a.** AvgBill = 31.4743 + 1.62536x; $55.855 **b.** AvgBill = 31.0071 + 1.93679x − 0.038929x^2; $51.300 **c.** quadratic, quadratic **18.67** I, 95.14; II, 89.69; III, 95.64; IV, 119.53 **18.69** I, 192.88; II, 63.87; III, 30.14; IV, 113.11 **18.71** I, 94.68; II, 93.54; III, 111.05; IV, 100.73 **18.73** I, 72.50; II, 83.11; III, 111.16; IV, 132.77 **18.75** I, 71; II, 91; III, 104; IV, 66 **18.77** Cost = 38.3268 + 0.794341x + 0.0538258x^2; Cost = 37.1427 + 1.38642x; quadratic **18.83** DW statistic = 0.48; positive autocorrelation **18.85** $CPI_t = 2.0216 + 1.01812{*}CPI_{t-1}$; MAD = 1.388; 213.08 **18.87** 6.36% increase

Chapter 19

19.9 0, 2000, 2000 shirts **19.11 a.** maximax **b.** maximax **c.** maximin **19.13** purchase, do not purchase, purchase **19.17** make claims, $600 **19.19** design B, $7.6 million **19.21** direct, 31.5 minutes, 0.5 minutes **19.25** A or C, $12,800, $12,800 **19.27** direct, 0.5 minutes; longer, 9.0 minutes **19.29** 296 cod **19.31** 3520 programs **19.33** 7 units **19.35** current; Dennis; Dennis **19.37** DiskWorth; ComTranDat; ComTranDat

Chapter 20

20.35 0.0026 **20.39** in control **20.41** out of control **20.43** out of control **20.45** LCL = 0, UCL = 0.088 **20.47** LCL = 0, UCL = 15.72 **20.49** centerline = 0.197; LCL = 0.113; UCL = 0.282; in control **20.51** centerline = 4.450; LCL = 0; UCL = 10.779; out of control **20.53** in control **20.55** no **20.57** 0.89; not capable **20.59** yes; mean chart failed test #1 at sample 6 **20.61** yes; mean chart failed tests #1 and #5 at sample 5 **20.63 a.** in control **b.** yes, 0.65; not capable **20.69** *c*-chart **20.71** 1.07; capable **20.73** centerline = 0.05225; LCL = 0.00504; UCL = 0.09946; in control **20.75** in control **20.77** centerline = 0.0988; LCL = 0.0355; UCL = 0.1621; in control **20.79** yes, 0.74; not capable

INDEX/GLOSSARY

J

K

L

M

P

Paired Observations Data collected from dependent samples, such as before-and-after measurements recorded for the same individuals or test units, 385

Parameter A characteristic of the population, such as the population mean (μ), standard deviation (σ), or proportion (π), 7, 117

Pareto Diagram Also known as the *Juran diagram,* a bar chart describing the relative frequencies with which various kinds of defects, or nonconformities, have occurred, 772

Partial Regression Coefficient In the multiple regression equation, the coefficient for each independent variable. It is the *slope* of the equation when all other x variables are held constant, 602
  interval estimation for, 620–621
  testing, 618–620

Payoff The outcome at the intersection of a decision alternative and a state of nature, 739–740
  expected, 745–746

Payoff Table A method of structuring the decision situation. The rows of the table are decision alternatives, the columns are states of nature, and the entry at each intersection of a row and column is a numerical consequence such as a profit or loss. The table may also include estimated probabilities for the states of nature, 739–740

***p*-Chart** A control chart that monitors a process by showing the proportion of defectives in each of a series of samples, 787–790

Percentiles Quantiles dividing data into 100 parts of equal size, with each comprising 1% of the observations, 70

Periodicity A potential problem with the systematic sample; this can occur when the order in which the population appears happens to include a cyclical variation in which the length of the cycle is the same as the value of k being used in selecting the sample, 123

Permutations The number of unique arrangements for n objects, r at a time, 157–158

Personal Interview A survey technique in which an interviewer personally administers the questionnaire to the respondent, 105

Pictogram A visual display using symbols to represent frequencies or other values, 32

Pie Chart A circular display divided into sections based on either the number of observations within, or the relative values of, the segments, 31

Point Estimate A single number (e.g., $\bar{x}$ or p) that estimates the value of a population parameter (e.g., μ or π), 271–272
  using multiple regression equation, 605
  using regression line, 558

Poisson Distribution A discrete probability distribution applied to events for which the probability of occurrence over a given span of time, space, or distance is extremely small. It can be used to approximate the binomial distribution when the number of trials is relatively large and the probability of a success in any given trial is very small, 187–193
  approximation to binomial distribution, 191–193
  description and applications of, 187–193
  exponential distribution and, 228

Polynomial models extend the linear regression model to include terms in which the independent variable(s) are raised to a power greater than 1. They enable the model to fit data from relationships that are curvilinear, 653–657
  with one quantitative predictor variable, 645–651
  with two quantitative predictor variables, 653–657

Population The set of all possible elements that could theoretically be observed or measured; sometimes referred to as the *universe,* 6, 117

Population Parameter The characteristic of the population that corresponds to the sample statistic. The true value of the population parameter is typically unknown, but is estimated by the sample statistic, 7, 117

Positively Skewed Distribution A nonsymmetrical distribution that tails off to the right; it has a mean that is greater than the median, 64

Posterior Probability A revised probability based on additional information, 151

Power Curve A graph in which the power $(1 - \beta)$ of the hypothesis test is plotted against a range of population parameter values for which the null hypothesis is false, 349–350

Power of a Test $(1 - \beta)$ The probability that the hypothesis test will correctly reject a null hypothesis that is false, 315–316, 346–350

Prediction Interval In regression analysis, the estimation interval for an individual y observation, 563–565
  multiple regression and, 612–614

Predictive Research An investigation for the purpose of obtaining a forecast or prediction of some value that will occur in the future, 103–104

Primary Data Data that have been collected by the researcher for the problem or decision at hand, 104 105–111
  from experimentation, 109–110
  from observation, 110–111
  from survey research, 105–108

Principle of Multiplication If, following a first event that can

S

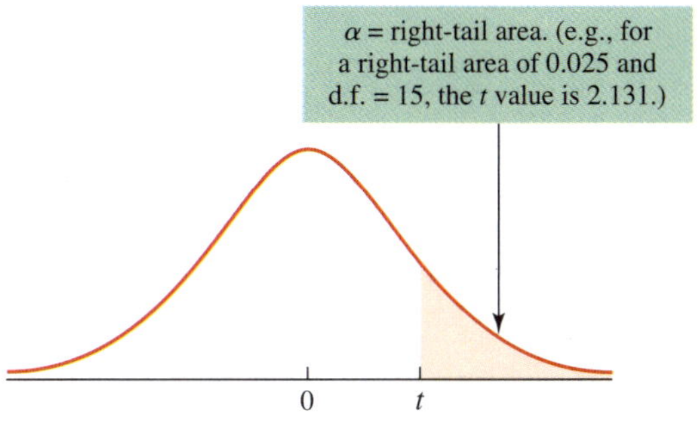

α:	0.10	0.05	0.025	0.01	0.005
d.f. = 1	3.078	6.314	12.706	31.821	63.657
2	1.886	2.920	4.303	6.965	9.925
3	1.638	2.353	3.182	4.541	5.841
4	1.533	2.132	2.776	3.747	4.604
5	1.476	2.015	2.571	3.365	4.032
6	1.440	1.943	2.447	3.143	3.707
7	1.415	1.895	2.365	2.998	3.499
8	1.397	1.860	2.306	2.896	3.355
9	1.383	1.833	2.262	2.821	3.250
10	1.372	1.812	2.228	2.764	3.169
11	1.363	1.796	2.201	2.718	3.106
12	1.356	1.782	2.179	2.681	3.055
13	1.350	1.771	2.160	2.650	3.012
14	1.345	1.761	2.145	2.624	2.977
15	1.341	1.753	2.131	2.602	2.947
16	1.337	1.746	2.120	2.583	2.921
17	1.333	1.740	2.110	2.567	2.898
18	1.330	1.734	2.101	2.552	2.878
19	1.328	1.729	2.093	2.539	2.861
20	1.325	1.725	2.086	2.528	2.845
21	1.323	1.721	2.080	2.518	2.831
22	1.321	1.717	2.074	2.508	2.819
23	1.319	1.714	2.069	2.500	2.807
24	1.318	1.711	2.064	2.492	2.797
25	1.316	1.708	2.060	2.485	2.787
26	1.315	1.706	2.056	2.479	2.779
27	1.314	1.703	2.052	2.473	2.771
28	1.313	1.701	2.048	2.467	2.763
29	1.311	1.699	2.045	2.462	2.756
30	1.310	1.697	2.042	2.457	2.750
31	1.309	1.696	2.040	2.453	2.744
32	1.309	1.694	2.037	2.449	2.738
33	1.308	1.692	2.035	2.445	2.733
34	1.307	1.691	2.032	2.441	2.728
35	1.306	1.690	2.030	2.438	2.724
36	1.306	1.688	2.028	2.435	2.719
37	1.305	1.687	2.026	2.431	2.715
38	1.304	1.686	2.024	2.429	2.712
39	1.304	1.685	2.023	2.426	2.708
40	1.303	1.684	2.021	2.423	2.704
41	1.303	1.683	2.020	2.421	2.701
42	1.302	1.682	2.018	2.418	2.698
43	1.302	1.681	2.017	2.416	2.695
44	1.301	1.680	2.015	2.414	2.692
45	1.301	1.679	2.014	2.412	2.690

The Standard Normal Distribu

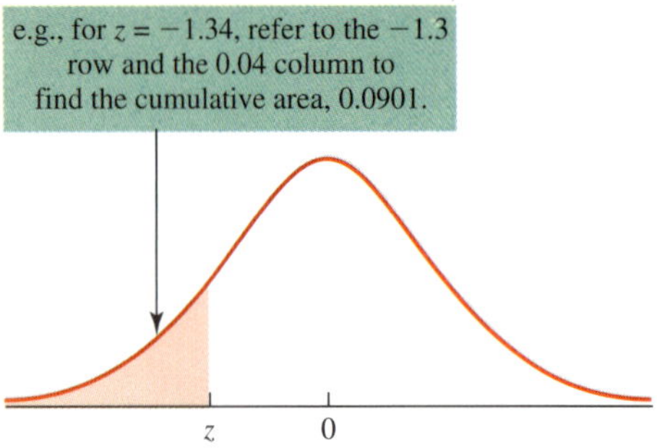

z	0.00	0.01	0.02	0.03	0.04	0.05	0.06	0.07	0.08	0.0
−3.0	0.0013	0.0013	0.0013	0.0012	0.0012	0.0011	0.0011	0.0011	0.0010	0.00
−2.9	0.0019	0.0018	0.0018	0.0017	0.0016	0.0016	0.0015	0.0015	0.0014	0.00
−2.8	0.0026	0.0025	0.0024	0.0023	0.0023	0.0022	0.0021	0.0021	0.0020	0.00
−2.7	0.0035	0.0034	0.0033	0.0032	0.0031	0.0030	0.0029	0.0028	0.0027	0.00
−2.6	0.0047	0.0045	0.0044	0.0043	0.0041	0.0040	0.0039	0.0038	0.0037	0.00
−2.5	0.0062	0.0060	0.0059	0.0057	0.0055	0.0054	0.0052	0.0051	0.0049	0.00
−2.4	0.0082	0.0080	0.0078	0.0075	0.0073	0.0071	0.0069	0.0068	0.0066	0.00
−2.3	0.0107	0.0104	0.0102	0.0099	0.0096	0.0094	0.0091	0.0089	0.0087	0.00
−2.2	0.0139	0.0136	0.0132	0.0129	0.0125	0.0122	0.0119	0.0116	0.0113	0.0
−2.1	0.0179	0.0174	0.0170	0.0166	0.0162	0.0158	0.0154	0.0150	0.0146	0.0
−2.0	0.0228	0.0222	0.0217	0.0212	0.0207	0.0202	0.0197	0.0192	0.0188	0.0
−1.9	0.0287	0.0281	0.0274	0.0268	0.0262	0.0256	0.0250	0.0244	0.0239	0.0
−1.8	0.0359	0.0351	0.0344	0.0336	0.0329	0.0322	0.0314	0.0307	0.0301	0.0
−1.7	0.0446	0.0436	0.0427	0.0418	0.0409	0.0401	0.0392	0.0384	0.0375	0.0
−1.6	0.0548	0.0537	0.0526	0.0516	0.0505	0.0495	0.0485	0.0475	0.0465	0.0
−1.5	0.0668	0.0655	0.0643	0.0630	0.0618	0.0606	0.0594	0.0582	0.0571	0.0
−1.4	0.0808	0.0793	0.0778	0.0764	0.0749	0.0735	0.0721	0.0708	0.0694	0.0
−1.3	0.0968	0.0951	0.0934	0.0918	0.0901	0.0885	0.0869	0.0853	0.0838	0.0
−1.2	0.1151	0.1131	0.1112	0.1093	0.1075	0.1056	0.1038	0.1020	0.1003	0.0
−1.1	0.1357	0.1335	0.1314	0.1292	0.1271	0.1251	0.1230	0.1210	0.1190	0.1
−1.0	0.1587	0.1562	0.1539	0.1515	0.1492	0.1469	0.1446	0.1423	0.1401	0.1
−0.9	0.1841	0.1814	0.1788	0.1762	0.1736	0.1711	0.1685	0.1660	0.1635	0.1
−0.8	0.2119	0.2090	0.2061	0.2033	0.2005	0.1977	0.1949	0.1922	0.1894	0.1
−0.7	0.2420	0.2389	0.2358	0.2327	0.2296	0.2266	0.2236	0.2206	0.2177	0.2
−0.6	0.2743	0.2709	0.2676	0.2643	0.2611	0.2578	0.2546	0.2514	0.2483	0.2
−0.5	0.3085	0.3050	0.3015	0.2981	0.2946	0.2912	0.2877	0.2843	0.2810	0.2
−0.4	0.3446	0.3409	0.3372	0.3336	0.3300	0.3264	0.3228	0.3192	0.3156	0.3
−0.3	0.3821	0.3783	0.3745	0.3707	0.3669	0.3632	0.3594	0.3557	0.3520	0.3
−0.2	0.4207	0.4168	0.4129	0.4090	0.4052	0.4013	0.3974	0.3936	0.3897	0.3
−0.1	0.4602	0.4562	0.4522	0.4483	0.4443	0.4404	0.4364	0.4325	0.4286	0.4
−0.0	0.5000	0.4960	0.4920	0.4880	0.4840	0.4801	0.4761	0.4721	0.4681	0.4

Source: Cumulative standard normal probabilities generated by Minitab, then rounded to four decimal places.

* Data Analysis Plus™ 7.0 add-in

Seeing Statistics Applets